MATERIALS, CIRCUITS & DEVICES SERIES 27

Nano-Scaled Semiconductor Devices

Other volumes in this series:

Volume 2 **Analogue IC Design: The current-mode approach** C. Toumazou, F.J. Lidgey and D.G. Haigh (Editors)
Volume 3 **Analogue-Digital ASICs: Circuit techniques, design tools and applications** R.S. Soin, F. Maloberti and J. France (Editors)
Volume 4 **Algorithmic and Knowledge-based CAD for VLSI** G.E. Taylor and G. Russell (Editors)
Volume 5 **Switched Currents: An analogue technique for digital technology** C. Toumazou, J.B.C. Hughes and N.C. Battersby (Editors)
Volume 6 **High-frequency Circuit Engineering** F. Nibler *et al.*
Volume 8 **Low-power High-frequency Microelectronics: A unified approach** G. Machado (Editor)
Volume 9 **VLSI Testing: Digital and mixed analogue/digital techniques** S.L. Hurst
Volume 10 **Distributed Feedback Semiconductor Lasers** J.E. Carroll, J.E.A. Whiteaway and R.G.S. Plumb
Volume 11 **Selected Topics in Advanced Solid State and Fibre Optic Sensors** S.M. Vaezi-Nejad (Editor)
Volume 12 **Strained Silicon Heterostructures: Materials and devices** C.K. Maiti, N.B. Chakrabarti and S.K. Ray
Volume 13 **RFIC and MMIC Design and Technology** I.D. Robertson and S. Lucyzyn (Editors)
Volume 14 **Design of High Frequency Integrated Analogue Filters** Y. Sun (Editor)
Volume 15 **Foundations of Digital Signal Processing: Theory, algorithms and hardware design** P. Gaydecki
Volume 16 **Wireless Communications Circuits and Systems** Y. Sun (Editor)
Volume 17 **The Switching Function: Analysis of power electronic circuits** C. Marouchos
Volume 18 **System on Chip: Next generation electronics** B. Al-Hashimi (Editor)
Volume 19 **Test and Diagnosis of Analogue, Mixed-signal and RF Integrated Circuits: The system on chip approach** Y. Sun (Editor)
Volume 20 **Low Power and Low Voltage Circuit Design with the FGMOS Transistor** E. Rodriguez-Villegas
Volume 21 **Technology Computer Aided Design for Si, SiGe and GaAs Integrated Circuits** C.K. Maiti and G.A. Armstrong
Volume 22 **Nanotechnologies** M. Wautelet *et al.*
Volume 23 **Understandable Electric Circuits** M. Wang
Volume 24 **Fundamentals of Electromagnetic Levitation: Engineering sustainability through efficiency** A.J. Sangster
Volume 25 **Optical MEMS for Chemical Analysis and Biomedicine** H. Jiang (Editor)
Volume 29 **Nano-CMOS and Post- CMOS Electronics: Devices and Modelling** Saraju P. Mohanty and Ashok Srivastava
Volume 30 **Nano-CMOS and Post- CMOS Electronics: Circuits and Design** Saraju P. Mohanty and Ashok Srivastava

Nano-Scaled Semiconductor Devices

Physics, Modelling, Characterisation, and Societal Impact

Edited by
Edmundo A. Gutiérrez-D

The Institution of Engineering and Technology

Published by The Institution of Engineering and Technology, London, United Kingdom

The Institution of Engineering and Technology is registered as a Charity in England & Wales (no. 211014) and Scotland (no. SC038698).

First published 2016

The Institution of Engineering and Technology
Michael Faraday House
Six Hills Way, Stevenage
Herts, SG1 2AY, United Kingdom

www.theiet.org

British Library Cataloguing in Publication Data
A catalogue record for this product is available from the British Library

ISBN 978-1-84919-930-8 (hardback)
ISBN 978-1-84919-931-5 (PDF)

Typeset in India by MPS Limited

Contents

Synopsis

The subject of nano-scaled semiconductor devices and technology is a strategic and emerging area of relevant societal importance in our ubiquitous electronics era. This book is intended for scientists, engineers, and graduate students involved in the research, technology development, education activities, and societal-related applications where nano-scaled semiconductor devices are involved. With the rapid evolution of the integrated circuit technology many new materials and fabrication processing steps have been incorporated. These new materials and processing steps enhance the electrical performance of nano-scaled semiconductor devices, resulting in faster, more functional and complex electronics. However, this boosted capacity comes with second-order effects that degrade device performance, impact its long-term reliability, energy-efficiency and ultimately contribute to worldwide energy waste and subsequent pollution. The characterization, modeling, and simulation prediction of these second-order effects is strongly correlated to each other. Therefore, this book, in a novel approach, attempts to blend the technology fabrication, with the characterization, physics, and modeling in a synergistic manner. For instance, the shrinkage of the device dimensions down to 14 nm and below pose a big challenge for the development of appropriate transport modeling mechanisms, which in turn require advanced characterization and modeling techniques. The characterization techniques, developed for studying transport in nano-scaled devices, is also a novel contribution of this book. Blending the characterization, modeling, physics, and technology is quite relevant for achieving energy-efficient electronics. Moreover, there is a benefit that device developers can get out from the complex nano-scaled semiconductor device technologies. This is also a topic that this book covers. The potential use of these new materials and structures for the development of new and alternative devices or functions, such as water decontamination, energy renewal, physiology research, energy-efficient milk and meat production, or optimal and energy-efficient crop irrigation. This book also correlates the fundamental physics and device electrical performance with energy efficiency, energy savings, and thus contributes to the development of a greener and more sustainable world.

Chapter 1
Introduction

Edmundo A. Gutiérrez-D.[1] and Fernando Guarín[2]

Semiconductor technology has profoundly changed the world influencing most aspects of today's society. As we review the evolution of the integrated circuit technology over the last three decades, from 1985 to 2015, we observe how the device dimension reduction evolves from the micrometer to the nanometer scale, but performance can no longer be improved by mere scale reduction and other innovations, such as the incorporation of new materials from the periodic table, new fabrication technology steps or structures, or the introduction of mechanical strain and crystal reorientation to boost carrier mobility. We notice the evolution from the classical two-dimensional metal-oxide-semiconductor field-effect transistor (MOSFET) toward the three-dimensional FinFET and further experimental device structures, such as nanowires and quantum dots. An overview of the potential new devices or functions, based on the new materials and processing steps, will also be introduced so the reader becomes aware of the wide and vast opportunities far beyond microprocessor-based electronics.

The MOSFET channel length has been reduced from 20 μm in 1985 down to 14 nm in 2015; this is 1428 times in an exponential trend, which implies going from 40,000 up to 4.3 billion transistors per chip. As a result of such dramatic technology evolution, there have been significant economic implications; the revenue, of the top 20 semiconductors companies, has linearly climbed from 22 billion USD in 1985 to 233 billion USD projected for 2015 as shown in Figure 1.1 [1.1]. Such an impressive technology–business evolution, over the last 30 years, has been possible thanks to the coordinated size scaling coupled with the introduction of ever more materials from the periodic table, as well as new and advanced fabrication processes and lithography.

The most prominent scientific and technology developmental findings, as well as their application, in the field of electronics and adjacent disciplines since 1985 will be reviewed. This overview is done through the material level up to the applications, passing through the fabrication technology, device structures, and device modeling and simulation.

Figure 1.2 shows a compilation of the different materials and FET-related device structures introduced since 1985.

[1]Department of Electronics, National Institute for Astrophysics, Optics, and Electronics, Puebla, Mexico
[2]GlobalFoundries, Inc., East Fishkill, NY, USA

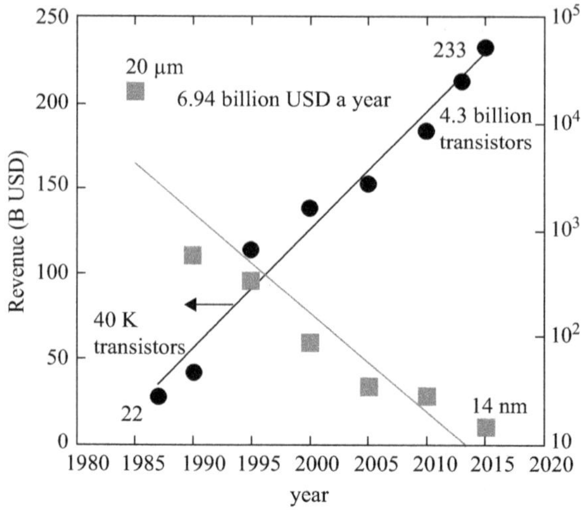

Figure 1.1 Business and device dimension evolution of MOSFET technology from 1985 to 2015

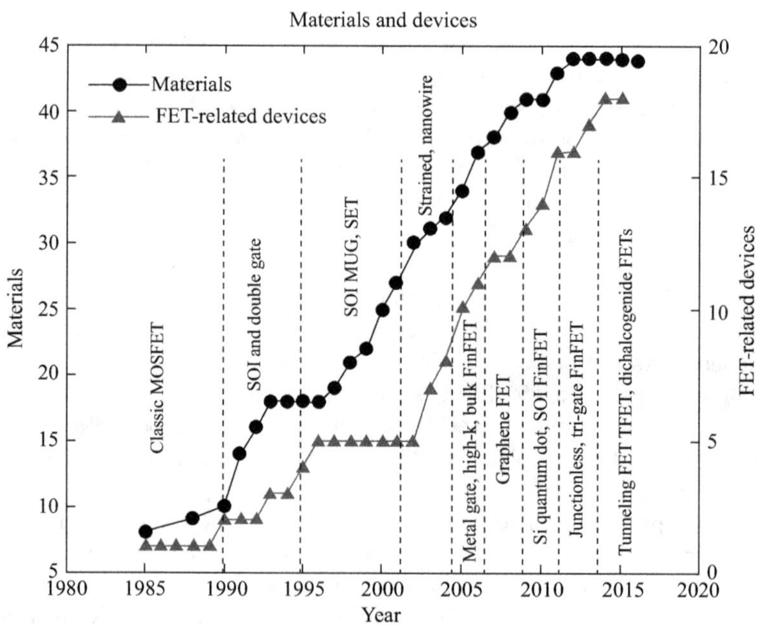

Figure 1.2 Introduction of materials and different FET-related device structures

The conventional planar MOSFET in 1985 used a basic number of around eight different materials including Si itself [1.2]–[1.5]. Then from the beginning of 1990, there is a linear increase in the number of materials incorporated in the fabrication process, which coincides with the introduction of the first commercial silicon-on-insulator (SOI) field-effect devices (FET) [1.6]. Since 1985 until around 1990 many materials were incorporated to reduce detrimental hot carrier effects [1.7], also to improve the quality of the semiconductor–oxide interface [1.8] that was also used to replace traditional polysilicon gates for metal gates. Germanium was also used to suppress hot carrier effects [1.9]. An elevated source/drain FET structure was proposed in 1990 [1.10] that was able to reduce the series resistance effects. The electrical contact has been found to be a big barrier for further down-scaling into the deep submicron scale. Therefore, Tungsten and Titanium were introduced to reduce contact resistance [1.11], [1.12]. Cobalt was introduced as a source/drain silicided contact in reference 1.13. Later on, in 1991 Tantalum was introduced as a diffusion barrier for Al or Cu in [1.14]. At the beginning of the 90s hot carrier, series resistance and self-heating were the two emerging research topics. Shallow trenches, for instance, replacing LOCOS was introduced in reference 1.15, which allowed the fabrication of 16 Mb DRAM cells with a low leakage condition.

In 1988 SOI MOSFETs device structures were introduced [1.16]. The double gate SOI MOSFET was introduced in 1992 [1.17]. Surrounding gate MOSFETs were subsequently introduced in 1991 [1.18]. The pocket implantation was introduced in 1992 [1.19]. Then the subquarter micrometer complementary metal-oxide-semiconductor (CMOS) on ultrathin SOI was introduced in 1992 [1.20]. Besides changes to the device structure, observed in the 1985–1990 period, other materials were introduced as the device dimension continued its way into the deep sub-micrometer range. Palladium, for instance, was incorporated as a Cu/Pd-silicide stack for metallization [1.21]. By 1989, Cooper had been introduced for VLSI inter-connects [1.22]. Then in 1992 a combination of bipolar and CMOS (BiCMOS) was developed as a potential technology for RF applications with a 60-GHz cutoff frequency [1.23]. In the same year the SiGe modulation-doped field-effect transistor was introduced [1.24]. This device is also known as the high electron mobility transistor.

In 1993 the introduction of monolithic Si RF amplifiers, for the 800–900 MHz band, brought the incorporation of CMOS inductors [1.25]. Other materials, such as Indium (In) and Antimony (Sb), were introduced as a method to improve short-channel behavior [1.26]. In order to reduce punchthrough, halo doping was introduced [1.27]. The first observation of process-induced mechanical strain impact on electrical performance for 0.5-μm transistors was observed in 1994 [1.28], where the trapping activation energy was changed with mechanical stress. A year before in 1993 the mechanical strain was introduced as a booster to enhance electron transport in MOSFETs [1.29]. The first fabricated 80-nm MOSFET was reported in 1994 [1.30].

By 1995 we saw the introduction of organic polymers, such as polymide (PI), which were used as low-capacitance dielectric for multilevel interconnection. The PI was used as reflowable organic spin-on glass (SOG). The SOG-filled dielectrics reduce interconnection capacitance down to a 70% respect to that of conventional structures [1.31]. As the CMOS continues scaling down below 0.1 μm (100 nm), RF applications begin to show up, and with that the need for high-Q reliable inductors, which are then introduced as four metal-level spiral inductors in reference 1.32. Then Chemical Vapor Deposition (CVD) stacked gate dielectrics are introduced to enhance "etch stop" and thus preventing diode leakage due to pitting of the substrate during the gate etch [1.33].

In 1997, the use of Deuterium (D) also known as heavy hydrogen (^2H) as a post-metal annealing to increase hot carrier reliability is seen [1.34]. A GaAs MESFET was transferred to a Si substrate by using a substrate removal process [1.35]. Silicon Carbide (SiC) showed promise as a wide-band material for MOS transistors. It was tested for P-type SiC and showed that a high-quality oxide could be grown on it. This opened the door for power and high-temperature electronics [1.36]. Amorphous Si (a-Si) is introduced as a component of a gate stacked W-polycide structure. This prevented the deterioration of the gate oxide caused by the fluorine ions in conventional W-polycide gated MOSFETs [1.37]. As the SiO_2 thickness was scaling down to 2.8 nm, it was necessary to incorporate oxynitrides [1.38] to obtain uniform oxide thickness introducing the new reliability concern for negative bias temperature instability (NBTI), which was first reported for advanced devices in 1997 [1.39]. This effect manifested itself as an increase in the threshold voltage for pFETs with a consequent decrease in the channel current [1.39]. NBTI optimization became a major integration challenge for new processes as Tox was further reduced with optimized nitrogen content. For 1.5 nm Vt stability represented a major reliability concern. Therefore, TiO_2 was introduced as a high-permittivity gate oxide dielectric [1.40]. W/TiN, as a way to control the threshold voltage, was introduced as a metal gate [1.41].

A 135-GHz cutoff frequency for a 70-nm NFET of CMOS technology was reported in reference 1.42. The NBTI is first introduced as a reliability issue. Oxide–nitride gate dielectrics thinner than 2.0 nm and Ti and Co silicidation were introduced for self-aligned T-shaped gate MOSFETs in references 1.43 and 1.44. This fabrication technology used chemical mechanical planarization, and the T-shaped gate structure helped in reducing the gate resistance, which in turn improved the thermal stability of gate structures. Zr-Sn-Ti-O amorphous dielectrics for potential application as storage capacitors in DRAM were introduced [1.45]. Tantalum (Ta) as Ta_2O_5 dielectric was proposed for future generation of MOS devices with reduced leakage current [1.46].

In 1999 Copper (Cu) as a metal gate was introduced [1.47]. Carbon (C) was introduced in vertical MOSFET by the use of thin SiGeC layers in the source and drain regions [1.48]. Zirconium as ZrO_2 was introduced as high-K dielectric for MOSFET application [1.49].

The above has been a rough overview of materials and device structures from 1985 to 2000. Many research works have not been cited thus far because the quantity is so large that there would not be enough room to accommodate all the references. Our primary intention was to acquaint the reader with a general overview.

The year 2000 began with the introduction of Indium (In) for threshold voltage control in 0.1 μm MOSFET [1.50]. La_2O_3 was tried as high-K dielectric [1.51]. FinFET showed up for the first time [1.52]. The electrical and reliability properties of ultrathin HfO_2 were investigated, and then proved to work as a gate dielectric in MOSFETs [1.53].

In 2001 Titanium (Ti) and Molybdenum (Mo) were introduced as metal gate in MOSFETs [1.54]. In 2002 Ruthenium (Ru) and Tantalum (Ta) were introduced as metal gate electrodes [1.55]. Strontium is tested as a high-K dielectric for MOS applications [1.56].

In 2003 different substrate orientations to enhance carrier mobility on high-k dielectric-based MOSFET was attempted for the first time [1.57]. Terbium (Tb) was introduced as doping material for high-density MIM capacitors [1.58].

In 2004 a CMOS nanotechnology with strained-silicon was introduced in a 90-nm technology [1.59]. The operation principle of the tunneling FET was introduced [1.60]. Ytterbium (Yb) introduced as a Schottky barrier source/drain in MOSFETs [1.61]. A 45-nm CMOS node 41-stage ring oscillator was designed, fabricated, and tested for a 25-nm FinFET technology [1.62]. A sub-1-nm effective oxide thickness CMOS process was developed [1.63].

In 2005 Terbium (Tb) was considered as $TaTb_xN$ metal gate electrode for n-MOSFETs [1.64]. Carbon nanotubes were considered as interconnects for gigascale integration [1.65]. Iridium (Ir) was introduced as gate dielectric in MOSFETs as the IrO_2-IrO_2-Hf-$LaAlO_3$ oxide [1.66]. A 3D stacked Fin-CMOS technology was proposed and implemented as the vertical stack of n- and p-type Fins on an SOI substrate. The proposed structure achieved a 50% area reduction [1.67]. Yttrium (Y) was introduced as a stacked Y_2O_3/HfO_2multimetal gate dielectric, which improved electron mobility and charge trapping characteristics [1.68].

In 2006 Iridium (Ir) was introduced as a silicided Ir_xSi-gated metal gate (4.9 eV work function) p-MOSFET with HfO_2 as a dielectric [1.69]. Chip cooling became a crucial issue with the high-density integration. For microprocessors a very aggressive and sophisticated cooling system will be required, such as liquid cooling. Then a wafer-level integration of microchannels at the wafer backside with microfluidic heat sink was proposed [1.70]. Scandium (Sc) was used to obtain a low work function in metal gate nMOSFETs [1.71]. The use of Dysprosium (Dy) as a stacked oxide for HfO_2-based n-MOSFETs, which resulted in low Vt, high drive current, and improved channel mobility was reported [1.72]. Gadolinium (Gd) was used as an oxide buffer layer on Hf-based dielectrics, which resulted in a reduction of the effective work function down to 4.05 eV [1.73]. Germanium (Ge) was introduced as strain-transfer structure, which served as a mobility booster [1.74].

In 2007 Zirconium (Zr) was introduced as a ZrO_2 dielectric with a dielectric constant of 18 and a bandgap range of 5.1 to 7.8 eV in [1.75]. Tantalum carbide (TaC) was tried as a metal gate on high-k dielectric MOSFETs [1.76].

In 2008 Selenium (Se) was used as a silicide contact for the S/D electrodes [1.77]. In 2010 the scaling down to 45 nm brought some effects such as line width or edge roughness that induced statistical variability was reported [1.78]. Another effect related to extremely small dimensions was the one concerning the metal granularity that induced threshold voltage variability was reported [1.79]. With the great advancement in downscaling and the incorporation of new materials came the three-dimensional stacking of chips that helped to increase the chip functionality [1.80].

The incorporation of new materials, different device structures, three-dimensionality, and further downscaling sparked the search for new CMOS-compatible materials, such as Lutetium (Lu) used as a composite $LaLuO_3$ with a dielectric k value of 30 for SOI MOSFETs in 2011 [1.81].

A new approach alternative to bulk MOSFETs is the junction-less transistors proposed for the sub-20-nm regime [1.82].

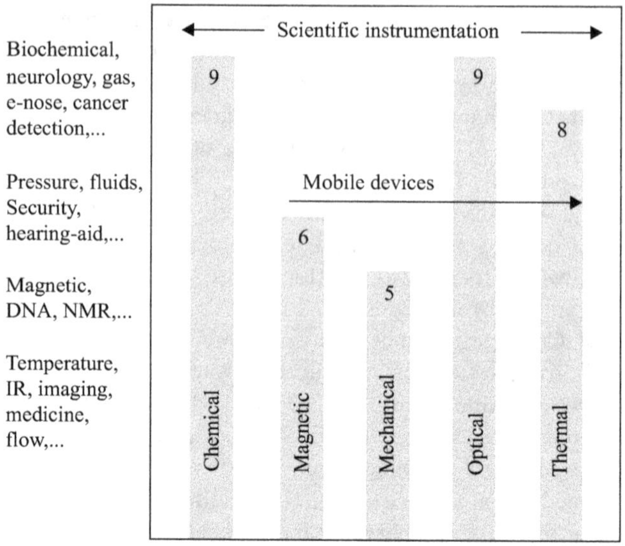

Figure 1.3 FET-related sensors and their applications into different fields

In the same 2011 the first atomistic studies on a 35 nm technology were introduced [1.83]. The Cerium (Ce) is now tried as a high-k dielectric that helps in improving the subthreshold characteristics of MOSFETs [1.84].

In 2012 Sulfur (S) was introduced as a molybdenum disulfide (MoS_2) in a monolayer channel MOSFET [1.85].

Later in 2014 the use of transition metal dichalcogenide materials, such as $MoTe_2$, $MoSe_2$, MoS_2, were being explored for potential MOSFET application in the Lg = 5 nm limit [1.86].

The incorporation of more than 40 different materials from the periodic table together with the introduction of different device structures and processing steps opened the door for the innovation and creation of a wide variety of FET-related sensors. This has resulted in more than 35 different chemical, magnetic, mechanical, optical, and thermal sensors as can be seen in Figure 1.3. The numbers for the different types of sensors are only a rough approximation based on the major FET-related sensor structures. There is a larger list of different sensors, but most of them are modifications to the major and original sensor structures. The applications of these sensors cover almost every aspect of human life, from the very obvious and natural as the temperature, light, pressure, flow, to the very sophisticated, such as; electronic nose (e-nose), DNA classification, and cancer detection, for instance. All these FET-related sensors are CMOS-compatible, which has permitted the creation of a wide variety of smart sensors. A more specific and detailed description of these CMOS-sensors and their applications will be given in Chapter 4. A visually schematic way to represent the impact of Si-based and FET-related devices is shown in Figure 1.4, using what we call the "technological tree" is shown. The

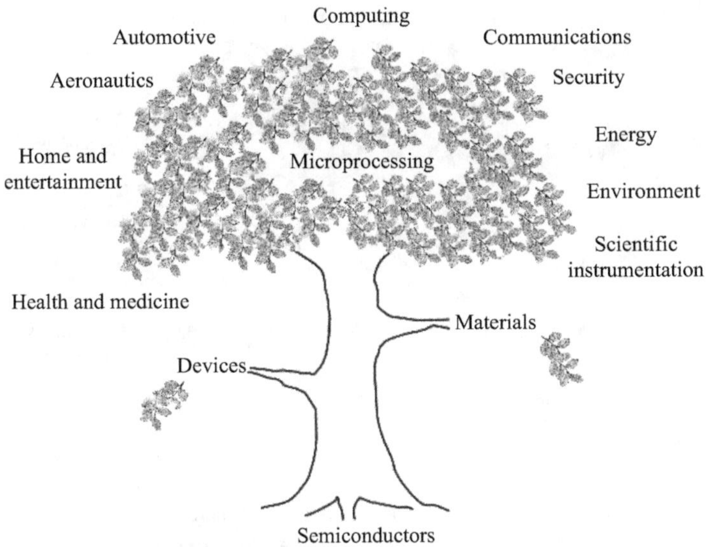

Figure 1.4 *The "technological tree" represented by its tree trunk and two main branches of materials and devices. The pervasiveness of semiconductor technology is shown through the plethora of applications*

microprocessing function represented by the computer processors is the tree trunk feeding from the semiconductor soil. While the microprocessor for computing is the main application, there are two important branches, one related to the different materials beyond the silicon used to boost the processing capabilities. The other branch is the one related to the different FET-related devices as the sensors above mentioned. The "technological tree" has grown so well that it has expanded to almost all practical aspects of human life.

The saga of the evolution of FET devices from the micrometer down to the nanometer size has not been free of complications. Many expected, an unexpected second-order effects have led to less energy-efficient devices, degradation, and inefficient electrical performance. These second-order effects are close to 20 different issues that will be described in detail on Chapter 2. In Chapter 3, the different experimental methodologies to measure the second-order effects will be described in detail. Finally in Chapter 5, we will deliver a summary of the previous chapters and write about beyond the silicon era and beyond earth mankind impact.

With Figure 1.5, we would like to add a final self-explanatory comment on the pervasiveness and ubiquity of the electronics–computing–communications–sensor/actuator feedback system.

The world and the way the human being interacts with it is analog by nature, but the way we electronically process the information, captured from our environment, is digital. The digital processing of information has been a key feature for

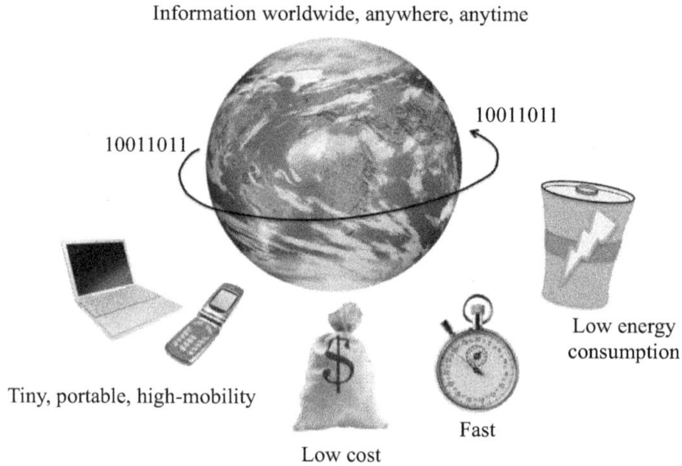

Figure 1.5 Pervasiveness and ubiquity of the joint assembly of electronics, computing, communications, and sensors and actuators

the worldwide rapid dissemination of information anywhere and anytime. This has enabled faster, portable, low-cost, and low-energy consumption systems. The wireless interconnectivity and the data storage and retrieval play an important role as they depend on the wireless network availability. However, as already mentioned, the world and the human being are analog, and thus the outcoming and incoming signals. Therefore, the development of high energy-efficient, low-cost, and portable sensors and actuators, and its corresponding conditioning circuitry, will play a key role in the future. Moreover the use of new materials and device structures compatible with the mainstream semiconductor technology will also enable the development of biomedical applications for health care, and the sustainable green environment, which are much needed for the future generations.

References

[1.1] Data collected from different sources, such as the SIA roadmap (http://www.semiconductors.org/), and the ITRS (http://www.itrs.net/).

[1.2] Moslehi, M. M. and Saraswat, K. C., "Thermal nitridation of Si and SiO2 for VLSI", IEEE Transactions on Electron Devices, Vol. ED-32, No. 2, 1985, pp. 106–123.

[1.3] Alperin, M. E., Holloway, T. C., Haken, R. A., Gosmeyer, C. D., Karnaugh, R. V., and Parmantie, W. D., "Development of the self-aligned titanium silicide process for VLSI applications", IEEE Transactions on Electron Devices, Vol. ED-32, No. 2, 1985, pp. 141–149.

[1.4] Butler, A. L., and Foster, D. J., "The formation of shallow low-resistance source-drain regions for VLSI CMOS technologies", IEEE Transactions on Electron Devices, Vol. ED-32, No. 2, 1985, pp. 150–155.

[1.5] Gardner, D. S., Michalka, T. L., Saraswat, K. C., Barbee, T. W., McVittie, J. P., and Meindl, J. D., "Layered and homogeneous films of aluminum and aluminum/silicon with titanium and tungsten for multilevel interconnects", IEEE Transactions on Electron Devices, Vol. ED-32, No. 2, 1985, pp. 174–183.

[1.6] Schwierz, F., Wong, H., and Liou, J. J., "Nanometer CMOS", Pan Stanford Publishing Pte. Ltd., Pan Stanford Publishing, Singapore, 2010, Chapter 2, p. 39.

[1.7] MacWilliams, K. P., Halle, L. F., and Zietlow, T. C., "Improved hot-carrier resistance with fluorinated gate oxides", IEEE Electron Device Letters, Vol. 11, No. 1, 1990, pp. 3–5.

[1.8] Schmidt, M. A., Raffel, J. I., Terry, F. L., and Senturia, S. D., "A metal-gate self-aligned MOSFET using nitride oxide", IEEE Transactions on Electron Devices, Vol. ED-32, No. 3, 1985, pp. 643–648.

[1.9] Ng, K. K., Pai, C.-S., Mansfield, W. M., and Clarke, G. A., "Suppression of hot-carrier degradation in Si MOSFET's by Germanium doping", IEEE Electron Device Letters, Vol. 11, No. 1, 1990, pp. 45–47.

[1.10] Pfiester, J. R., Sivan, R. D., Liaw, H. M., Seelbach, C. A., and C. D. Gunderson, "A self-aligned elevated source/drain MOSFET", IEEE Electron Device Letters, Vol. 11, No. 9, 1990, pp. 365–367.

[1.11] Dori, L., Megdanis, A., Brodsky, S. B., Arienzo, M., and Cohen, S. A., "Sputtered tungsten for deep submicron complementary metal-oxide-semiconductor technology," Thin Solid Films , Vols. 193/194, 1990, p. 501.

[1.12] Kim, K. T., Kang, L. G., Park, T. S., *et al.*, "Tungsten silicide/titanium nitride compound gate for submicron CMOSFET," In: Proc. IEEE Symp. VLSI Technology, 1990, p. 115.

[1.13] Ronkainen, H., Drozdy, G., and Franssila, S., "The use of disposable double spacer and self-aligned cobalt silicide for LDD MOSFET fabrication", IEEE Electron Device Letters, Vol. 12, No 3, 1991, pp. 125–127.

[1.14] Kolawa, E., Pokela, P. J., Reid, J. S., Chen, J. S., Ruiz, R. P., and Nicolet, M. A., "Sputtered Ta-Si-N diffusion barriers in Cu metallization for Si", IEEE Electron Device Letters, Vol. 12, No. 6, 1991, pp. 321–323.

[1.15] Davari, B., Koburger, C., Furukawa, T., *et al.*, "A variable-size shallow trench isolation (STI) technology with diffused sidewall doping for submicron CMOS", In: IEDM Tech. Dig, 1988, p. 92.

[1.16] Sturm, J. C., "Performance advantages of submicron silicon-on-insulator devices for ULSI", Materials Research Society Symposium Proceedings, Vol. 107, 1988, pp. 295–307.

[1.17] Venkatesan, S., Neudeck, G. W., and Pierret, R. F., "Dual-gate operation and volume inversion in n-channel SOI MOSFET's", IEEE Electron Device Letters, Vol. 13, No. 1, 1992, pp. 44–46.

[1.18] Takato, H., Sunouchi, K., Okabe, N., *et al.*, "Impact of surrounding gate transistor (SGT) for ultra-high-density LSI's", IEEE Transactions on Electron Devices, Vol. 38, 1991, pp. 573–578.

[1.19] Hori, A., Segawa, M., Shimomura, H., and Kameyama, S., "A self-aligned pocket implantation (SPI) technology for 0.2 μm dual-gate CMOS", IEEE Electron Device Letters, Vol. 13, No 4, 1992, pp. 174–176.

[1.20] Kistler, N., VerPloeg, E., Woo, J., and Plummer, J., "Sub-quarter-micrometer CMOS on ultrathin (400 A) SOI", IEEE Electron Device Letters, Vol. 13, No. 5, 1992, pp. 235–237.

[1.21] Tao, J., Cheung, N. W., Hu, C., Kang, H.-K., and Wong, S., "Electromigration performance of electroless plated Cooper/Pd-Silicide metallization", IEEE Electron Device Letters, Vol. 13, No. 8, 1992, pp. 433–435.

[1.22] Pai, P. L. and Ting, C. H., "Selective electroless cooper for VLSI interconnection", IEEE Electron Device Letters, Vol. 10, No. 9, 1989, pp. 423–425.

[1.23] Warnock, J., Shahidi, G. G., Davari, B., *et al.*, "BiCMOS technology with 60-GHz n-p-n bipolar and 0.25 μm CMOS", IEEE Electron Device Letters, Vol. 13, No. 11, 1992, pp. 578–580.

[1.24] König, U., Boers, A. J., Schäffler, F., and Kasper, E., "Enhancement mode n-channel Si/SiGe MOSFET with high intrinsic transconductance", Electronics Letters, Vol. 28, 1992, p. 160.

[1.25] Chang, J. Y.-C., Abidi, A. A., and Gaitan, M., "Large suspended inductors on silicon and their use in a 2-μm CMOS RF amplifier", IEEE Electron Device Letters, Vol. 14, No. 5, 1993, pp. 246–248.

[1.26] Shahidi, G. G., Davari, B., Bucelot, T. J., *et al.*, "Indium channel implant for improved short-channel behavior of submicrometer NMOSFET's", IEEE Electron Device Letters, Vol. 14, No. 8, 1993, pp. 409–411.

[1.27] Lee, W.-H., Park, Y.-J., and Lee, J. D., "A new 0.25-μm recessed-channel MOSFET with selectively halo-doped channel an deep graded source/drain", IEEE Electron Device Letters, Vol. 14, No. 12, 1993, pp. 578–580.

[1.28] Hamada, A. and Takeda, E., "Hot-electron trapping activation energy in PMOSFET's under mechanical stress", IEEE Electron Device Letters, Vol. 15, No. 1, 1994, pp. 31–33.

[1.29] Vogelsang, T. and Hofmann, K. R., "Electron transport in strained Si layers on $Si_{1-x}Ge_x$ substrates", Applied Physics Letters, Vol. 63, 1993, p. 186.

[1.30] Mii, Y., Rishton, S., Taur, Y., *et al.*, "Experimental high performance sub-0.1 μm channel nMOSFET's", IEEE Electron Device Letters, Vol. 15, No. 1, 1994, pp. 28–30.

[1.31] Furusawa, T., and Homma, Y., "Low capacitance multilevel interconnection using low-k organic spin-on glass for quarter-micron high speed ULSIs", 1995 Symposium on VLSI technology Digest of Technical papers, pp. 59–60.

[1.32] Burghartz, J. N., Jenkins, K. A., and Soyuer, M., "Multilevel-spiral inductors using VLSI interconnect technology", IEEE Electron Device Letters, Vol. 17, No. 9, 1996, pp. 428–430.

[1.33] Perera, A. H., and Tseng, H.-H., "Gate etch induced diode leakage prevention with 7-nm CVD stacked dielectric", IEEE Electron Device Letters, Vol. 17, No. 11, 1996, pp. 528–530.

[1.34] Kizilyalli, I. C., Lyding, J. W., and Hess, K., "Deuterium post-metal annealing of MOSFET's for improved hot carrier reliability", IEEE Electron Device Letters, Vol. 18, No. 3, 1997, pp. 81–83.

[1.35] Wheeler, C. B., Mathine, D. L., Johnson, S. R., Maracas, G. N., and Allee, D. R., "Selectively oxidized GaAs MESFET's transferred to a Si substrate", IEEE Electron Device Letters, Vol. 18, No. 4, 1997, pp. 138–140.

[1.36] Dimitrijev, S., Li, H.-F., Harrison, H. B., and Sweatman, D., "Nitridation of silicon-dioxide films grown on 6H silicon carbide", IEEE Electron Device Letters, Vol. 18, No. 5, 1997, 175–177.

[1.37] Lee, K.-Y., Fang, Y.-K., Chen, C.-W., Liang, M.-S., and Hiseh, J.-C., "A novel stack structure to improve degradation of W-polycide gated MOS device with undoped a-Si/heavily-doped poly-Si multilayer", IEEE Electron Device Letters, Vol. 18, No. 5, 1997, pp. 181–183.

[1.38] Morimoto, T., Momose, H. S., Ozawa, Y., Yamabe, K., and Iwai, H., "Effects of boron penetration and resultant limitations in ultra thin pure-oxide and nitride-oxide gate-films", IEDM Proc., 1990, pp. 429–432.

[1.39] La Rosa, G., Guarin, F., Rauch, S., Acovic, A., Lukaitis, J., and Crabbé, E., "NBTI-channel hot carrier effects in pMOSFETs in advanced CMOS technologies", IEEE International Reliability Physics Symposium, 1997, 35th Annual Proceedings, pp. 282–286.

[1.40] Kim, H.-S., Campbell, S. A., and Gilmer, D. C., "Charge trapping and degradation in high-permittivity TiO2 dielectric films", IEEE Electron Device Letters, Vol. 18, No. 10, 1997, pp. 465–467.

[1.41] Lee, D. H., Yeom, K. H., Cho, M. H., Kang, N. S., and Shim, T. E., "Gate oxide integrity of MOS transistors with W/TiN stacked gate", In: Symp. VLSI Technol. Dig. Tech. Papers, 1996, pp. 208–209.

[1.42] Wann, C., Assaderaghi, F., Shi, L., *et al.*, "High-performance 0.7-μm CMOS with 9.5-ps gate delay and 150 GHz fT", IEEE Electron Device Letters, Vol. 18, No. 12, 1997, pp. 625–627.

[1.43] Parker, C. G., Lucovsky, G., and Hauser, J. R., "Ultrathin oxide-nitride gate dielectric MOSFET's", IEEE Electron Device Letters, Vol. 19, No. 4, 1998, pp. 106–108.

[1.44] Lin, H.-C., Lin, R., Wu, W.-F., *et al.*, "A novel self-aligned T-shaped gate process for deep submicron Si MOSFET's fabrication", IEEE Electron Device Letters, Vol. 19, No. 1, 1998, pp. 26–28.

[1.45] van Dover, R., and Schneemeyer, L. F., "Deposition of uniform Zr-Sn-TiO films by on-axis reactive sputtering", IEEE Electron Device Letters, Vol. 19, No. 9, 1998, pp. 329–331.

[1.46] Lu, Q., Park, D., Kalnitsky, A., *et al.*, "Leakage current comparison between ultra-thin Ta2O5 films and conventional gate dielectrics", IEEE Electron Device Letters, Vol. 19, No. 9, 1998, pp. 341–343.

[1.47] Ma, Y., Evans, D. R., Nguyen, T., Ono, Y., and Hsu. S. T., "Fabrication and characterization of sub-quarter-micron MOSFET's with a Copper gate electrode", IEEE Electron Device Letters, Vol. 20, No. 5, 1999, pp. 254–256.

[1.48] Yang, M., Chang, C.-L, Carroll, M., and Sturm, J. C., "25-nm p-channel vertical MOSFET's with SiGeC source-drains", IEEE Electron Device Letters, Vol. 20, No. 6, 1999, pp. 301–303.

[1.49] Qi, W.-J., Nieh, R., Lee, B. H., *et al.*, "MOSCAP and MOSFET characteristic using ZrO2 gate dielectric deposited directly on Si", Proceedings of the IEEE IEDM, 1999, pp. 145–148.

[1.50] Chang, S.-J., Chang, C.-Y., Chao, T.-S., and Huang, T.-Y., "High performance 0.1 μm dynamic threshold MOSFET using Indium channel implantation", IEEE Electron Device Letters, Vol. 21, No. 3, 2000, pp. 127–129.

[1.51] Chin, A., Wu, Y. H., Chen, S. B., Liao, C. C., and Chen, W. J., "High quality La2O3 and Al2O3 gate dielectrics with equivalent oxide thickness 5-10 Å", 2000 Symposium on VLSI Technology Digest of Technical Papers, 2000, pp. 16–17.

[1.52] Hisamoto, D., Lee, W.-C., Kedzierski, J., *et al.*, "FINFET—a self-aligned double-gate MOSFET scalable to 20 nm", IEEE Transactions on Electron Devices, Vol. 47, No. 12, 2000, pp. 2320–2325.

[1.53] Kang, L., Lee, B. H., Qi, W.-J., *et al.*, "Electrical characteristics of highly reliable ultrathin hafnium oxide gate dielectric", IEEE Electron Device Letters, Vol. 21, No. 4, 2000, pp. 181–183.

[1.54] Yeo, Y.-C., Lu, Q., Ranade, P., *et al.*, "Dual-metal gate CMOS technology with ultrathin silicon nitride gate dielectric", IEEE Electron Device Letters, Vol. 22, No. 5, 2001, pp. 227–229.

[1.55] Misra, V., Zhong, H., and Lazar, H., "Electrical properties of Ru-based alloy gate electrodes for dual metal gate Si-CMOS", IEEE Electron Device Letters, Vol. 23, No. 6, 2002, pp. 354–356.

[1.56] Lue, H.-T., Liu, C.-Y., and Tseng, T.-Y., "An improved two-frequency method of capacitance measurement for $SrTiO_3$ as high-k gate dielectric", IEEE Electron Device Letters, Vol. 23, No. 9, 2002, pp. 553–555.

[1.57] Onishi, K., Seok, C., Choi, R., *et al.*, "Performance of polysilicon gate HfO2 MOSFETs on (100) and (111) silicon substrates", IEEE Electron Device Letters, Vol. 24, No. 4, 2003, pp. 254–256.

[1.58] Kim, S. J., Cho, B. J., Li, M.-F., Zhu, C., Chin, A., and Kwong, D.-L., "Lanthanide (Tb)-doped HfO2 for high-density MIM capacitors", IEEE Electron Device Letters, Vol. 24, No. 7, 2003, pp. 442–444.

[1.59] Thompson, S. E., Armstrong, M., Auth, C., *et al.*, "A logic nanotechnology featured strained-silicon", IEEE Electron Device Letters, Vol. 25, No. 4, 2004, pp. 191–193.

[1.60] Kim, K. R., Kim, D. H., Song, K.-W., *et al.*, "Silicon-based field-induced band-to-band tunneling effect transistor", Electron Device Letters, Vol. 25, No. 6, 2004, pp. 439–441.

[1.61] Zhu, S., Chen, J., Li, M.-F., *et al.*, "N-type Schottky source/drain MOSFET using Ytterbium silicide", IEEE Electron Device Letters, Vol. 25, No. 8, 2004, pp. 565–567.

[1.62] Collaert, N., Dixit, A., Goodwin, M., *et al.*, "A functional 41-stage ring oscillator using scaled FinFET devices with 25-nm gate lengths and 10-nm Fin widths applicable for the 45-nm CMOS node", IEEE Electron Device Letters, Vol. 25, No. 8, 2004, 568–570.

[1.63] Ren, C., Yu, H. Y., Kang, J. F., *et al.*, "A dual-metal gate integration process for CMOS with sub-1-nm EOT HfO_2 by using HfN replacement gate", IEEE Electron Device Letters, Vol. 25, No. 8, 2004, pp. 580–582.

[1.64] Ren, C., Yu, H. Y., Wang, X. P., *et al.*, "Thermally robust TaTbxN metal gate electrode for n-MOSFETs applications", IEEE Electron Device Letters, Vol. 26, No. 2, 2005, pp. 75–77.

[1.65] Naeemi, A., Sarvari, R., and Meindl, J. D., "Performance comparison between carbon nanotube and copper interconnects for gigascale integration (GSI)", IEEE Electron Device Letters, Vol. 26, No. 2, 2005, pp. 84–86.

[1.66] Yu, D. S., Liao, C. C., Cheng, C. F., Chin, A., Li, M. F., and McAlister, S. P., "The effect of IrO2-IrO2-Hf-LaAlO3 gate dielectric on the bias-temperature instability of 3-D GOI CMOSFETs", IEEE Electron Device Letters, Vol. 26, No. 6, 2005, pp. 407–409.

[1.67] Wu, X., Chan, P. C. H., Zhang, S., Feng, C., and Chan, M., "Stacked 3-D Fin-CMOS technology", IEEE Electron Device Letters, Vol. 26, No. 6, 2005, pp. 416–418.

[1.68] Zhu, F., Rhee, S. J., Kang, C. Y., *et al.*, "Improving channel carrier mobility and immunity to charge trapping of high-k nMOSFET by using stacked Y_2O_3/HfO_2 gate dielectric", IEEE Electron Device Letters, Vol. 26, No. 12, 2005, pp. 876–878.

[1.69] Wu, C. H., Yu, D. S., Wang, S. J., *et al.*, "High work function Ir_xSi gates on HfAlON p-MOSFETs", IEEE Electron Device Letters, Vol. 27, No. 2. 2006, pp. 90–92.

[1.70] Dang, B., Bakir, M. S., and Meindl, J. D., "Integrated thermal-fluidic I/O interconnects for an on-chip microchannel heat sink", IEEE Electron Device Letters, Vol. 27, No. 2, 2006, pp. 117–119.

[1.71] Yang, H., Lee, D., Rahman, M. S., Hasan, M., Jung, H.-S., and Hwang, H., "ScN_x gate on atomic layer deposited HfO_2 and effect of high-pressure wet post deposition annealing", IEEE Electron Device Letters, Vol. 27, No. 6, 2006, pp. 435–437.

[1.72] Lee, T., Rhee, S. J., Kang, C. Y., *et al.*, "Structural advantage for the EOT scaling and improved electron channel mobility by incorporating Dysprosium oxide (Dy_2O_3) into HfO_2 n-MOSFETs", IEEE Electron Device Letters, Vol. 27, No. 8, 2006, pp. 640–642.

[1.73] Thareja, G., Wen, H. C., Harris, R., Majhi, P., Lee, B. H., and Lee, J. C., "NMOS compatible work function of TaN metal gate with Gadolinium oxide buffer layer on Hf-based dielectrics", IEEE Electron Device Letters, Vol. 27, No. 10, 2006, pp. 802–804.

[1.74] Cea, S. M., Soman, R., Nagisetty, R., Tyagi, S., and Natarajan, S., "Transistor with strain-inducing structure in channel," U.S. Patent 7 019 326, 2006.

[1.75] Liu, C.-H. and Chiu, F.-C., "Electrical characterization of ZrO_2/Si interface properties in MOSFETs with ZrO_2 gate dielectrics", IEEE Electron Device Letters, Vol. 26, No. 1, 2007, pp. 62–64.

[1.76] Yen, F. Y., Hung, C. L., Hou, Y. T., *et al.*, "Effective work function engineering of TaxCy metal gate on Hf-based dielectrics", IEEE Electron Device Letters, Vol. 28, No. 3, 2007, pp. 201–203.

[1.77] Wong, H.-S., Liu, F.-Y., Ang, K.-W., Samudra, G., and Yeo, Y.-C., "Novel nickel silicide contact technology using selenium segregation for SO N-FETs with silicon-carbon source/drain stressors", IEEE Electron Device Letters, Vol. 28, No, 8, 2008, pp. 841–843.

[1.78] Fukutome, H., Yoshida, E., Hosaka, K., Tajima, M., Momiyama, Y., and Satoh, S., "Effects of gate line width roughness on threshold-voltage fluctuation among short-channel transistors at high drain voltage", IEEE Electron Device Letters, Vol. 31, No. 3, 2010, pp. 240–242.

[1.79] Brown, A. R., Idris, N. M., Watling, J. R., and Asenov, A., "Impact of metal gate granularity on threshold voltage variability: A full-scale three-dimensional statistical simulation study", IEEE Electron Device Letters, Vol. 3, No. 11, 2010, pp. 1199–1201.

[1.80] Dang, B., Shapiro, M., Andry, P., *et al.*, "Three-dimensional chip stack with integrated decoupling capacitors and thru-Si via interconnects", IEEE Electron Device Letters, Vol. 31, No. 12, 2010., pp. 1461–1463.

[1.81] Özben, E. D., Lopes, J. M. J., Nichau, A., *et al.*, "Integration of LaLuO3 (k~30) as high-k dielectric on strained and unstrained SOI MOSFETs with a replacement gate process", IEEE Electron Device Letters, Vol. 32, No. 1, 2011, pp. 15–17.

[1.82] Gundapaneni, S., Ganguly, S., and Kottantharayil, A., "Bulk planar junctionless transistor (BPJLT): an attractive device alternative for scaling", IEEE Electron Device Letters, Vol. 32, No. 3, 2011, pp. 261–263.

[1.83] Wang, X., Roy, S., Brown, A. R., and Asenov, A., "Impact of STI on statistical variability and reliability of decanometer MOSFETs", IEEE Electron Device Letters, Vol. 32, No. 4, 2011, pp. 479–481.

[1.84] Wong, H., Yang, B. L., Kakushima, K., and Iwai, H., "Subthreshold characteristics of MOS transistors with CeO2/La2O3 stacked gate dielectric", IEEE Electron Device Letters, Vol. 32, No. 8, 2011, pp. 1002–1004.

[1.85] Liu, H. and Ye, P. D., "MoS_2 dual-gate MOSFET with atomic-layer-deposited Al2O3 as top-gate dielectric", IEEE Electron Device Letters, Vol. 33, No. 4, 2012, pp. 546–548.

[1.86] Majumdar, K., Hobbs, C., and Kirsch, P. D., "Benchmarking transition metal dichalcogenide MOSFET in the ultimate physical scaling limit", IEEE Electron Device Letters, Vol. 35, No. 3, 2014, pp. 402–404.

Chapter 2

Device physics, modeling, and technology for nano-scaled semiconductor devices

E. A. Gutiérrez-D.[1]*, F. Gámiz*[2]*, V. Sverdlov*[3]*, S. Selberherr*[3] *and A. Torres-J.*[1]

This chapter introduces the device physics, modeling, and technology for the different silicon-based device structures. Quantum-mechanical treatment for the device physics is done as well as the different and alternative approaches for advanced device simulation. The last section takes over the potential use that can be given to new materials and device structures. A preliminary set of applications are reviewed, such as Si-based materials with nanostructured properties, amorphous SiGe alloys applications such as thermal and photodetector sensors. Furthermore, the possibility to make use of CMOS fabrication steps for 3D Si die stacking is also reviewed.

2.1 Bulk MOSFETs and related device structures

The Metal-Oxide-Semiconductor Field-Effect Transistor (MOSFET) has been and still is the workhorse device behind the semiconductor industry. The MOSFET was born in the United States under the fatherhood of Atalla and Kahng in 1960 [2.1], but its birth certificate was issued until 1963 as a patent number 3,102,230 [2.2]. Once the MOSFET showed good and reproducible electrical characteristics, the idea to form a circuit composed of multiple field-effect transistors (FETs) flourished with the realization of a commercial integrated circuit (IC) in 1963. Later in 1968 RCA produced the first commercial IC made of 120 p-type FETs as a 20-bit shift register [2.3]. Later in 1968 RCA produced the first commercial complementary metal-oxide-semiconductor (CMOS) IC, the 4000 series of CMOS logic gates [2.4]. Of course the concept of the IC is attributed to Jack Kilby who in 1958 patented the idea [2.5], and later in 2000 was granted the Nobel Prize. In 1963 Wanlass and Sah also introduced [2.6], in an experimental way, a set of CMOS logical gates, flip-flops, and ring oscillator with a maximum operating voltage of 40 V.

[1]National Institute for Astrophysics, Optics and Electronics, INAOE, Puebla, Mexico
[2]Department of Electronics, University of Granada, Spain
[3]Institute for Microelectronics, Technical University Vienna, Austria

This technology showed a transconductance $g_m = 5 \times 10^{-4}$ A/V, and a leakage current for a single inverter that would amount to 1 W of standby power for 10^7 inverters, which is about 100 nW per inverter, or 2.5 nA leakage current per inverter. The propagation delay time is about 100 ns. While today a 28 nm CMOS technology, for instance, can have a $g_m = 7 \times 10^{-3}$ A/V, and a delay time in the order of ps.

In the two following sections a description of the conventional FET physics and its limitations will be introduced, and then alternative FET-related structures that alleviate the limited electrical performance of traditional FET devices are introduced in the second section. In the subsequent sections advanced FET-related devices, quantum effects, different approaches for semiconductor modeling and simulation, alternative materials and device structures, are reviewed.

2.1.1 Conventional field-effect transistor (FET) physics and limitations

As already mentioned the MOSFET concept was demonstrated in early 1960s based on the surface semiconductor charge controlled by the electric field as shown in Figure 2.1.

The charge Q_s at the semiconductor surface, close to the semiconductor–gate oxide interface, is controlled by the electric field E_y in the y-axis as shown at the right side, which in turn is controlled by the gate (G) voltage. However, when a potential difference between the drain (D) and source (S) electrodes increases (in the x-axis direction), the charge Q_s becomes dependent of both electric field components, E_y and E_x. Basically what we have is a cloud of charges, either electrons or holes, that modulates the drain-source conductance controlled by the four potentials V_G, V_S, V_D, and V_B. In a conventional MOSFET of the 1980s–1990s, with a typical channel length L from 10 to 0.5 μm, a gate oxide T_{ox} from some 60 nm down to 14 nm, the electrical characteristics are like those depicted in Figure 2.2. Both are conventional, non-LDD, polysilicon gate technologies, with a SiO_2 gate oxide.

These are well-behaved MOSFETs with electrical characteristics close to the ideal behavior for non-low-voltage analog applications. The 10 μm transistor has been fabricated with an experimental process at INAOE [2.7], while the 0.5 μm

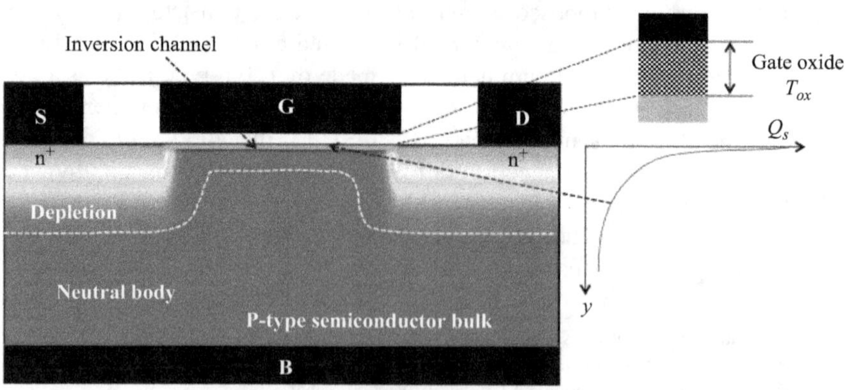

Figure 2.1 A conventional n-type MOSFET device

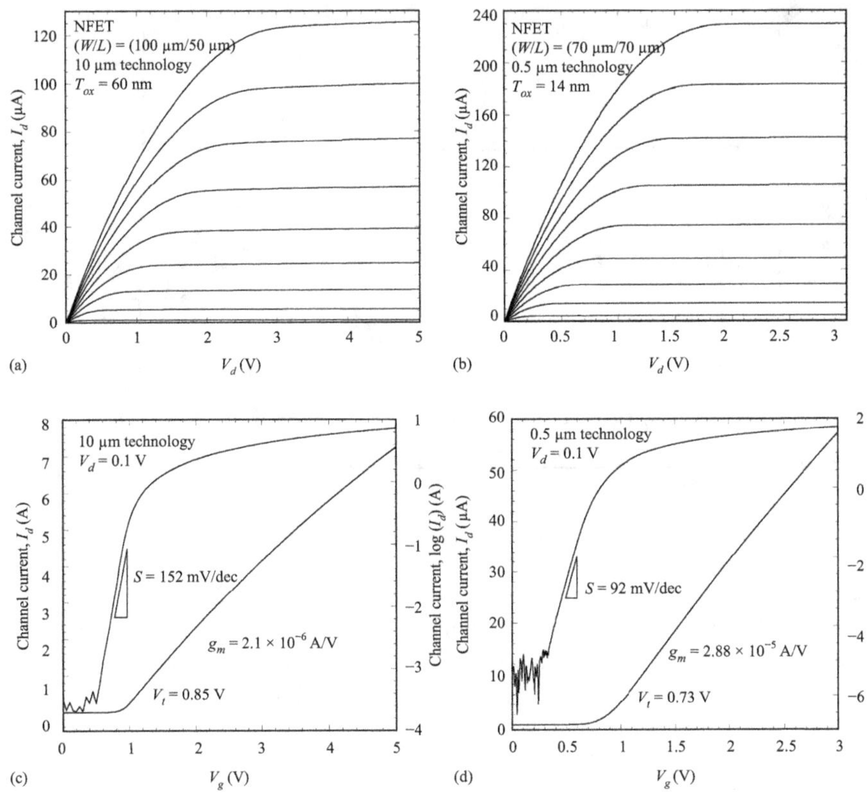

Figure 2.2 Electrical characteristics of two conventional MOSFET technologies of 10 μm and 0.5 μm. Plots (a) and (b) have V_g as a parameter from low to high values, while plots (c) and (d) correspond to I_d–V_g characteristics in both log and linear scales

transistor is from a commercial AMI technology [2.8]. From the I_d–V_d characteristics a desirable high output resistance is observed for the 0.5 μm technology. In both cases the saturation, or high output resistance region (a and b), is well defined. The subthreshold regime (c and d) in both cases is also well defined. The 0.5 μm transistor, because of the thinner gate oxide, has a larger current drive capability than that of the 10 μm technology. The subthreshold slope *S*, a speed-switching feature, is smaller for a 10 μm technology, while the threshold voltage V_t is smaller for the 0.5 μm technology with respect to that of 10 μm.

Smaller transistors with advanced technology features, such as thinner gate oxide, show better performance, faster switching, smaller threshold voltage V_t, higher current drive capability, and higher transconductance g_m ($\partial I_d/\partial V_g$), which result in lower bias voltage, faster circuits, and larger transistor density integration. The device physics and technology behind the enhanced electrical performance, as well as the limitations for a further down scaling, are described as follows.

At the semiconductor–oxide interface the energy bands bend as shown in Figure 2.3.

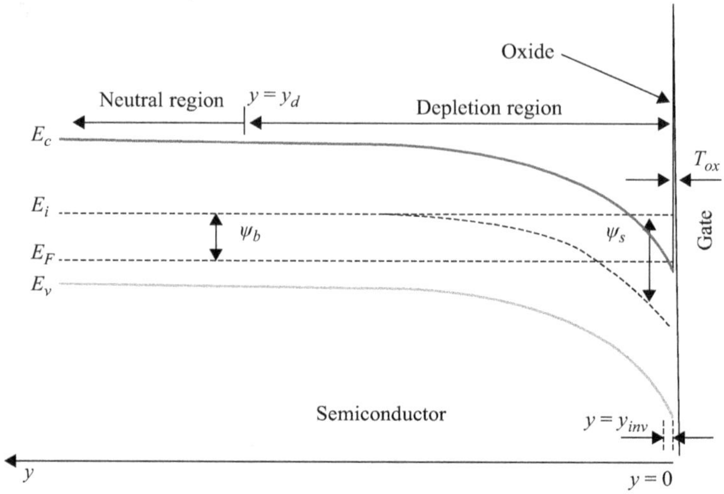

Figure 2.3 Energy band bending at the semiconductor–oxide interface for a p-type semiconductor case

Several features are observed, which are fundamental for the electrical performance of a MOSFET. Under the influence of a positive potential V_g applied to the gate terminal, the energy bands, E_c and E_v, bend down at the semiconductor-oxide interface. The energy band bending produces, in the semiconductor region, a region depleted of holes with an extension y_d, and very close the surface, within an extension y_{inv}, an inverted region populated with electrons. Such an inversion layer populated with electrons connects electrically the source with the drain terminal. The conductance of the MOSFET channel depends on both the gate (V_g) and the drain voltage (V_d) in such a way that it produces the electrical characteristics shown in Figure 2.2. A potential barrier at the semiconductor-oxide, and through the gate oxide thickness T_{ox}, develops as well. Such a potential barrier at the oxide prevents electrons from the inversion layer to flow to the gate. However, if the gate oxide T_{ox} is thinner than 10 nm, a considerable amount of electrons can tunnel through the gate oxide to the gate electrode. Another relevant feature is the thickness of the inversion layer y_{inv}, which is in the range within the 10 nm as shown in Figure 2.4.

Figure 2.4 shows the simulated current density for an n-type MOSFET with a gate oxide $T_{ox} = 14$ nm for a particular bias condition $V_g = 0.3$ V and $V_d = 0.1$ V. Most of the current that flows from source-to-drain is confined into a well potential of a thickness y_{inv} of about 5 nm. Such a thickness, which determines the channel conductivity, is bias and technology dependent. It is uniform along the channel if the drain-to-source potential is zero. The gate voltage dependence of y_{inv}, the inversion charge n, the surface potential ψ_s, and the channel current I_d are shown in Figure 2.5. The inversion channel gets thinner as the V_g voltage increases, while the inversion charge in the channel n increases exponentially, which compensates the thinning of the inversion channel. The surface potential ψ_s first increases linearly in the subthreshold regime, $0 < V_g < 1.0$ V, and then tends

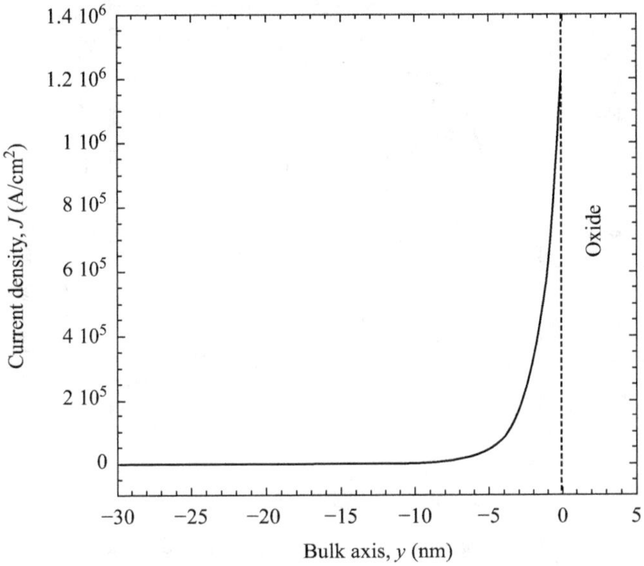

Figure 2.4 Current density J distribution from the semiconductor surface into the bulk axis y

to saturate at higher V_g voltages. The ψ_s potential is a measure of the energy band bending at the surface, and because both n and I_d are exponentially dependent on ψ_s, they follow the ψ_s–V_g shape.

It follows that the subthreshold slope S, as well as the transconductance g_m, is tightly related to ψ_s, and its derivative with respect to V_g.

The optimization of the electrical performance of a MOSFET, i.e., a high current drive capability, a small subthreshold slope S, a high transconductance g_m, and a low threshold voltage V_t, is intimately linked to the ability of the gate electrode to control the energy band bending at the semiconductor surface, and the capability of the channel to drive the charge from the source to the drain electrode. The impact of the gate oxide thickness T_{ox} on the subthreshold slope S, the transconductance g_m, the threshold voltage V_t, and the current drive capability I_d at the gate overdrive voltage $V_g + 1.0$ V is shown in Figure 2.6. The thinning of T_{ox} makes the MOSFET faster as the subthreshold slope S becomes smaller, which implies a faster digital off-on-off commutation. A larger transconductance g_m means the MOSFET is able to charge and discharge a capacitive load in a shorter time. For very thin gate oxides the reduction of the threshold voltage with T_{ox}, makes the MOSFET very leaky at low V_g voltages. The channel current I_d at the same gate overdrive voltage ($V_g = V_t + 1.0$ V), for the different gate oxide thicknesses, increases for thinner T_{ox}, which is good in terms of an enhanced current drive capability. It is not only the drastic reduction of the threshold voltage V_t, which degrades the electrical performance, but two other effects like tunneling through the gate oxide [2.9] and polysilicon depletion at the gate electrode [2.10].

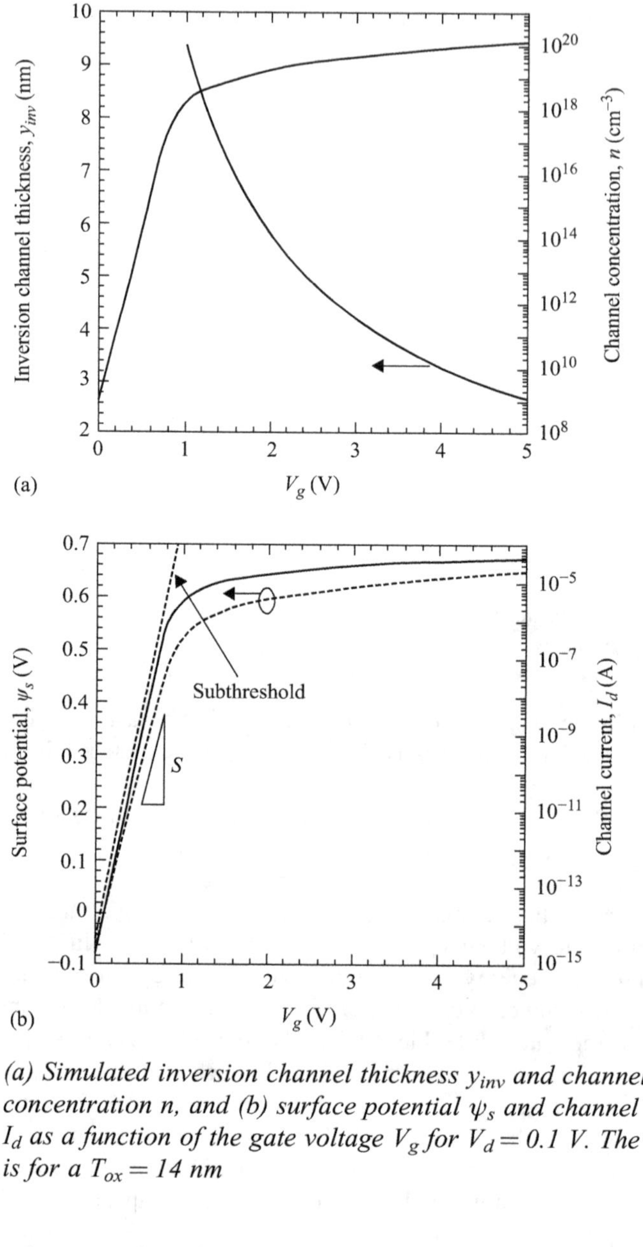

Figure 2.5 *(a) Simulated inversion channel thickness y_{inv} and channel concentration n, and (b) surface potential ψ_s and channel current I_d as a function of the gate voltage V_g for $V_d = 0.1$ V. The simulation is for a $T_{ox} = 14$ nm*

These two performance degradation mechanisms are shown in Figure 2.7. As the gate oxide thickness T_{ox} becomes thinner, the probability for carriers to tunnel through the oxide potential barrier increases and becomes evident for $T_{ox} < 10$ nm as shown in Figure 2.7b. For oxides thinner than 4 nm the direct tunneling (DT) [2.11] dominates, while for thicker oxides the tunneling is assisted by trap-assisted tunneling (TAT) [2.12]. For T_{ox} in the order of 2 nm the gate oxide leakage current can be in the range of 300 pA.

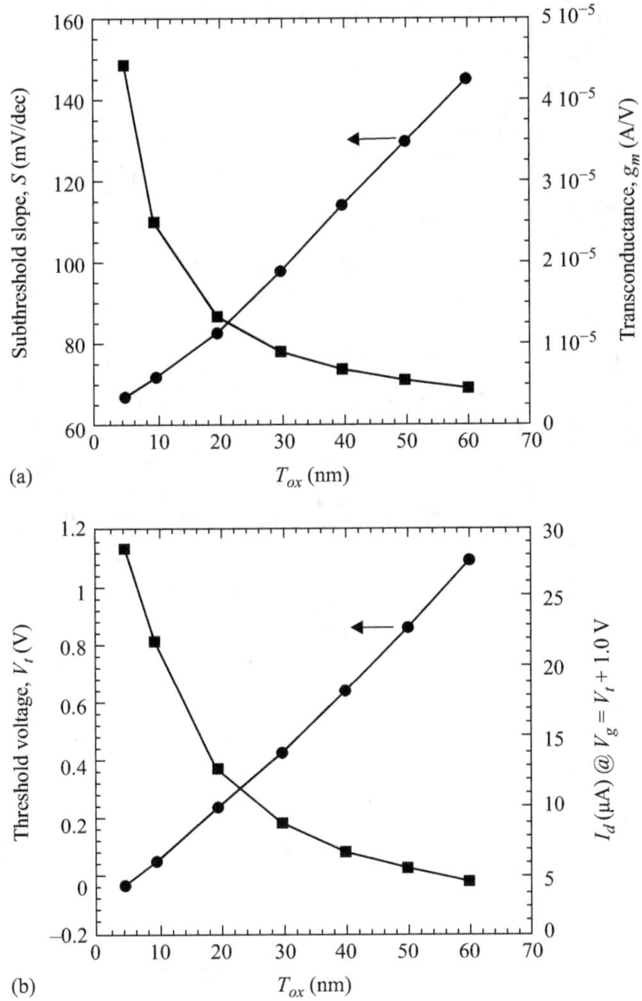

Figure 2.6 Simulated subthreshold slope S ((a) left axis), transconductance g_m ((a) right axis), threshold voltage V_t ((B) left axis), and current drive capability I_d at $V_g = V_t + 1.0$ V ((b) right axis), versus gate oxide thickness Tox

With the increase of the V_g voltage the lower part of the polysilicon gate at the interface with the oxide depletes. This mechanism is more pronounced for thinner gate oxides where the polysilicon gate is not uniformly doped, but has a lower doping profile at the gate–oxide interface. This makes the gate oxide effectively to increase from its nominal value T_{ox} to $T_{oxeff} = (T_{ox} + W_{pd})$, where W_{pd} is the thickness of the polysilicon depletion region. The straightforward effect is a reduction of current drive capability at high V_g values and a reduction of the transconductance as shown in Figure 2.7b. However, when the MOSFET drain voltage V_d increases, the distribution of the oxide electric field is not homogeneous anymore as shown in Figure 2.8.

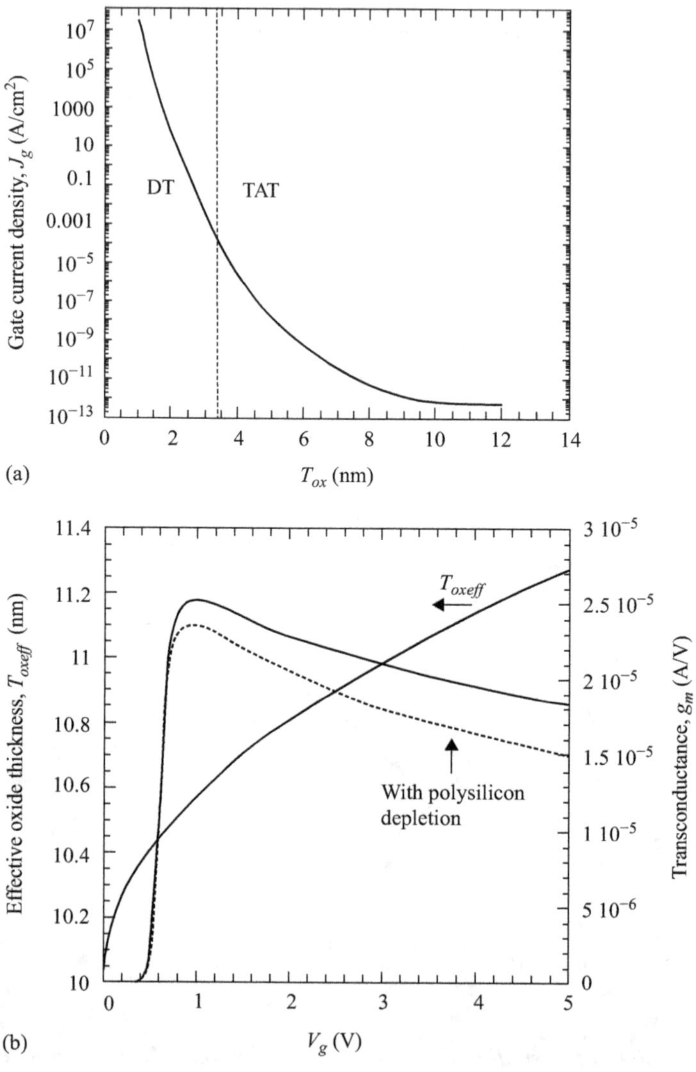

Figure 2.7 (a) Simulated gate current density J_g versus oxide thickness T_{ox}, and (b) effective oxide thickness T_{oxeff} and transconductance g_m versus V_g

The depletion of the polysilicon gate at the bottom side of the gate–oxide interface makes the oxide to virtually extend a little bit into the polysilicon material, which results in a thicker effective oxide $T_{oxeff} = (T_{ox} + W_{pd})$. The non-uniform polysilicon doping, in the y-axis, as shown at the upper right side, is responsible for the polysilicon depletion effect. However, when the drain voltage V_d increases, the distribution of the oxide electric field E_{ox} along the channel axis (x-axis) is not uniform anymore, and thus the polysilicon depletion effect becomes x-dependent through the bias condition as shown by the dotted line drawn through the oxide. The polysilicon

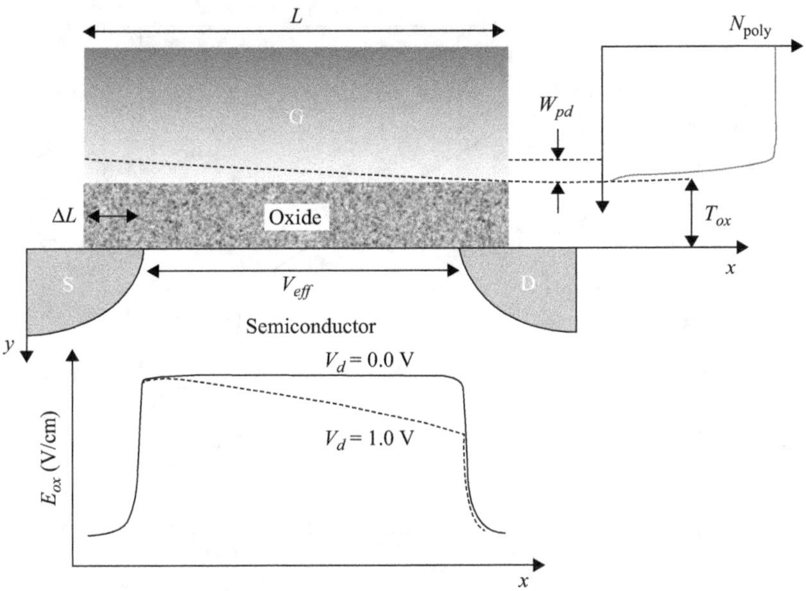

Figure 2.8 Schematic representation of the polysilicon depletion effect at the bottom side of the gate electrode (G), and simulation of the oxide electric field as a function of the x-axis for $V_d = 0.0$ V and 1.0 V

becomes less depleted at the drain than at the source side. The physics inside of the MOSFET becomes two-dimensional (2D). The 2D nature of the physics inside of the MOSFET becomes more evident when looking at the conduction band bending E_c and channel electric field $-E$ for two MOSFET with channel length L of 1.0 μm and 30 nm (see Figure 2.9). Three relevant differences are readily observed. The magnitude and the shape of the longitudinal or channel electric field $-E$, which for an $L =$ 30 nm MOSFET is wide and extends over almost all the channel length with a magnitude above 1×10^5 V/cm. For the $L = 1$ μm MOSFET the electric field $-E$ is narrow in shape with a magnitude above 1×10^5 V/cm extending only about 10 nm close to the drain side. The maximum electric field for an $L = 30$ nm peaks at 1.5×10^6 V/cm, while for $L = 1.0$ μm peaks at 2.5×10^5 V/cm. The conduction band edge E_c for $L = 1.0$ μm is linear in the channel, while that for $L = 30$ nm is never linear, which indicates a strong penetration of the longitudinal electric field lines into the channel. This implies that the charge in the inversion channel, for short channel devices, is not only controlled by the gate voltage (transversal electric field) but by the longitudinal electric field as well.

The second relevant feature for a short channel MOSFET is the generation of hot carriers [2.13]. A high electric field means the carriers, either electrons for an n-type or holes for a p-type MOSFET, acquire enough energy above the lattice temperature producing electron–hole pair ionization along the channel where they multiply by the avalanche mechanism. This is measured as an impact ionization bulk current I_b as shown in Figure 2.10a.

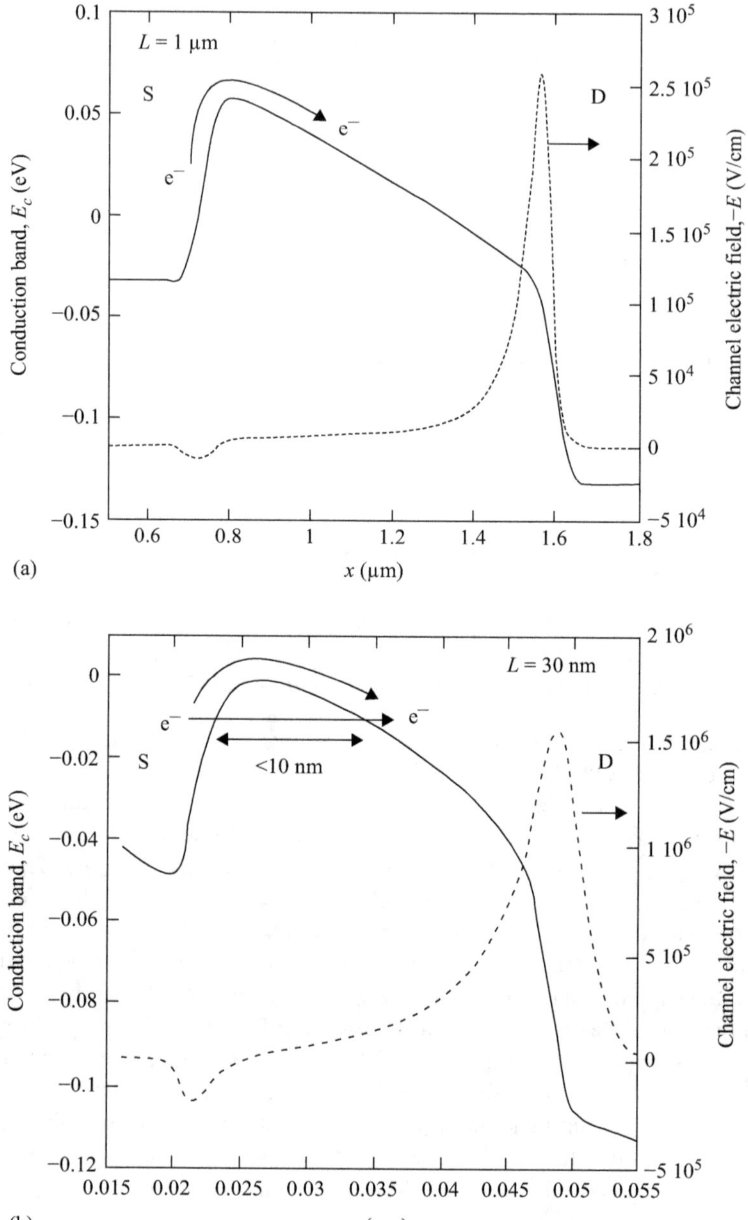

Figure 2.9 Simulated conduction band energy E_c and channel electric field versus channel axis x for two MOSFET of (a) L = 1 μm, and (b) L = 30 nm

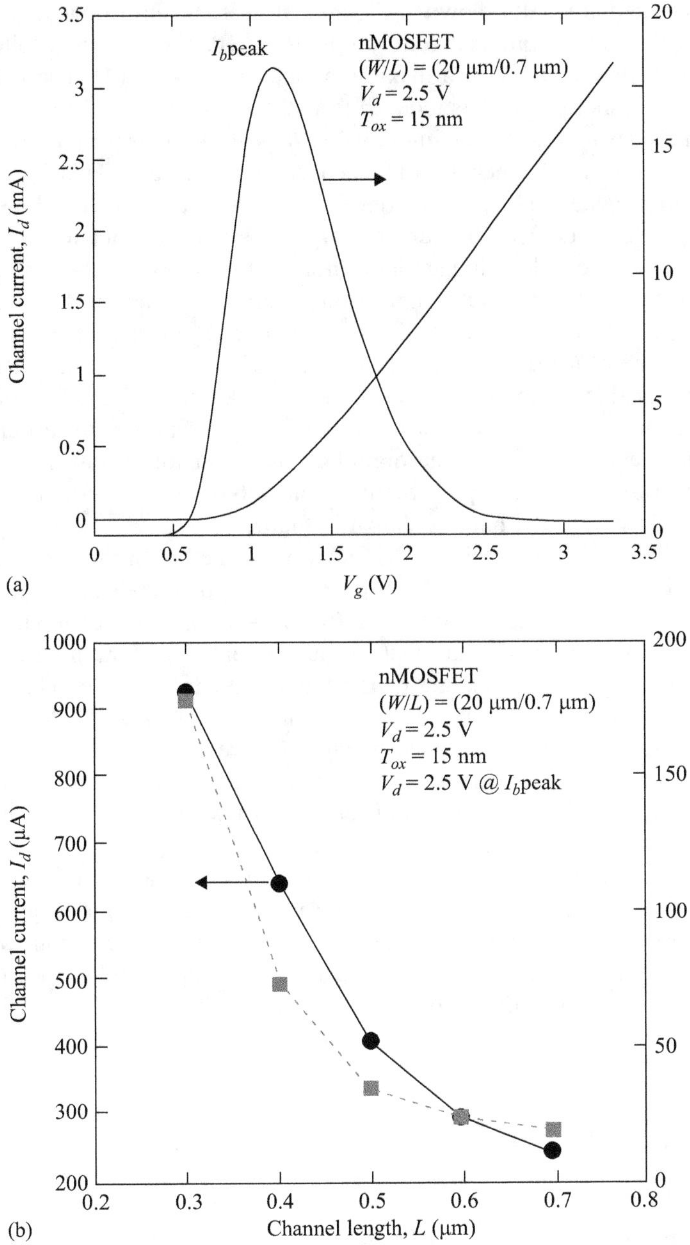

Figure 2.10 (a) Experimental I_d and bulk current I_b versus V_g, and (b) I_d and I_b versus channel length L

The experimental results shown in Figure 2.10 were taken from a 0.7 μm LDD technology with $T_{ox} = 15$ nm. As seen from Figure 2.10a the I_b current follows the classical bell-shaped curve, with a maximum value I_b peak [2.14]. The I_d follows the expected ($1/L$) increase as L is reduced from 0.7 μm down to 0.3 μm. However, the I_b current, measured at its maximum value I_b peak, follows a rather more pronounced increase ($1/L^{3.5}$) when L is reduced down to 0.3 μm. This implies that impact ionization generated by hot carriers worsens at shorter device dimensions, which prompts for a device structure redesign to reduce impact ionization. An alternative structure to reduce impact ionization will be shown in the next section.

Together with hot carrier and impact ionization comes the third relevant feature, which is self-heating [2.15]. Self-heating is a deleterious effect of the electrical performance of MOSFETs. It reduces the carrier mobility in the channel by increasing the local temperature and the phonon population, and adversely impacts its reliability. The self-heating originates at the device region where carriers reach their maximum energy, close to the drain-bulk junction for a conventional bulk MOSFET. And because of the poor thermal conductivity of the gate oxide thickness, the heat flows into the bulk as shown in Figure 2.11.

The simulation shown in Figure 2.11 was performed with the gtsFramework tool [2.16]. The x–y profile of the differential temperature ΔT is shown in Figure 2.12. The differential temperature $\Delta T = (T_{Si} - T_{amb})$ is the difference of the internal device temperature T_{Si} and that of the ambient T_{amb}. The increase of the local temperature due to self-heating concentrates close to the drain/bulk junction and spreads out toward the source side and into the bulk. The gate oxide blocks the heat flow toward the gate electrode as shown in Figure 2.12b.

The differential temperature ΔT or the device temperature above the room temperature (300 K) follows an almost linear behavior with respect to the electrical power consumed by the device (see Figure 2.13a).

As shown by the simulated results in Figure 2.13b the rate of increase of the local temperature per mW of device power consumption ($\mathrm{d}\Delta T/\mathrm{d}P$) is proportionally larger for shorter devices. Combining the use of the gtsFramework simulation tool and data published in References 2.17 and 2.18, the thermal conductance G_{th} as a function of the channel length is shown in Figure 2.14a. It is seen that the ability of

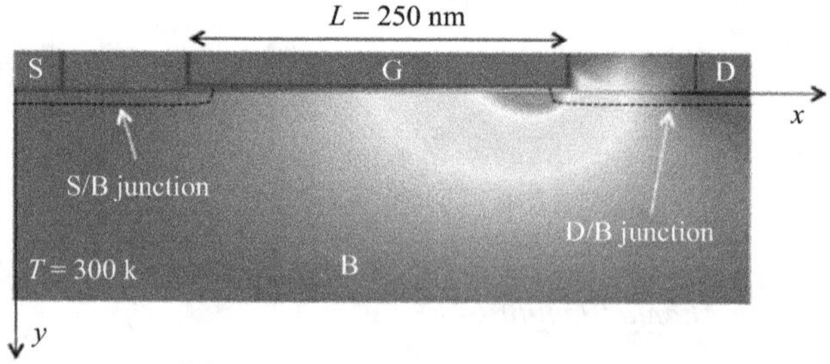

Figure 2.11 Simulated lattice temperature for a (W/L) = (10 μm/250 nm) nMOSFET biased at $V_g = 2.0$ V and $V_d = 5.0$ V

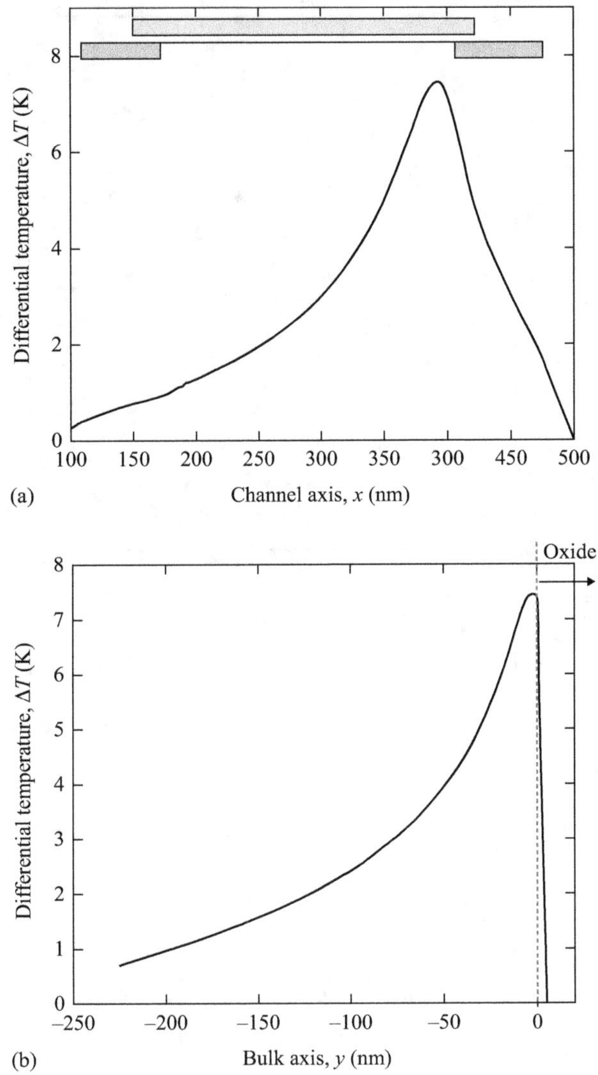

Figure 2.12 *(a) Simulated differential temperature ΔT along the channel axis, and (b) differential temperature ΔT along the bulk axis*

a MOSFET to conduct heat decreases as the channel length becomes shorter. For the gtsFramework simulation case a nMOSFET with a bulk thickness $T_b = 10$ μm has been assumed, while the simulation is performed for different channel lengths with a constant width $W = 1.0$ μm.

The self-heating effect worsens as the device length reduces from the micro- to the nanometer scale size. The thermal conductance G_{th} is also less dependent on temperature as the device dimensions shrinks down as shown in Figure 2.14b. The

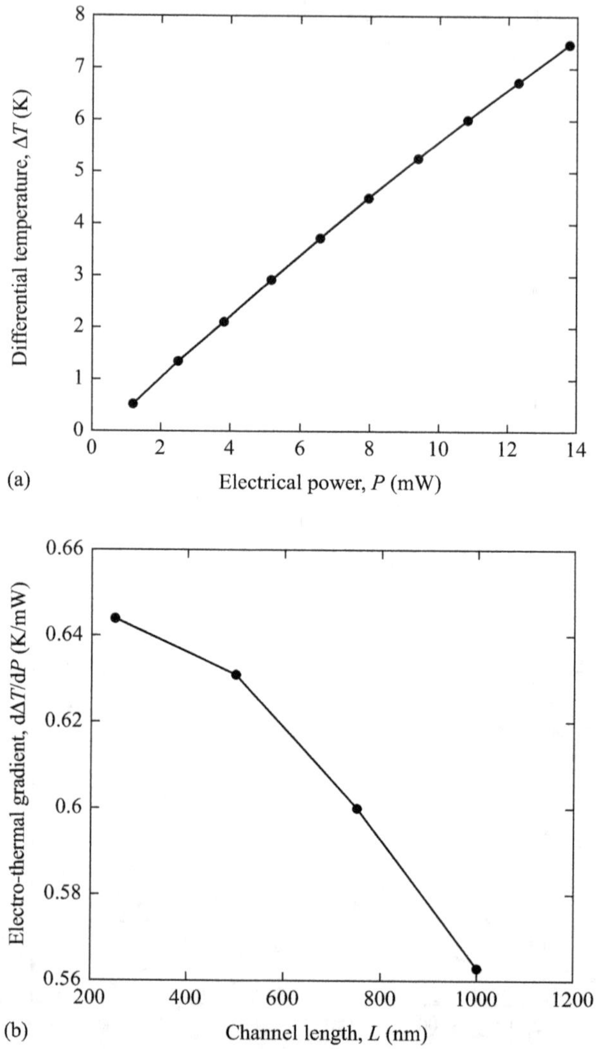

Figure 2.13 (a) Simulated ΔT temperature versus the device electrical power consumption, and (b) electro-thermal gradient $d\Delta T/dP$ versus channel length L

modeling and prediction of self-heating in nano-scaled semiconductor devices becomes quite complex and requires solving the lattice heat flow equation.

$$\nabla \cdot (G_{th} \nabla T_L) = \rho C_{th} \cdot \frac{\partial T_L}{\partial T} - H \tag{2.1}$$

The coefficient ρ is the mass density, while C_{th} is the specific heat or the equivalent thermal capacity, and H is the heat generation. However, the model for

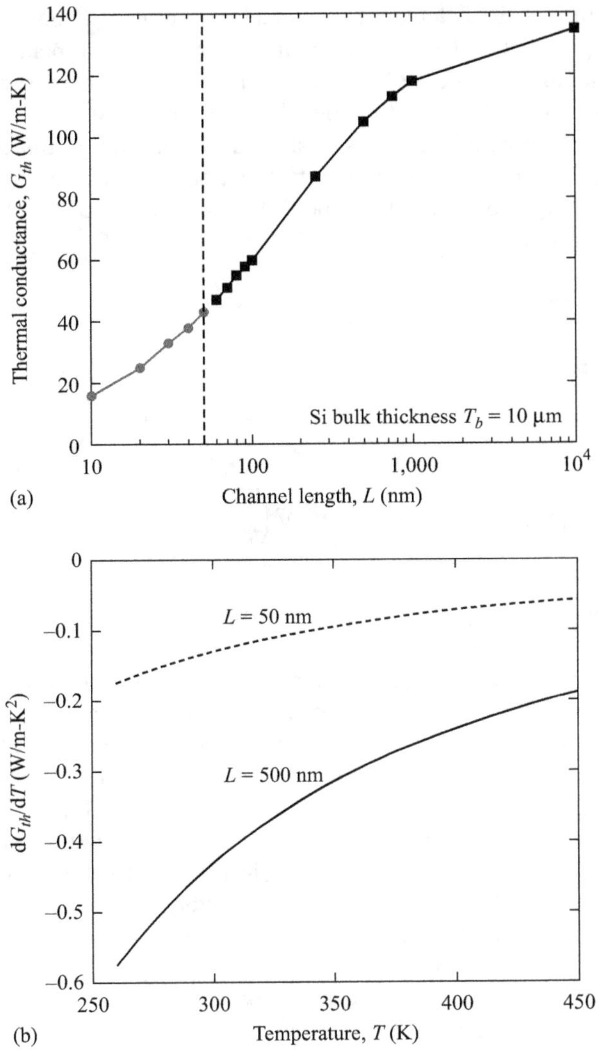

Figure 2.14 (a) Simulated thermal conductance G_{th} versus channel length L, and (b) simulated temperature gradient of G_{th} versus temperature for two different lengths L

the heat generation H depends on the charge transport model to be used, which can be Drift-Diffusion (DD), Hydrodynamic (HD), Monte-Carlo (MC), or Density Gradient (DG). The classical Drift-Diffusion (DD) and semiclassical approach considering HD transport are expected to fail at the sub 0.1-μm regime [2.19]. In order to cover quantum-mechanical effects, such as quantum confinement or charge tunneling, in a computationally economical approach, the DD model is modified through the DG approach, which recreates the inversion centroid charge

[2.20] and improves the calculation of tunneling through the gate oxide. For carriers under high electric field, they might reach carrier temperatures above the lattice temperature. In this case HD must be used for carrier transport simulation.

The conventional way of treating charge transport in large semiconductor devices, in the order of micrometers in size, has been through the combined solution of the Poisson equation, the continuity equation for electrons and holes, and the charge transport equations, which can be DD or any other approach as shown in the next set of equations.

$$\varepsilon\Delta\varphi = q(n - p - C) \tag{2.2}$$

$$qR = \nabla \cdot \vec{J}_n - q\frac{\partial n}{\partial t} \tag{2.3}$$

$$-qR = \nabla \cdot \vec{J}_p + q\frac{\partial p}{\partial t} \tag{2.4}$$

$$\vec{J}_n = q\mu_n n\left[\nabla\left(\frac{E_c}{q} - \varphi\right) + \frac{kT_L}{q}\cdot\frac{N_c}{n}\nabla\left(\frac{n}{N_c}\right)\right] \tag{2.5}$$

$$\vec{J}_p = -q\mu_p p\left[\nabla\left(\frac{E_v}{q} - \varphi\right) - \frac{kT_L}{q}\cdot\frac{N_v}{p}\nabla\left(\frac{p}{N_v}\right)\right] \tag{2.6}$$

The relation between charges and potential φ is captured with the Poisson equation (2.2), where n and p are the electron and hole charge concentrations in the semiconductor, respectively. C represents the concentration of fixed ionized charges. The mass conservation is captured with the continuity equations (2.3) and (2.4), where R represents the net generation of electron–hole pairs minus the net recombination rate of electron–hole pairs. Under thermal equilibrium R is neglected. $\vec{J}_n$ and $\vec{J}_p$ give the current density for electrons and holes, respectively. This couple of equations states that the semiconductor system will contain neither charge sources nor sinks. Under the presence of hot electrons, which causes either impact ionization or gate-oxide tunneling current, the mass conservation is violated, which requires the model to be modified.

Finally, when the charges, either electrons or holes n and p, are exposed to a gradient of the electric potential φ, or a gradient of the charges, which represents forces upon the charges, the charges will move or transport from an initial to a final position. This charge displacement or transport is conventionally represented by (2.5) and (2.6). In these two equations *drift* is represented by the first term, and *diffusion* by the second term between brackets. The gradient of the internal potential is referred to the conduction band edge E_c for electrons, and to the valence band edge E_v for holes. N_c and N_v refer to the effective density of states for electrons and holes, respectively. T_L refers to the lattice temperature, which may be different to that of the carriers $T_{n(p)}$. However, for the DD case $T_{n(p)} = T_L$. The carrier mobility for electrons and holes are represented by μ_n and μ_p. The force of an externally applied electric field $\boldsymbol{E}$ contributes to the charge drift. The charge flux originated by the externally applied electric field leads to an internal gradient of the charge distribution, which contributes to the charge diffusion. In general, the DD approach states that a charge can move either by drift or diffusion. However, under the action of a high electric field carrier temperature T_n or T_p may differ from that of the lattice T_L. In this case the hydrodynamic model (HD) must be used to calculate the current density as shown by the next set of equations [2.21].

$$\vec{J}_n = q\mu_n n\left[\nabla\left(\frac{E_c}{q} - \varphi\right) + \frac{k}{q}\cdot\frac{N_c}{n}\nabla\left(\frac{n\cdot T_n}{N_c}\right)\right] \tag{2.7}$$

$$\vec{J}_p = q\mu_p p\left[\nabla\left(\frac{E_v}{q} - \varphi\right) + \frac{k}{q}\cdot\frac{N_v}{p}\nabla\left(\frac{p\cdot T_p}{N_v}\right)\right] \tag{2.8}$$

If average carrier energies are to be conserved, then the energy balance equations stay as follows [2.22].

$$\nabla\cdot\left(-G_{thn}\cdot\nabla T_n - \frac{5}{2}\cdot\frac{kT_n}{q}\cdot\vec{J}_n\right) = \nabla\left(\frac{E_c}{q} - \varphi\right)\cdot\vec{J}_n - \frac{3k}{2}\left[\frac{\partial(n\cdot T_n)}{\partial t} + R\cdot T_n + n\cdot\frac{T_n - T_L}{\tau_{rn}}\right] \tag{2.9}$$

$$\nabla\cdot\left(-G_{thp}\cdot\nabla T_p + \frac{5}{2}\cdot\frac{kT_p}{q}\cdot\vec{J}_p\right) = \nabla\left(\frac{E_v}{q} - \varphi\right)\cdot\vec{J}_p - \frac{3k}{2}\left[\frac{\partial(p\cdot T_p)}{\partial t} + R\cdot T_p + p\cdot\frac{T_p - T_L}{\tau_{rp}}\right] \tag{2.10}$$

Here, τ_{rn} and τ_{rp} refer to the energy relaxation times, while the term between brackets at the left side of the equation refers to the energy fluxes. G_{thn} and G_{thp} are the thermal conductivities for electrons and holes, which follow the Wiedemann-Franz law [2.23] that represents the ratio of the electronic contribution of the thermal conductivity to the electrical conductivity.

$$G_{thn} = \frac{5}{2}\cdot\frac{k^2}{q}\cdot T_n\cdot\mu_n\cdot n \tag{2.11}$$

$$G_{thp} = \frac{5}{2}\cdot\frac{k^2}{q}\cdot T_p\cdot\mu_p\cdot p \tag{2.12}$$

In the case of DD the heat generation H in the heat flow equation (2.1), becomes

$$H = \nabla\left(\frac{E_c}{q} - \varphi\right)\cdot\vec{J}_n + \nabla\left(\frac{E_v}{q} - \varphi\right)\cdot\vec{J}_p + R(E_c - E_v) \tag{2.13}$$

While for the HD case, the heat generation H stays as

$$H = \frac{3}{2}k \cdot \left(n\frac{T_n - T_L}{\tau_{rn}} + p\frac{T_p - T_L}{\tau_{rp}} \right) \tag{2.14}$$

As mentioned earlier DG is a computationally inexpensive an alternative way to incorporate quantum corrections without the need of self-consistently solving the coupled Schrodinger-Poisson equation system. The quantum corrections in the DG approach are introduced through the use of quantum correction potentials γ_n and γ_p for electrons and holes, which results in a modified transport equations as follows [2.24]:

$$\vec{J}_n = q\mu_n n \left[\nabla \left(\frac{E_c}{q} - \varphi - \gamma_n \right) + \frac{kT_L}{q} \cdot \frac{N_c}{n} \nabla \left(\frac{n}{N_c} \right) \right] \tag{2.15}$$

$$\vec{J}_p = q\mu_p p \left[\nabla \left(\frac{E_v}{q} - \varphi - \gamma_p \right) - \frac{kT_L}{q} \cdot \frac{N_v}{p} \nabla \left(\frac{p}{N_c} \right) \right] \tag{2.16}$$

The quantum correction potentials are given by [2.25]

$$\gamma_n = \frac{\hbar^2}{m_0 \lambda_n q} \frac{\nabla^2 \sqrt{n}}{\sqrt{n}} \tag{2.17}$$

$$\gamma_{np} = \frac{\hbar^2}{m_0 \lambda_p q} \frac{\nabla^2 \sqrt{p}}{\sqrt{p}} \tag{2.18}$$

where the $\lambda_{n,p}$ parameters can take values of 3 for the non-degenerate semiconductor case, or 9 for the degenerate case. In practice this is fitting parameter that can be adjusted to values between 3 and 9 depending on the doping level.

Overall, for devices at or below the 100 nm size-scale the incorporation of quantum effects is a must, and requires the combined solution of the Schrodinger and Poisson equations together with those of transport, and the adoption of the MC method for their multi-dimensional solution. A detailed discussion of advanced modeling and simulation techniques will be described in Section 2.1.3.

2.1.2 *Alternative FET-related structures on silicon bulk and other materials*

The conventional MOSFETs built with gate oxides thicker than 10 nm and polysilicon gates, show electrical performance limitations as shown in the previous section. Such a limitations are a large threshold voltage V_t, which inhibits the use of low-voltage supplies, a large subthreshold slope S, which makes transistor for high-speed digital electronics unsuitable. The small transconductance g_m is also another limitation for high-speed electronics. The polysilicon depletion and the series source/drain resistance are also two additional mechanisms, which negatively impact the electrical performance. A way to circumvent all these limitations is through the reduction of the channel length L and gate oxide thickness T_{ox}. However, as the channel length and the gate oxide are reduced, the energy of the carriers increases giving rise to hot carrier effects, such as impact ionization measured as a

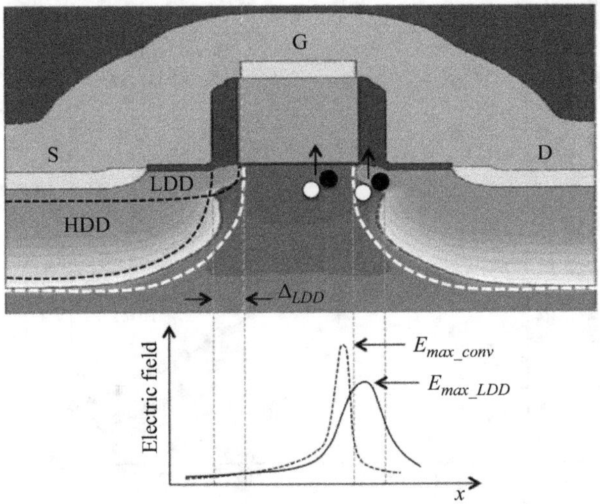

Figure 2.15 LDD MOSFET structure and electric field distribution for a single implanted conventional structure (dashed line), and for an LDD structure (continuous line)

bulk current, oxide tunneling measured as gate leakage current, channel self-heating measured as an increase of the local temperature, drain-induced barrier lowering (DIBL) [2.26] measured as a reduction of the threshold voltage, or increase of the channel leakage current.

The first approach to alleviate hot carrier effects has been the introduction of the lightly doped drain/source (LDD) MOSFET structure [2.27]. The LDD structure is shown in Figure 2.15. The structure is composed of a double implantation with a highly doped region (HDD) and a lightly doped section (LDD). A conventional single implanted S/D MOSFET structure is represented by the thick white dashed contour. The combination of the HDD and the LDD doping profiles results in a buffer region Δ_{LDD}, where the electric field reduces in magnitude and its maximum peak value E_{max_LDD} moves away from the active channel region. In the case of a single S/D implanted conventional structure, the electric field E_{max_conv} is larger than for the LDD and peaks in the active channel region. As the hot carrier generation is exponentially dependent on the electric field, a drastic reduction of hot carriers happens for the LDD structure. Moreover, as the electric field peaks, in the case of the LDD structure, outside the active channel region, thus the amount of hot carriers injected into the gate oxide diminishes, which minimizes degradation of the threshold voltage and improves gate oxide reliability.

The reduction of hot carrier generation by the use of a LDD structure is not for free, it comes with a bias-dependent series resistance effect [2.28]. Because of the light doping level at the Δ_{LDD} region, the charges at the gate-to-source/drain overlapping regions ΔL becomes modulated by the gate voltage, which results in a gate voltage-dependent series resistance as shown in Figure 2.16. The extracted gate

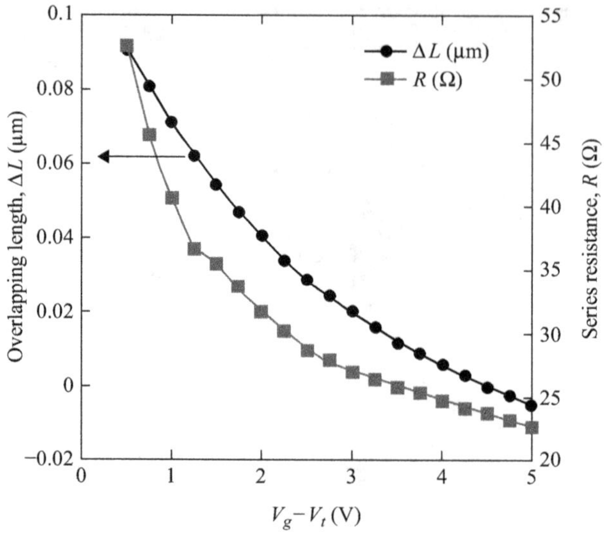

Figure 2.16 Measured overlapping length ΔL and series resistance R as a function of the gate voltage, for a 0.7 μm n-type MOSFET

overlapping length ΔL becomes also gate voltage dependent. At high gate voltages ΔL becomes even negative, which means the fringe gate electric field extends beyond the gate edge, controlling not only the charges under the gate-source/drain overlap but also the charges beyond the gate edge. Despite the LDD structure has been the workhorse for the MOSFET-based electronics for a long period of time since its inception in the late 1970s when used for a 5-mm MOSFET generation [2.29]. The reduction of the supply voltage and the contact resistance [2.30], has required a modification to its structure in order to be ready for a further miniaturization down to the 0.1 μm generation [2.31]. The 0.1 μm generation is a 1.5 V power supply technology with dual n^+/p^+ polysilicon gates on a 3.5 nm gate SiO_2 oxide.

This 0.1 μm technology still makes use of local oxidation (LOCOS) as an insulator between n- and p-type MOSFETs. The realized 50 nm shallow source/drain junctions in this 0.1 μm technology implies the consumption of silicon during the silicidation process, a problem avoided using the source-drain extension, a modification to the original LDD structure. A counter doping pocket implant (halo) is used to increase the doping level at the extensions, which further helps in suppressing short-channel effects. The halo doping allows for heavier channel doping, which in turn results in shallower source/drain junctions and thus in a reduced gate length. However, transistors with a high uniform doping level, from the surface through the bulk, this results in a degradation of carrier mobility due to combined action of Coulombic and phonon scattering. Thus the incorporation of super-steep retrograde channel doping profiles mitigates the carrier mobility degradation [2.32], and helps keeping a threshold voltage V_t with low variability from device to device. Gate oxides thinner than 10 nm brought the need to modify once again the

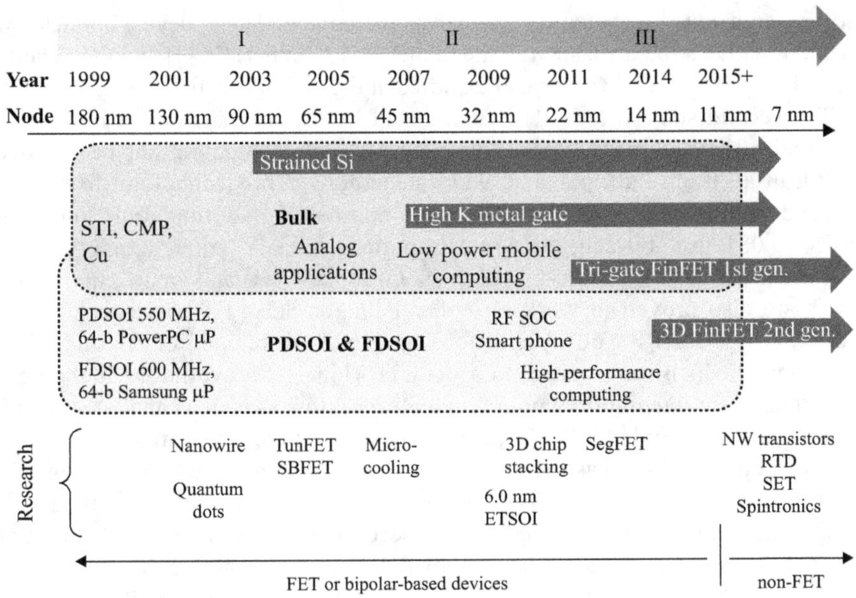

Figure 2.17 Technology evolution for Si- and FET-based devices. The two upper blocks Bulk and PDSOI and FDSOI shows the technologies already introduced for production. The lower part shows some of the alternative research approaches. The upper part shows the node definition and yearly progress

original LDD structure by making use of nitrided oxide films in place of pure SiO_2 [2.33]. The use of stacked nitride films prevents the boron penetration through the gate oxide into the channel, which in turn, alleviates the threshold voltage variation caused by the boron penetration into the channel.

In general the evolution of the FET-based IC technology can be split into three stages and three blocks as shown in Figure 2.17.

Stage I goes from the 180 nm to approximately the 65 nm technology node, where the LDD device structure incorporates many fabrication processes changes, such as mechanical strain to boost carrier mobility, retrograde channel doping, elevated source/drain regions [2.34] with shallow junctions and silicide contacts. Shallow trench isolation (STI) is introduced as a way to prevent current leakage between adjacent devices [2.35]. Chemical mechanical planarization (CMP) together with Cooper interconnects are introduced as a way to flat and smooth the inter-level dielectric layers in ICs [2.36]. In this first stage two popular microprocessor for computing application, the 550 MHz 64-b Power PC and the 600 MHz 64-b Samsung microprocessors, technologies are commercially introduced on Silicon-on-insulator (SOI) technologies [2.37, 2.38]. These microprocessors are based on partially depleted (PDSOI) or fully depleted (FDSOI) SOI technologies [2.39]. Alternative structures, such as nanowires (NWs) [2.40] and quantum dots [2.41], are researched and for the first time envisioned as potential alternative for semiconductor device applications. Other alternative structures, such as the tunnel FET (TunFet) [2.42] and the Schottky Barrier FET (SBFet) [2.43], are also explored at the end of this first stage.

The second stage (Stage II) is dominated by the introduction of high-k gate oxide dielectrics and metal gates. Low Power Mobile computing has been proved to work in a 20 nm bulk planar CMOS technology. This technology features an advanced high-k metal gate process, strain engineering. A fully functional high density (0.081 μm^2 bit-cell) SRAM is reported at 0.9 V voltage supply [2.44]. Radio frequency (RF) System-On-Chip (SoC) for wireless and smart phone applications are also proven to work on either bulk- or SOI-FET-based technologies [2.45, 2.46]. These RF SoC applications have been developed on 32 and 28 nm technologies, which have resulted on 4G/LTE (long-term evolution) mobile SoC chips suitable for integration with multicore processors as well as analog and audio mixed-signal front ends [2.47]. With increasing levels of integration the need for 3D silicon chip stacking has increased. The 3D silicon chip stacking is required to meet the demand for high-density interconnection at medium and high frequencies [2.48]. This approach is explored in this second stage. As a result of the high-density integration high heat flux removal becomes a major consideration. An alternative to dissipate the heat flux out of an IC chip is the use of multiple drainage micro-trenches as proposed in Reference 2.49. Extremely Thin SOI (ETSOI) devices, with low GIDL and low V_t variability, are researched as candidates for low power applications. These ETSOI devices are built on 6 nm SOI layers [2.50].

Finally at the third stage comes the commercial introduction of both bulk and SOI FinFETs for high-performance computing, which marks an evident transition from the evolved planar MOSFET toward a 3D FinFET structure. In 2012 a 22 nm generation featuring a fully depleted (FD) tri-gate transistor is first introduced for SRAM and microprocessor applications [2.51]. Later in 2014 the second-generation 14 nm FinFET technology featuring with air-gapped interconnects is introduced for mass production of high-performance microprocessors and high-density SRAM [2.52]. Also in 2014 IBM introduced a 14 nm SOI FinFET technology with 0.0174 μm^2 embedded DRAM and 15 levels of Cu metallization [2.53]. An alternative to FinFET devices, called Segmented-Channel MOSFET (SegFET), is researched as an option to enhance electrostatic integrity and reduce short-channel effects using a still planar-silicon substrate [2.54]. This device structure has the advantage of a simplified planar MOSFET fabrication process with electrical performance as good as that of thin-body SOI devices.

In an exploratory stage there are various devices among them a vertically stacked gate-all-around silicon NW FET [2.55], which enables dynamic configuration of the device polarity to be n- or p-type. Si NW transistors with gate lengths in the range of 20 nm have been shown a high I_{on}/I_{off} ratio larger than 10^6. This NW transistor is fabricated with a simpler process without junctions [2.56], which results in improved short-channel characteristics compared to the inversion-mode devices. A more advanced option is the silicon single-electron quantum-dot transistor, based on the Coulomb blockade effect, which has been demonstrated to work at room temperature for a silicon dot of about 12 nm of diameter [2.57]. Another option for the post-CMOS or post-FET era is the control of spin transport in Silicon [2.58]. Spin-based devices are non-volatile information by nature. So they are highly efficient in storing and processing information. They are also quite efficient in transferring information as well, which in addition to other outstanding features such as injection of spin-polarized currents into silicon makes them very attractive for a CMOS-spin circuit combination [2.59].

Another option for high-speed electronics is the Resonant Tunneling Diode (RTD) device [2.60]. Maximum frequencies of up to 2.2 THz have been demonstrated with RTDs, which outperform that of Si-based FET devices [2.61]. The RTDs may found application in different areas, such as high-resolution imaging systems or wide-band secure communication systems.

Other than Silicon materials such as graphene, have been also researched as a material with excellent electrical transport properties. Graphene can be grown as a dimensional mono-atomic layer with excellent semimetal properties, such as concentrations as high as 10^{13} cm^2 with room-temperature mobilities of around 1×10^4 cm^2/Vs [2.62].

The advanced semiconductor devices, such as FinFETs, SOIs, TunFets, NWs, quantum dots, and spin-based devices, will be treated in detail in the subsequent sections.

2.2 Advanced FET-related devices

2.2.1 Introduction

Power-Performance-Area-Cost (PPAC) trade-off relationship is the "*leitmotif*" that moves today's semiconductor industry. Since early 1960s, the reduction of MOSFET dimensions with some punctual modifications to control short channel effects (SCEs) has been enough to fulfill the above balance equation while following Moore's Law (the number of devices in an IC will double every 18 months) [2.63]. Meanwhile, the power consumption has been kept under control, at the same time that the drain current and therefore the performance were largely increased. Finally, the cost-per-device of each node was smaller than that in the just previous one, or equivalently, the number of transistors that can be bought per dollar was monotonically increasing node after node (Figure 2.18).

The major factors that control the drain current and therefore the performance of a MOSFET transistor are the channel length, the carrier mobility, and gate capacitance. Although the reduction of channel length would theoretically increase the drain current and reduce the switching times of the device making it faster, when this length approaches nanometric dimensions, the arising of physical and essential limitations makes this picture much more complicated. These limitations are related to SCEs, degradation of carrier transport, and increase of leakage current and/or power consumption. SCEs (i.e., charge sharing between source/drain depletion regions and the channel) [2.64] mean the lack of electrostatic control of

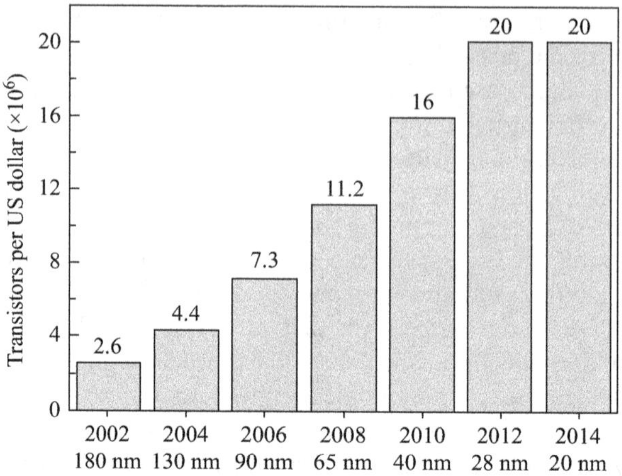

Figure 2.18 Transistors that can be bought per dollar at each technological node
© 2014 Semiconductor Engineering. Adapted with permission from Brian Bailey http://semiemgineering.com/the-tail-of-moores-law-nolonger-wagging-the-dog/

the channel by the gate, and the degradation of the proper switching capabilities of the transistor. To avoid SCEs, substrate doping level has to be increased. However, there is a limit to how much the doping level can be increased before junction breakdown becomes a problem. In addition, the increase in the substrate doping produces an increase in the transversal electric field and a degradation of the carrier mobility [2.65, 2.66], and also degradation in the gate-to-channel capacitance, both facts producing a reduction of the on-current.

Keeping the gate-to-channel capacitance, while downsizing the gate length, requires an equal factor of decrease in the gate oxide thickness. In the 65 nm technology node with gate length of 32 nm, the thickness of the silicon oxide should be about 0.9 nm, which is very close to the theoretical limit (around 0.7nm) for bulk silicon dioxide [2.67]. Further device downsizing is only possible by introducing high-dielectric constant (high-k) materials [2.68, 2.69]. By using gate insulators with higher permittivity a larger value of the gate capacitance can be achieved with a thicker film, thus not only resolving the physical constraint of the oxide thickness for further downsizing but also helping to suppress the leakage tunnel current through the gate.

Degradation of mobility because of high doping substrates can be compensated by mobility or transport boosters [2.70]. The most straightforward way to improve the carrier transport in MOSFETs is to induce uniaxial or biaxial tensile strain in NMOSFETs and uniaxial compressive strain in PMOSFET channels [2.71]. Tensile strain produces an energy band splitting of the silicon conduction band, and a repopulation of the carriers, which originates a decrease of the transport effective mass and a reduction of the phonon scattering. Biaxial strain can be induced by substrate engineering, i.e., from epitaxial growth of tensile silicon on relaxed SiGe layers, and uniaxial tensile or compressive strain can be obtained from Contact

Table 2.1 Mobility boosters for CMOS VLSI technology

	Electrons	**Holes**
Channel direction	<100>	<100> on (100) <110> on (110)
Surface orientation	(100)	(110)
Strain	biaxial tensile	biaxial tensile uniaxial compressive
Materials	(III–V)	SiGe/Ge

Etch Stop Layer (CESL) approach. Another efficient way to improve the carrier mobility is to choose the appropriate crystalline orientation and the transport direction (Table 2.1). Finally, other materials presenting higher electron and/or hole mobilities, such as III-Vs, SiGe, or Ge, can be used to improve the transport in the channel.

For very short channel devices, the resistance of the source/drain and the extension regions can be comparable to that of the channel, thus becoming in a showstopper limiting the performance enhancement obtained by the channel length reduction. The source/drain engineering is therefore another technology booster that implies the optimal design of source impurity profiles and Schottky metal source structures.

But, performance is only one term in the equation. Power dissipation is also an important ingredient. The power dissipation of a MOSFET is due to static and dynamic contributions. Static power consumption is the product of the supply voltage times the off-state current. Therefore, the reduction of static power consumption requires an excellent control of the off-state current (including leakage currents), and a low value of the threshold voltage, which would allow the scaling down of supply voltage. However, a minimum value of the threshold voltage is required due to subthreshold swing degradation, short channel threshold voltage roll-off, and DIBL [2.72]. Another important effect of scaling is related to the increased current that flows through the device in its off state. In addition to the subthreshold current, several parasitic currents can contribute to the static power consumption: (i) first, for nanometer size gate dielectrics, the probability for carriers to tunnel between the gate and the channel is non-negligible [2.73]. An efficient way to reduce the gate leakage is to increase the physical dielectric thickness, while keeping a sufficient potential barrier height between the channel and the dielectric. In order to keep at the same time a large capacitive coupling between the gate and the channel high-k materials have to be used, as mentioned earlier. Another source of current leakage is the junction leakage current between source/drain to body current of the PN junction. In addition to this reverse PN junction current, a drain to body leakage can be induced when a high gate to drain voltage is applied, as a consequence of the band-to-band tunneling (BTBT) of electrons between the conduction band in the drain region and the valence band in the accumulated region below the gate oxide. Finally, for channel lengths below 10 nm, a significant amount of carriers can tunnel directly from source-to-drain through the barrier potential [2.74].

With regard to dynamic power consumption, it can be kept under control by reducing parasitic capacitances as much as possible [2.75].

In summary, so far PPAC trade-off condition has been achieved by geometrically scaling the gate length of the transistors to nanometric dimensions, without involving any major change in the bulk MOSFET structure except a complexity increase in the fabrication process using technology boosters which can be classified into three categories: (i) gate-stack engineering with the use of high-k insulators and metal gate electrodes; (ii) source-drain engineering which includes Schottky barrier source structures and optimal design of source impurity profiles; and (iii) channel engineering, which includes the use of strain (global or local) and the use of alternative high-mobility channel materials.

Thus channel lengths in the 30 nm range have been achieved using classical bulk-Si MOSFETs [2.76]. However, a scaling limit for this classical CMOS bulk structures beyond $L_g = 30$ nm is now evident. The complex doping profiles required to control SCEs, and the variability issues arising by the randomness of the dopant atoms, stop bulk-Si MOSFETs from being scaled with reliable results further than 30 nm.

The increasing variability in the device characteristics is among the major challenges to scaling and integration for the present and next generation of nano-CMOS transistors and circuits [2.77]. The statistical variability of transistor characteristics, which has been previously concern only in the analog design domain, has become a major concern associated with CMOS transistors scaling and integration [2.78, 2.79]. It already critically affects SRAM scaling [2.80] and introduces leakage and timing issues in digital logic circuits [2.81]. The statistical variability in modern CMOS transistors is introduced by the inevitable discreteness of charge and matter, the atomic scale non-uniformity of the interfaces and the granularity of the materials used in the fabrication of ICs. The granularity introduces significant variability when the characteristic size of the grains and irregularities become comparable to the transistor dimensions. For conventional bulk MOSFETs Random discrete dopants (RDD) is the main source of statistical variability [2.82]. Random dopants are introduced predominantly by ion implantation and redistributed during high temperature annealing. Apart from special correlation in the dopant distribution imposed by the silicon crystal lattice, there may be also correlations introduced by the Coulomb interactions during the diffusion process. For a complete study of variability consider the work of Prof Asenov's group [2.77, 2.83, 2.84].

No additional performance "boosters," such as strained-Si channels and metal gate stacks [2.76, 2.85], will enable a reliable classical technology beyond channel lengths shorter than 30 nm. As the channel area underneath the gate is getting very short, the gate is no longer powerful enough to control it properly. It could control the top part of the channel but the further from the gate the less the control. In particular, when the gate is off there are paths between source and drain that remain on and so there is very high leakage. It is clear that new transistor architectures are required.

The basic constraint is that the entire channel needs to be close to the gate so that it can be controlled properly. Therefore, if a thin channel is put on top of an insulator, and the gate is built on top of that then there is once again good control and low leakage (Figure 2.19). There are simply no paths through the channel that

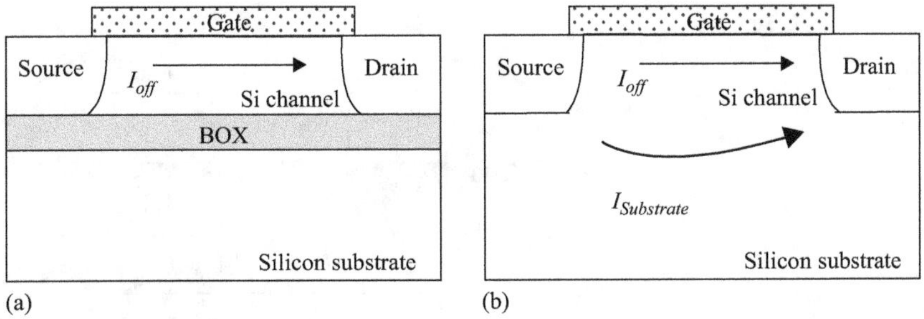

Figure 2.19 Schematic of (a) a SOI MOSFET transistor and (b) a bulk MOSFET showing different components of the off-current

are far from the gate and so poorly controlled because the insulator isolates the source and drain from the substrate. These devices are called SOI transistors. That is thick-box SOI (box just stands for buried oxide, the insulator underneath the channel). If, however, the box is very thin then the substrate itself becomes a sort of second back gate and can further be used to control the channel, not to turn it on and off but to affect its performance.

Another alternative to do this is to make the channel into a thin vertical fin and wrap the gate around it. This structure is known as FinFET. Since the fin is thin, it is never far from the gate and control is good and leakage is low. The following parts of this chapter are devoted to review these interesting devices.

2.2.2 Silicon-on-insulator FET technology

Typical silicon wafers have a total thickness of less than 1 mm (usually 775 μm for 300 mm wafers [2.86]). Only a tiny slice of several micrometers at the top is used for fabricating nanoelectronic devices, Figure 2.20. The unused bottom region of monocrystalline silicon is necessary to ensure the structural feasibility of the wafer and the devices, providing strength and avoiding breaks [2.87].

But using a semiconductor substrate also contributes to some undesirable parasitic effects such as:

- Implicit electrical connection between different devices. It is usually solved using lateral isolation techniques like STI, or channel stop implantations. These techniques consume area and require more fabrication steps.
- Appearance of parasitic capacitances and devices. An NPN parasite BJT can appear using the N^+ diffusion of the source or drain of a N-MOSFET as emitter, the P-type well acts as the base and the N-type well as collector. A PNP BJT can also appear the same way next to a P-MOSFET.
- Lower control over SCEs. The gate looses the electrostatic control over the channel due to source and drain region influence. This limits the minimum channel length of the device that can be fabricated. To overcome this problem, high body doping to reduce the depletion regions has been used. However, this increase in the body doping has a negative effect on mobility, threshold voltage, transconductance, and device variability.

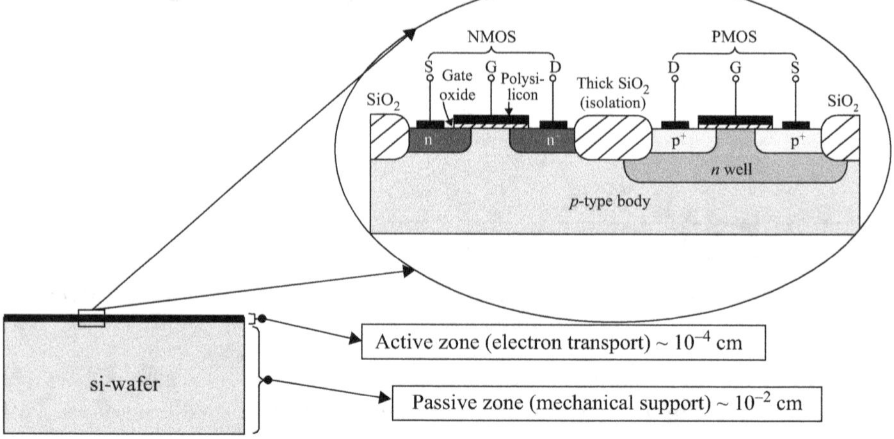

Figure 2.20 Details of the useful layer thickness on a bulk Silicon wafer. Only the top region is used for fabricating devices, while the thick bottom region remains unused providing mechanical strength and avoiding breaks

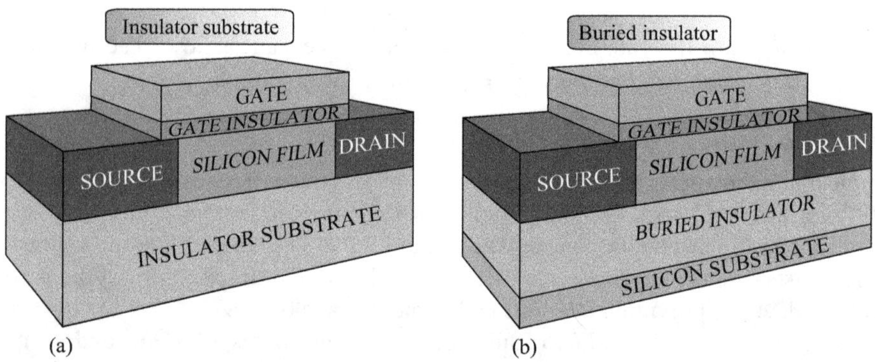

Figure 2.21 Different SOI wafer configuration depending on the buried oxide thickness. (a) In the panel A, a thin Si-film layer is grown or deposited on top of an insulator substrate. (b) In the panel B, only a buried thin layer of insulator separates the active area to the silicon substrate

In order to solve these problems, the SOI technology proposes the use of a new film made of dielectric. This insulator layer isolates the active region where devices will be fabricated from the silicon substrate. There are two different alternatives: using a thin insulator layer buried in the silicon wafer or using a wafer made of an insulator material and growing a thin layer of silicon on its top. Both types of approaches are depicted in Figure 2.21.

The most common option is to use the buried layer called BOX (Buried OXide). This enables substrate biasing when the insulator layer is not very thick, less than 200 nm. Nevertheless, to achieve a useful SOI wafer is very challenging.

A thin monocrystalline layer of silicon needs to be grown on top of the insulator. The Si-film must feature the same quality and properties as those grown in volume:

- low density of interface states
- uniform thin layers
- good performance of the dielectric

Figure 2.22 summarizes the main techniques for fabrication of SOI wafers.

Among all of them, the most important and used today is the Unibond technique or *Smart-Cut*. This is one of the most effective processes to produce commercial SOI wafers. The Smart-Cut technique combines an ion implantation and wafer bonding to transfer a thin slice of one wafer to another wafer or insulator substrate. It was first developed in 1992 at CEA-LETI [2.88]. The main steps are illustrated in Figure 2.23 and summarized below:

1. Initially, two bulk silicon wafers, "A" and "B," are used. The surface of one of them should be completely oxidized, for example, wafer "A." This wafer will be used for the active layer on top of the BOX. The next step consists on implanting positively charged hydrogen ions (protons) on the wafer "A." The implantation's energy will define the silicon film thickness.
2. After this, the wafer is cleaned and then bonded onto wafer "B" using the SiO_2 previously grown to maintain them together.
3. Then, a first heat treatment is performed at around 400–600 °C. With this, the micro-cavities produced by the implantation of protons split up the wafer into two pieces, one of them a bulk wafer, "A," and the other an SOI wafer, "B." After this, a second heat treatment to improve the connection between the BOX layer and the wafer itself is carried out.

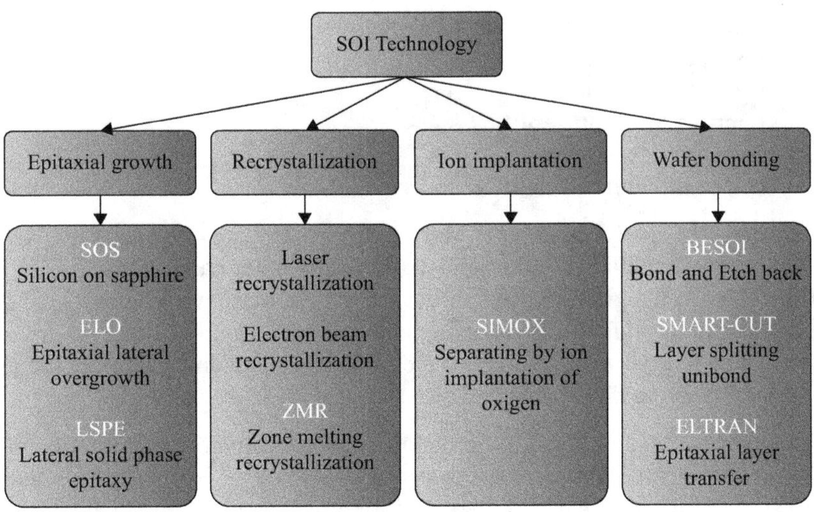

Figure 2.22 Main techniques for SOI wafers fabrication

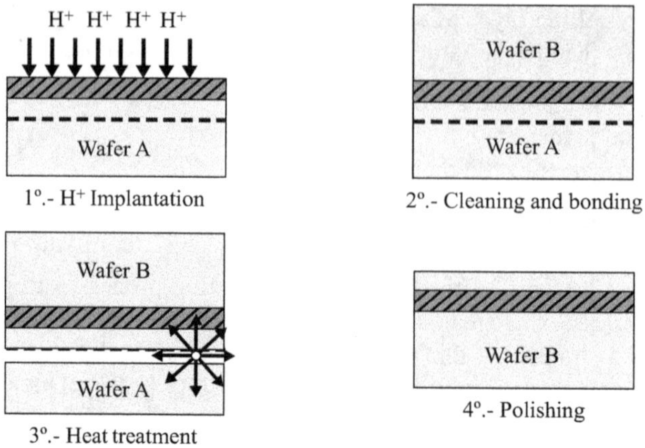

Figure 2.23 Smart-Cut SOI wafer fabrication technique based on wafer bonding

4. Finally a chemical mechanical polishing (CMP) process is performed to achieve a uniform wafer surface and also to thin the Si-film to the desired final thickness.

The advantages that SOI technology provides compared to bulk are numerous. Some of the improvements are [2.89]

- Avoid the connection between device and substrate:
    - area consumption reduction
    - avoid the latch-up
    - lower power consumption
    - better subthreshold swing
    - lower leakage currents
    - better transconductance
    - improve the control of SCEs
    - lower power voltage
    - better immunity to ionizing radiation
- No need of body doping: lower Random Dopant Fluctuation (RDFs), decreasing variability on threshold voltage, and other parameters.
- Capability of integrating different devices on the same wafer such as, CMOS, power devices, bulk designs, and optoelectronics.
- Possibility to stack more than one layer of devices (the drawbacks are the interconnection issues that appear and also self-heating-related problems [2.89]).

The main disadvantage of SOI technology seems to be the higher price of the wafers due to the extra fabrication steps needed. However, it is worth noting that the final cost of an IC in SOI technology may be lower due to a simpler fabrication process with less manufacturing steps. Together with the extra price, there is another important drawback of SOI. The self-heating effect (SHE) is more remarkable in SOI wafers since the SiO_2 is about two orders of magnitude less thermal conductive than silicon. Thin-film SOI MOSFETs are therefore prone to accumulate heat because it cannot be dissipated rapidly through the BOX. However, recent studies show that the SHE is present but it does not represent a limiting factor for the reliability in ultra-thin silicon films transistors, especially for fast switch operations [2.90]. Other constraints that SOI wafers may include are [2.91–2.93]:

(i) The uniformity of the BOX and silicon layers is crucial to avoid high variability.
(ii) Floating-body effects (FBE) [2.94–2.96]. Although this is not always a drawback, it may lead to an abnormal behavior.
(iii) Inter-gate coupling effects between front and back-channels. As the FBE, it may be also an advantage.
(iv) Large series resistance in thin film devices. This is normally reduced by raising the source and drain regions, RSD (raised source drain also known as ESD, elevated source drain) technique [2.97, 2.98].

2.2.2.1 Partially depleted SOI MOSFETs

SOl CMOS involves building more or less conventional MOSFETs on a thin layer of crystalline silicon, as illustrated in Figure 2.24. The thin layer of silicon is separated from the substrate by a thick layer (typically 100 nm or more) of buried SiO_2 film, thus electrically isolating the devices from the underlying silicon substrate and from each other. SOl CMOS process can be readily developed due to the compatibility with established bulk processing technology.

SOl MOSFETs are often distinguished as partially depleted (PD) and fully depleted (FD) [2.99, 2.100].

In case of PD transistors, the silicon film under the gate is not completely depleted of mobile charges. In contrast, for FD MOSFETs no charge accumulation or storage can be achieved without biasing the gates: there is no quasi-neutral region, which serves as a potential well to store majority carriers. In general, the thickness of the silicon film that determines whether it is PD or FD depends mainly

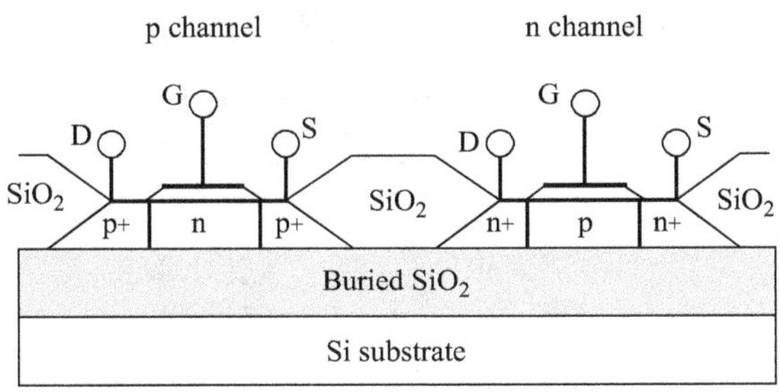

Figure 2.24 Schematics of p-channel and n-channel transistors in SOI CMOS technology

on the body doping concentration, $N_{A,D}$. The following equation indicates the depletion depth for one interface with no lateral influence (1D model):

$$x_D = \sqrt{\frac{2\epsilon_0 \epsilon_{Si} \Phi_s}{qN_{A,D}}} \tag{2.19}$$

where ε_0 and ε_{Si} are the permittivity of vacuum-free space and silicon relative permittivity, respectively, q is the electron charge, and Φ_S is the surface potential at the interface. The maximum depletion width, $X_{D,max}$, corresponds to $\Phi_S = 2\Phi_F$, where Φ_F is the Fermi potential, equal to $\frac{kT}{q} ln(\frac{N_{A,D}}{n_i})$.

Typically, commercial silicon film thicknesses, t_{Si}, larger than 70 nm correspond to PD, and below 40 nm to FD [2.101]. The main differences between both types of transistors are resumed in Table 2.2 and Figure 2.25 [2.102–2.105].

In a PDSOI device there is no interaction between the front and the back depletion zones because $t_{Si} > X_{D,max}$. Therefore, the threshold voltage in a PDSOI transistor is the same as in a bulk MOSFET:

$$V_{TH} = V_{FB} + 2\Phi_F + \frac{qN_A x_{D,max}}{C_{ox}} \tag{2.20}$$

where $V_{FB} = \Phi_{MS} - \frac{Q_{ox}}{C_{ox}}$ is the flatband voltage.

As previously mentioned, in a PDSOI device, the body is thick enough such that only part of the body region is depleted across the bias range of operation. This means that part of the silicon film remains neutral, populated with mobile majority carriers and surrounded by the buried oxide at the bottom, and laterally by source/body and drain/body PN junctions. Therefore, this neutral area remains isolated from any contact, as observed in Figure 2.26, i.e., the body of the transistor remains floating and its potential is determined by capacitive coupling through the gate oxide, C_g, the sidewall source and drain junctions, C_s and C_d, the BOX interface C_b, and the leakage current through the junctions and the oxides. In Si-bulk devices, if majority carriers are created during the operation of the transistor, they can be evacuated through the substrate since this is directly connected to the substrate

Table 2.2 Partially depleted, fully depleted SOI, and bulk transistors comparison

Parameter	**Partially depleted**	**Fully depleted**	**Bulk**
Silicon thickness	>70 nm (typ.)	<40 nm (typ.)	—
Source/drain resistance	Moderate	High	Low
I_{on}	Moderate	High	High
I_{off}	Very low	Low	Moderate
DIBL	Low	Very low	Moderate
Subthreshold swing	Moderate	Very low	High
Floating-body effects	✓	✖	✖
Kink effect	✓	✖	✖
History effect	✓	✖	✖
Coupling channel	✖	✓	✖

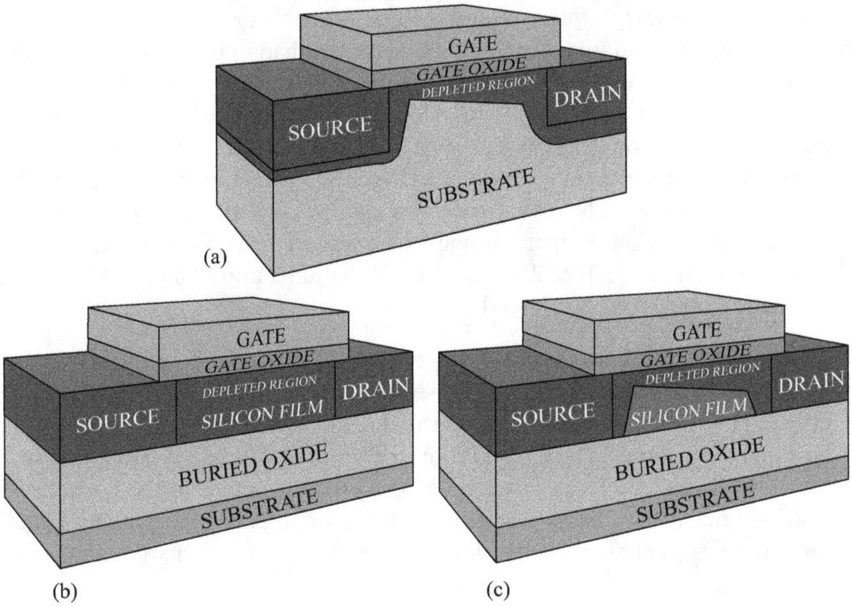

Figure 2.25 Depleted region comparison between (a) bulk MOSFET, (b) fully depleted SOI transistor, and (c) partially depleted SOI transistor. The depletion region induced by the back-gate is not represented

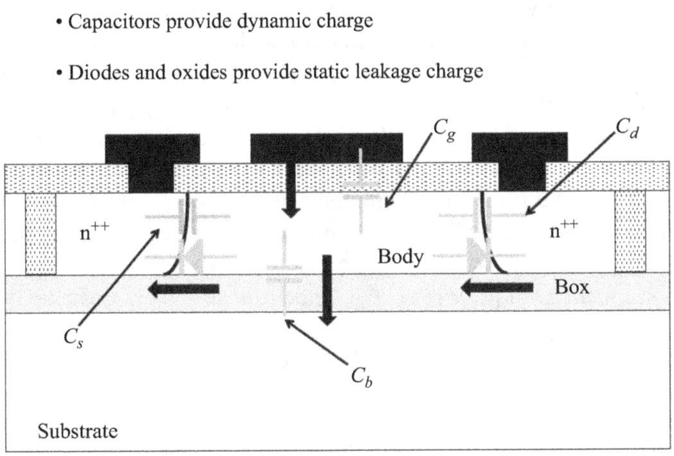

Figure 2.26 Schematic cross-section of a partially depleted SOI transistor showing the capacitive coupling of the neutral body of the transistor through the gate oxide, C_g, the sidewall source and drain junctions, C_s and C_d, the BOX interface C_b, and the leakages current through the junctions and the oxides

contact. On the contrary, in the case of a PDSOI transistor, the excess of majority carriers accumulates in the body of the device, changing its potential, which therefore is not fixed but depends on the history of the device. This fact is the origin of a set of effects named FBE [2.89]. All these effects have their origin in the charging-discharging of the floating-body by currents coming from the source or the drain and in the capacitive coupling between the gate and the floating-body.

Although these effects may be detrimental, it also allows the MOSFET to exhibit very interesting properties such as memory capabilities, the so-called floating-body memory [2.106–2.108]. Some of the most important aspects of FBEs are summarized below [2.89, 2.109]:

1. **History effect and threshold voltage variability**
   The most prominent electrical property of the PDSOI device is the history effect. The *I–V* characteristics (and the threshold voltage) of the transistor are no longer constant, but dependent on the amount of charge contained in the body of the device at any given time. The charge content of the body and the distribution of that charge caused by gate, source, and drain potentials determine the final behavior of the device. The magnitude of charge contained in the body depends on several factors that include:

   - the previous state of the transistor
   - the device architecture (length, width, Si-film thickness, etc.)
   - the applied biases
   - the frequency of operation
   - the temperature

   Figure 2.27 shows an example of history effect in a PDSOI transistor. To do so, we have considered a PDSOI transistor with a channel length of $L_{ch} = 1$ μm. Initially a voltage ramp is applied to the gate of the transistor during 10 ns. During this time, the drain to source voltage is set to $V_{DS} = 0.1$ V. Figure 2.27c shows the hole concentration in the body of the transistor at the end of the gate voltage ramp ($t = 10$ ns). At that moment, a high voltage pulse is applied to the drain ($V_{DS} = 3$ V) during 5 ns, while the gate voltage is kept at the value reached at the end of the ramp. In these conditions, a high value of the electric field and a high value of drain current occur near the drain edge of the channel; impact ionization mechanism [2.98] produces electron–hole pairs near the drain. While electrons are drifted to the drain, holes are stored in the neutral body of the transistor, producing the concentration increase shown in Figure 2.27d at the end of the pulse. The increase of hole concentration in the body of the transistor produces a decrease of the threshold voltage. This is the reason why the same gate voltage ramp is now applied to the transistor, the measured I_D is higher (Figure 2.27b). At the end of the new voltage ramp the hole concentration in the body of the transistor is shown in Figure 2.27e. At $t = 35$ ns, a negative voltage pulse is applied to the drain ($V_{DS} = -1.5$ V) while the gate is kept grounded. In these conditions, the body-drain junction is forward-biased, allowing the holes to escape through the drain. At the end of the negative pulse, the concentration of

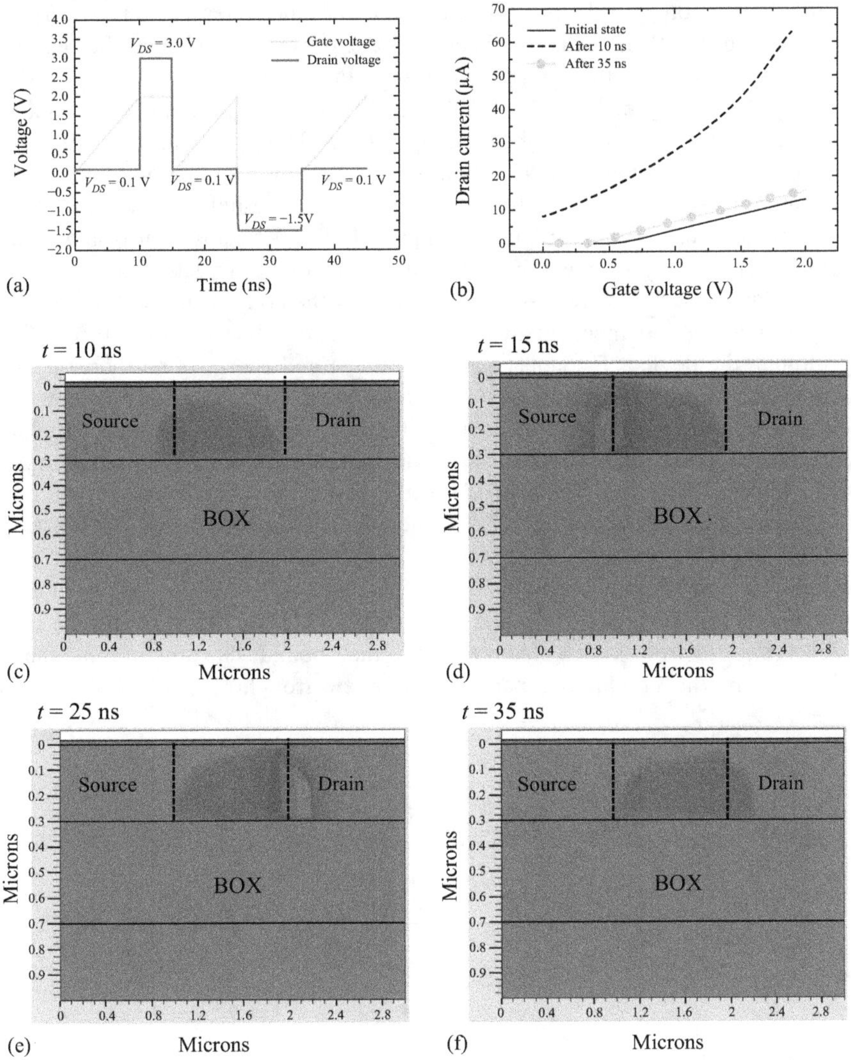

Figure 2.27 History effect in a PDSOI transistor. (a) Voltage patterns applied to the gate and drain of the transistor. (b) I_D–V_{GS} characteristics measured at different times. (c)–(f) Hole concentration in the transistor at different times

holes in the body (Figure 2.27f) is now similar to the one at $t = 0$ s. Therefore, the threshold voltage has returned to almost its initial value and when the I_D–V_{GS} characteristic is measured again, a similar curve to the first one is obtained.

The pronounced hysteresis effect observed in PDSOI transistors has been an intense matter of investigation [2.110–2.113]. Since any kind of hysteresis in a transistor entails a memory effect, much attention started to be paid to the design and fabrication of new memory cells on SOI substrates [2.114, 2.115]. The research of the FBE in PDSOI devices rose exponentially when at the beginning of the last decade the start-up company, Innovative Silicon, introduced the Z-RAM memory cell [2.94]. The basic principle of operation of the cell was the shift induced in the threshold voltage of the PDSOI MOSFET, caused by the injection of holes in the floating-body (a transient overpopulation of holes in the floating-body) [2.116]. This transitory shift of the threshold voltage leads to two different current levels at a given bias point [2.94]. At equilibrium (stable state), the floating-body of the transistor remains neutral; this situation defines the "0" state. The "1" state is forced by charging the body with holes produced by impact ionization mechanism [2.93] that occurs when a large current flows through the device. The consequence of the stored charge is an increase in the potential of the body of the device and a decrease in the threshold voltage: for the same bias a larger current is then obtained. Figure 2.28 presents simulation results obtained, under the drift-diffusion approximation, with calibrated models for impact ionization. The top-side of the figure shows an example of the bias pattern used to demonstrate the 1T-DRAM functionality of the PDSOI MOSFET. In the bottom side, the driven current is monitored. Initially, the holes are injected in the floating-body by impact ionization due to the large current driven by the device (V_{DS} = 1.6 V while $V_{GS} > V_{TH}$, W "1" in Figure 2.28): the highly energetic electrons knock electrons out of their bound state and promote them to a state in the conduction band, creating electron–hole pairs. Electrons are

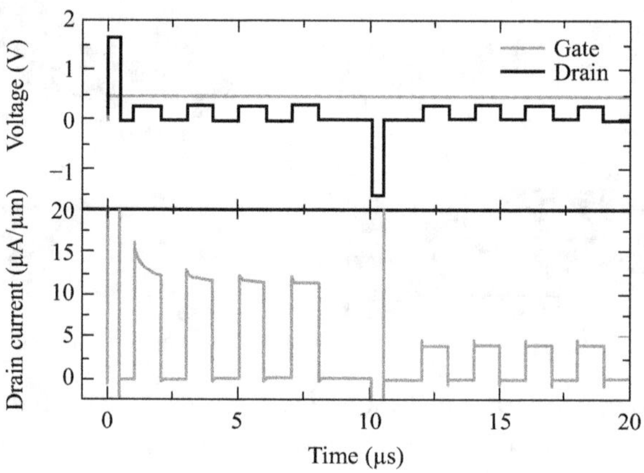

Figure 2.28 Simulation results for the operation of a PD 1T-RAM based on the Z-RAM approach. The picture shows the bias pattern (top) and the driven current (bottom). For simplicity the gate bias is maintained constant ($V_{GS} > V_{TH}$). The floating-body is initially charged with holes generated by impact ionization (W "1"), and the cell state is read four times by using a small drain bias. At t = 10 μs, the cell is purged (write "0" state: W "0"), and read again four times reflecting the difference with respect to the previous "1" states. $L = 1\ \mu m$, $t_{Si} = 300$ nm, $t_{ox} = 3$nm, $t_{BOX} = 400$ nm, $N_A = 10^{17}$ cm^{-3}

evacuated through the drain, while holes are trapped into the neutral body of the silicon film. The hole overpopulation of the body of the device leads to a decrease in the threshold voltage and therefore a transitory increase of the drain current. The cell can be purged of charge by forward biasing the drain-to-body junction (negative drain bias, W "0" in Figure 2.28).

In this process, holes are evacuated from the floating-body through the channel-to-drain p–n junction. If the cell state is read again, the current level remains in the stable level (lower current). For simplicity, the gate bias has been maintained constant and larger than the threshold voltage ($V_{GS} > V_{TH}$) during the whole simulation period in order to have always a conductive channel.

2. **Kink effect**

The kink effect is a direct consequence of the FBE. It makes the drain current, I_{DS}, to show overshoot when applying a large V_{DS} voltage [2.89]. The charge stored in the body modifies the potential (increases in case of N-MOS), reducing therefore the threshold voltage and leading to a sudden increase in the drain current. This effect tends to appear for a high drain bias where the charge is injected in the body by impact ionization. If the large drain bias is hold for a long time, more impact ionization would be obtained leading at the end to a higher increase in drain current (positive feedback effect). An example of this effect is represented in Figure 2.29. I_D–V_{DS} curves have been calculated with Silvaco Atlas [2.117] in a PDSOI transistor for different values of the gate voltage. For the sake of comparison, dashed lines show the drain current data when impact ionization is ignored in the simulation. Solid lines (with impact ionization taken into account) show an overshoot in the drain current produced

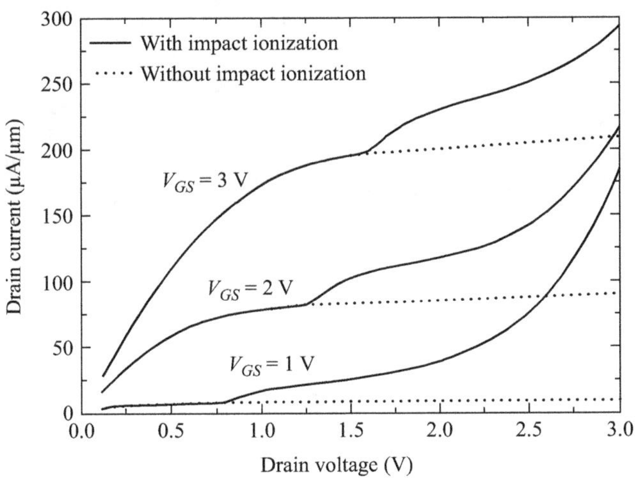

Figure 2.29 Simulated drain current versus drain voltage for different values of the gate voltage in a PDSOI transistor showing kink effect when impact ionization is considered (solid line). Kink effect disappears when impact ionization is ignored (dashed lines): $L_{ch} = 0.8\ \mu m$, $t_{Si} = 300\ nm$, $t_{ox} = 17\ nm$, $t_{BOX} = 400\ nm$, $N_A = 10^{17} cm^{-3}$

by the hole accumulation in the body of the transistor. This effect disappears when impact ionization is ignored (dashed lines).

If the minority carrier lifetime in the silicon film is high enough, the kink effect can be reinforced by the NPN bipolar parasitic transistor structure present in the device (second kink).

3. **Transient effects**

Due to the existence of the floating-body in PDSOI devices, transient effects are significant when the body is not tied to a fixed voltage [2.118].

The drain current experiences a long transient delay before reaching a stable value, thus influencing the drain current just after gate switching and in steady state. This subsequently causes different subthreshold slope and threshold voltage. When the gate is switched on, majority carriers are expelled from the depletion region (instantly formed by capacitive coupling, Figure 2.26) and collected in the neutral body, giving rise to a drain current "overshoot." The drain current decreases gradually with time during electron–hole recombination. A reciprocal "undershoot" occurs when the gate is switched from strong to weak inversion: the drain current increases with time as the majority carriers are generated and allow the depletion depth to shrink. The amplitude of current overshoot or undershoot is proportional to the difference between the final and initial body charges, and the transient duration depends on the generation-recombination rate in the film volume, at interfaces and on the edges. Figure 2.30 shows the turn-on and turn-off transients of drain current in a PD body SOI device, showing the corresponding transient effects.

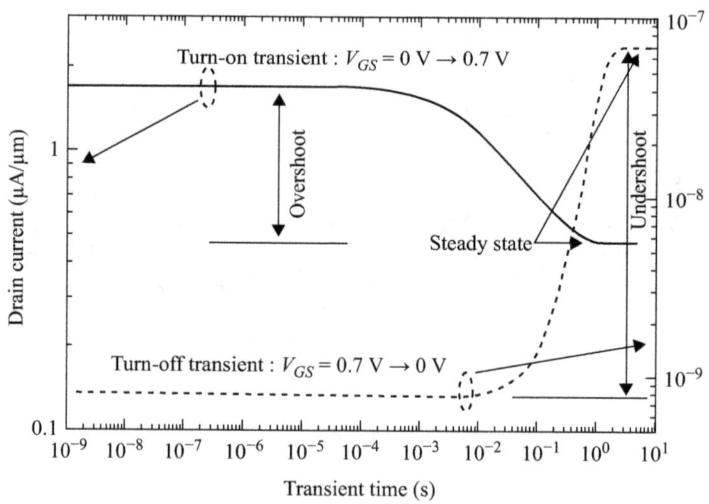

Figure 2.30 Turn-on (solid line) and turn-off (dashed line) in a PDSOI transistor. At $t = 0$ s the gate voltage is switched from 0 to 0.7 V (resp. 0.7 to 0 V) while $V_{DS} = 0.1$ V. Drain current is calculated using Silvaco Atlas until the steady state is reached. $L_{ch} = 0.8$ μm, $t_{Si} = 300$ nm, $t_{ox} = 17$ nm, $t_{BOX} = 400$ nm, $N_A = 10^{17}$ cm^{-3}

Due to the capacitance coupling between the front gate and the floating-body, switching of the front gate voltage changes the body potential momentarily. If the body of a PDSOI device is not tied to a fixed potential, carriers need to be generated or recombine through the front and back interfaces as well as through the PN junctions in order to reach steady state. This may take several seconds. Before carriers are generated or recombine, the potential is lower or higher than its final value, resulting in a change of threshold voltage and drain current. As time elapses during the turn-off transient, holes are generated and the potential in the device rises. The drain current, therefore, will also rise from a low value to its final, stable value. During the turn-on transient, holes recombine, and the potential is high initially and then returns to its final, stable value. Thus the drain current experiences a high to low transition, as shown in Figure 2.30.

These drain current transients in a PD floating-body SOI device can affect the circuit operation in various ways. The circuit may not behave similarly under different frequencies [2.119–2.121].

4. **Parasitic bipolar transistor**
If we consider an n-channel PDSOI device, the N^+-source, the P-type body, and the N^+-drain also form the emitter, the base, and the collector of an NPN bipolar transistor. As in a PDSOI transistor the body is floating (unless a contact is provided) the accumulation of the majority carriers in the body of the transistor produced by impact ionization or BTBT at the drain edge of the channel makes the body potential become high enough so that the PN (body-source) junction turns on. The NPN (Source/film/Drain) bipolar transistor is activated. The parasitic bipolar transistor (PBT) is responsible for different effects; for example, it can amplify impact ionization current and produce a reinforcement of kink effect as mentioned earlier, or also can trigger an extremely low inverse subthreshold slope and reduce the drain breakdown voltage [2.93].

Anomalous subthreshold slope, hysteresis, and single transistor latch
As mentioned in the explanation of kink effect, the generation of majority carriers by impact ionization near the drain can give rise to an increase of the body potential and a related decrease of the threshold voltage. Sometimes, a similar effect can occur at gate voltages lower than the threshold voltage. If the drain voltage is high enough, impact ionization can occur in the subthreshold region, even though the drain current is very small [2.93]. Body charging is particularly effective in weak inversion, where the current depends exponentially on potential. Figure 2.31 compares the drain current versus the gate voltage in a PDSOI MOSFET when impact ionization is considered (solid) and when impact ionization is ignored (dashed). As observed, impact ionization (solid) helps to improve the subthreshold behavior of the device. When the device is turned off, there is no impact ionization (both curves coincide) and the body potential is equal to zero. When the gate voltage is increased the weak inversion current can induce impact ionization in the high electric field region near the drain (if V_{DS} is high enough), holes are generated, the body potential increases, and the threshold voltage is reduced. Consequently, the whole

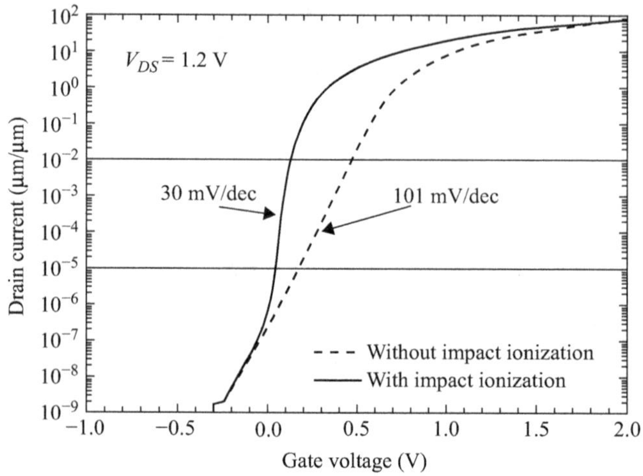

Figure 2.31 Simulated I_D–V_{GS} characteristics of a PDSOI n-channel transistor taking into account impact ionization (solid line) and ignoring impact ionization (dashed line). An anomalous subthreshold slope lower than the theoretical limit of 60 mV/dec can be observed. L_{ch} = 0.8 μm, t_{Si} = 300 nm, t_{ox} = 17 nm, t_{BOX} = 400 nm, $N_A = 10^{17}$ cm^{-3}

I_D/V_{GS} curve shifts to the left and the current can increase with gate voltage at a rate larger than 60 mV/decade, i.e., an anomalous subthreshold slope lower than the theoretical limit of 60 mV/dec can be observed [2.93].

This increase of drain current in weak inversion constitute a positive feedback loop: the higher the drain current, the higher the impact ionization, the higher the body potential, the lower the threshold voltage, and the higher the drain current, which suddenly increases with a subthreshold slope equal to zero millivolt per decade (transistor latch [2.122]) (Figure 2.32).

In Figure 2.32, for V_{DS} = 1.8 V, the impact ionization current that is generated near the drain during the forward gate voltage scan ($V_{GS} = -1.5 \rightarrow 2$ V) raises the body potential. The increased body bias in turn reduces the threshold voltage of the SOI MOSFET leading to an increase in I_{DS} and more impact ionization current. This positive feedback, which occurs when the impact ionization current is larger than the body-to-drain diode leakage current leads to the abrupt increase of the subthreshold current and the body potential. The positive feedback is self-limiting: increased body bias also increases the drain saturation voltage which results in lower channel electric field and smaller impact ionization current. Also, as the drain current increases, the effective potential across the channel decreases due to resistive voltage drops across the source and drain regions. During the descending V_{GS} scan, the impact ionization current under the large drain bias keeps the body potential high, which in turn keeps the MOSFET threshold voltage low, to sustain the inversion layer and a high I_{DS} until this positive feedback cannot be maintained and I_{DS} and the

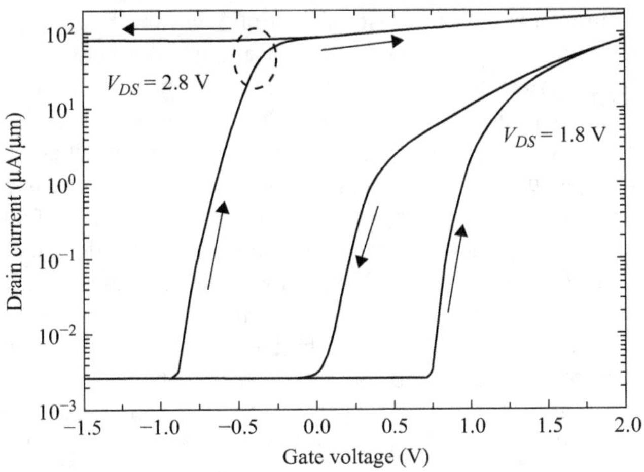

Figure 2.32 Simulated I_D current for ascending and descending scans of the gate voltage in a PDSOI transistor for two values of the drain voltage. If V_{DS} is large enough, a hysteresis cycle is observed. For even higher V_{DS} values, single transistor latch-up occurs: the inversion channel is maintained, although the gate is turned-off. $L_{ch} = 0.8$ μm, $t_{Si} = 300$ nm, $t_{ox} = 17$ nm, $t_{BOX} = 400$ nm, $N_A = 10^{17}$ cm^{-3}

body potential collapse suddenly. It can be noticed that the gate voltage at which the current suddenly raises during a forward scan is higher than the voltage at which the current suddenly falls during a reverse scan. As a consequence, hysteresis is observed in the I_D–V_{GS} curve. For $V_{DS} = 2.8$ V, the positive feedback loop cannot be turned off once it has been triggered, and the device does not turn on (at least for reasonable V_{GS} values) [2.123].

Leakage current

Another effect originated by the parasitic bipolar effect in SOI transistors is related to the subthreshold leakage current (I_{OFF}). The main involved OFF leakage currents are BTBT, impact ionization (II), and direct gate tunneling currents [2.124–2.126]. In short-channel SOI devices, these three contributions can all be amplified by the inherent PBT. When front gate is negative and drain is positive, holes are generated either by BTBT or impact ionization at the drain side and they are driven into body by the lateral electric filed. The body potential increases and turns on the source–body junction (here, playing the role of base–emitter junction); consequently electrons are injected from source and collected by the drain as collector current, I_C, which is an amplified value of the base current, I_B.

5. **Gate-induced floating-body effect**

The gate-induced floating-body effect (GIFBE) [2.127] occurs for very thin gate oxides and at strong gate voltage; the leakage currents by tunnel effect can be important, leading to body charging, a variation of the film potential (even for low drain current) and an increase of the drain current. Experiments and simulations show that the gate-to-body current charges the body causing an unexpected "kink" effect to occur at low drain voltage. The drain current input characteristics measured at low drain voltage show a sudden increase of the drain current. This unexpected "kink" on the drain current gives rise to a second peak in transconductance, which can exceed by up to 40% the normal peak.

The classical remedy to floating-body problems is the use of a body contact in PD devices [2.93]. Using body contacts can restore the device characteristics of SOl MOSFETs back to the bulk-MOSFET-like characteristics [2.128]. Body contact, however, carries a delay penalty and loses the body effect advantage of SOl devices. Advantage can be taken of current overshoot effects and increased current drive due to the kink effect to boost switching speed and therefore, circuit performance [2.129]. However, elaborate circuit models need to be used to avoid unexpected device behavior due to FBE [2.93, 2.129].

2.2.2.2 Fully depleted SOI transistor

FBE can be largely avoided in FD SOl devices in which either the silicon film is thin enough or the doping is light enough that the entire film is depleted, i.e., there is no neutral region in the body. In fact, the entire silicon film can be undoped because FDSOl MOSFETs scale by the silicon film thickness, not by the gate depletion width (x_D) as bulk and PDSOl CMOS do. The inverse subthreshold slope of a long-channel FDSOl MOSFET can be near the ideal 60 mV/decade number at 300 K. Figure 2.33 compares the charge distribution in a PDSOI transistor (left, $T_{Si} = 400$ nm) and an FDSOI transistor (right, $T_{Si} = 100$ nm) with the same channel length. Doping concentration was fixed to $N_A = 10^{17}$ cm^{-3} and a $V_G = 1$ V is applied to the front gate.

In the case of FDSOI device, the whole channel is depleted (charged), while in the PDSOI device, there is a neutral region (dark area) under the channel. The lack of neutral region in an FDSOI transistor makes that the front and back surface

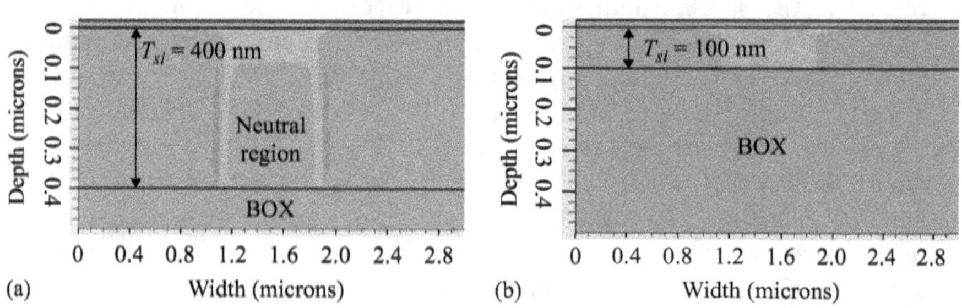

Figure 2.33 Total charge concentration in a PDSOI transistor (a) and in an FDSOI transistor (b) Dark gray indicates zero total charge. The neutral region in the body is clearly observed in the PDSOI transistor, while the body of the FDSOI transistor is FD and therefore charged

potentials become inter-related [2.87], and FDSOI technology presents numerous advantages:

- The possibility to non-dope the conduction channel enables increasing the mobility compared with a PDSOI transistor.
- Short-channel effects are largely reduced if T_{Si} is very small [2.130, 2.131].
- Lower threshold voltage compared to Si bulk technologies allows low power consumption applications.
- The subthreshold slope (S) is reduced due to the replacement of the dynamic depletion capacitance (C_{dep}) by a fixed Si film capacitance ($C_{Si} = \varepsilon_{Si}/t_{Si}$).

Interface coupling means that the electrical characteristics of one channel vary with the bias applied to the opposite gate. In FDSOI MOSFETs, two inversion channels can be activated, one at the front Si–gate oxide interface and the other at the back Si–BOX interface. A better coupling is then obtained between the gate bias and the inversion charge, leading in particular to an increase of the drain current. Because the front and back interface can either be in accumulation, depletion, or inversion, there are nine possible modes of operation in an FDSOI transistor, as a function of the front gate voltage and back gate voltage. In particular, the threshold voltage of the front gate depends on the bias of the substrate or back-gate voltage, V_{G2}.

Figure 2.34 represents I_D–V_G characteristics of an FDSOI device with T_{Si} = 50nm for different conditions of the back interface: solid line corresponds to the case when the back interface is in depletion; dash-dot line corresponds to the case when the back interface is in accumulation. If a positive voltage is applied to the substrate or back gate (V_{G2}), the back interface becomes inverted (dashed line).

Figure 2.35 shows the transconductance of the FDSOI device with a $T_{Si} = 50$ nm, for different states of the back interface. In general, the transconductance of FDSOI devices is complicated function of the gate voltage because of the influence of the back interface on the front channel threshold voltage. As observed in Figure 2.18 the transconductance has its maximum value when the back channel is in depletion. When the back interface is in inversion, the transconductance curve shows a plateau, originated by the activation of the back channel before the front channel [2.132].

Threshold voltage

The threshold voltage of an FDSOI device can be obtained by solving the Poisson equation [2.93]. If ϕ_{S1} and ϕ_{S2} are the surface potential at both interfaces (front and back) (Figure 2.36), the front and back gate voltages, V_{G1} and V_{G2}, respectively, can be expressed as

$$V_{G1} = \phi_{s1} + \phi_{ox1} + \phi_{MS1} \tag{2.21}$$

$$V_{G2} = \phi_{s2} + \phi_{ox2} + \phi_{MS2} \tag{2.22}$$

where ϕ_{oxi} and ϕ_{MSi} are the potential drop across the front (resp. back) gate oxide and the front (resp. back) workfunction differences.

Figure 2.36 shows the potential distribution in an FDSOI device with $t_{Si} = 50$ nm for two conditions of the back interface, inversion (solid line), and accumulation (dashed line). Electron and hole concentrations in the direction perpendicular to the channel are also shown in Figure 2.37.

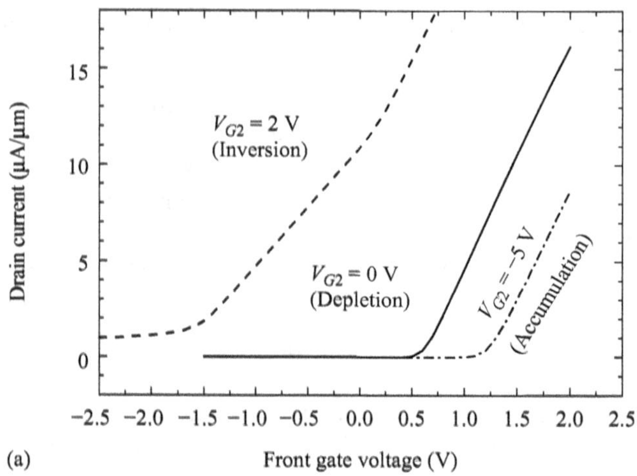

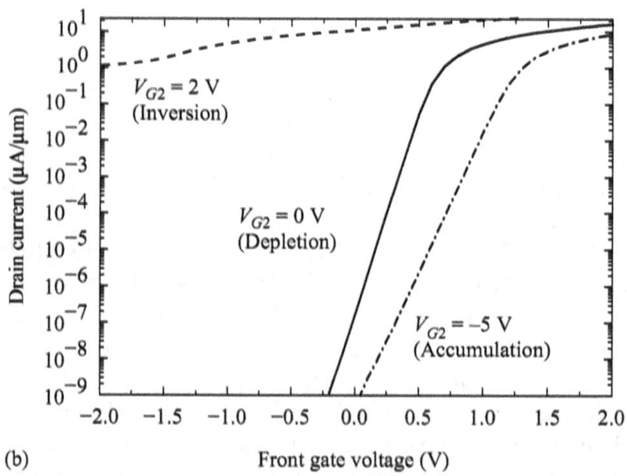

Figure 2.34 Calculated I_D–V_{GS} characteristics of an FDSOI device with $T_{Si} = 50$ nm for different conditions of the back interface (a) linear scale, (b) log scale

We can obtain a relationship between the front gate voltage and the surface potentials [2.91, 2.93]:

$$V_{G1} = \phi_{MS1} - \frac{Q_{ox1}}{C_{ox1}} + \left(1 + \frac{C_{Si}}{C_{ox1}}\right)\phi_{s1} - \frac{C_{Si}}{C_{ox1}}\phi_{s2} - \frac{\frac{1}{2}Q_{depl} + Q_{inv1}}{C_{ox1}} \tag{2.23}$$

$$V_{TH1,depl2} = V_{TH1,acc2} - \frac{C_{Si}C_{ox2}}{C_{ox1}(C_{Si} + C_{ox2})}\left(V_{G2} - V_{G2,acc}\right) \tag{2.24}$$

$$V_{TH1,inv2} = \phi_{MS1} - \frac{Q_{ox1}}{C_{ox1}} + 2\phi_F - \frac{Q_{depl}}{2C_{ox1}} \tag{2.25}$$

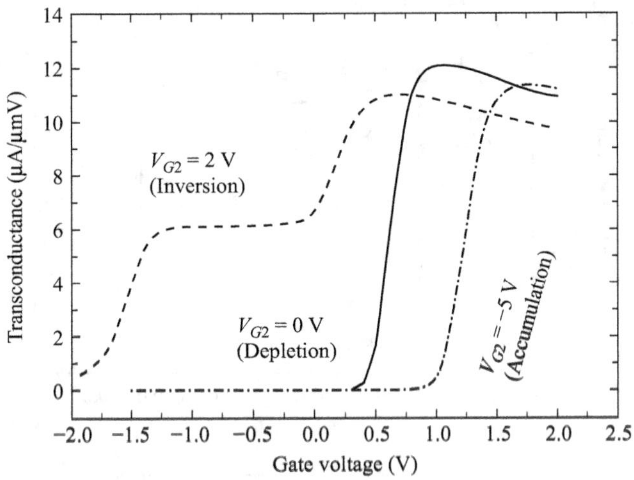

Figure 2.35 Calculated transconductance for the device of Figure 2.34 in different conditions of the back interface (accumulation, depletion, and inversion)

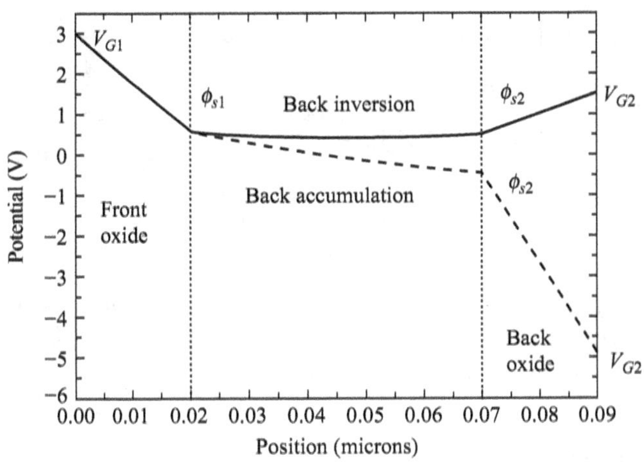

Figure 2.36 Potential distribution in an FDSOI device with t_{Si} = 50 nm for two conditions of the back interface: inversion (solid line) and accumulation (dashed line)

where Φ_F is the Fermi potential, equal to $\frac{kT}{q}ln(\frac{N_A}{n_i})$, and $V_{G2,acc}$ is the value of the back gate voltage for which the back interface reaches accumulation and is given by

$$V_{G2,acc} = \phi_{MS2} - \frac{Q_{ox2}}{C_{ox2}} + 2\left(1 + \frac{C_{Si}}{C_{ox1}}\right)\phi_F - \frac{Q_{depl}}{2C_{ox2}} \tag{2.26}$$

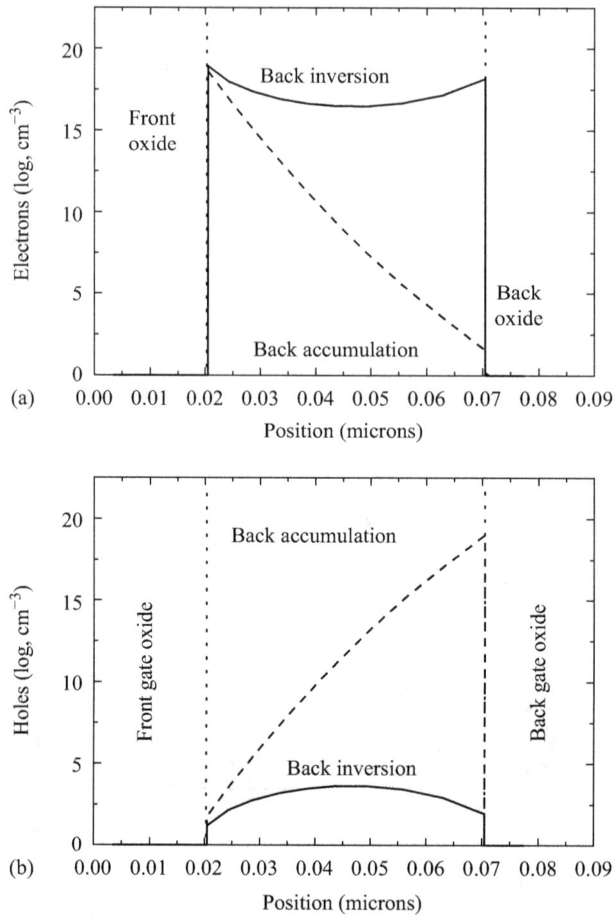

Figure 2.37 *(a) and (b) Electron and hole concentrations in an FDSOI device with $t_{Si} = 50$ nm for two conditions of the back interface: inversion (solid line) and accumulation (dashed line)*

Figure 2.38 shows the evolution of the front gate threshold voltage with the back gate voltage. When the back interface is in accumulation (or inversion), the front threshold voltage slightly increases (or decreases).

The dependence of the threshold voltage with the silicon thickness has been studied by different authors [2.133, 2.134]. If the device is PD, the threshold voltage does not depend on the silicon thickness. In an FD transistor, the threshold voltage decreases with the silicon thickness. This is due to the reduction of the depletion charge as the silicon thickness decreases (Figure 2.39). When the film thickness is below 10 nm, a threshold voltage rebound is observed: the conduction band splits into subbands and the minimum energy of the conduction band increases as the film thickness decreases [2.134].

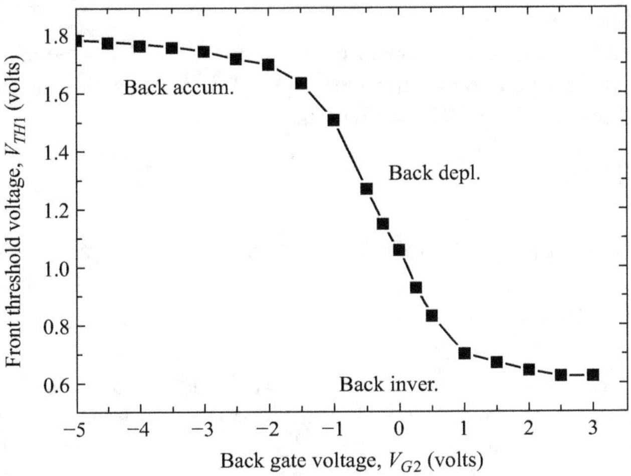

Figure 2.38 Dependence of the threshold voltage with the back gate bias in an FDSOI transistor

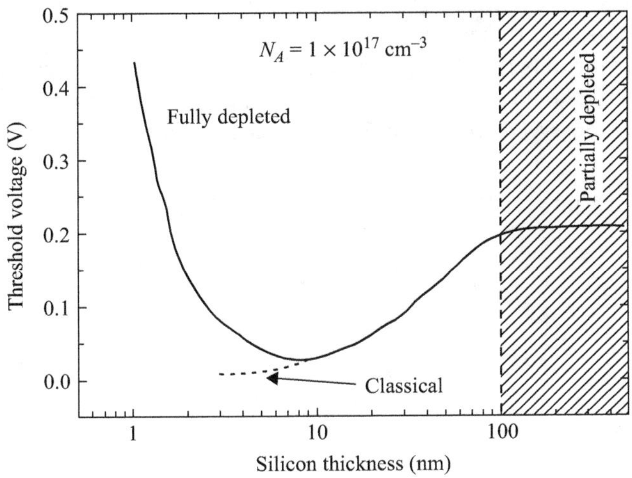

Figure 2.39 Evolution of the threshold voltage of a SOI transistor with the silicon thickness. When the device is PD, threshold voltage does not depend on the silicon thickness. As the silicon thickness decreases and the device become FDSOI, V_{TH} decreases. Below 10 nm, quantum effects make V_{TH} to increase

Subthreshold slope

The fact that the depletion region is constant in an FDSOI device improves the subthreshold slope when comparing with PDSOI devices as shown in Figure 2.40:

The subthreshold slope S is defined as

$$S = \frac{dV_G}{dlog(I_D)} \tag{2.27}$$

which in the case of a bulk MOSFET or a PDSOI transistor can be written as

$$S = \frac{kT}{q} ln(10)\left(1 + \frac{C_D + C_{it}}{C_{ox}}\right) \tag{2.28}$$

where C_D is the depletion capacitance, $C_D = dQ_D/d\phi_s$, and C_{it} is the capacitance related to the interface traps. In an FD transistor with depleted back interface, the subthreshold slope can be expressed as [2.135]

$$S^{dep} = \frac{kT}{q} \ln(10)\left(1 + \frac{C_{it1}}{C_{ox1}} + \alpha_1 \frac{C_{Si}}{C_{ox1}}\right) \tag{2.29}$$

α_1 is the interface coupling coefficient, given by

$$\alpha_1 = \frac{C_{ox2} + C_{it2}}{C_{Si} + C_{ox2} + C_{it2}} \tag{2.30}$$

which accounts for the influence of the back interface traps and the buried oxide (BOX) thickness, and it is always smaller than 1.

The comparison between (2.28) and (2.29) shows that the inverse subthreshold slope of an FDSOI device is smaller than that of a PDSOI transistor, with the same parameters, as observed in Figure 2.40. By accumulating the back channel, the

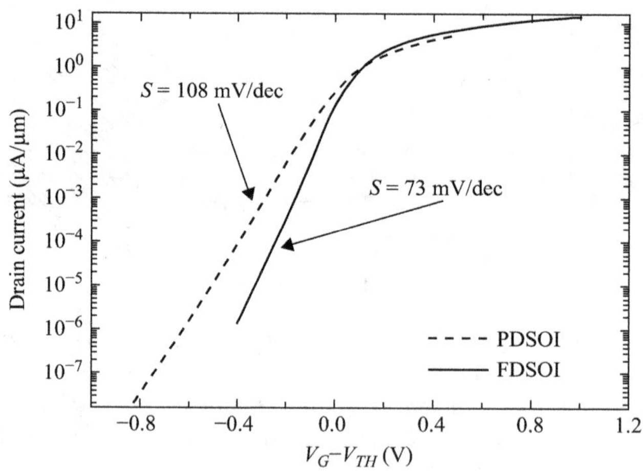

Figure 2.40 Comparison of subthreshold characteristics of PDSOI and FDSOI transistors

front inversion channel becomes decoupled from the defects in the back interface and α_1 tends to 1 [2.135], causing an overall degradation of the subthreshold slope, as shown in Figure 2.40. As a result, S has the lowest value in an FD device, it is larger in the bulk of PDSOI device, and even larger in the FD device with back accumulation.

The subthreshold slope normally improves for thinner silicon films and thicker BOXs, only in the case of few states at the silicon layer/BOX interface [2.133]. On the contrary, a high concentration of interface traps at the buried interface strongly degrades the subthreshold swing.

Short-channel effects in FDSOI devices

In a bulk MOSFET, the reduction of the channel length produces numerous effects so-called short channel effects and whose last origin is the loss of control by the gate of the depletion zone below it, i.e., the depletion charge under the gate it is not totally controlled by the gate because of the encroachment from the source and drain. This results in a roll-off of the threshold voltage as the channel length decreases [2.76] [2.76]. In bulk-Si MOSFETs, increasing the substrate doping and forming shallow junctions enable length scaling. However, ultra-large doping degrades the carrier mobility and promotes BTBT current thus increasing leakage currents.

In FDSOI devices, the fraction of depletion charge controlled by the gate is larger than that in a bulk MOSFET for the same channel length, and the situation improves as the silicon thickness decreases. Figure 2.41 shows the threshold voltage roll-off in bulk and SOI devices for different values of the silicon thickness, t_{Si}.

DIBL is another SCE also due to charge sharing between the gate and the source/drain junctions. It occurs in both bulk and SOI devices, although as in the case of threshold voltage roll-off, in SOI, it is better controlled by reducing the film thickness [2.100]. However, the major SCE in SOI is due to the penetration of the electric field from the drain into the BOX and the substrate (Figure 2.42) [2.135].

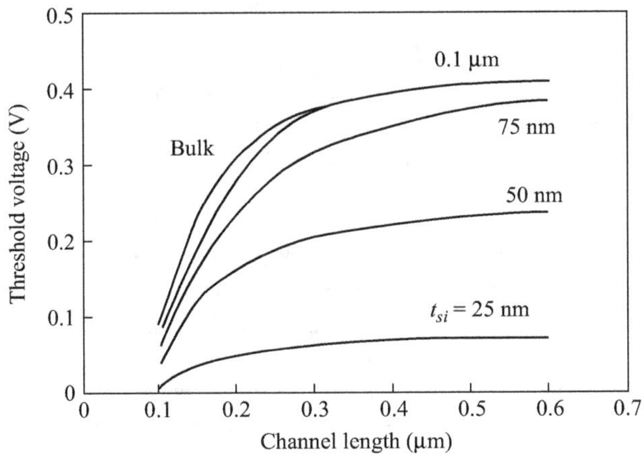

Figure 2.41 Variation of the threshold voltage as a function of the channel length for different values of the silicon thickness in a SOI transistor

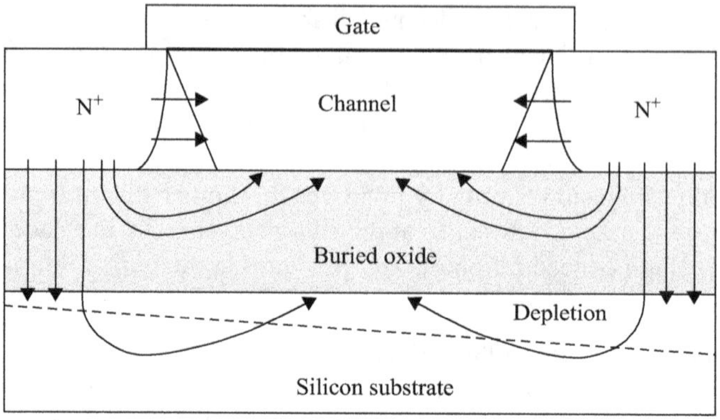

Figure 2.42 Drain-induced virtual substrate biasing (DIVSB) effect

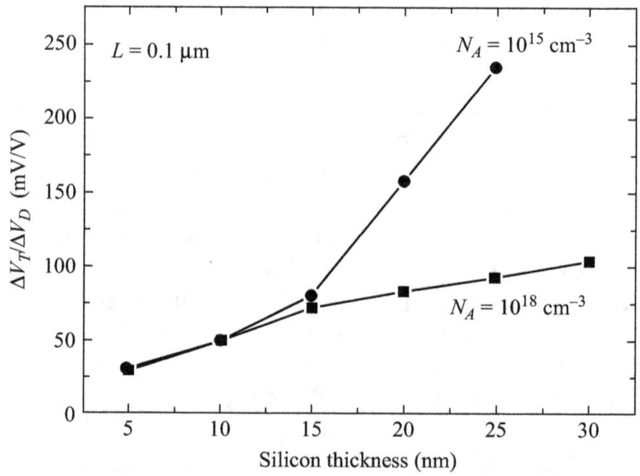

Figure 2.43 Threshold voltage lowering induced by DIBL and DIVSB in highly doped and undoped SOI MOSFET as a function of the silicon thickness

The fringing field tends to increase the surface potential at the film–BOX interface. As mentioned earlier, front and back interfaces are coupled in FDSOI devices; therefore the front-channel properties become degraded. For example, the threshold voltage is lowered with increasing drain bias, as in DIBL, but caused by a different effect.

Figure 2.43 compares the threshold voltage lowering induced by DIBL and DIVSB in highly doped and undoped SOI MOSFETs as a function of the silicon thickness. For silicon thicknesses below 15 nm, ΔV_T becomes very small even for undoped substrates. It is concluded that an undoped and ultra-thin silicon layer is extremely robust to SCEs. Further improvements of SCEs could be obtained reducing the effect of the fringing field, for example, using thinner BOX, multigate devices, or using a ground plane (GP) or back plane (BP), i.e., a highly doped region or metal layer underneath the BOX [2.137].

Self-heating

SOI transistors are thermally insulated from the substrate by the buried insulator. As a result, the heat generated inside the device is not efficiently removed and the temperature of the device increases up to more than 150 °C, and a mobility reduction is observed [2.138, 2.139]. Due to self-heating and the mobility degradation as V_{DS} increases, a negative resistance can be seen in the output characteristics of SOI MOSFETs.

Figure 2.44 shows the output characteristics of a SOI transistor with a channel length of 1 μm calculated with Silvaco ATLAS [2.117] under continuous conditions (solid line) and under pulsed conditions (symbols, 0.1% duty cycle, 1 ms period). When the device is measured under continuous conditions, the effect of self-heating is evident. Drain current decreases as drain voltage increases because of the increase of the lattice temperature that produces mobility degradation.

Figure 2.45(a) shows the lattice temperature in the device at $V_{GS} = 5$ V and $V_{DS} = 3.5$ V calculated in DC conditions. The lattice temperature highly increases in the body of the SOI transistor, which produces a strong degradation in the mobility. The poor heat conductivity of SiO_2 prevents the sink of the heat generated

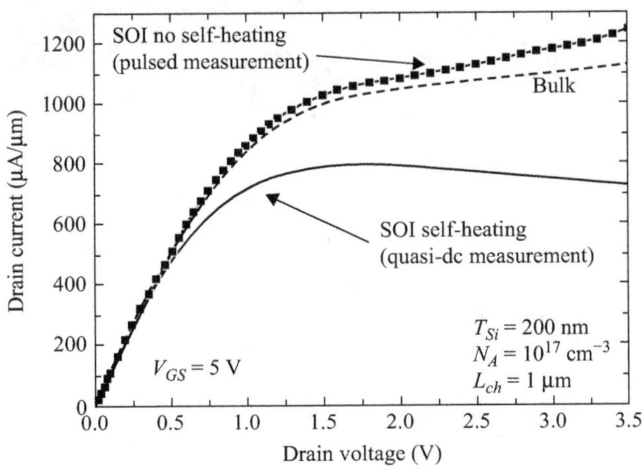

Figure 2.44 Calculated output characteristics of a SOI transistor: (a) under quasi-DC conditions (solid line) the effect of self-heating and the negative resistance can be observed, (b) under pulsed-conditions (symbols) (0.1% duty cycle, 1 ms period) no self-heating effect is observed. (c) For the sake of comparison, the output characteristics of a bulk transistor with the same parameters as the SOI transistor have been simulated (dashed line). No degradation of the output conductance is observed

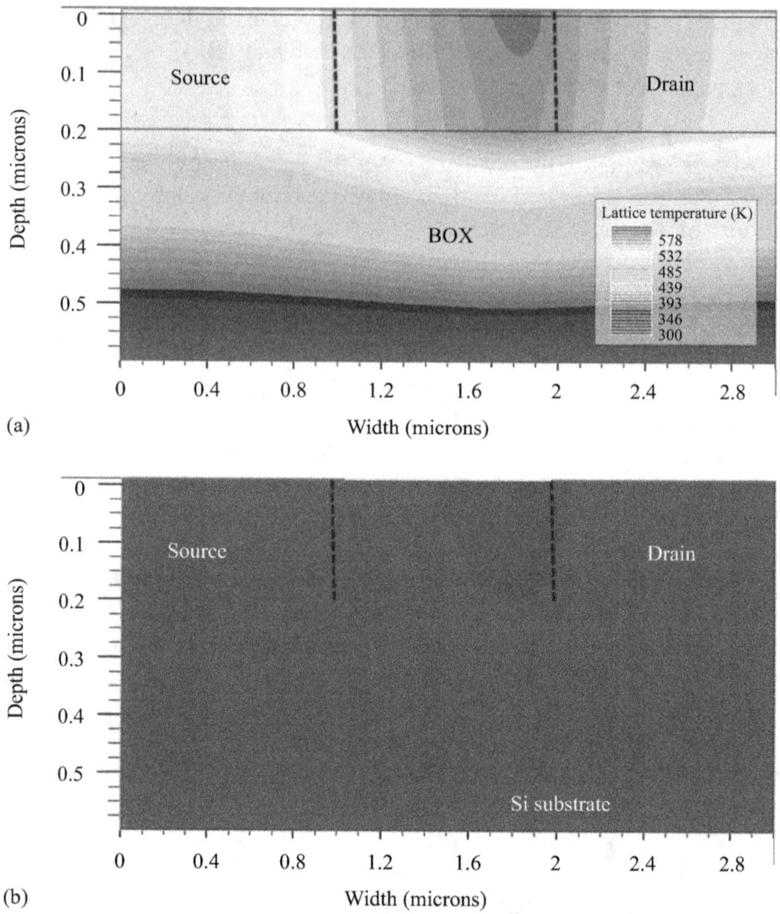

Figure 2.45 (a) Lattice temperature in the SOI device of Figure 2.44 for $V_{GS} = 5$ V and $V_{DS} = 3.5$ V calculated under DC conditions. (b) Lattice temperature in a bulk device with the same characteristics and calculated in the same conditions

in the device. For the sake of comparison a similar bulk device has been considered working in the same conditions as the SOI transistor. Figure 2.45(b) shows that the increase of the temperature is much lower in the case of the bulk device, and therefore the degradation of the mobility because the increase of the temperature is very weak. The output characteristics for the bulk device calculated for the same conditions as the SOI device are shown in dashed line in Figure 2.44.

As the silicon layer is made thinner, self-heating is accentuated [2.100]; this is why FD SOI MOSFETs are more affected. The channel temperature is also raised with increasing the BOX thickness and the channel-to-contact separation. Fortunately, self-heating is highly reduced under dynamic and/or low-voltage operation. As discussed in Reference 2.93, self-heating takes place as power is dissipated in the device, as is the case when the device is measured in quasi-DC mode, but not in operating digital CMOS circuits, since there is virtually no current flowing through the devices in the standby mode, and power is dissipated in the devices only during switching for short periods of time normally. However, if the duty cycle or the frequency are large, although the negative resistance would not be a problem for digital circuits, the increase of the overall local temperature should be taken into consideration, since mobility could be modified [2.90].

For analog circuits, heating effects are more serious: the output conductance of a transistor becomes frequency dependent because of self-heating. Analog designers must take into account self-heating and use circuit simulation models in which this effect is included [2.140, 2.141].

2.2.3 Ultra-thin FDSOI MOSFETs

As discussed previously, for ultimate scaling, FDSOI devices consist of an undoped ultra-thin silicon layer ($T_{Si} < 15$ nm) sandwiched between two oxide layers. In this situation, electrons are quantized in the direction perpendicular to both Si/SiO_2 interfaces, but they can move freely in the plane parallel to them [2.142]. This confinement of the carriers in the direction perpendicular to the channel greatly modifies their transport behavior, and as a consequence, in addition to the advantages mentioned earlier, ultra-thin SOI devices specifically present other benefits, as highlighted below.

2.2.3.1 Electron distribution and quantum effects

In these ultra-thin devices, the maximum extension of the electrons in the direction perpendicular to both Si/SiO_2 interfaces is limited by the silicon thickness that is comparable to the De Broglie wavelength of the carriers. As a consequence, to accurately evaluate the electron distribution in these structures we must self-consistently solve the 1D-Schrödinger and Poisson equations in the direction perpendicular to both Si/SiO_2 interfaces [2.143, 2.144]. The reduction of the silicon film thickness sandwiched between the two oxide layers causes important effects on electron distribution and on electron transport properties, such as the *subband modulation effect*.

The size quantization in the silicon inversion layer produces a redistribution of the carriers between the two subband ladders, which arise from the split of the degeneracy of the six equivalent valleys of bulk silicon [2.142, 2.145]. In SOI inversion layers, the redistribution of the inversion electrons is more acute as the silicon layer shrinks and the following consequences are observed [2.146]:

1. A reduction in the conduction effective mass of electrons in the inversion layers as T_{Si} is reduced. A lower effective conduction mass means a greater electron velocity for the same drift field value, and thus a greater mobility [2.144]. The reduction of the conduction effective mass of electrons as the silicon layer thickness decreases can clearly be observed in Figure 2.46, where the average conduction effective mass versus the total electron concentration for different silicon layer thicknesses is shown.

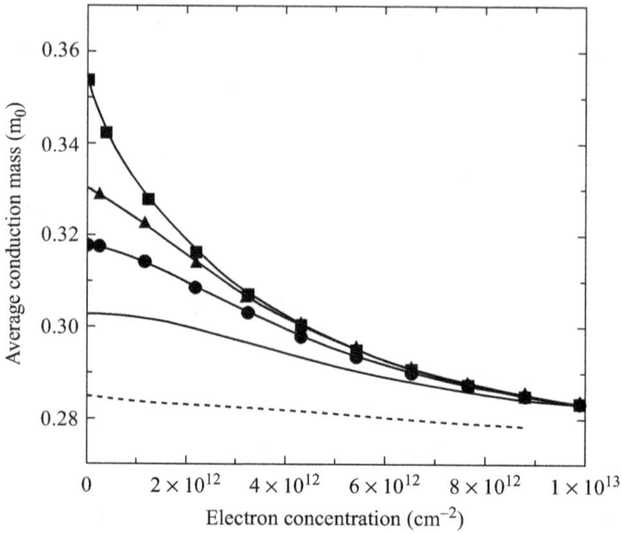

Figure 2.46 Average conduction effective mass versus the inversion charge concentration for different silicon layer thicknesses in a SGSOI inversion layer. (■ $T_{Si} = 50$ nm, ▲ $T_{Si} = 25$ nm, ● $T_{Si} = 15$ nm, (solid) $T_{Si} = 10$ nm, (dashed) $T_{Si} = 5$ nm)

2. A reduction of the intervalley scattering rate between non-equivalent valleys (f-scattering, [2.144, 2.146]) due to the greater separation of the energy levels related to prime subbands with respect to the non-prime ones as the silicon film thickness is reduced.

Both effects, 1 and 2, simultaneously contribute to an electron mobility increase.

Phonon-scattering increase

Another important effect that appears in SOI-inversion layers as the silicon layer thickness is reduced is an increase in the phonon-scattering rate [2.147]. The electron confinement in ultra-thin SOI-inversion layers is greater than in bulk-inversion layers [2.144, 2.146]. Thus, the uncertainty concerning the location of the electrons in the direction perpendicular to the interface is less in SOI samples than in bulk samples. In accordance with the uncertainty principle, there is a wider distribution of the electron's momentum perpendicular to the interface. In other words, due to size quantization, the electrons interface-directed momentum does not have a single value (as in 3D electrons), but a distribution of possible values that expands as the silicon layer thickness is reduced. Taking into account the momentum conservation principle, there are more bulk phonons available that can assist in transitions between electron states, and therefore an increase in the phonon-scattering rate is expected. As a consequence, for the same inversion-charge concentration, the phonon-scattering rate is greater in thinner films than in thicker ones (since the confinement is greater), and therefore a mobility reduction can be expected [2.146].

Subband modulation effect and phonon-scattering increase, discussed earlier, indicate that two opposing trends appear in electron mobility as the silicon layer thickness is reduced. However, the contribution of these two trends varies depending on the considered structure and on the temperature. It would be interesting to know which of these trends, if either, is dominant, and whether there is a critical silicon layer thickness at which a change in mobility behavior is observed: in other words, if a change is produced in the trend of electron mobility as the silicon layer thickness is reduced.

2.2.3.2 Electron mobility in ultra-thin FDSOI transistors

Electron mobility behavior in ultra-thin SOI inversion layers has been calculated using a one-electron MC simulator. Phonon and surface-roughness scattering have been taken into account. In ultra-thin FDSOI structures, the silicon layer is usually left undoped, and if, in addition, the Si/SiO2 interfaces are of good quality (which is desirable), the interface charge concentration is also quite small. As a consequence, Coulomb scattering is much less important than phonon and surface-roughness scattering, especially at intermediate and high inversion charge concentrations. Bulk electron–phonon scattering models considering acoustic deformation potential scattering and intervalley scattering (between both equivalent and non-equivalent valleys [2.143, 2.148]) have been considered. If the scattering mechanisms related to the presence of the Si/SiO_2 interface significantly affect the electron transport properties in bulk silicon inversion layers, one can easily understand that this effect should, at least a priori, be taken into account in those physical systems where electrons are simultaneously affected by two such Si–SiO_2 interfaces. This is the case of the ultra-thin SOI inversion layers: the presence of a second interface plays a very important role, both by modifying the surface-roughness scattering rate due to the gate interface, and by itself providing a non-negligible scattering rate. In addition, we have shown that the usual surface-roughness scattering model in bulk silicon inversion layers overestimates the effect of surface-roughness scattering arising from one of the interfaces as a consequence of the presence of the other [2.145, 2.149]. Therefore, it was necessary to improve the surface-roughness model in order to calculate the scattering rate due to both interfaces (which are assumed not to be correlated) [2.143, 2.145, 2.148].

Figure 2.47a shows mobility curves versus the electron concentration for different values of silicon layer thicknesses, t_{Si}, at room temperature. Phonon scattering and surface-roughness scattering were taken into account.

The following surface-roughness parameters were assumed: $\Delta_{m1} = \Delta_{m2} = 0.1$ nm, $L_1 = L_2 = 1.5$ nm, where Δ_{mi} and L_i are the rms value and the autocovariance length of the roughness fluctuations, respectively. The effect of surface-roughness scattering is more acute at high inversion charge concentrations and in the thinnest samples. In the latter case, the effect of surface-roughness scattering is noticeable even at very low transverse effective fields. There is more than one trend in the electron mobility as the silicon thickness is reduced, and this behavior strongly depends on the electron concentration. To see this more clearly, the evolution of

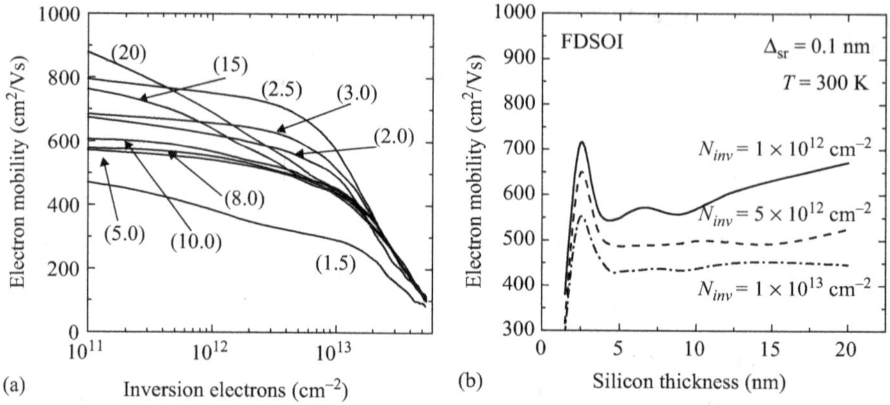

Figure 2.47 *(a) Electron mobility curves in an ultra-thin FDSOI MOSFET at room temperature versus the inversion electron concentration for different values of the silicon layer thickness (T_{Si}). Phonon scattering and surface-roughness scattering due to both interfaces ($\Delta_{m1} = \Delta_{m2} = 0.1$ nm), ($L_1 = L_2 = 1.5$ nm) have been considered. The thickness of the silicon layer is expressed in nanometers between parentheses. (b) Evolution of electron mobility in an FDSOI MOSFET with the silicon layer thickness for different values of the inversion charge concentration at room temperature*

electron mobility with the silicon layer thickness for different values of the transverse effective field is shown in Figure 2.47b. Three regions can be distinguished:

1. For high silicon thickness values, electron mobility tends toward the mobility value in bulk inversion layers.
2. As T_{Si} decreases, electron mobility gradually decreases until a minimum is reached around 5 nm.
3. Then it quickly increases until a maximum at 3 nm is reached, and then abruptly falls.

This behavior is a direct consequence of the opposing trends mentioned earlier.

2.2.3.3 Ultra-thin BOX

From the point of view of the control of SCEs, ultra-thin SOI technology presents another important advantage compared to standard bulk devices: the addition of new degrees of freedom in the definition of the device geometry, i.e., the thicknesses of the silicon and buried oxide layers. In this way, the scaling is carried out thanks to a channel thickness reduction instead of the implantation of complicated doping profiles with very high doping densities. Following the classical scaling rules for bulk MOSFETS, the maximum depth for source and drain implants must be close to $L_G/5$ [2.150]. Therefore, the fabrication of decanano bulk devices demands an important reduction in the implant depth of source and drain regions with the subsequent challenge of obtaining higher doping densities stopping the dopant diffusion to reach the targeted depth. This issue disappears in ultra-thin SOI technology since the BOX constitutes a natural barrier to dopant diffusion. The scaling rules stand also different for bulk and SOI devices. In the case of FDSOI devices, the silicon thickness plays the role of the implant depth, however the conventional design rule can be modified to a less restrictive $L_G/T_{Si} = 4$ [2.151]. In this way, a 4 nm silicon slab would be necessary for a 16 nm channel-length device. However, for T_{Si} smaller than 5 nm two effects limit the use of FDSOI devices: On the one hand, it is very difficult to keep good enough thickness uniformity at wafer level to avoid V_{TH} fluctuations. On the other hand, electron mobility is dramatically reduced as a consequence of confinement effects [2.152].

The condition $L_G/4$ can be relaxed for Multiple-Gate FET devices (as will be discussed later) where the recommended T_{Si} to minimize SCEs follows $N_G L_G/4$ being N_G the effective number of gates [2.153]. Therefore, if we use a double gate structure ($N_G = 2$), the necessary thickness of the silicon slab in the previous example will be increased to 8 nm; if we use a three-gates structure, the necessary silicon thickness will be 12 nm. In both cases, the thickness of the silicon layer is large enough to avoid the above problems of degradation of the mobility and variability. In most cases, the use of multiple-gate devices means the use of 3D architectures whose mass production implies important efforts from an economical and technological point of view.

Therefore, it is necessary to relax the channel length to thickness constraint to extend the use of ultra-thin single-gate FDSOI (SGSOI) transistors to the future nodes. There is still an unexploited way to improve SCEs control for SGSOI devices and, thus, to extend the use of FDSOI devices beyond the limit given by standard design rules considering a thick BOX, as it has been the standard up to now. The use of ultra-thin BOX (UTBOX) and the addition of a GP, i.e., a highly doped region underneath the BOX, can improve the behavior of the device. This fact has been already experimentally demonstrated for gate lengths of 33 nm as shown in Reference 2.154.

Different authors have studied the possibilities that the combined use of UTBOX + GP offers to determine whether the scaling of planar FDSOI devices makes possible to fulfil the requirements of sub-32 nm nodes reducing the impact of thickness fluctuation effects [2.154–2.157]. As the channel length is reduced, the control of DIBL and V_{TH} roll-off is one of the biggest challenges from the point of view of device optimization. Figure 2.48 shows I_D–V_{GS} curves for an FDSOI transistor with a channel of 18 nm with different BOX configurations. These current curves have been calculated using a multi-subband MC simulator described elsewhere [2.158–2.160].

The gate stack structure includes a midgap metal and an HfO_2/SiO_2 dielectric bilayer with an EOT of 1.2 nm, which is in the order of the recommended for a well-tempered decanano device [2.161]. For all the cases T_{Si} is fixed to 6 nm, which

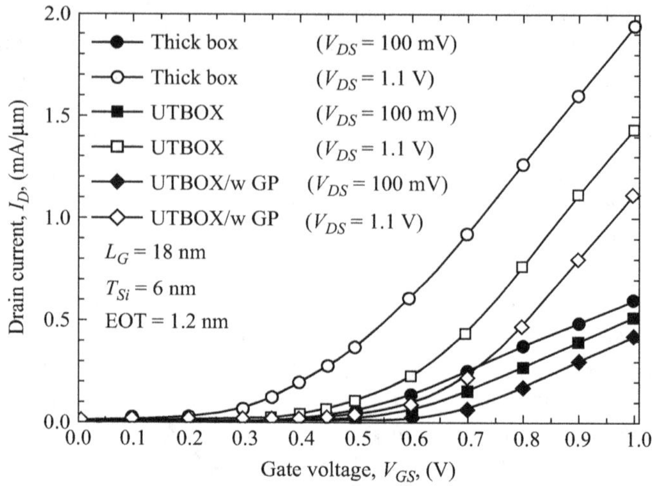

Figure 2.48 I_D versus V_{GS} curves for the 18 nm FDSOI devices calculated at V_{DS} = 100 mV (closed symbols) and V_{DS} = 1.1 V (open symbols). It can be observed the important variation of the characteristics for the standard thick BOX devices demonstrating that the scaling for such architecture cannot be extended beyond the standard rules

fits the requirements for a 24 nm channel length following the standard rule [2.151]. Concerning the BOX, three configurations were studied:

1. a standard thick BOX (T_{BOX} = 145 nm),
2. an Ultra-thin BOX (UTBOX, T_{BOX} = 10 nm),
3. the same UTBOX (T_{BOX} = 10 nm) but including a GP contact with N_A = 3.0×10^{18} cm^{-3}.

In all the cases, source and drain regions are doped with $N_D = 5.2 \times 10^{19}$ cm^{-3}, where a Gaussian transition profile is considered into the channel with no variation of the doping in the transversal direction.

As can be observed in Figure 2.48, the different BOX configurations have a considerable impact especially on the threshold voltage at both high and moderate drain bias. From the results it can be inferred that the use of thick BOX is not recommended due to the important variation on the V_{TH} requiring a channel thinner than 5 nm in order to fulfill the scaling rules. However, this will not make things much better since the aforementioned problems from the point of view of performance and variability will start to play an important role. Focusing on UTBOX options, the variation of the characteristics from high to moderate V_{DS} remains under control and a closest study is necessary to determine whether UTBOX or UTBOX + GP could meet the minimum requirements on SCEs

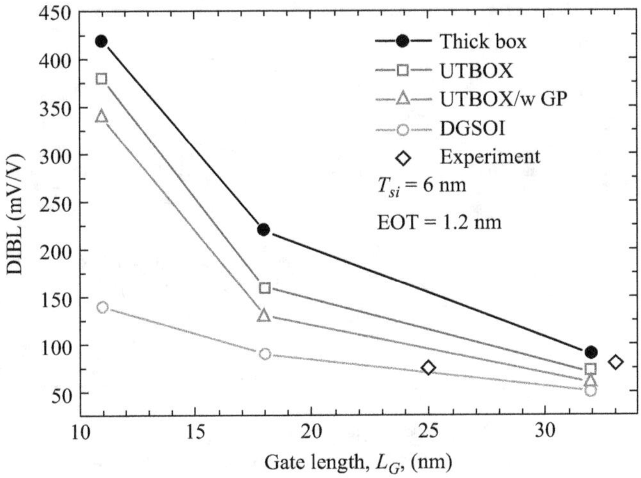

Figure 2.49 DIBL as a function of the gate length including the different BOX configurations for FDSOI, and the corresponding DGSOI device (open circles). For the sake of comparison, experimental results obtained for $L_G = 33$ nm and $L_G = 25$ nm are also shown (diamonds)

control. Figure 2.49 shows the channel length dependence of the DIBL for the different devices under study.

As expected, there is an important increase of DIBL as L_G is reduced; however, the UTBOX + GP device keeps it under control (130 mV/V) for the 18 nm device (triangles). For the sake of comparison, results corresponding to a Double-Gate SOI transistor (discussed later) with $T_{Si} = 6$ nm are also shown. DGSOI devices show smaller values of DIBL as a consequence of the higher electrostatic control as will be discussed later. Diamond symbols represent experimental results for $L_G = 33$ nm [2.154] and $L_G = 25$ nm [2.162] ultra-thin FDSOI devices.

Concerning another important consequence of scaling, V_{TH} roll-off as a function of the gate length, Figure 2.50a shows the evolution of threshold voltage as a function of L_G for the three considered BOX configurations. As the gate length is decreased, a reduction of V_{TH} is observed for all the cases; however the smallest variation among the FDSOI devices occurs again for the UTBOX configurations (triangles and squares). The use of UTBOX also increases V_{TH} respect to the standard thick BOX case for a given length due to the electrostatic influence of the BOX. This fact allows keeping a single metal gate for both p and n devices, which is not possible if V_{TH} becomes very small since the noise margins are dramatically reduced. The impact of channel thickness in UTBOX devices has been also studied. Figure 2.50b shows I_D–V_{GS} curves for $V_{DS} = 1.1$ V (open symbols) and $V_{DS} = 100$ mV (closed symbols) corresponding to the 18 nm device and silicon thickness ranging

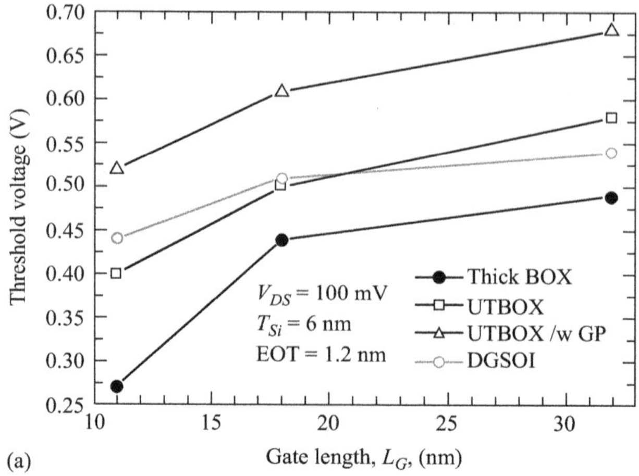

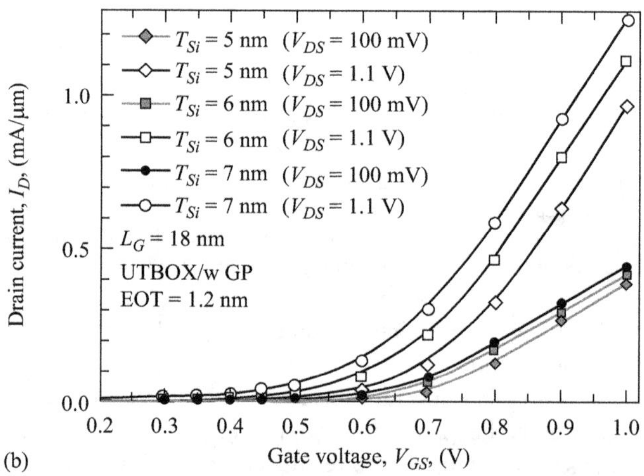

Figure 2.50 (a) Threshold voltage as a function of the gate length for all the BOX configurations and the DGSOI reference device (open circles). For all the cases the same midgap metal gate is used. (b) I_D versus V_{GS} curves for the 18 nm FDSOI devices calculated at $V_{DS} = 100$ mV (closed symbols) and $V_{DS} = 1.1$ V (open symbols) and T_{Si} ranging from 5 to 7 nm. The variation of the characteristics can be specially noticed for saturation bias conditions

from 5 to 7 nm. As observed, there is a V_{TH} increase as the T_{Si} is reduced especially for the case of saturation bias conditions. As a consequence, there is an important variation in the DIBL value specially between the $T_{Si} = 6$ nm and the $T_{Si} = 7$ nm thick devices: $DIBL_{(@TSi=6nm)} = 130$ mV/V and $DIBL_{(@TSi=7nm)} = 210$ mV/V.

However there is a small variation (only 10 mV/V) when the channel thickness is reduced to $T_{Si} = 5$ nm: $DIBL_{(@TSi=5nm)} = 120$ mV/V. Therefore, there could be an important influence of the thickness fluctuations on the SCEs control that cannot be neglected when extremely thin film devices are used.

All these features obtained from the use of FDSOI devices combined with UTBOX + GP allow extending the use of SGSOI transistors for sub-32 nm nodes. The simulations show that 18 nm gate length devices with a channel thickness of 6 nm present good performance and excellent SCEs control. Therefore, the standard design rule which relates gate length and channel thickness can be relaxed to give an extra technological node for a given T_{Si} and delaying the aforementioned end of the scaling capabilities based on channel thickness. Following the considerations presented in Reference 2.153, the proposed design rule corresponds to $N_G \approx 1.5$, giving an idea of the extra electrostatic control obtained from the UTBOX + GP which could be represented as an additional half-gate. The main advantage of this configuration in between of single and double gate structures is the compatibility with the standard ultra-thin FDSOI fabrication flow. However, further studies are necessary in order to evaluate the impact of channel thickness variability on the performance of sub-32 nm node devices.

2.2.3.4 Multi-V_T ultra-thin body and buried oxide FDSOI platform

Multi-V_T CMOS design platforms are commonly used to continue increasing the speed of low-power (LP) applications while keeping adequate static power consumption. High-V_T (HVT) (500 mV $\leq$ HVT $\leq$ 650 mV) transistors are used in noncritical paths to keep low leakage currents, whereas standard-VT (SVT) (350 mV $\leq$ SVT $\leq$ 500 mV) and low-VT (LVT) (200 mV $\leq$ LVT $\leq$ 350 mV) transistors are used in critical paths to meet timing constraints [2.163, 2.164].

In contrast to bulk technology, V_{TH} is primarily set by the gate material work function (WF) in FDSOI devices. Therefore, setting up multi-V_{TH} devices in FDSOI technology is then very challenging. Although today, the cointegration of two gate materials has been demonstrated [2.165–2.168], cointegrating more than two gate materials prohibitively complicates the process. Researchers from LETI and STMicroelectronics have demonstrated recently that using undoped ultra-thin body and ultra-thin BOX (UTBB) devices it is possible to develop a multiple threshold voltage VT platform for digital circuits compatible with bulk CMOS. To do so, various technology options, such as gate materials, buried oxide thickness, BP doping type, and back biasing, were conveniently combined in order to achieve a technology platform that offers at least three distinct VT options (high-VT, standard-VT, and low-VT). In References 2.169 and 2.170, it has been shown that integrating a doped BP below an ultra-thin buried oxide (BOX) (<30 nm) allows setting up three distinct VT options with a single metal gate. Figure 2.51 summarizes the different BP and back bias (V_b) configurations needed to achieve three different VT values in an n-channel UTBB device.

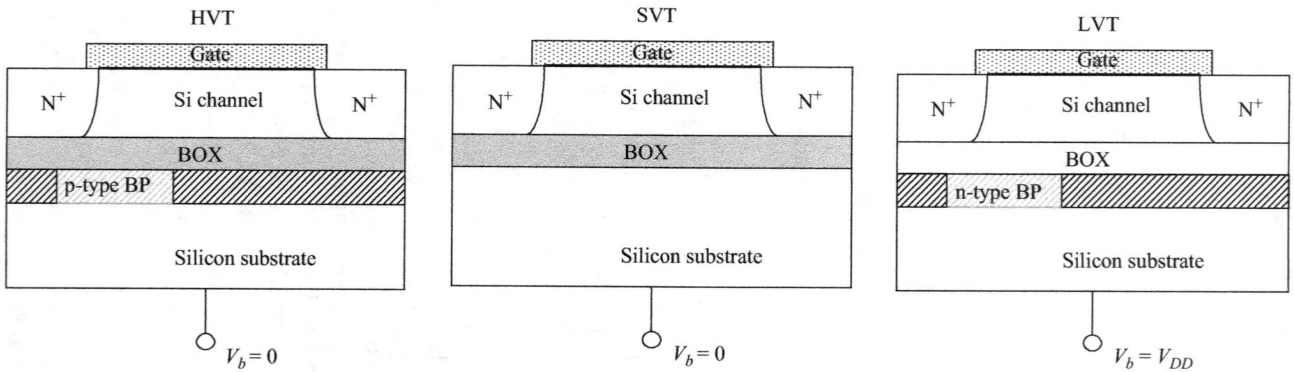

Figure 2.51 NMOS FDSOI multi-V_T devices

Table 2.3 Threshold voltages for UTBB SOI devices with different back plane (BP) and back gate (V_B) configurations

	NMOS			PMOS		
	HVT p-type BP, $V_B=0$	SVT No BP $V_B=0$	LVT n-type BP, $V_B=V_{DD}$	HVT n-type BP, $V_B=0$	SVT No BP $V_B=0$	LVT n-type BP $V_B=-V_{DD}$
V_{TH}(mV) @ $V_D=0.1$ V	604	506	253	−427	−302	−153

1. For standard-VT (350 mV < SVT < 500 mV), no BP is necessary and the back bias is set to zero.
2. Low-VT option (200 mV < LVT <350 mV) is obtained with an n-type BP set to a back bias of $V_b=V_{DD}$.
3. High-VT option (500 mV < HVT < 650 mV) is obtained with a p-type BP biased to $V_b=0$ V.

For p-type devices, complementary BP doping type and biasing are applied. Using $T_{Si}=8$ nm, $T_{BOX}=10$ nm, and a nominal gate length of $L_G=40$ nm, the values obtained for the threshold voltage are shown in Table 2.3 [2.171].

2.2.4 Double-gate and FinFETs

As shown in the previous section, the use of UTBB devices together with BPs/GPs and back biases allows a better electrostatic control of the channel, and therefore a better control of SCEs in ultra-thin FDSOI transistors. This fact suggests that addition of more than one gate to the transistors will enhance their performance and functionality. Several versions of multigate device are discussed extensively in the literature [2.87, 2.100, 2.93, 2.151]. There are two varieties: planar and vertical structures [2.87] (Figure 2.52). The former group contains GP and back-gate devices, which are derivatives of the SOI device [2.93]. The vertical structures contain the FinFET [2.172–2.174], the Omega FET [2.175], the Tri-Gate [2.176], the Gate-All-Around FET [2.177], and the junctionless FET [2.178].

The main difficulty with multiple-gate transistors is the realistic fabrication of such devices. Manufacturing a self-aligned double-gate MOSFET has been the goal of device engineers and researchers ever since it was proposed by Sekigawa and Hayashi in 1984 [2.179]. Different approaches for the fabrication of DGSOI transistors have been discussed since then.

(i) The planar solution (Figure 2.52a) is suitable but it does not guarantee the self-alignment of the two gates [2.180–2.182], although some advantages can still be obtained using asymmetrical and misaligned DGSOI devices [2.181].
(ii) A totally different approach is to adopt a non-planar technology. In fully vertical DG-MOSFET (Figure 2.52b), the source-body-drain stack, and therefore the current is perpendicular to the Si-wafer. These devices are attractive because the channel length (source-to-drain distance) is controlled by epitaxy, instead of lithography. They suffer, however, from the

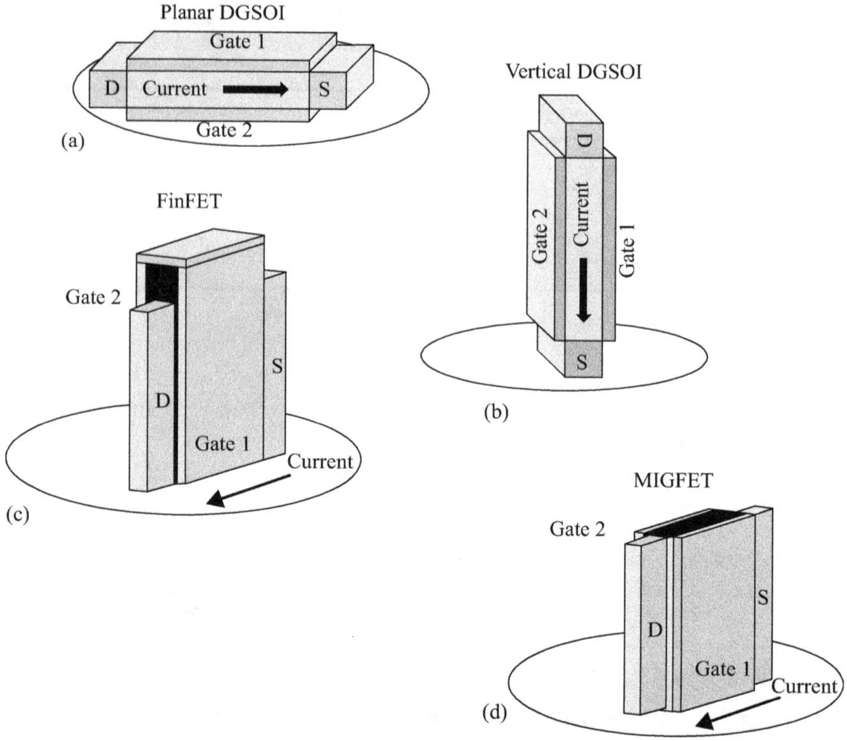

Figure 2.52 (a) Planar DGSOI (b) Vertical DGSOI (c) FinFET (d) Multiple Independent Gate FET

asymmetry of the source and drain terminals and from the difficulty of achieving tiny pillars with ultra-small integrate distances [2.183].

(iii) The third option is the FinFET. In this case, source and drain are set at both edges of the silicon fin (Figure 2.52c). The current is controlled by the two vertical gates and flows horizontally along the body sidewalls. On top of the silicon fin, a dielectric layer called "hard mask" deactivate the formation of an inversion channel at the top corners of the device. Another alternative of the FinFET is the MIGFET (Multiple Independent Gate FET). In this device, the top gate is etched, thus the lateral gates become independent, and therefore they can play different roles (Figure 2.52d).

(iv) If the top gate is made active by reducing the top dielectric layer, the device is named triple-gate MOSFET, or tri-gate MOSFET, although in fact one single gate controls three different sections of the channel: two vertical and one horizontal (Figure 2.53). The addition of the third gate improves the electrostatical integrity of the device. This electrostatical integrity can be improved still more by extending the sidewall portions of the gate electrode to some extent inside the buried oxide (Pi-Gate transistor) or/and underneath the channel region (Omega-Gate transistor). In fact, these devices will behave, from the electrostatic point of view, as having a number of gates between 3 and 4 [2.151, 2.184,–2.186].

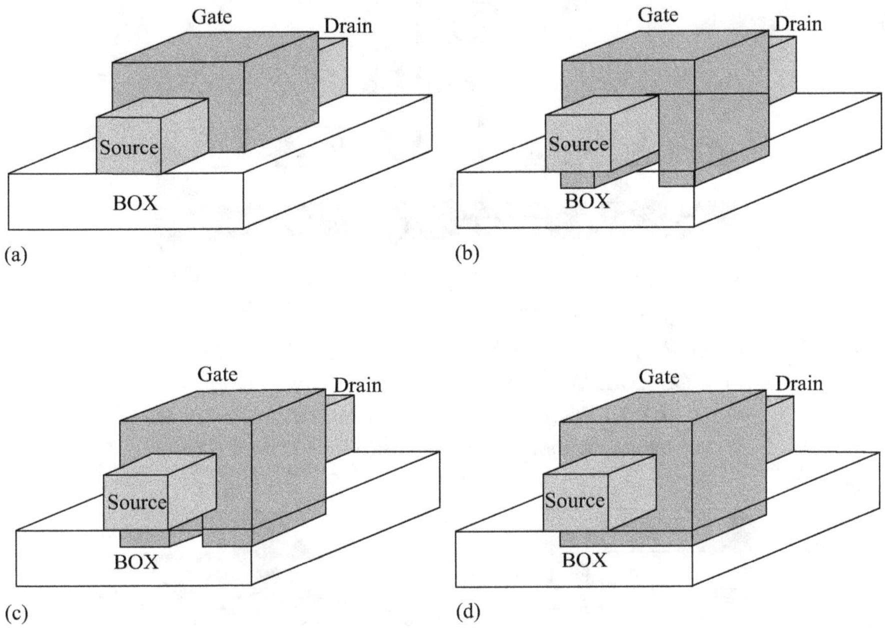

Figure 2.53 Multigate transistors. (a) Tri-gate FET, (b) Pi-gate FET, (c) Omega-gate FET, and (d) GAA-FET

(v) The device with better control of the channel region by the gate is the surrounding gate MOSFET, or Gate-all-around (GAA) MOSFET [2.187]. In these devices, the gate electrode is wrapped around all sides of the channel region.

2.2.4.1 Fabrication technology for multigate transistors

The first modern self-aligned vertical multigate MOSFET was called DELTA (DEpleted Lean channel TrAnsistor). This device was proposed by D. Hisamoto *et al.* in 1989 [2.173]. Figure 2.54 shows a cross section of the DELTA MOSFET.

The critical fabrication steps in the front-end processing of a multigate MOSFET include, sequentially: (1) fin formation, (2) gate stack formation, (3) source and drain extension implant, (4) spacer formation, (5) epitaxial raised source/drain formation, and (6) deep source/drain implantation and activation anneal. The fabrication flow of a tri-gate MOSFET on a SOI substrate is shown in Figure 2.55:

(a) The SOI silicon top layer (T_{Si}) thickness defines the fin height (FinHEIGHT).
(b) The fin pattern and the critical dimension of fin width (FinWIDTH) can be defined by optical lithography or by spacer image transfer (SIT) [2.189, 2.190], followed by plasma etching. After fin etch, the fin sidewall surfaces are rough. Therefore, oxidation and H_2 annealing are often used to smooth the sidewalls [2.191, 2.192].
(c) Next, the gate dielectric is grown and metal gate is deposited. It is suitable to tune the threshold voltage (V_{TH}) of the MOSFET by using a gate material that has the appropriate effective workfunction rather than by doping the channel,

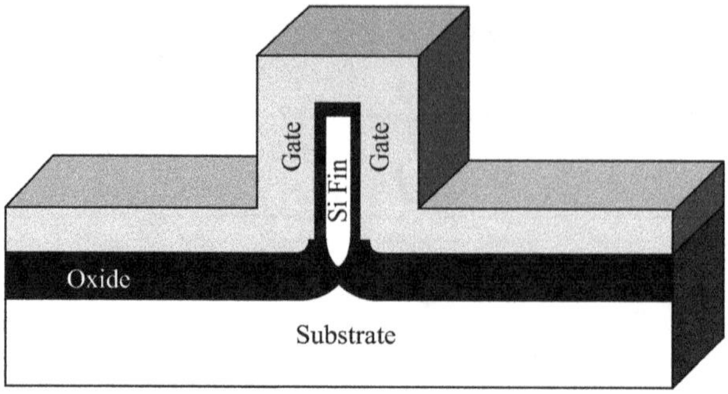

Figure 2.54 DELTA MOSFET. The oxide under the Si fin was formed through LOCOS oxidation, while the Si fin was protected by the nitride hard mask and a nitride spacer prior to the oxidation process

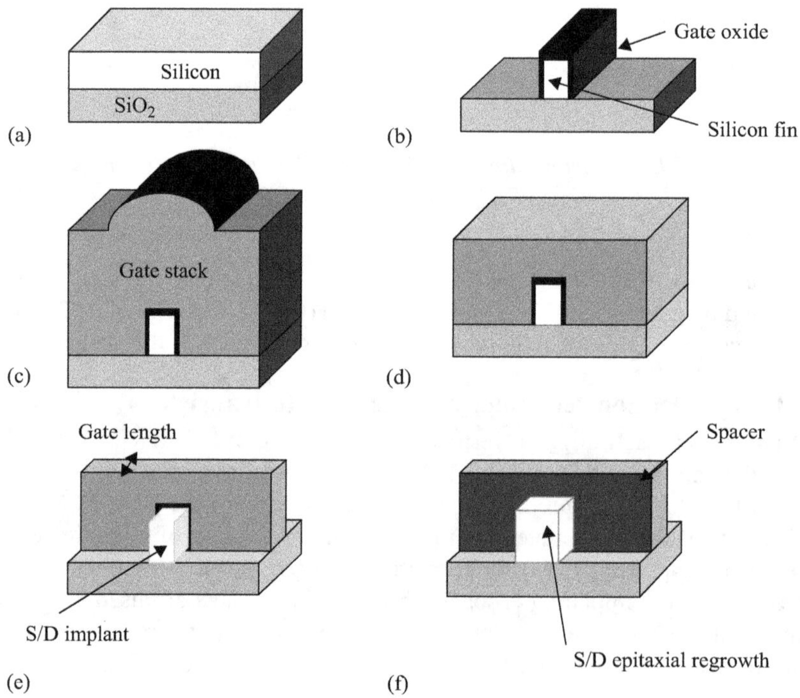

Figure 2.55 Fabrication sequence of a FinFET transistor

since as mentioned earlier, it is highly desirable to have intrinsic or lightly doped channels.

(d) Since the gate stack is over the fin topography, a planarization step is desirable to flatten the gate surface, which reduces the burden on photolithography and gate etch. Significant overetch of the gate material is required to clear the bottom of the fins. As a result, the gate etch must have a high selectivity to the gate dielectric on top of the fin, if one wants to avoid damage to the fin during gate etch.

(e) Source and drain (S/D) extensions are formed after gate patterning using low-energy and large-tilt angled implants [2.193–2.195]. Next, S/D offset spacers are formed along the sidewalls of the gate and fin.

(f) The sidewall spacers on the fins are subsequently removed to expose the fin to grow raised source and drain using selective epitaxy [2.193–2.195]. The raised source and drain structure helps to reduce the parasitic resistance associated with thin fins [2.193].

For a complete treatment of multigate MOSFET technology, the reader can follow Reference 2.188.

2.2.4.2 Multigate transistors and short channel effects

Starting from Poisson equation, Yan and coworkers [2.196] and Lee and coworkers [2.186] introduced a powerful concept, the "natural length," λ, which is a measure of SCEs in multigate transistors. It represents the distance of penetration of the drain electric field into the channel [2.185, 2.186]. A device will be free of SCEs if the channel length is at least six times the natural length [2.186]. The natural length depends on the gate oxide thickness, the silicon film thickness, and the geometry of the gate, i.e., the number of gates. Table 2.4 summarizes the expression of λ, for different device geometries.

Suzuki *et al.* [2.198] proposed an accurate expression for the natural length in a double-gate device:

$$\lambda_2 = \sqrt{\frac{\varepsilon_{Si}}{2\varepsilon_{ox}}\left(1 + \frac{\varepsilon_{ox}}{4\varepsilon_{Si}}\frac{t_{Si}}{t_{ox}}\right)t_{Si}t_{ox}} \tag{2.31}$$

Table 2.4 Natural length for different multigate FETs

Device	Natural length	Ref.
Single gate	$\lambda_1 = \sqrt{\frac{\varepsilon_{Si}}{1\varepsilon_{ox}}t_{Si}t_{ox}}$	[2.185]
Double gate	$\lambda_2 = \sqrt{\frac{\varepsilon_{Si}}{2\varepsilon_{ox}}t_{Si}t_{ox}}$	[2.186]
Triple gate	$\lambda_3 = \sqrt{\frac{\varepsilon_{Si}}{3\varepsilon_{ox}}t_{Si}t_{ox}}$	[2.186]
Surrounding gate (square cross-section)	$\lambda_4 = \sqrt{\frac{\varepsilon_{Si}}{4\varepsilon_{ox}}t_{Si}t_{ox}}$	[2.186]
Surrounding gate (circular cross-section)	$\lambda_o = \sqrt{\frac{2\varepsilon_{Si}t_{Si}^2\ln\left(1+\frac{2t_{ox}}{t_{Si}}\right)+\varepsilon_{ox}t_{Si}^2}{16\varepsilon_{ox}}}$	[2.197]

Colinge and coworkers [2.184] also introduced the concept of effective gate number, N, to generalize the expression of the natural length to an arbitrary number of gates.

$$\lambda_N = \sqrt{\frac{\varepsilon_{Si}}{N\varepsilon_{ox}}\left(1 + \frac{\varepsilon_{ox}}{4\varepsilon_{Si}}\frac{t_{Si}}{t_{ox}}\right)t_{Si}t_{ox}} \tag{2.32}$$

The value of N can be extracted experimentally from the dependence of threshold voltage on the silicon film thickness, which in a multigate device is given by

$$V_{TH,N} = V_{FB} + 2\phi_F + \frac{qN_A}{C_{ox}}\frac{t_{Si}}{N} \tag{2.33}$$

Thus, for a Pi-gate transistor, N takes the value of 3.14, and for an Omega-gate FET, N is between 3 and 4, depending on the penetration of the fourth gate underneath the channel [2.186].

The natural length depends on the silicon thickness, the oxide thickness, and the number of gates. Based on the "natural length" concept, Suzuki *et al.* [2.198] defined a scaling parameter, α_n, which allows one to estimate the short-channel sensitivity of devices with different gate structures.

$$\alpha_N = \frac{L_{eff}}{2\lambda_N} \tag{2.34}$$

Given a silicon thickness, t_{Si}, and oxide thickness t_{ox}, the minimum gate length avoiding SCE can be estimated imposing a value of $\alpha \approx 2.2$ [2.186].

2.2.4.3 Corner effects

Despite their benefits controlling SCEs, when multiple gate devices are considered, new coupling effects appear due to their 3D architecture. Design studies of multiple gate SOI MOSFETs have revealed that the corners of the silicon body can significantly affect their I–V characteristics [2.176, 2.199–2.201]. This phenomenon is commonly referred to as corner effects, and they are due to the formation of independent channels with different threshold voltages next to the corners as compared to the top or the sidewall gates. The corner components of the total current reflect a lower threshold voltage than in the rest of the device, giving rise to a higher I_{off}, which degrades the I_{on}/I_{off} ratio [2.202, 2.203]. In addition, the radius of curvature of the corners has a significant impact on the device electrical characteristics and can determine if the corner sections of the channel and the planar interfaces of the channel will have a different threshold voltage. A comprehensive study of corner effects in Pi-gate SOI transistors was performed in Reference 2.201. It was observed that the extension of corner regions has an inverse dependence on doping concentration. Thus, a reduction of doping density would help to prevent the presence of undesirable double threshold voltages. However, it has also been demonstrated that, even when highly doped substrates are used, corner effects can be suppressed as long as the device dimensions are small enough. Moreover, the influence of corner rounding and the reduction of the gate oxide thickness were also analyzed. For these cases, the elimination of the corner effects is based on the reduction of electrostatically favorable regions and, therefore, of potential variations along the Si–oxide interface. This would avoid the possibility that different regions of the device become inverted at different gate voltages. Therefore, in accordance with these results, corner effects are not expected to play an important role in ultra-small structures. These conclusions, reached from the study of Pi-gate MOSFETs, can be extended to similar structures, such as tri-gate FETs, Ω-gate FETS, or GAA SOI MOSFETs.

2.2.4.4 Bulk FinFETs

The multigate devices described so far are fabricated on SOI wafers. However, multiple gate devices can also be made on bulk silicon wafers [2.204, 2.205]. Indeed, Intel introduced bulk FinFETs in their 22 nm CMOS technology [2.206] (Figure 2.56)

The main advantages of using bulk silicon over SOI substrates are basically two:

1. lower wafer cost, and
2. better substrate heat transfer rate.

The disadvantages seem to be much more numerous:

1. Bulk multigate FETs require for the fabrication additional isolation steps, which increase the number of steps in the fabrication process and therefore the final cost per device. Figure 2.57 compares the fabrication processes for a bulk tri-gate transistor and for a SOI tri-gate transistor (see for instance soiconsortium.org).

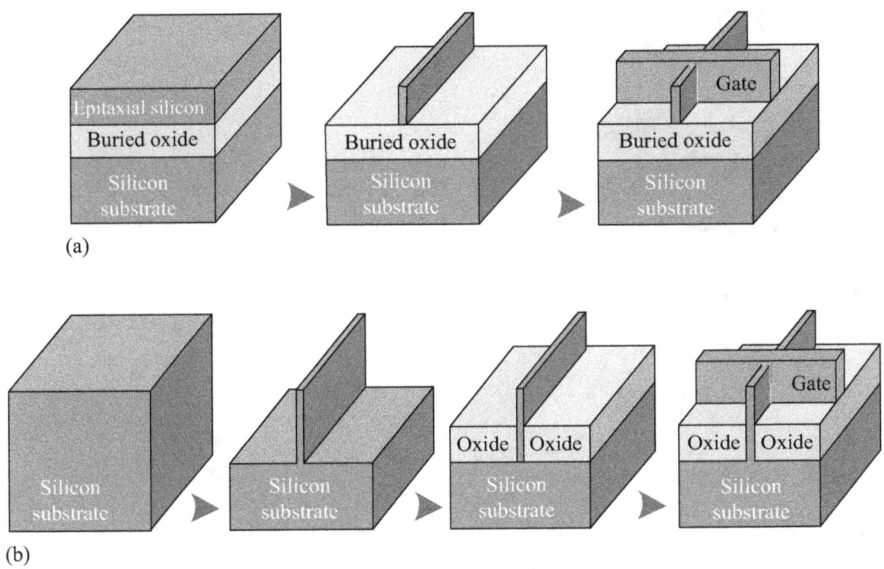

Figure 2.56 Comparison of (a) SOI FinFET and (b) bulk FinFET

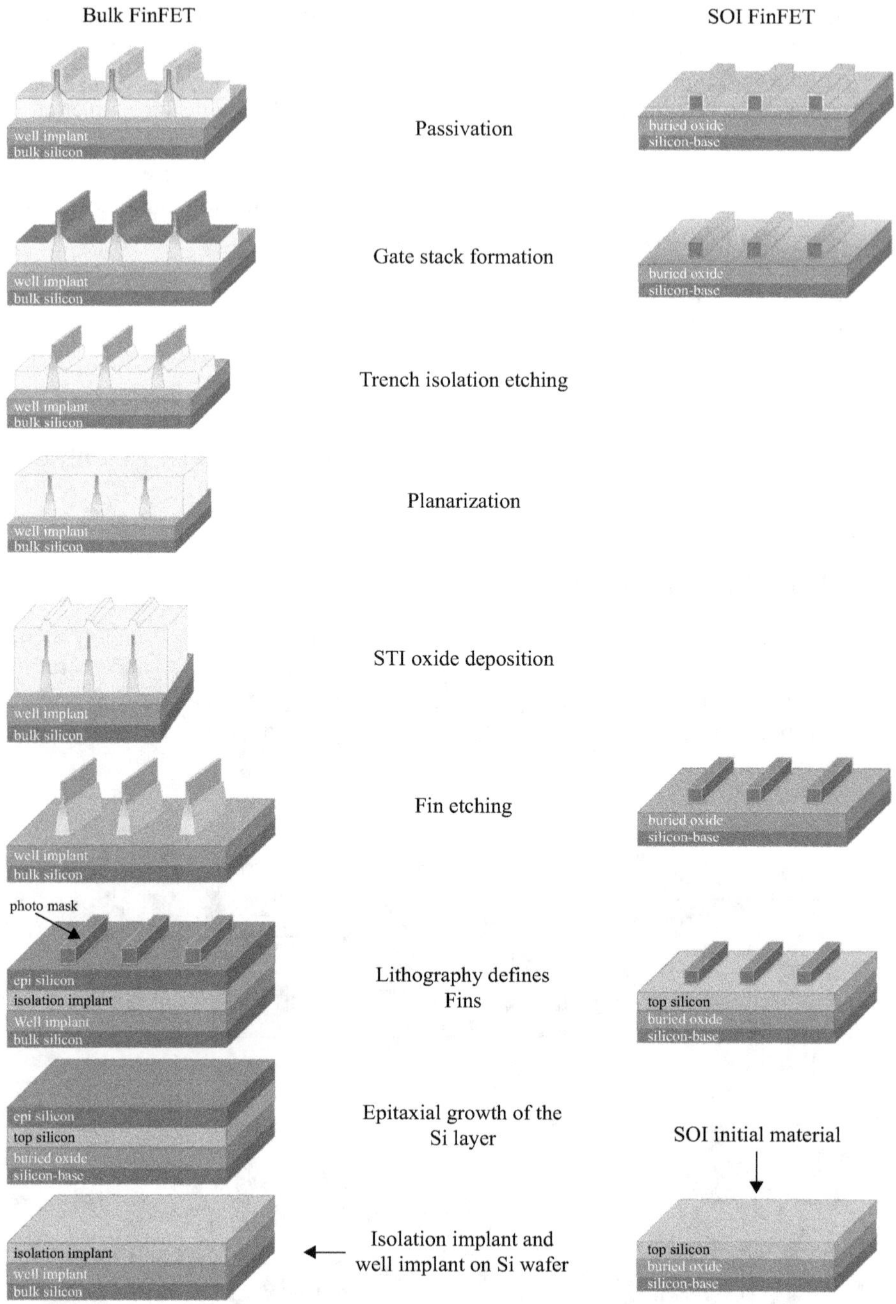

Figure 2.57 Comparison of fabrication processes for bulk and SOI multigate transistors. Adapted from soiconsortium.org

2. There are in addition some issues related to geometrical variability that are better solved in SOI FinFETs than in bulk FinFETs. The fin height in a bulk FinFET is entirely determined by the fin etch step (not by the Si layer thickness as in the case of SOI FinFET). This puts more pressure on etch variation control, since any fin height variation translates into transistor width variation.
3. Although the control of the electrostatics of the channel by the gate in bulk FinFETs is as good as in their SOI counterparts, underneath the fin, in bulk devices, the source and drain have to be separated by heavy channel stop implants to prevent sub-surface punch through.
4. The bulk silicon under the fin can be accessed by a body contact. But the body factor of a bulk FinFET is very low, since the electrostatic potential inside the fins is dominate by the gate, not by the body. Therefore, body bias is no effective in changing the threshold voltage of a multigate bulk transistor.

Asenov's group at University of Glasgow performed a comprehensive simulation study comparing the performance of bulk multigate transistors and SOI-based devices [2.84]. This study indicates that for a tri-gate FinFET, following the normal design practice, the SOI device can introduce more than 6% performance and I_{on}/I_{off} ratio advantage compared to bulk FinFET, or can provide more than two times reduction in leakage current at the same drive current. This is thanks to the BOX isolation compared to the junction isolation which depletes the bottom part of the bulk fin. For a DG FinFET, the advantage of SOI FinFET over its bulk counterpart will be more pronounced: over 10% improvement on drive current and I_{on}/I_{off} ratio can be expected in SOI architecture for devices with the same leakage current, and more than five times reduction of leakage current in SOI can be achieved if both devices have the same drive current.

Although SOI FinFETs have slightly worse SCEs compared to bulk FinFETs, to some extent it can be mitigated by BOX and substrate doping optimization. SOI technology can efficiently help to reduce the process-induced FinFET variability. In SOI FinFETs, there is no obvious degradation on statistical variability performance compared to bulk FinFETs that have the best possible RDD performance. Considering the larger process variation associated with the fin formation in bulk technology, the SOI FinFETs can have better overall variability performance compared to bulk FinFETs.

For a comprehensive discussion about bulk Si versus SOI FinFETs, the reader can follow the work by Fossum *et al.* [2.85].

2.2.4.5 Quantum effects in quantum-well-based multigate transistors

The thickness and/or width of the multigate transistors have to be reduced to values below 10 nm in order to control SCEs as discussed previously. Under these conditions, the carriers in the channel become quantized in one dimension (in quantum-well-based devices, as DGSOI and FinFETs) or even in two dimensions (in quantum wire-based devices as tri-gate, four-gate, or GAA devices). This results in the formation of energy subbands and in electron distributions in the silicon film

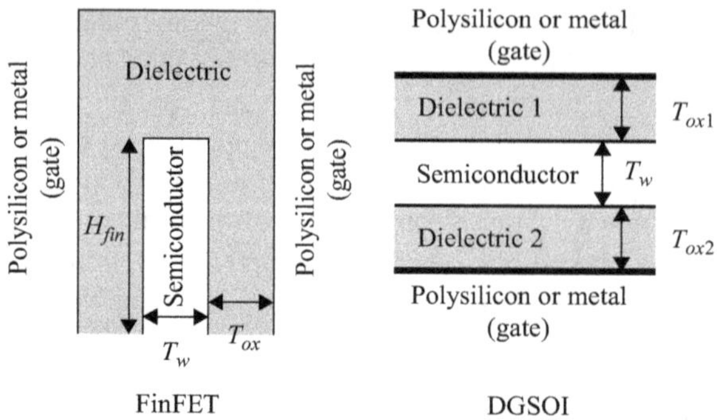

Figure 2.58 Schematic representation of a FinFET (left) and a Double-Gate transistor (right)

that can be significantly different from what is predicted by the classical theory. The confinement of carriers is responsible for different behavior of carrier mobility and threshold voltage.

We will start analyzing quantum effects and electron mobility in quantum-well-based devices, i.e., those devices as planar DGSOI and FinFETs in which carriers are confined only in the direction perpendicular to both gates (2DEG). We will consider the case of quantum wire based devices (1DEG) in a next section of this chapter.

In a planar DGSOI structure the silicon slab is sandwiched between two oxide layers. A metal or a polysilicon film contacts each oxide (Figure 2.58). Each of these films acts as a gate electrode (front and back gate), which can generate an inversion region near the Si–SiO_2 interfaces if an appropriate bias is applied. Thus, we would have two MOSFETs sharing the substrate, source and drain. The outstanding feature of these structures lies in the concept of *volume inversion*, introduced by Balestra *et al.* [2.206] if the Si film is thicker than the sum of the depletion regions induced by the two gates, no interaction is produced between the two inversion layers. The operation of this device is similar to that of two conventional MOSFETs connected in parallel (Figure 2.59b).

However, if the Si thickness is reduced, the whole silicon film is depleted and an important interaction occurs between the two potential wells. In such conditions the inversion layer is formed not only at the top and bottom of the silicon slab (i.e., near the two silicon–oxide interfaces) but throughout the entire silicon film thickness. It is then said that the device operates in "*volume inversion*," i.e., carriers are no longer confined at one interface, but distributed throughout the entire silicon volume (Figure 2.59a).

Several authors have claimed that *volume inversion* presents a significant number of advantages, such as: (i) enhancement of the number of minority carriers; (ii) increase in carrier mobility and velocity due to the reduced influence of scattering associated with oxide and interface charges and surface roughness; (iii) as a consequence of the latter, an increase in drain current and transconductance; (iv) a decrease in low-frequency noise; and (v) a large reduction in hot carrier effects [2.206].

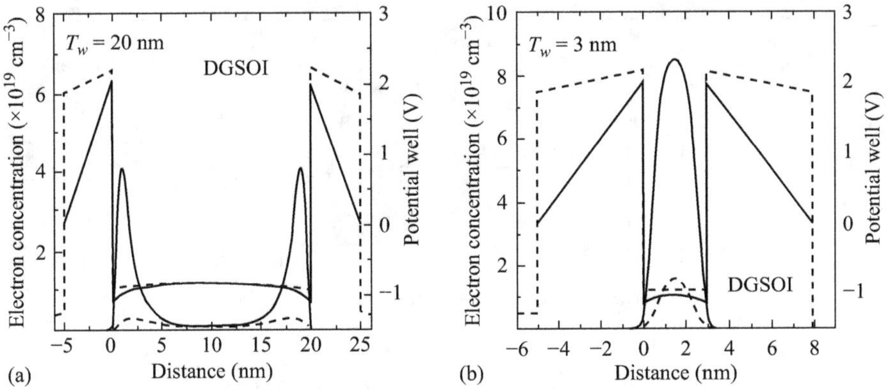

Figure 2.59 Electron distribution and potential well for two DGSOI devices with different silicon-layer thicknesses and (a) and (b) two values of the inversion charge concentration. Dashed lines correspond to $N_{inv} = 1 \times 10^{12}$ cm^{-2}, and solid lines correspond to $N_{inv} = 8 \times 10^{12}$ cm^{-2}

To evaluate accurately the electron distribution in a DGSOI structure we must self-consistently solve the Schrödinger and Poisson equations [2.146, 2.208–2.211]. Figure 2.59 shows the potential distribution in the structure and the electron concentration for two silicon thicknesses $T_{Si} = 20$ nm (a) and $T_{Si} = 3$ nm (b) and for two electron concentrations (solid line: $N_{inv} = 1 \times 10^{12}$ cm^{-2}; and dashed line: $N_{inv} = 8 \times 10^{12}$ cm^{-2}). In this case, N_{inv} is defined as

$$N_{inv} = \int_0^{\frac{t_{Si}}{2}} n(x)dx \tag{2.35}$$

where $n(x)$ is the electron distribution.

From the self-consistent solution of the Poisson and Schrodinger equations, the following conclusion can be drawn:

1. As in ultra-thin FDSOI discussed earlier, the subband modulation effect is an important effect caused by the reduction of the silicon film thickness [2.210, 2.211]. This effect is related to the redistribution of the carriers among the different electric subbands originated by the size quantization. The direct consequence of a decrease of the conduction effective mass as the silicon thickness decreases. Although the picture in DGSOI devices is slightly more complicated than that observed in FDSOI structures, Figure 2.60a shows that, also in DGSOI devices, the average conduction effective mass decreases as the silicon layer thickness is reduced.
2. Another important quantum effect that appears in ultra-thin FDSOI inversion as silicon thickness decreases is an increase of the phonon-scattering rate, i.e., the uncertainty in the location of the electrons in the direction perpendicular to

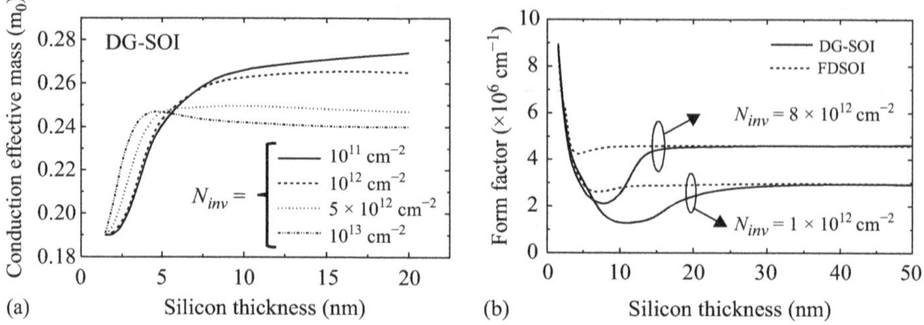

Figure 2.60 *(a) Evolution of the average conduction effective mass with the silicon slab thickness, for different inversion charge concentrations. (b) Evolution of phonon-scattering rate for the ground subband as a function of the silicon thickness for two values of the inversion charge concentration. Solid line: for a DGSOI inversion layer; dashed line: for an ultra-thin FDSOI inversion layer*

the interface is lower as T_{Si} decreases. By the uncertainty principle, there is a wider distribution of the electron's momentum perpendicular to the interface. In other words, due to size quantization, the electron's interface-directed momentum does not have a single value (as in three-dimensional electrons), but rather a distribution of likely values that expands as the silicon layer thickness is reduced. Taking into account momentum conservation, there are more phonons available that can assist the transitions between electronic states, and therefore the phonon-scattering increases. Thus for the same inversion charge concentration, the phonon-scattering rate is greater in thinner films than in thicker ones (since the confinement is greater), and therefore we expect a reduction of the mobility. Phonon-scattering rate is proportional to the form factor displayed in Figure 2.60b. The form factor corresponding to ultra-thin FDSOI inversion layers is also shown (dashed line). For thinner samples the form factor is very large due to the geometrical confinement of electrons in a very narrow space (no differences are observed between DGSOI and FDSOI). As the silicon slab thickness increases, the form factor is quickly reduced, until a minimum is reached in the region between 5 and 15 nm. Then, it increases to approach, for thick samples ($T_{Si} \sim 20$ nm), the value presented in ultra-thin FDSOI inversion layers and in bulk inversion layers. As can be seen in the figure, in the range $T_{Si} = 5$–15 nm the phonon-scattering rate for DGSOI is lower than the one corresponding to FDSOI for the same thicknesses, and even lower than the one corresponding to bulk inversion layers. Consequently, in intermediate ranges of the silicon thickness (which depend on the inversion charge concentration) the phonon-scattering rate in the DGSOI inversion layer decreases, instead of increasing as expected. This is an important result, a direct consequence of the volume inversion effect. This fact is important in explaining the behavior of the transport properties.

2.2.4.6 Electron mobility in quantum-well-based transistors

Phonon scattering is not the only scattering mechanism present in multigate devices. Although other scattering mechanisms (namely, those associated with the Coulomb interaction with oxide and interface charges and with the roughness of the silicon–oxide interfaces) are likely to be weakened by a volume inversion operation [2.212–2.216], their contribution has to be taken into account. The weakness of these scattering mechanisms is justified, at least a priori, by the spread of the electrons throughout the whole silicon region. Nevertheless, we must not forget that in order to achieve volume inversion, both channels must interact strongly, and this only happens in the medium-high transverse electric field range when the silicon slab between the two oxides is thin enough (below 20 nm as pointed out by [2.144, 2.212–2.216]). In these thin devices, although electrons are certainly spread along the whole silicon layer, they may not be far enough from the interfaces and may therefore be significantly affected by surface scattering mechanisms; much more so, in fact, than in bulk MOSFETs, since they are now interacting with two interfaces. This means that scattering mechanisms may play a very important role in the electron mobility in ultra-thin DGMOSFETs, contrary to what was previously believed. This imposes a serious limitation on the minimum silicon thicknesses that can be used in these devices, in addition to the limitations already presented by other physical and technological issues, as detailed elsewhere [2.208, 2.212].

Using a one-electron MC method the stationary electron transport properties in DGSOI and FinFET inversion layers have been evaluated. Electron quantization in the inversion layer was taken into account in an appropriate manner, self-consistently solving Poisson's and Schrödinger's equations assuming a simple non-parabolic band model for the silicon. Once the electron distribution in the silicon layer was determined, the Boltzmann transport equation was solved by the MC method, simultaneously taking into account phonon, surface-roughness, and Coulomb scattering. To do this, it was necessary to improve on existing scattering models. The presence of two close silicon–oxide interfaces in a DGMOSFET makes it significantly different from its standard-bulk counterparts. Figure 2.61 shows the total electron mobility for two values of the total inversion charge as a function of the silicon thickness in a planar DGSOI transistor. For the sake of comparison, the electron mobility in an ultra-thin FDSOI transistor with the same parameters is also shown.

Figure 2.61 reveals the existence of three regions with different behavior in the DGSOI electron mobility:

(i) The first region corresponds to thick silicon slabs. In DGSOI inversion layers the two channels are sufficiently separated and no interaction appears between them. This situation corresponds to two conventional inversion layers in parallel, separated by a large potential barrier. The behavior of electrons in each of these inversion layers is the same as that observed in a bulk silicon inversion layer. As the silicon thickness is reduced, the interaction between the two inversion layers causes the electrons to occupy the entire silicon volume. This is the beginning of the second region, which

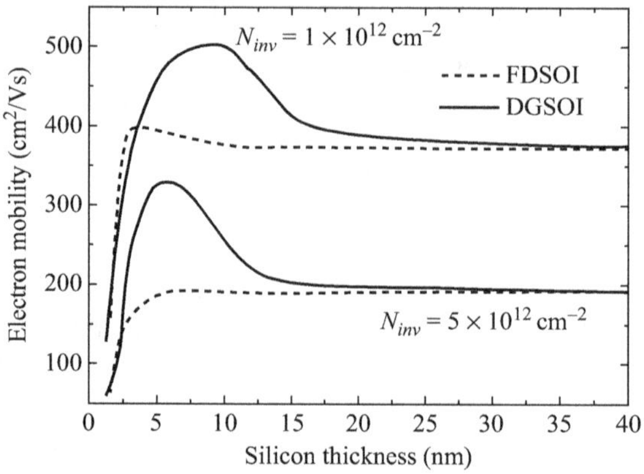

Figure 2.61 Evolution of electron mobility in DGSOI and FDSOI with the silicon thickness

strongly depends on the value of the transverse effective field, since for high electric fields a potential barrier, which obstructs the mutual influence of the two channels, is formed in the middle of the silicon slab.

(ii) In the second region, the electron mobility in DGSOI inversion layers is up to 20% larger than the mobility in FDSOI inversion layers. The limits of this region and the values of the mobility depend on the inversion charge concentration. This is the region in which volume inversion occurs. In this region of silicon thickness, both subband energy levels and wave functions vary significantly as a consequence of the two channels interacting. It is for this reason that the phonon-scattering rate decreases [2.146] compared to its value in conventional bulk. This happens down to a certain value of silicon thickness. For lower thicknesses, although the electrons are distributed throughout the entire silicon layer, their confinement is greater (due to the geometrical confinement), and therefore, the phonon-scattering rate increases, as shown in Figure 2.61. This marks the beginning of the third region.

(iii) In the third and last region ($t_{Si} < 4$ nm), the mobility for DGSOI falls abruptly. In this zone, mobility is limited by the thickness of the silicon slab, and therefore electron mobility falls abruptly.

For a complete and comprehensive study of the electron mobility in multigate devices, the reader can follow Reference 2.217.

2.2.5 Silicon multigate nanowires

Multiple-gate FETs provide a good electrostatic control of the channel and therefore the possibility of a higher reduction of the channel length compared to traditional bulk MOSFETs. It has been demonstrated that for DGSOI transistors to operate correctly is necessary silicon thicknesses t_{Si} lower than half of the channel length $L_{ch}/2$. The use of several gates can relax this condition without degradation of the device performance. Therefore, since the characteristic dimensions for the next node generations is well below 100 nm, the silicon fin cross-section will reach the nanometer scale, confining the carriers in the two directions perpendicular to the transport one. As a consequence, the 2D electron gas is transformed in a 1D one. The density of states experiences an important transformation due to their different energy dependence. Under these circumstances, it has been experimentally observed a change of fundamental electrical parameters such as the threshold voltage [2.218].

The need to scale the active channel region below 30 nm requires a silicon body width and thickness of the same dimensions or even lower in order to maintain an acceptable gate electrostatic control of the channel potential. For these reduced dimensions, carriers are confined in the directions perpendicular to the transport, such devices being called NWs. The potential application of semiconductor NW field- FETs as potential building blocks for highly downscaled electronic devices with superior performance is attracting considerable attention [2.219, 2.220].

With regard to their manufacturing process, it is possible to establish two different approaches, bottom-up and top-down:

(i) The bottom-up approach refers to the methodology that employs chemistry to promote the self-assembly of complex mesoscopic architectures. One of the most important discoveries in recent years has been the growth of single-crystal nanostructured materials at low temperatures using different nanometer-sized metallic nanoparticles (e.g., Ni, Au, Fe) as catalysts [2.221]. A wide variety of semiconductor materials such as Si, Ge, GaAs, GaN, and InP can be synthesized employing this technique [2.222, 2.223]. Different applications, such as laser action, photoluminescence, sensing, p-n junction, and FETs, have already been demonstrated. Currently, the FETs fabricated from Vapor-Liquid-Solid (VLS) grown NWs may offer better size uniformity than etching for very small diameters (<5 nm) due to controlled chemical synthesis.

(ii) The top-down approach refers to those devices with dimensions in the nanometer range, fabricated using the standard techniques employed for CMOS processing, namely, photolithography, thin-film deposition, etching, and metallization, to obtain multigate SOI FETs with very small dimensions [2.224, 2.225].

As has been established previously in this chapter, Multiple-Gate SOI MOSFETs are considered an attractive alternative to traditional bulk MOSFETs since they have proved to give better electrostatic control of the channel, allowing a greater reduction of the channel length (L_g) while the SCEs are kept under control. Moreover, the use of two, three, or even four gates allows a relaxation of the width (W_{Si}) and height (H_{Si}) of the silicon fin compared with L_g. When the dimensions of the semiconductor fin, W_{Si} and H_{Si} (see Figure 2.62), reach the nanometer scale, the carriers are confined in two dimensions, in the plane perpendicular to the transport direction. Therefore, the bulk crystal symmetry is not preserved and fundamental magnitudes, such as the density of states, and the band structure, experience

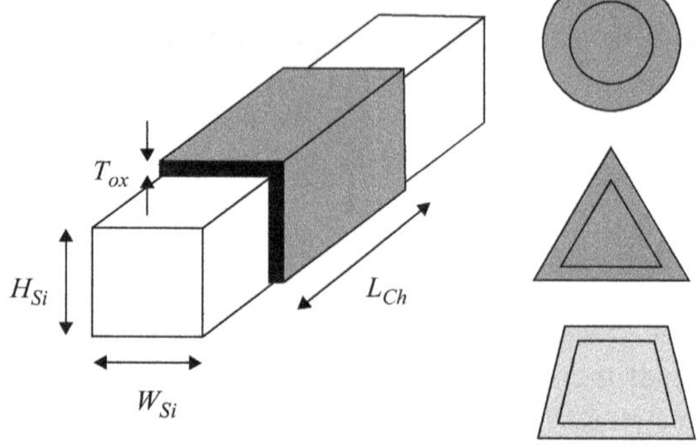

Figure 2.62 Representation of a SiNW where W_{Si} and H_{Si} represent the semiconductor fin width and height respectively, L_g the gate length, and T_{ox} the gate oxide thickness. Different cross-sections, such as triangular, circular, or trapezoidal can be considered in addition to the rectangular one

important modifications, which will influence the carrier transport properties of these new devices.

2.2.5.1 Quantum effects in Si nanowires

It has been demonstrated that band structure effects begin to manifest in silicon NWs with diameters below 5 nm. For higher dimensions, the simple parabolic effective-mass approach with bulk effective-masses is the optimum solution due to its reduced computational cost. Moreover, with very small dimensions (<5 nm) this method can be still used when appropriate tuning parameters are employed [2.226]. Therefore, the modified semiconductor band structure should be taken into account when the device dimensions are below the limit of 5 nm since other important parameters, such as the threshold voltage and the gate-channel capacitance, suffer considerable modifications [2.227]. In order to understand in depth the transport properties of these one-dimensional (1D) devices, detailed knowledge of the electron density and the electrostatic potential is necessary. Obviously this requires the solution of the Poisson equation. Two different approximations can be carried out to achieve this goal. On the one hand, whether or not the whole device is considered, the solution of the three-dimensional (3D) Poisson equation for the electrostatic potential has to be carried out [2.228–2.230].

On the other hand, if a very long device is considered, this equation can be restricted to a plane perpendicular to the transport direction and the influence of the source and drain contacts neglected. In this case, a 2D Poisson equation must be solved:

$$\nabla(\varepsilon\nabla\phi) = -q\left(p - n + N_D^+ - N_A^-\right) \tag{2.36}$$

where ϕ is the electrostatic potential, ε is the dielectric constant, q is the electric unit, n and p are the electron and hole concentrations, and $N_D{}^+$ and $N_A{}^-$ are the ionized donor and acceptor concentrations. Use of the finite element method allows the simulation of different geometries, such as triangular, cylindrical, or rectangular cross-sections. Due to the reduced dimensions of the devices under study, it is mandatory to include the quantum effects in the simulation, through the self-consistent solution of the Schrödinger equation. The most common approximation is the solution of the equation in two dimensions in the channel cross-section. If the whole device is studied, this solution is carried out in an arbitrary number of slices along the device length and then coupled with the corresponding transport equation [2.228, 2.229].

It can be assumed that confinement is produced in the y- and z-directions and transport in the x-direction and, as a first approximation, the effect of source and drain contacts can be neglected. Therefore, the 2D Schrödinger equation can be written as

$$-\frac{\hbar^2}{2}\frac{\partial}{\partial y}\left(\frac{1}{m_y}\frac{\partial \Psi_v}{\partial y}\right)-\frac{\hbar^2}{2}\frac{\partial}{\partial z}\left(\frac{1}{m_z}\frac{\partial \Psi_v}{\partial z}\right)+V(y,z)\Psi_v=E_v\Psi_v \tag{2.37}$$

where m_y and m_z are the effective electron masses along the y- and z-axis, respectively, and Ψ_v the wave function belonging to energy level E_v.

In order to self-consistently solve the two equations, different algorithms can be employed. However, the predictor-corrector scheme proposed by Trellakis *et al.* [2.231] has been tested by several authors [2.232, 2.233] on different semiconductor structures with excellent results in every case. In order to show the correct operation of the above procedure, it has been applied to a silicon Gate All-Around (GAA) MOSFET, where the semiconductor is completely surrounded by the insulator and the gate contact. In all the calculations, we assumed a square cross-section ($W_{Si}=H_{Si}$), substrate doping of 10^{15} cm^{-3}, $T_{ox}=1$ nm, and a midgap workfunction metal gate ($\phi_m=4.61$ eV).

Figure 2.63a,c represents the electron distribution in a silicon GAA with $W_{Si}=H_{Si}=15$ nm while Figure 2.63b corresponds to a 4 nm lateral size. Figure 2.63d was calculated using a classical solution of the structure and an applied gate voltage (V_G) equal to 1 V. The maximum electron density is located at the Si–SiO_2 interface, right in the corners, and its value is clearly overestimated when compared with the corresponding quantum simulation shown in Figure 2.63c. Figure 2.63a was calculated for the same device shown in Figure 2.63c but with a gate voltage reduced to 0.25 V. It shows how the electrons are spread throughout the whole silicon body with a peak density at the centre of the structure due to the so-called volume inversion effect [2.207]. Figure 2.63b corresponds to a device with reduced silicon fin dimensions ($W_{Si}=H_{Si}=4$ nm) and shows that for the same gate voltage as Figure 2.63c ($V_G=1$ V), the maxima of the electron density are again located at the center of the semiconductor due to the volume inversion effect.

2.2.5.2 Electron transport in Si nanowires

The electron transport properties in Si NWs can be studied from different approximations such as the Kubo-Greenwood formula [2.234, 2.235] modified for 1D transport, the nonequilibrium Green's function (NEGF) formalism [2.236, 2.237] or an MC simulation where the quantum effects have been taken into account

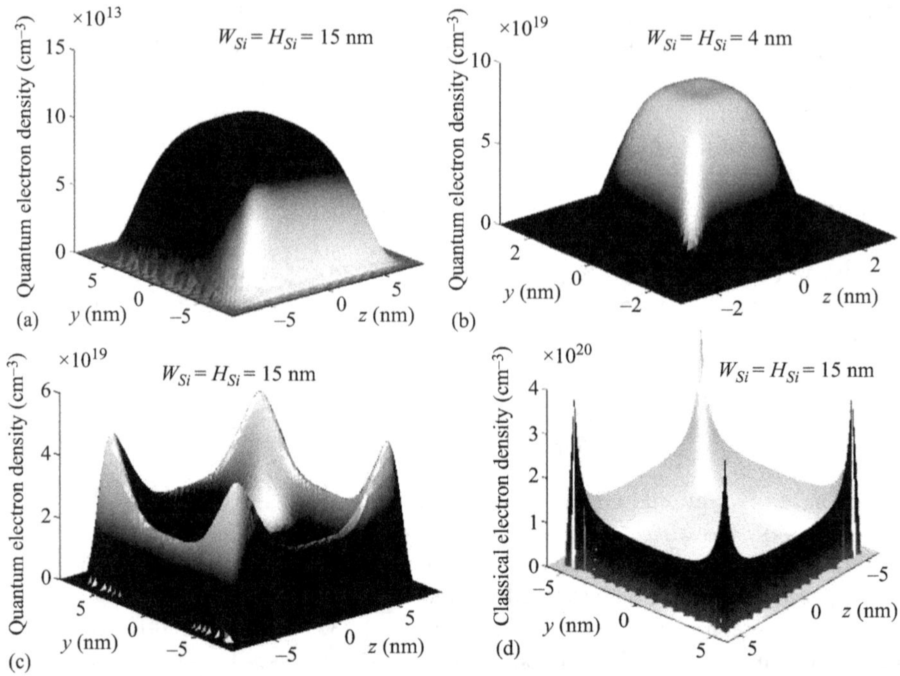

Figure 2.63 (a)–(c) Calculated quantum electron distributions in a silicon GAA nanowire with $W_{Si} = H_{Si} = 15$ nm (a) and (c), and $W_{Si} = H_{Si} = 4$ nm (b). (d) Calculated classical electron distribution in a silicon GAA nanowire with $W_{Si} = H_{Si} = 15$ nm and the same gate bias as (c)

[2.238, 2.239]. The MC procedure has been quite popular in recent decades and the scattering models have been tested in a large number of semiconductor structures [2.240]. However, the MC algorithm normally used to calculate the electron transport properties in silicon inversion layers has to be modified to take into account the special characteristics of carrier confinement in two dimensions. Among the different scattering mechanisms, which can influence carrier mobility in a SiNW, it has been demonstrated that phonon scattering is the dominant mechanism in mobility degradation in a Si MOSFET under operating conditions at room temperature. Therefore, phonon mobility is dominant in low effective fields and its value is determined by the acoustic and intervalley scattering rates. The results obtained from MC simulation are shown in Figure 2.64 where the calculated values of phonon-limited mobility are depicted as a function of the gate voltage for three different values of the square cross-section ($W_{Si} = H_{Si} = 15$ nm; 10 nm; 5 nm). As can be observed, phonon-limited mobility is quite similar for the larger cross-sections (15 and 10 nm of lateral size). However, if the silicon fin is reduced to 5 nm, a significant degradation is found [2.239, 2.241].

Electron mobility in multigate NWs based in III–V materials have been widely studied in References 2.242–2.246.

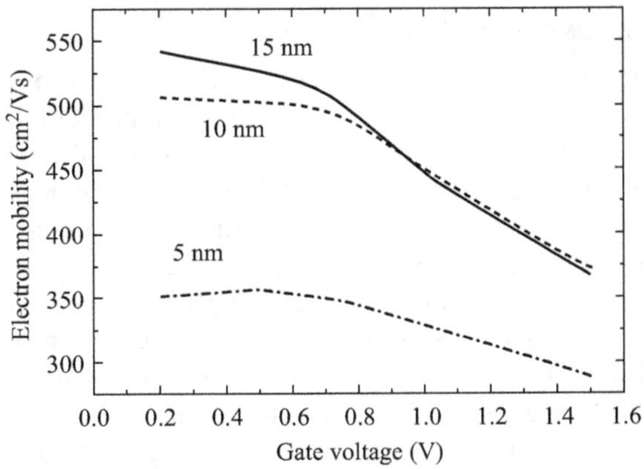

Figure 2.64 Phonon-limited electron mobility in a silicon GAA as a function of the gate voltage. Three different cross-sections are considered ($W_{Si} = H_{Si}$), 15 nm, 10 nm, and 5 nm

Different research groups are currently working on the manufacture and characterization of Si NWs and a wide variety of experimental results have been presented. As an example, we could mention the studies published by Lieber's group [2.247, 2.248], which have shown that the average transconductance and mobility show substantial advantages for SiNWs obtained from vapour-liquid-solid (VLS) synthesis compared with more conventional multigate FETs. However, a number of issues including device performance, reproducibility, and high-quality ohmic contacts must be addressed if such systems are to be implemented in the future. More recently, electron mobility as high as $\approx$1,000 cm^2/Vs has been reported for n-channel SiNW FETs made following conventional semiconductor manufacturing techniques, from a p-type SOI wafer with n$^+$ source and drain. Moreover, the authors of this work conclude that as the channel width decreases, the inversion layer mobility of the SiNW increases to approximately twice the mobility of the larger channel-width FETs [2.249].

Singh *et al.* [2.225] reported the fabrication of GAA n- and p-FETs on a SOI wafer with a diameter smaller than 5 nm and lengths around 1,000 nm where the estimated electron and hole mobilities are $\sim$750 and $\sim$325 cm^2/Vs at high fields. These experimental results cover a wide range of mobility values and, to date, a clear explanation is not available for all of them. Various reasons have been put forward to justify these discrepancies, such as strain in the semiconductor due to the oxidation process or the possible suppression of intervalley phonon scattering, volume inversion, and reduced surface roughness at high fields [2.225].

The unique properties observed in 1D electron gases can be appreciated only when lateral dimensions are well below 10 nm since otherwise their behavior much more closely resembles that found in traditional silicon inversion layers. This conclusion can be relaxed when very low temperatures are applied, as was demonstrated by Colinge *et al.* [2.250].

2.2.6 Junctionless transistors

All the devices presented so far along this chapter are based on the formation of junctions. Junctions are capable of both blocking current and allowing it to flow, depending on an applied bias. For example, the MOSFET transistors are made using two p–n junctions: the source junction and the drain junction. Therefore, an n-channel MOSFET transistor is an N-P-N structure, while a p-channel transistor is a P-N-P structure. As detailed in previous sections, trends in the electronic industry require smaller and smaller components resulting in transistor sizes down to the nanoscale. This is starting to pose significant manufacturing problems. In classical very small transistors one has to form two junctions, since source and drain regions are separated by channel area with opposite doping polarity. The diffusion of source and drain doping atoms is difficult to control in very short-channel transistors. In all transistors, the scattering and diffusion of source and drain impurities into the channel region becomes a bottleneck to the fabrication of very short channel devices, and very low thermal budget processing techniques must be used [2.251]. Very costly techniques are used to minimize this diffusion, but even in the absence of diffusion the statistical variation of the impurity concentration due to ion implantation or other doping techniques can cause device parameter variation problems. To overcome these problems, in 2010 J. P. Colinge proposed [2.252] the junctionless transistors (JNTs). These devices are fabricated without source and drain formation process, as the doping type and concentration in the channel region is essentially equal to that in the source and drain, or at least to that in the source and drain extensions [2.253–2.255]. A JNT is basically an FD accumulation-mode device, consisting of a heavily doped SOI NW resistor with an MOS gate to control current flow. Doping concentration is constant and uniform throughout the device and typically ranges from 10^{18} and 10^{20} cm^{-3}. The JNT device can be tuned to normal-off state when the gate workfunction is properly chosen and the highly doped channel can be FD with no gate bias. As gate voltage is increased, the JNT enters into partially depletion state, and current conducts in the centre of the NW when V_D is supplied, and then at flatband voltage, the depletion region is completely gone. The accumulation starts at the NW surface when further raises the V_G, which additionally offers an accumulation current, in spite of the bulk current.

Figure 2.65 shows calculated I_D–V_G curves for a cylindrical GAA JNT with a gate length of $L_{ch} = 20$ nm and a radius, $R_i = 3$ nm. Doping concentration was set to $N_D = 8 \times 10^{19}$ cm^{-3}. The metal gate was assumed to have a workfunction of 5.5 eV and the oxide thickness is set to $t_{ox} = 2$ nm. The curves were calculated by self-consistently solving Poisson equation and drift-diffusion and continuity equations. In this device, the subthreshold slope results to be $S = 70$ mV/dec.

In subthreshold operation ($V_G = -0.2$ V in Figure 2.66), the silicon is FD. Threshold voltage is reached when a portion of the silicon becomes neutral. At that point the device is PD. The bulk current flows in this neutral channel (that is not depleted silicon). Therefore carriers in a JNT transistors see a zero electric field in the directions perpendicular to the current flow. This is a big difference with inversion mode GAA or tri-gate transistors. As gate voltage increases, depletion decreases and the diameter of the neutral channel increases. When the gate voltage reaches flatband

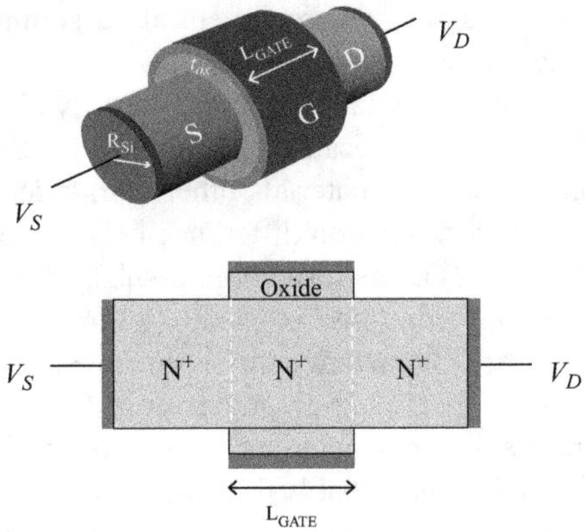

Figure 2.65 Schematics of a cylindrical GAA junctionless nanowire

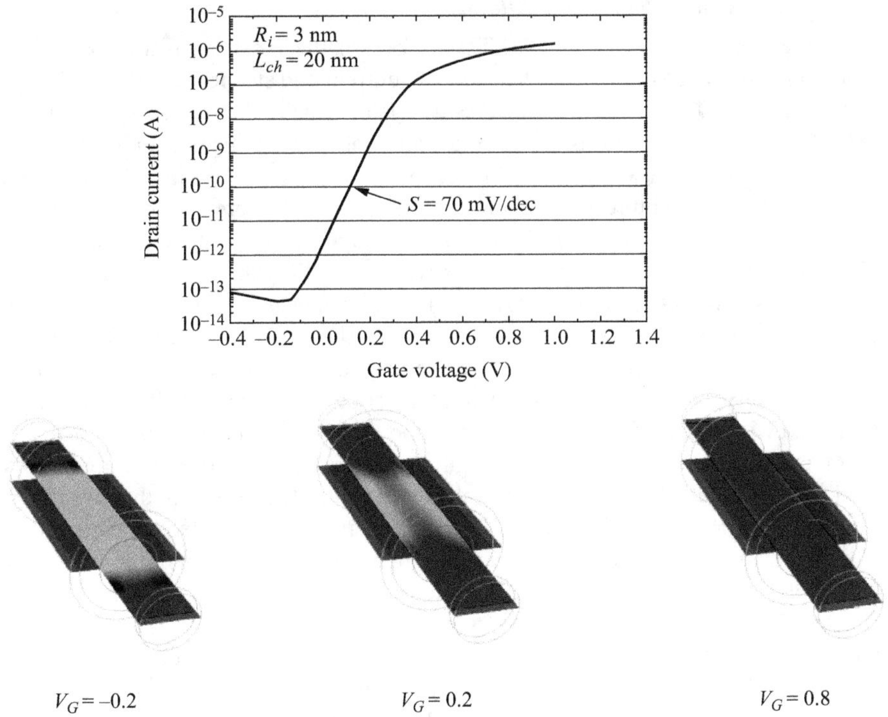

Figure 2.66 I_D–V_G curve for a cylindrical junctionless transistor. V_{DS} = 50 mV. Electron concentration contour plots for different gate voltage V_G. $V_G = -0.2$ V (below threshold): the channel is completely depleted of electrons; $V_G = 0.2$ V: the channel starts to be populated with electrons and the current starts to increase; $V_G = 0.8$ V (above threshold): the channel is almost completely populated with electrons

voltage the entire channel region becomes neutral (assuming low V_{DS}). Further increasing the gate voltage brings about the formation of an accumulation layer.

In conclusion, JNTs are unipolar, thin-film, heavily doped (typically in the 10^{19} cm^{-3} range) MOS transistors. Because of its simple design, the JNT architecture has been adapted to semiconductor materials other than silicon. In the off-state the channel is FD owing to the workfunction difference between the semiconductor and the gate material. When the device is turned on, a substantial part of the current is carried in the bulk of the thin film, and is usually augmented by an accumulation current contribution. JNTs are characterized by reduced SCEs and present excellent subthreshold slope and low DIBL [2.256]. As a result, CMOS junctionless devices with outstanding short-channel characteristics have been demonstrated for gate lengths down to 13 nm. The JNT is probably the most scalable of all FET structures, as demonstrated by both ab initio simulations and experimental devices with gate lengths as small as 3 nm.

2.2.7 Tunnel field-effect transistor

Tunnel Field-Effect Transistors (TFETs) are one of the most promising devices to replace conventional MOSFETs. Their low off-current and steeper subthreshold slope overcoming the 60 mV/dec limit of MOS transistors, make them enormously attractive for low-power applications. One of the main problems arising in MOSFETs is that when they are scaled down, so as to do their power supply voltage in order to reduce power density. The subthreshold swing limit of 60 mV/dec present in conventional MOSFETs, imposes a severe roadblock to reduce the supply voltage plateau of 1 V and maintain high ON-state currents along with low OFF-state leakages. TFETs, on the other hand, are based in the so-called BTBT mechanism which makes the carrier injection into the channel essentially dependent on the quantum process of tunneling across an energy barrier. This fact allows extremely low subthreshold swings when the device turns on due to the quasi-exponential dependence of the current on the barrier width. Likewise, when the transistor is off, the tunneling barrier keeps the leakage current extremely low. In MOSFET scaling, tunneling phenomena from heavily doped junctions resulted in parasitic leakage currents. However, as this process is precisely the working principle of TFETs, it is no longer an unwanted parasitic effect. Furthermore, since tunneling only takes place in a very small region of these devices, this may allow significant gate scaling up to the distance of BTBT which in silicon represents less than 10 nm. As source-to-drain DT is negligible for channel lengths greater than that value [2.257], TFETs could in principle be scaled to very small dimensions without relevant leakage current degradation.

Since the discovery of BTBT in 1957 by Esaki [2.258] when studying very narrow germanium p–n junctions, this phenomenon based on the tunneling injection of carriers from occupied states in the valence (conduction) band to empty states in the conduction (valence) band has been demonstrated in many devices. As indicated earlier, such a mechanism has been shown, for example, in MOSFETs (both lateral and vertical). The first gated p-i-n structure was proposed in 1978 at Brown University [2.259] suggesting it for spectroscopy. Transistors based on it (like

B2T–MOSFETs [2.260] or others replacing the i-region under the gate by a p$^-$-region [2.261]) were investigated, showing the lack of V_{TH} roll-off and temperature dependence of the device characteristics when scaling. However, the first gated p-i-n diodes operating as Surface Tunnel Transistors were proposed on III–V compounds [2.262]. Similar tunneling transistor operation was developed in silicon at Cambridge [2.263], and later at Toshiba [2.264]. The interest of these first results was limited until the experimental results presented in 2000 by W. Hansch and I. Eisele on vertical p-i-n diodes [2.265–2.267]. In 2004, a lateral gated p–n junction diode (without intrinsic region) on silicon-on-insulator was fabricated [2.268] at Brown University. In this last case, the lack of intrinsic region reduced gate capacitance but did not significantly improved ON-current – which was still very low – and also produced an increase in leakage current. Also in 2004, Appenzeller *et al.* [2.269] reported for the first time a subthreshold swing under 60 mV/dec in carbon nanotube FETs (CNFET). A back gate and a top gate were employed to achieve the necessary band configuration to trigger BTBT. In 2005, the same authors [2.270] compared several CNFETs concluding that the single gate configuration presented the best performance. Despite the obtained results for CNFETs, the research on silicon-based FETs offered in principle a more immediate possibility to industrial applications due to the greater development of this technology [2.271, 2.272].

2.2.7.1 Structure and operation

The device structure of a TFET essentially differs from that of the MOSFET in the nature of the dopants used in the source and the drain. While MOSFETs have the same type of dopants, in TFETs, source and drain are of opposite types. The basic constituent of TFETs is thus a gated p-i-n diode, or less frequently – as previously mentioned – a gated p–n diode [2.268]. The name of the terminals is chosen to resemble the MOSFET biasing. To switch the device ON, the diode has to be reverse biased and a voltage applied to the gate. Therefore, an n-type TFET would require a positive voltage in the gate and also in the n-doped region, which would play the role of the drain if one recalls the analogy with the NMOS. The other p$^+$ region would act as the source and the intrinsic region as the channel. Figure 2.67 shows a schematic of a lateral single gate n-type silicon TFET where the dielectric

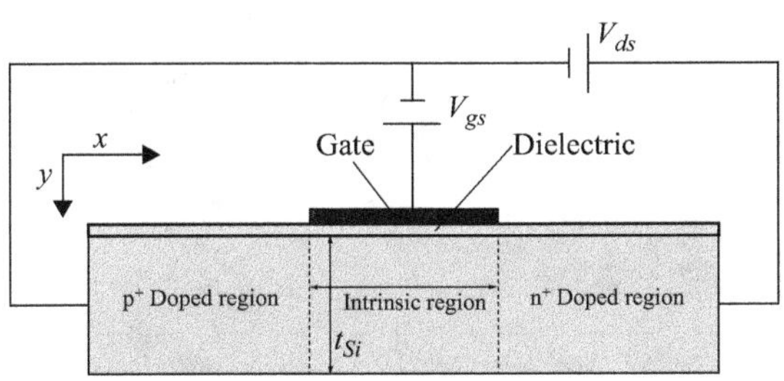

Figure 2.67 Single gate n-type TFET. The p$^+$-region acts as the source and the n$^+$-region as the drain

covers the source, the drain and the channel. In a p-type TFET, the dopings would be the opposite: the source would be n^+ and the drain, p^+.

Prior to study the operation regimes of these transistors, and in order to better understand them, it may result very useful to analyze the qualitative behavior of the p–n tunnel diodes, in which p–i and n–i junctions of the TFET are based. Tunnel diodes consist of a p–n junction in which both p and n sides are degenerate (i.e., very heavily doped).

To illustrate this, let us consider the tunnel diode configurations depicted in Figure 2.68 along with their corresponding points in the $I-V$ curve, and discuss the

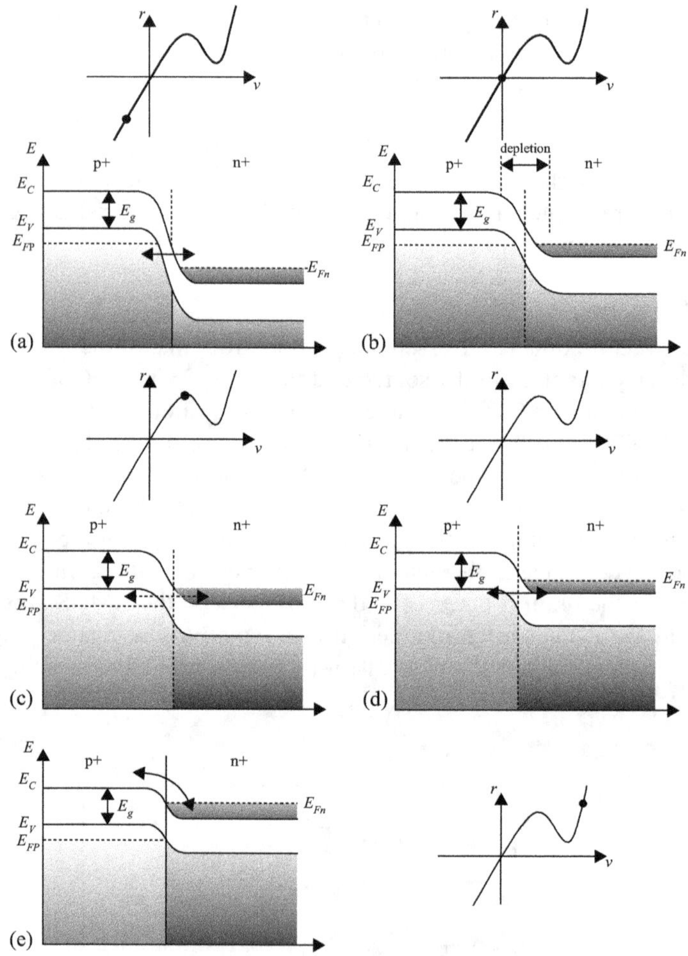

Figure 2.68 Energy band diagrams and I−V characteristics of a tunnel diode at (a) reverse bias with increasing tunneling current; (b) thermal equilibrium, zero bias; (c) forward bias V such that peak current is obtained; (d) forward bias approaching valley current; and (e) forward bias with diffusion current and no tunneling

different processes taking place in them at absolute zero temperature. Each of the different figures corresponds to:

(a) When the tunnel diode is reverse biased (p-side negative bias with respect to n-side), the BTBT current increases monotonically and indefinitely.
(b) At thermal equilibrium, no voltage is applied and the Fermi levels are aligned. That means that above the Fermi level there are almost no filled states on either side of the junction, and below it there are almost no empty states available on either side of the junction. Hence, net tunneling current at zero bias is zero.
(c) In the forward direction (positive voltage at the p-side with respect to n-side), the current first increases to a maximum because electrons can tunnel from the conduction band to the valence band. Tunneling is possible as there is a common band of energies with filled states on the n-side and unoccupied states on the p-side.
(d) If forward voltage is further increased, this range of common energies decreases and so does the current until the bands are uncrossed and there are no available states aligned with filled states.
(e) Once tunneling current becomes zero, normal diffusion current begins to dominate and current increases again exponentially.

However, in Figure 2.68e there also exists another type of current contribution called excess current. The excess current arises from a BTBT process that takes place indirectly through energy states within the forbidden gap. Several possible routes followed by carriers can be seen in Figure 2.69 [2.273]. As an example, an electron could drop down from A to an empty level at B, decreasing its energy, from which it might tunnel to the final state D in the valence band. Alternative trajectories would be ACD, or even a staircase route formed by several tunneling transitions followed by vertical losses of energy. This last one is by far less probable and requires a sufficiently high concentration of intermediate states.

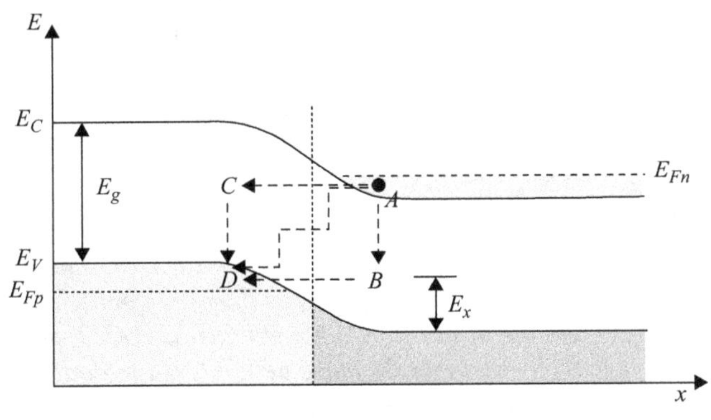

Figure 2.69 Example of different BTBT trajectories through intermediate states in the forbidden gap. E_x is the energy through which the electron must tunnel in B

2.2.7.2 Operating regimes of the TFET

In Figure 2.70, we show the band diagrams for the OFF and ON states of the TFET. There are two configurations that correspond to the OFF state. The first is when the band structure is in equilibrium and no bias applied, Figure 2.70a, which corresponds to a situation as that shown in Figure 2.68b. In this initial state, the built-in potentials of the p–i and n–i junctions determine the characteristic staircase-like profile that can be observed. When we only apply a positive voltage to the drain, $V_{DS} > 0$, the current flow is not allowed through the device in absence of gate bias because electron and hole currents are blocked by the built-in barriers (Figure 2.70b with $V_{GS} = 0$). In this situation, only reverse biased p-i-n diode leakage current flows between the source and the drain. This leakage current is extremely low (may result of order fA/μm).

When we apply positive voltage to the gate, $V_{GS} > 0$, the conduction band inside the channel is pushed down until it is aligned with the top of the valence band of the source. From that point onward, BTBT begins to be possible and carriers (in this case, electrons) are injected from the source into the channel (Figure 2.70b with $V_{GS} > 0$). This operating mode is the n-channel ON-state with source/channel junction resembling that of Figure 2.68a. We clearly see how the gate controls the band bending inside the channel and, consequently, the BTBT mechanism. TFETs designed with symmetry between the p- and n-doped regions, may show ambipolar behavior provided that adequate voltages are applied to the terminals. In our case, if we change the sign of the voltage applied to the gate,

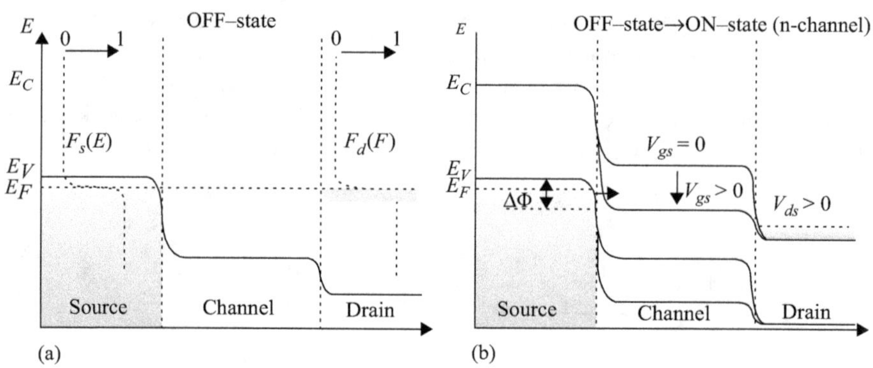

Figure 2.70 Energy band diagrams of the TFET taken horizontally along the channel close to the dielectric interface. (a) OFF-state showing the equilibrium configuration of energy bands when no bias is applied at the gate and the drain. (b) Combined ON- and OFF-states. When $V_{GS} = 0$, BTBT cannot take place and only p-i-n diode leakage current exists: OFF-state. If a big enough $V_{GS} > 0$ is applied, the conduction band in the channel is pushed down and BTBT may appear: ON-state. ΔΦ represents the difference between the top of the valence band in the source and the bottom of the conduction band in the channel

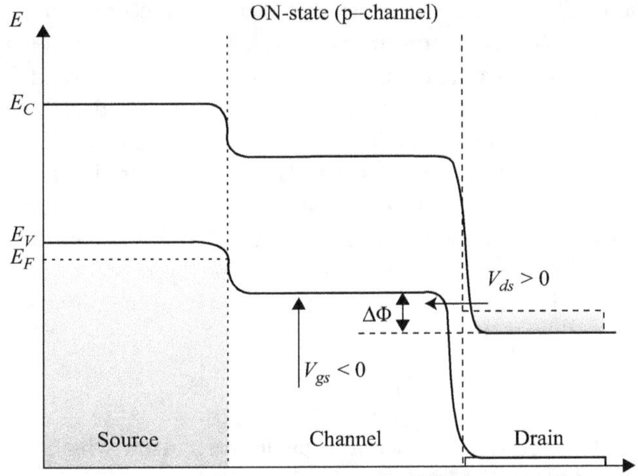

Figure 2.71 Band diagram configuration of the p-channel ON-state of the TFET. Carriers are injected from the drain into the channel once the top of the valence band inside the channel is raised over the bottom of the conduction band in the drain

$V_{GS} < 0$ – while keeping $V_{DS} > 0$ – the bands inside the channel move up and carriers can tunnel through the drain/channel junction as soon as the valence band inside the channel is lifted above the bottom of the conduction band in the drain. This can be seen in Figure 2.71. In this case, we again recall the situation of Figure 2.68a, but now between the drain and the channel. Obviously, this ambipolar behavior would imply a reassignation of "source" and "drain" labels for the p-channel ON-state if one wants to keep the analogy with the MOSFET operation.

To complete the description of the operating modes of the n-type TFET, let us consider $V_{DS} < 0$ and sufficiently large. In that case, the p-i-n structure is forward biased which means that carriers can flow with $V_{GS} = 0$ and results in exponential diode characteristics. The application of gate bias may block either the electrons or the holes by means of a potential barrier but not both. This configuration is not appropriate for switching purposes. Hence, while the drain bias switches the device characteristics from that of a forward-biased p-i-n diode to that of a TFET, the gate bias switches the TFET characteristics from an n-channel to a p-channel mode of operation, when the diode is reverse-biased.

TFETs are essentially based on tunneling rather than thermal emission. In this sense, they clearly differ from the normal operation pattern of conventional MOSFETs. In the subthreshold regime, we can use the following expressions [2.274] to describe the BTBT current in a tunnel diode assuming DT where the momentum is conserved in direct bandgap

$$I_{BTBT} = C_1 \int_{E_{Cn}}^{E_{Vp}} [f_C(E) - f_V(E)]T(E)N_C(E)N_V(E)dE \tag{2.38}$$

C_1 is a constant, $N_{C,V}(E)$ are the density of states in the conduction and valence bands respectively, $T(E)$ is the transmission probability across the energy tunneling barrier width – which is assumed to be equal for both directions – and $f_{C,V}(E)$ are the occupation probabilities of the bands described by the Fermi distribution functions.

Let us focus now on the tunneling process that happens at the source/channel junction in the n-channel mode. The bands profile is similar to that shown in Figure 2.68a. An accurate way to describe the transmission probabilities through the barrier was developed by Sze [2.275] using the WKB approximation and a triangularly shaped potential barrier, as depicted in Figure 2.72.

The 1D expression for the tunneling transmission probability is given by

$$T(E) \approx exp\left[-2\int_{x_{start}}^{x_{end}} k(x)dx\right] \tag{2.39}$$

with $k(x)$ the wave vector of the electron inside the barrier, which, using the parabolic band approximation is given by

$$k(x) = \sqrt{\frac{2m^*}{\hbar^2}(U(x) - E)} \tag{2.40}$$

where m^* is the electron effective mass. If we assume that, according to Figure 2.71, the electron approaches the barrier at the bottom of the triangle, then $E = 0$ for it. The linear equation for the potential energy reads as

$$U(x) = \frac{E_g}{2} - qF(x)x \tag{2.41}$$

With these assumptions, the transmission probability reads as

$$T(E) \approx exp\left[\frac{4}{3}\frac{\sqrt{2m^*}}{qF\hbar}\left(\frac{E_g}{2} - qFx\right)^{\frac{3}{2}}\right]\Bigg|_{x_{start}}^{x_{end}} = exp\left(-\frac{4\sqrt{2m^*}E_g^{\frac{3}{2}}}{3qF\hbar}\right) \tag{2.42}$$

To perform the integration we have assumed that the electric field, $F(x)$, is uniform along the integration path ($F(x) = F$) which is consistent with the approximated shape of the barrier of Figure 2.72. This assumption is made in the so-called

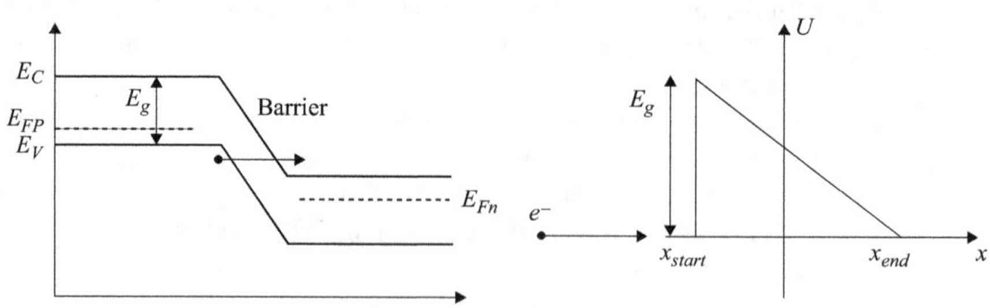

Figure 2.72 BTBT process approximated by a triangular barrier with the electron tunneling at the base of the triangle

semiclassical local models and allows obtain an analytical expression for the transmission probability. A more realistic approximation is incorporated in non-local models, which explicitly regard F as a function of x given its dependence on the band bending at every point inside the barrier. As a consequence, in non-local models, $T(E)$ is more accurately described. Nonetheless, unlike what happens in local models, the integration cannot be solved analytically but only numerically. The electric field may be replaced by $F = \frac{E_g}{qw}$ in the case of the situation shown in Figure 2.72 (with $w = x_{end} - x_{start}$), thus leaving $T(E)$ as a function of the width and height of the barrier

$$T(E) \approx exp\left(-\frac{4w\sqrt{2m^* E_g}}{3\hbar}\right) \tag{2.43}$$

Finally the BTBT current can be expressed as

$$I_{BTBT} = \frac{Aq^2}{36\pi\hbar^2}\sqrt{\frac{2m^*}{E_g}}Dexp\left(-\frac{4\sqrt{2m^*}E_g^{\frac{3}{2}}}{3qF\hbar}\right) \tag{2.44}$$

where the integral D is

$$D = \int [f_C(E) - f_V(E)]\left[1 - exp\left(-\frac{2E_S}{\overline{E}}\right)\right]dE \tag{2.45}$$

with $\overline{E}$ and E_S given by

$$\overline{E} = \frac{\sqrt{2}q\hbar F}{\pi\sqrt{m^* E_g}} \tag{2.46}$$

$$E_S = min\left(E_{Vp} - E, E - E_{Cn}\right) \tag{2.47}$$

The BTBT current is finally given by

$$I_{btbt} \propto exp\left[-\frac{4\lambda\sqrt{2m^*}E_g^{\frac{3}{2}}}{3\hbar\left(E_g + \Delta\Phi\left(V_{gs}\right)\right)}\right]\Delta\Phi\left(V_{gs}\right) \tag{2.48}$$

Here, λ is the screening tunneling length and describes the spatial extent of the transition region at the source–channel interface; it depends on the specific device geometry. In a TFET, at constant drain voltage, V_{DS}, the V_{GS} increase reduces λ and increases the energetic difference between the conduction band in the source and the valence band in the channel ($\Delta\Phi$), so that in a first approximation the drain current is a super-exponential function of V_{GS}. As a result, in contrast to the MOSFET, the point subthreshold swing of the TFET is no longer a constant but strongly depends on V_{GS}. This behavior clearly differs from that of conventional MOSFETs, in which SS is a constant and does not vary with V_{GS}.

Figure 2.73 compares the subthreshold characteristics of a TFET and an ideal MOSFET transistor (SS = 60 mV/dec). Swing is smallest at the lowest V_{GS} for which BTBT occurs, and increases as V_{GS} does likewise. As a consequence of this

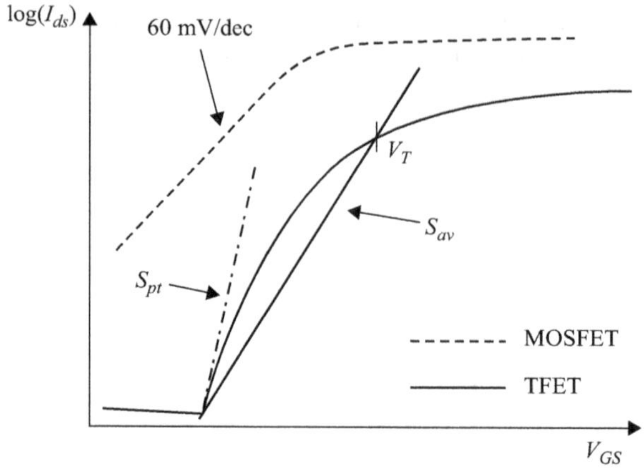

Figure 2.73 Qualitative comparison between and ideal MOSFET and TFET subthreshold characteristics. S_{pt} and S_{av} are depicted for TFETs showing the nonuniformity of subthreshold slope (SS)

variation, two different SS are defined in TFETs: the point swing, S_{pt}, and the average swing, S_{av}. S_{pt} is the smallest swing anywhere in the I_{DS}–V_{GS} curve, and in most cases coincides with the point where BTBT starts. On the other hand, S_{av} is the swing taken from the point where BTBT begins, up to the threshold voltage. There is not a unified definition of the threshold voltage in TFETs. Some authors use the constant current technique (usually 10^{-7} A/μm). Other authors choose a more physically based definition and regard the threshold voltage as the voltage at which the control that the corresponding electrode exerts over the current changes from quasi-exponential to linear.

These two swings are qualitatively shown in Figure 2.73 along with the conventional MOSFET characteristics. Note that typically the TFET has a lower ON-state and OFF-state current. S_{av} is the most important swing for switch performance. Unlike conventional MOSFETs where SS is a function of the thermal factor kT/q, SS does not depend on temperature to a first approximation in TFETs. This fact is not surprising since tunneling currents are weakly dependent on temperature. However, this does not imply that there is no degradation in S_{pt} or S_{av}. Degradation with T indeed exists given that rising temperatures clearly affect leakage current by increasing it, and making the steepest region of the curves disappear. Figure 2.74 shows simulated I_{ds}–V_{gs} curves for a single gate and a double gate n-channel TFETs (schematically shown in Figure 2.75).

Today TFETs represent the most promising steep-slope switch candidate, having the potential to use a supply voltage significantly below 0.5 V and thereby offering significant power dissipation savings. Because of their low off currents, they are ideally suited for low-power and low-standby-power logic applications operating at moderate frequencies (several hundred MHz). Other promising applications of TFETs include ultra-low-power specialized analog ICs with improved temperature stability and low-power SRAM.

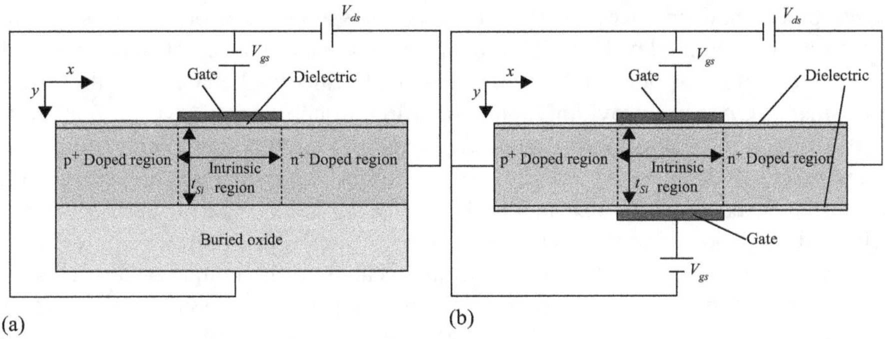

Figure 2.74 Schematic cross-section (not to scale) of the single gate (a) and double gate (b) n-channel TFETs considered. Dopings: p^+ region: 10^{20} cm^{-3}; intrinsic/lightly doped region: 10^{17} cm^{-3} (n-type); n^+ region: 10^{20} cm^{-3}; $t_{ox} = 1$ nm; $L_{source} = L_{drain} = 100$ nm; $L_g = 20$ nm; $t_{Si} = 3$ nm

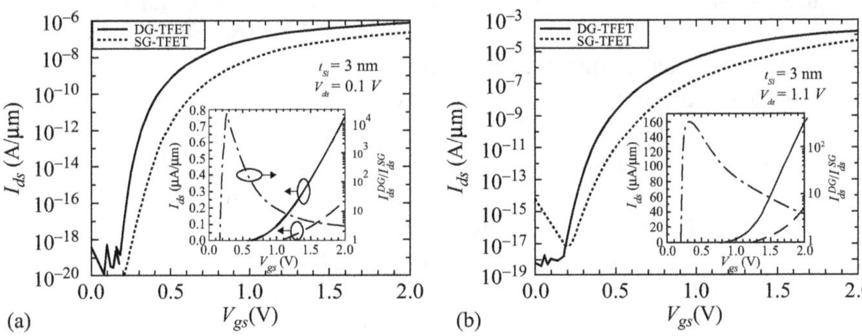

Figure 2.75 Transfer characteristics for (a) single gate and (b) double gate TFETs with $t_{Si} = 3$ nm for $V_{DS} = 0.1$ and 1.1 V. The inset shows the current on linear scale (left axis) and the ratio $\frac{I_{ds}^{DG}}{I_{ds}^{SG}}$ (right axis)

The biggest challenge is to achieve high performance (high I_{ON}) without degrading I_{OFF}, combined with an SS of less than 60 mV per decade over more than 4 decades of drain current. This requires the additive combination of the many technology boosters specific to complementary heterostructure TFETs, which are available or under research on advanced SOI CMOS platforms.

2.3 Different approaches for semiconductor device modeling and simulation

2.3.1 TCAD tools: technological motivation and general outlook

The IC plays a key role in our modern digital information society. Sophisticated technology computer-aided design (TCAD) tools are currently used to assist in IC development and engineering at practically all stages from process design and definition to circuit development and optimization. The level of success which microelectronics has achieved by now was predestined and enabled by the extensive smart TCAD use providing and supporting an appropriate level of mathematical abstraction necessary to design elements and circuits containing billions of interconnected devices. At present, TCAD tools allow to reduce research and development costs by approximately 35%–40% [2.276]. Superior performance is achieved by making transistors smaller, faster, cheaper, and more energy efficient and by assembling more elements on a die. Thanks to the unique MOSFET scalability, the minimum feature size of a transistor has been successfully reduced for more than four decades, allowing to effectively double the number of transistors on a chip by every second year. Today, with the 14 nm CMOS technology introduced [2.59, 2.66], the physical transistor gate length is shorter than 20 nm. Nowadays, this trend which is institutionalized as the famous Moore's law is showing signs of saturation as Intel has recently announced a delay in bringing to the market the next generation of devices manufactured within the 10 nm technology node [2.315]. However, supported by a demonstration of MOSFETs with a gate length as short as 6 nm already in 2002 [2.290], the International Technology Roadmap for Semiconductors [2.276] predicts further scaling beyond the10 nm node in the coming decade. This demands extension and possibly a redesign of the known TCAD tools for modeling the transport behavior in a given device down to sub-10 nm scale. Many new aspects must be taken into consideration at these dimensions. For example, the famous three-dimensional tri-gate architecture [2.341] employed by Intel for the 22 nm and 14 nm technology nodes require fully three-dimensional device modeling. The cutting-edge devices fabricated with the 14 nm process are so small that the fluctuations of the number of dopants inside the channel alter their characteristics significantly [2.282]. Even more, a particular positioning of an impurity in the channel affects the current thus demanding TCAD to take this randomness into account.

The first suggestion of a fully numerical transport description in a one-dimensional bipolar transistor was available already in 1964 [2.311]. The approach was further developed and extended to describe the carrier transport in *PN*-junctions [2.288] and impact ionization avalanche diodes by Scharfetter and Gummel in 1969 [2.354]. The first application of a solution of the two-dimensional Poisson equation to address electrostatics of metal-oxide-semiconductor (MOS) structures was performed by Loeb [2.335] and Schroeder and Muller [2.355]. A simultaneous solution of the coupled continuity and Poisson equations to describe the transport in junction gate field effect [2.325] and bipolar transistors [2.361] goes back to 1969.

Since the pioneering work on transport modeling numerical approaches have been successfully developed and applied to practically all important devices prompting the number of papers in the field to grow exponentially. Today modeling of transport in modern ICs has matured into a well-established field with vast commercial applications and intense software development. Numerous textbooks, monographs, and reviews devoted to theoretical and computational aspects of transport modeling in ICs are available. Not pretending to cover all the literature we would like to mention one of the first monographs [2.356], which addresses practically all aspects from modeling and discretization to applications, the textbooks describing various semiclassical transport models [2.336] and modern quantum mechanics-based approaches to electron transport [2.287], and a monograph [2.363] investigating the role of mechanical stress to boost the performance of modern MOSFETs.

As the development and maintenance costs of modern sophisticated TCAD have increased significantly, only large semiconductor companies can afford to support their own TCAD development team. Fortunately, there exist a fairly large number of commercial TCAD software products available on the market, e.g., [2.360, 2.366], which serve most of the industrial demands. Numerous TCAD tools developed at the universities have an advantage of being open-source licensed software [2.316, 2.345, 2.368]. These tools gain their popularity due to the need of multiscale approaches to simulations combining different levels of complexity and precision. It makes these tools valuable not only for pure educational or research purposes but also satisfies the demand from semiconductor manufacturing companies for more refined simulations of complex phenomena. This urges a creation of new successful spin-off companies, e.g., [2.26, 2.78], focusing on the development and commercialization of specialized tools.

Regardless of the small transistor size, even today most TCAD tools are based on semi-classical macroscopic transport models. The celebrated drift-diffusion transport model has enjoyed an amazing success due to its relative simplicity, numerical robustness, and an ability to perform two- and three-dimensional simulations on large unstructured meshes [2.356]. However, with device size dramatically reduced and new technology elements and materials introduced, the TCAD tools based on the standard semi-classical transport description are becoming less accurate.

From the viewpoint of transport modeling in nanoscale transistors, the problem is two-fold. First, with downscaling the driving field and its gradient increase dramatically inside the short channel. As a result, the carrier distribution along the channel can no longer be described by even a heated shifted Maxwellian distribution typically assumed in current and energy transport. In order to properly account for the hot carrier and non-local high-field effects, the drift-diffusion and even the energy transport model have to be improved to incorporate higher-order corrections beyond the heated Maxwellian in the carrier distribution function. This leads to a more complicated and computationally involved macroscopic transport model equations set.

The second reason for semi-classical modeling tools to become inadequate is a growing importance of quantum-mechanical effects, especially the under-the-barrier tunneling which increases the leakage and causes parasitic power dissipation in the off-state. The band-to-band quantum-mechanical tunneling is the phenomenon, which defines the functionality of a TFET, the device with a potentially very steep subthreshold characteristic. Although less pronounced in the on-state current due to the averaging over many states at different energies involved in transport, the quantum-mechanical effects are believed to play a role in determining the current as the device size is getting comparable to the De Broglie electron wave length, where the transport is becoming more ballistic. However, since the devices operate at room temperature, the carrier scattering in silicon-based FETs is still important [2.348] and a crossover from diffusive to ballistic transport may occur at much shorter channel lengths [2.300]. Therefore, an adequate transport model of an ultra-scaled MOSFET must account for both quantum-mechanical effects and scattering simultaneously.

In the direction perpendicular to transport the most important quantum-mechanical effect is the quantization of carrier motion in the potential well. This results in the formation of subbands described with the corresponding wave functions. The subband wave functions nearly do not penetrate in the gate dielectric, which results in the rapid decay of the corresponding subband charge densities close to the gate–oxide interface. Classically, however, the charge density is characterized by the maximum value at the interface. As a consequence, the transport can no longer be accurately described by the classical three-dimensional equations, and a new description based on two- or even one-dimensional subbands in tri-gate structures and FinFETs must be adopted. In addition, the subbands are characterized by the transport effective masses which, apart from strain, depend strongly on the channel orientation, and, most importantly, on the confinement strength.

Modern TCAD tools must be flexible enough to address challenges of the upcoming technological changes due to the use of new materials and structures. They have to describe properly transport in silicon and new channel materials depending on strain and confinement and must be prepared to adequately address the quantum-mechanical phenomena which are expected to determine the transport properties in ultra-scaled CMOS and post-CMOS devices.

The electron spin attracts at present much attention as a possible candidate for complementing or even replacing the charge degree of freedom in future devices. The electron spin state is characterized by one of two of its possible projections on a given axis and could be potentially used in digital information processing. In addition, it takes an amazingly small amount of energy to invert the spin orientation, which is necessary for low power applications. The electron spin may also point in any direction on a unit Bloch sphere, which opens new directions in storing and processing information by initializing, manipulating, and detecting the spin orientation. Therefore, the TCAD tools must be ready to help foreseeing and guiding the future device development based on new principles of operation.

Without pretending to cover the large field of transport modeling research here, we present several important examples and outline some difficulties and challenges regarding the transport description in modern MOSFETs. We begin with the semi-classical description of carrier dynamics inside the device and outline methods to solve the Boltzmann equation. A particular emphasis will be put on the inclusion of strain effects into the transport simulations. We also demonstrate how the spin lifetime can be evaluated by methods similar to those employed for the low-field mobility calculations. In ultra-scaled devices quantum-mechanical effects start playing an important role. Different types of quantum potential and density gradient corrections can be introduced. We conclude with briefly mentioning fully quantum-mechanical approaches dealing with the dissipative quantum transport in ultra-scaled devices.

2.3.2 Drift-diffusion transport model

In order to analyze a semiconductor device under general operating conditions, a mathematical model has to be formulated first. Regardless of the complexity of carrier dynamics inside the semiconductor, two equations are the most important ingredients of any particular model. These are the Poisson equation and the current continuity equation [2.356]. The Poisson equation relates the carrier charge density ρ to the electrostatic potential V as

$$\nabla \varepsilon \nabla V = -\rho \tag{2.49}$$

where ε is the dielectric permittivity. The charge density is related to the electron n and hole p via

$$\rho = q(p - n + C) \tag{2.50}$$

where q is the electron charge value and C is the concentration of fixed ionized charges.

The continuity equation relates the charge current density

$$\vec{j}(\vec{r}, t) = \vec{j}_n(\vec{r}, t) + \vec{j}_p(\vec{r}, t) \tag{2.51}$$

to the time derivative of the charge density $q(p–n)$

$$\frac{\partial(n - p)}{\partial t} = \frac{1}{q} \nabla \vec{j} \tag{2.52}$$

In case $\frac{\partial C}{\partial t = 0}$ the electron and hole contributions are described by two separate equations.

$$\frac{\partial n}{\partial t} = \frac{1}{q} \nabla \vec{j}_n + R \tag{2.53}$$

$$\frac{\partial p}{\partial t} = -\frac{1}{q} \nabla \vec{j}_p + R \tag{2.54}$$

Here R is the electron–hole generation-recombination rate.

In order to complete the formulation of the mathematical model at this level, the system of these two equations must be supplemented with the current density expressed via the electron and hole concentrations and the electric field. By assuming a constant temperature in the device, the electron $\vec{j}_n$ and hole $\vec{j}_p$ current densities can be, in the simplest form, written as

$$\vec{j}_n = qn\mu_n \vec{E} + qD_n \nabla n \tag{2.55}$$

$$\vec{j}_p = qp\mu_p \vec{E} - qD_p \nabla p \tag{2.56}$$

$\mu_{n(p)}$ is the electron (hole) mobility and $D_{n(p)}$ is the electron (hole) diffusion coefficient.

The electron mobility and diffusion coefficient are material dependent. In order to calculate their values an accurate description of the carrier dynamics in the material is required.

Equations (2.49–2.56) form the framework of the transport description in the drift-diffusion approximation. Although more accurate transport models have been extensively developed, implemented, and used, the drift-diffusion-based simulations continue to be mainstay. One reason is the relative simplicity of the model, which allows obtaining timely yet accurate results [2.337]. Another reason is that the drift-diffusion simulations are surprisingly predictive even in situations, where they are not expected to work at all. Several examples [2.337], for which the drift-diffusion provides amazingly good results, include the diffusion through a thin base of a bipolar transistor near the ballistic limit and an emission-diffusion theory through the Schottky barrier, which reproduces the Landauer theory. Even a quasi-ballistic transient heat phonon transport can be well modeled with the drift-diffusion model [2.337].

Although the drift-diffusion formalism can be applied to describe a wide range of phenomena, it is instructive to understand the formal limitations of the approach and ways to go beyond the approximations. We demonstrate the common generalizations of the drift-diffusion approximation by considering first the semi-classical transport model and then transport models including higher moments of the carrier distribution function.

2.3.3 Semi-classical transport and higher moments transport models

An ensemble of classical particles is conveniently described by the single-particle distribution function $f(\vec{r}, \vec{k}, t)$ in phase space. The distribution function satisfies the Boltzmann equation.

$$\frac{\partial f}{\partial t} + \vec{v} \cdot \nabla_r f + \frac{S_v q}{\hbar} \vec{E} \cdot \nabla_{\vec{k}} f = \left(\frac{\partial f}{\partial t} \right)_{coll} \tag{2.57}$$

$\vec{v} = \nabla_{\vec{k}} E(\vec{k})$ is the carrier velocity.

The sign function s_v distinguishes between negatively charged electrons, $s_n = -1$, and positively charged holes, $s_p = 1$. The right-hand side in (2.57) represents the collision operator due to phonons, impurities, interfaces, and other scattering sources.

For realistic structures, a direct numerical solution of this equation by discretization of the phase space is computationally expensive. Therefore, the Boltzmann transport equation is not solved directly in the TCAD tools. Usually an approximate solution is obtained by using the method of moments of the distribution function. By defining the moments of the distribution function $f(\vec{r}, \vec{k}, t)$, one consecutively obtains the drift-diffusion model [2.311], the energy transport model [2.362], or the six-moments transport model [2.305].

The drift-diffusion mode has several shortcomings, when it is applied to miniaturized devices. Hot carrier effects can be partly addressed by a mobility dependence on the driving field. However, non-local effects such as velocity overshoot are completely neglected within the drift-diffusion approach. Higher-order transport hydrodynamic model [2.285] and the energy transport model [2.362] are designed to overcome some of these shortcomings. However, the energy transport model tends to overestimate the non-local effects and thus the on-current in a device. This also results in unacceptable errors in the hot carrier induced gate tunneling current [2.299]. Another example is a particularly poor description of transport in partially depleted SOI devices, where, because of the overestimated hot carrier diffusion into the floating-body of the device, the energy transport model fails completely in predicting the device characteristics [2.308–2.310], and the application of transport models including higher-order moments is required. The six-moments transport model includes additional information about the shape of the distribution function and is predestine to overcome the above-mentioned limitations [2.305, 2.307, 2.308, 2.329].

The derivation of the transport models is based on equations of the moments-statistical averages as

$$\langle\vec{\Phi}\rangle = \frac{1}{4\pi^3}\int \vec{\Phi}\left(\vec{k}\right) f\left(\vec{r},\vec{k},t\right) d^3k \tag{2.58}$$

where $\langle\vec{\Phi}\rangle$ is a weight function in $\vec{k}$ space. In order to derive a particular model, the following weight functions are considered.

$$\begin{aligned} &\Phi_0 = 1, \quad \Phi_2 = E\left(\vec{k}\right), \quad \Phi_4 = E^2\left(\vec{k}\right) \\ &\vec{\Phi}_1 = \hbar\vec{k}, \quad \vec{\Phi}_3 = \vec{v}E\left(\vec{k}\right), \quad \vec{\Phi}_5 = \vec{v}E^2\left(\vec{k}\right) \end{aligned} \tag{2.59}$$

Here an isotropic parabolic dispersion $E(\vec{k}) = \frac{\vec{k}^2}{m_v}$ with the effective mass m_v is assumed; however, a generalization to non-parabolic energy band dispersions is straightforward [2.303]. Taking the moment of the Boltzmann equation gives the following general moment equation

$$\frac{\partial\langle\vec{\Phi}_j\rangle}{\partial t} + \nabla_{\vec{r}}\cdot\langle\vec{v}\otimes\vec{\Phi}_j\rangle - s_v q\vec{E}\cdot\langle\nabla_p\otimes\vec{\Phi}_j\rangle = \int d^3k\vec{\Phi}_j\left(\frac{\partial f}{\partial t}\right)_{coll} \tag{2.60}$$

where $\otimes$ denotes the tensor product. In order to obtain a closed set of equations for moments several approximations have to be introduced. One is concerned with the moments of the scattering integral, which are frequently approximated using a macroscopic relaxation time expression.

$$\int d^3k\vec{\Phi}_j\left(\frac{\partial f}{\partial t}\right)_{coll} \cong -\frac{\langle\vec{\Phi}_j\rangle - \langle\vec{\Phi}_j\rangle_0}{\tau_\Phi} \tag{2.61}$$

The distribution function can be split into a symmetric and an antisymmetric part, where the symmetric part of the distribution function depends only on the absolute value of $\vec{k}$.

$$f\left(\vec{k}\right) = f_S\left(\left|\vec{k}\right|\right) + f_A\left(\vec{k}\right) \tag{2.62}$$

The carrier concentration ν, the carrier temperature T_ν, the current density $\vec{J}$, the energy flux density $\vec{S}$, the "second-order" temperature Θ_ν, the moment of the sixth order M_6, and the flux $\vec{K}_\nu$ (related to the kurtosis of the distribution) are defined by

$$\nabla \cdot \vec{J}_\nu = -s_\nu q\left(\frac{\partial \nu}{\partial t} + R_\nu\right) \tag{2.63}$$

$$\nabla \cdot \vec{S}_\nu = -C_4 \frac{\partial(\nu T_\nu)}{\partial t} + \vec{E} \cdot \vec{J}_\nu - C_4 \nu \frac{T_\nu - T_L}{\tau_{E(k)}} + G_{E(k)\nu} \tag{2.64}$$

$$\nabla \cdot \vec{K}_\nu = -C_5 \frac{\partial(\nu T_\nu \Theta_\nu)}{\partial t} + 2 s_\nu q \vec{E} \cdot \vec{S}_\nu - C_5 \nu \frac{T_\nu \Theta_\nu - T_L^2}{\tau_\Theta} + G_{\Theta_\nu} \tag{2.65}$$

$$C_4 = \frac{3}{2} k_B, C_5 = \frac{15}{4} k_B^2 \tag{2.66}$$

The system of balance equations for the densities is completed with the equations for fluxes

$$\vec{J}_\nu = -C_1\left(\nabla(\nu T_\nu) - s_\nu \frac{q}{k_B} \vec{E}_\nu\right), C_1 = s_\nu k_B \mu_\nu \tag{2.67}$$

$$\vec{S}_\nu = -C_2\left(\nabla(\nu T_\nu \Theta_\nu) - s_\nu \frac{q}{k_B} \vec{E}_\nu T_\nu\right), C_2 = \frac{5}{2} \frac{k_B^2}{q} \frac{\tau_S}{\tau_m} \mu_\nu \tag{2.68}$$

$$\vec{K}_\nu = -C_3\left(\nabla(\nu M_6) - s_\nu \frac{q}{k_B} \vec{E} \nu T_\nu \Theta_\nu\right), C_3 = \frac{35}{4} \frac{k_B^3}{q} \frac{\tau_K}{\tau_m} \mu_\nu \tag{2.69}$$

where the mobility μ_ν is defined as

$$\mu_\nu = \frac{q \tau_m}{m_\nu} \tag{2.70}$$

We note that generation-recombination terms may depend on both electron and hole distribution functions in an integral, non-local manner [2.356], which makes the task of solving the corresponding equations extremely difficult. Therefore, generation-recombination terms have to be modeled carefully using knowledge from solid-state physics of semiconductors, which may represent a significant challenge [2.306].

Let us now derive the hierarchy of TCAD transport models. The drift-diffusion transport model consists of the continuity equations (2.63) and the current relations

(2.67). The latter is decoupled from the higher-order equation by introducing a closure assumption for the second order moment

$$T_v = T_L \tag{2.71}$$

The physical meaning of this assumption is that the carrier gas is in equilibrium with the lattice. The energy transport model additionally takes into account the carrier energy balance equation (2.64) and the energy flux equation (2.68). To close the system of equations, an assumption on the fourth-order moment has to be introduced. The assumption of a heated Maxwellian distribution for the symmetric part of the distribution function gives the closure relation

$$\Theta_v = T_v \tag{2.72}$$

Going one step further in the model hierarchy, the balance equation for the average squared energy (1.16) and the related flux equation (2.69) are added. To close the equation system, the moment of sixth-order M_6 has to be approximated using the lower-order moments. An empirical closure relation that accounts for the "second-order" temperature Θ_v is the best choice [2.304].

$$M_6 = T_v^3 \left(\frac{\Theta_v}{T_v}\right)^c \tag{2.73}$$

From MC simulations, which are an accurate reference, the value of $c = 2.7$ has been estimated [2.302] for the simulation of miniaturized MOSFETs. Compared to the energy transport models, the six-moments model requires two additional relaxation times, namely, the relaxation time of the second-order temperature τ_Θ and the kurtosis flux relaxation time τ_K. Since analytical models for these new parameters are not available, tabulated values obtained from bulk MC simulations [2.302] can be used.

2.3.3.1 Deterministic solution of the Boltzmann transport equation

In order to evaluate the accuracy of the transport models the solution of the Boltzmann transport equation for the distribution function with subsequent calculations for the moments and the current density is required. The current density obtained with this method, as an integral of the velocity with the distribution function, is free from the assumptions used to close the set of the equations for the moments and fluxes, and can be used for benchmarking the transport models.

With computers getting more powerful, a deterministic numerical solution of the Boltzmann equation for the distribution function can be found by expanding the angular dependence of the distribution function $f(\vec{r}, \vec{k})$ on $\vec{k}$ using a complete set of spherical harmonics $Y_{lm}(\theta, \phi)$.

$$f\left(\vec{r}, \vec{k}\right) = \sum\nolimits_{lm} f_{lm}\left(\vec{r}, \vec{k}\right) Y_{lm}(\theta, \phi) \tag{2.74}$$

θ and ϕ are the polar angles between the electric field $\vec{E}$ and $\vec{k}$. In the low-field limit one can truncate the expansion (2.74) after the terms with $l = 1$. Importantly, this truncation results in a drift-diffusion transport model for the current under the assumptions of parabolic isotropic bands and randomizing elastic scattering. It turns out that in silicon, where the valleys are anisotropic, this approximation gives good results [2.313] in the limit of a weak electric field.

For general scattering processes and realistic band structures as well as at higher driving fields, more terms in the expansion (2.74) are required [2.349]; however, a limited number (less or around ten) of spherical harmonics is needed to achieve good accuracy. Because the angular dependence is accounted for by a small number of harmonics, the number of discretization points in momentum space is reduced with respect to the full three-dimensional discretization procedure. This speeds up the computation significantly.

The knowledge of the distribution function allows evaluate numerically all the moments and fluxes needed for the formulation of the TCAD transport models. Although typically requiring more computer resources and longer simulation time than corresponding TCAD simulations, the deterministic solution of the Boltzmann equation by using the spherical harmonics expansion is essential in verifying the accuracy of the TCAD transport models.

Current–voltage characteristics computed with the spherical harmonic expansion method and using the macroscopic transport models based on the moments of the distribution function are shown in Figure 2.76. It is demonstrated that for a short-channel device the drift-diffusion model underestimates the current. Indeed, since the carrier temperature is assumed constant, the drift-diffusion model cannot

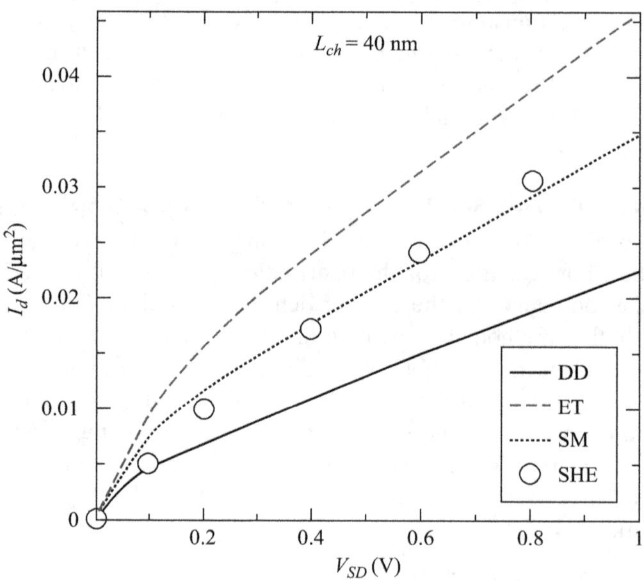

Figure 2.76 I–V characteristics computed with the drift-diffusion (DD), energy transport (ET), six moments (SM) transport models, and with the spherical harmonic expansion method. The six-moments model gives results closest to the results of the spherical harmonic expansion method

account for the non-local effects inside the short-channel device properly. Thus, for short devices the restriction of constant carrier temperature must be relaxed.

When the temperature is not constant, a temperature gradient causes heat flow and thermal diffusion appears. The drift-diffusion transport model must be augmented to allow the energy flow. The energy transport model additionally takes into account the energy flux and the carrier energy balance. The model, however, overestimates the drive current. Figure 2.77 illustrates the average velocity profile in a 40 nm long channel. The drift-diffusion model underestimates the average velocity, while the energy transport model overestimates it. In order to reduce this spurious velocity overshoot effect the next moments should be included for devices with L_{ch} shorter than about 40 nm. This is accomplished by introducing a transport model of sixth order. Balance equations for the average squared energy and the related flux are added. To close the equation system, the moment of sixth order has to be approximated by using the lower order moments (2.73).

The inclusion of higher moments improves the quality of the transport model significantly. The current–voltage characteristics are reproduced fairly well even for a short device as shown in Figure 2.76, because of the more accurate results for the averaged velocity (Figure 2.77). The solution of the Boltzmann equation is required in describing transport and hot carriers effects in power devices at very

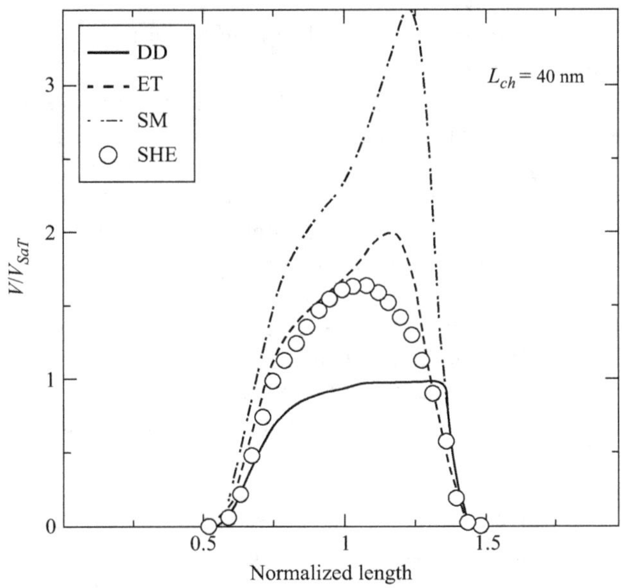

Figure 2.77 Average velocity along a device with 40 nm channel length computed with the macroscopic transport models and with the spherical harmonic expansion method. While the drift-diffusion model underestimates the velocity and current, the energy transport model overestimates the velocity in short-channel devices. The six-moments model gives best results as compared to the velocity obtained by the spherical harmonic expansion method

high-applied voltages, when even the use of the six-moments transport model may become insufficient.

The energy transport model requires knowing the mobilities for the current density and the energy flux, for each carrier type, one relaxation time, and the non-parabolicity factor for non-parabolic bands as the material parameters. The six-moments model requires two additional relaxation times for the second order temperature and the kurtosis flux. Having too many adjustable parameters is a particular inconvenience of the six-moments model. Solving the Boltzmann transport equation using the spherical harmonics expansion method or an MC algorithm usually tabulates these parameters. The parameter dependences on temperature, doping, and the electric field are determined from the conditions that the six-moments transport model mimics exactly all the moments obtained from the MC simulator under homogeneous conditions. The full-band MC method of the solution of the Boltzmann equation is required, when the band structure is strongly non-parabolic and is known only numerically.

We present an example of the mobility evaluation in stressed silicon where the band structure depends on strain in a complex manner. Alternatively, a Kubo-Greenwood approach can be applied to find the low-field mobility [2.298] as we show later on an example of ultra-thin silicon films with stress. The Kubo-Greenwood method can also be used to evaluate the electron spin lifetime and its dependence on strain.

2.3.3.2 Monte Carlo methods for the Boltzmann equation solution

Methods based on MC techniques are well established for studying transport in semiconductors [2.317]. The motion of charge carriers is simulated in the appropriate phase space formed by position and momentum. In the presence of external fields, the carriers which are considered as point-like objects with well-defined positions and momentum move according to Newton's law on classical trajectories. A dispersion relation expressing the carrier energy dependence on the crystal momentum is determined by the band structure. The free flight of carriers along the trajectory is interrupted by scattering events, which are assumed local in space and instantaneous in time. Scattering is modeled as a random process. The duration of a free flight, the type of scattering mechanism, and the state after the scattering are selected randomly from the given probability distributions characteristic to the microscopic scattering processes [2.314, 2.320, 2.339, 2.352]. The method of generating sequences of free flights and scattering events appears to be so intuitively transparent that it is frequently interpreted as a direct emulation of the physical transport process rather than a numerical method. For these reasons the MC methods of transport evaluations are quite time consuming. Being statistical by nature, the method provides an error inversely proportional to the square root of the total number of scattering events. The MC method for solving the Boltzmann transport equation can be useful when the deterministic solution is too expensive, as in strained silicon.

Mathematically one can reformulate the transport equation as an integral equation and then develop an MC algorithm for its solution [2.319, 2.343]. When the Boltzmann equation is transformed to an integral equation, which is then iteratively solved [2.319], the iteration series results in the MC backward technique. This algorithm is useful, if rare events have to be simulated or the distribution function is needed only in a small phase space domain [2.328]. If the Boltzmann equation is reformulated in an adjoin integral form [2.326], a link between the physically based MC method and the iterative procedure of the solution of the integral equation is established.

Already in 1966 Kurosawa [2.332] applied the MC method to high-field transport in semiconductors. In References 2.292 and 2.293 GaAs and Ge were studied, correspondingly. In the mid-1970s a physical model capable of explaining the major macroscopic transport characteristics in silicon was developed [2.286, 2.318]. Considerable improvement of the method and application to a variety of materials was reported in Reference 2.317.

With an increase of the carrier energy the need of an accurate energy band structure description has been realized [2.297, 2.331, 2.358, 2.372]. For electrons in silicon, the most thoroughly investigated case, the "standard model" [2.295], provides a description of all scattering mechanisms. Transport analyzes considering the accurate band structure were performed in [2.296, 2.298, 2.323].

2.3.4 Stress- and orientation-dependent mobility in silicon

To compute the low-field electron mobility in strained Si, one can use the VMC simulator [2.369]. It includes a comprehensive set of scattering models with phonons, ionized impurities, alloy scattering, as well as impact ionization for both electrons and holes.

For electrons in n-silicon, a phonon-scattering model based on Jacoboni and Reggiani [2.320] is used. The model takes into account long-wavelength acoustic phonons causing intravalley transitions as well as optical phonons triggering intervalley transitions. Intravalley scattering is treated as an elastic process. Following [2.289] we adjusted the original values for the coupling constants for intervalley phonon scattering to achieve a bulk mobility enhancement factor of 70% in biaxially strained silicon layers. The coupling constants for acoustic and optical intervalley phonons, as well as the phonon energies are listed in Table 2.5. In full-band simulations the scattering rates are proportional to the density of states calculated from the band structure [2.323].

Figure 2.78 demonstrates the orientation-dependent electron mobility enhancement factor for tensile stress applied in [111] direction obtained from the full-band MC simulations.

Table 2.5 Modes, coupling constants, phonon energies, and selection rule of inelastic phonon scattering

Mode	**Δ (MeV/cm)**	**$\hbar\omega$ (meV)**	**Selection rule**
Transverse acoustic	47.2	12.1	f
Longitudinal acoustic	75.5	18.5	f
Longitudinal optical	1042.0	62.0	f
Transversal acoustic	34.8	19.0	g
Longitudinal acoustic	232.0	47.4	g
Transversal optical	232.0	58.6	g

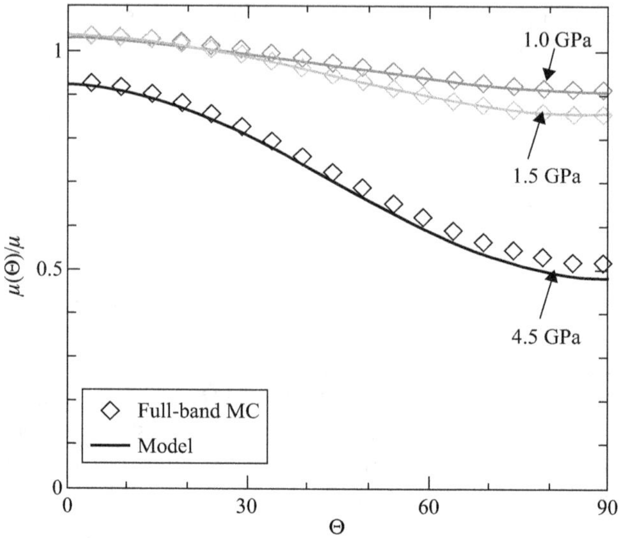

Figure 2.78 Orientation-dependent electron mobility variation in silicon under tensile stress along [111] direction, for several stress values. Results of full-band MC are well explained by the dependence of the effective masses on shear strain. Note the mobility degradation at high stress values caused by an increase of scattering due to the stress-dependent increase of the density-of-states mass

The current direction is defined by the unity vector $\hat{n}$.

$$\hat{n} = \left[\frac{cos(\Theta + \Phi_0)}{\sqrt{2}}, \frac{cos(\Theta + \Phi_0)}{\sqrt{2}}, sin(\Theta + \Phi_0)\right] \tag{2.75}$$

where $\Phi_0 \approx 36^\circ$ is the angle between [111] and [110] directions. The angle Θ determines the current direction in the plane defined by the [111] and [110] axes and is counted from the [111] direction. Results of the full-band simulations coincide well with an analytical model for the mobility described below.

Stress along the [111] direction does not produce a relative energy shift between the valleys and thus affects all six valleys in a similar way. The mobility of an ellipsoidal valley in an arbitrary orientation $\hat{n}$ with respect to the crystallographic coordinate system is determined by the inverse of the conductivity mass $1/m_{\mathbf{n}}$ along this direction computed as

$$m_n^{-1} = \hat{n}^T \hat{m}^{-1}(\eta)\hat{n} \tag{2.76}$$

where $m^{-1}(\eta)$ is the inverse effective mass tensor for the valley in the crystallographic system. $\eta = 2D\varepsilon_{xy}/\Delta$ is a dimensionless strain with $D = 14$ eV being the shear strain deformation potential, ε_{xy} the shear strain component, and $\Delta = 0.5$ eV [2.363].

Taking into account the dependence of the density-of-state effective mass $m_{dos}(\eta) = [m_l(\eta) \times m_{t1}(\eta) m_{t2}(\eta)]^{\frac{1}{3}}$ on dimensionless strain η, the mobility dependence on [111] strain in a direction $\hat{n}$ is obtained as

$$\frac{\mu(\eta, \hat{n})}{\mu_0} = \frac{m_c}{m_n} \left[\frac{m_{dos}}{m_{dos}(\eta)} \right]^{\frac{3}{2}} \tag{2.77}$$

μ_0 is the mobility of relaxed silicon, $m_c = 3\left(\frac{1}{m_l} + \frac{2}{m_t}\right)^{-1}$, and $m_l = 0.91 m_0$ and $m_t = 0.19 m_0$ are the longitudinal and transversal effective masses. The analytical curves for the mobility dependence on $\hat{n}$ in silicon in Figure 2.78 are obtained by substituting in (2.77) the strain-dependent longitudinal and transversal masses m_{t1} (η), $m_{t2}(\eta)$, and m_l (η). For the [001] oriented valleys the dependences are [2.363]

$$\frac{m_t}{m_{t1}(\eta)} = \begin{cases} \left(1 - \eta \frac{m_t}{M}\right), |\eta| < 1 \\ \left(1 - sgn(\eta) \frac{m_t}{M}\right), |\eta| > 1 \end{cases} \tag{2.78}$$

$$\frac{m_t}{m_{t2}(\eta)} = \begin{cases} \left(1 + \eta \frac{m_t}{M}\right), |\eta| < 1 \\ \left(1 + sgn(\eta) \frac{m_t}{M}\right), |\eta| > 1 \end{cases} \tag{2.79}$$

$$\frac{m_l(\eta)}{m_l} = \begin{cases} (1 - \eta^2)^{-1}, |\eta| < 1 \\ \left(1 - \frac{1}{|\eta|}\right)^{-1}, |\eta| > 1 \end{cases} \tag{2.80}$$

$sgn(\eta)$ denotes the sign function.

2.3.5 *Mobility in ultra-thin body of strained SOI transistors*

In thin silicon films the three-dimensional band structure splits into a number of two-dimensional subbands. A special subband MC method has to be implemented in order to evaluate the transport properties. Fortunately, to find the low-field mobility it is not necessary to run the MC simulations. Instead, following the Kubo-Greenwood approach [2.298], one can linearize the Boltzmann transport equation with respect to the proportionally small electric field and perturbation to the distribution function and obtain an approximate expression for the mobility

$$\mu = \frac{q}{4\pi^2 \hbar^2 k_B T n_S} \sum_i \int_0^{2\pi} d\phi \int_{E_1^{(0)}}^{\infty} dE \frac{|\vec{k}_i|}{\left|\frac{\partial E(\vec{k}_i)}{\partial \vec{k}_i}\right|} \times \left(\frac{\partial E\left(\vec{k}_i\right)}{\partial \vec{k}_i}\right)^2_{\varphi = \frac{\pi}{4}, E} \tau^{(i)} f(E)(1 - f(E)) \tag{2.81}$$

$n_s = \sum_i n_i$, n_i is the population of the subband i, and $\tau^{(i)}$ is the scattering rate in the *i-th* subband. It turns out that the mobility value evaluated in this simplified way coincides well with the one computed with the help of full quantum-mechanical non-equilibrium Green's function-based solvers employed to study transport in 14 nm FDSOI structures [2.366].

Scattering mechanisms are the most important physics ingredients determining the relaxation rates. The interface between silicon and oxide plays an important role in determining carrier transport. Small perturbations at the interface contribute to scattering. In ultra-thin films the importance of surface-roughness scattering increases due to the presence of two interfaces. The surface-roughness momentum relaxation rate is calculated after [2.298] as

$$\frac{1}{\tau_i^{SR}\left(\vec{k}_i\right)} = \frac{2\pi}{\hbar(2\pi)^2}\sum_j \int_0^{2\pi} \pi\Delta^2 L^2 \frac{1}{\varepsilon_{ij}^2\left(\vec{k}_i - \vec{k}_j\right)} \frac{\hbar^4}{4m_l^2} \frac{\left|\vec{k}_j\right|}{\left|\frac{\partial E\left(\vec{k}_j\right)}{\partial \vec{k}_j}\right|}$$

$$\times \left[\left(\frac{d\Psi_{i\vec{k}_i\sigma}}{dz}\right)^* \left(\frac{d\Psi_{j\vec{k}_j\sigma}}{dz}\right)\right]^2_{z=\pm\frac{t}{2}} exp\left(\frac{-\left(\vec{k}_j - \vec{k}_i\right)L^2}{4}\right) d\varphi \tag{2.82}$$

$\vec{k}_i$ and $\vec{k}_j$ are the in-plane wave vectors before and after scattering, φ is the angle between $\vec{k}_i$ and $\vec{k}_j$, ε is the dielectric permittivity, L is the autocorrelation length, Δ is the mean square value of the surface-roughness fluctuations, $\Psi_{i\vec{k}_i}$ and $\Psi_{j\vec{k}_j}$ are the wave functions, $f(\vec{E})$ is the Fermi function, and $\sigma = +/-1$ is the spin projection to the [001] axis.

The momentum scattering relaxation rate is calculated as

$$\frac{1}{\tau_i^{PH}\left(\vec{k}_i\right)} = \frac{2\pi k_B T}{\hbar\rho v_{PH}^2}\sum_j \int_0^{2\pi} \frac{d\phi}{2\pi}\frac{1}{(2\pi)^2}\frac{\vec{k}_j}{\left|\frac{\partial \vec{E}_j}{\partial \vec{k}_j}\right|}\left[1 - \frac{\left|\frac{\partial E\left(\vec{k}_j\right)}{\partial \vec{k}_j}\right| f\left(E\left(\vec{k}_j\right)\right)}{\left|\frac{\partial E\left(\vec{k}_i\right)}{\partial \vec{k}_i}\right| f\left(E\left(\vec{k}_i\right)\right)}\right]$$

$$\times 2\pi \int_0^t \left[\Psi^\dagger_{j\vec{k}_{j\sigma}}(z) M^{PH} \Psi_{i\vec{k}_{i\sigma}}(z)\right]\left[\Psi^\dagger_{j\vec{k}_{j\sigma}}(z) M^{PH} \Psi_{i\vec{k}_{i\sigma}}(z)\right] dz \times \theta(\vec{E}_i - \vec{E}_j) \tag{2.83}$$

where the phonon velocity $v_{PH} = \frac{2v_{TA}+2v_{LA}}{3}$ [2.320], and the deformation potential $M^{PH} = 12$ eV [2.346].

The electron mobility enhancement along tensile stress in [110] direction is shown in Figure 2.79 as a function of the shear strain component. The mobility in

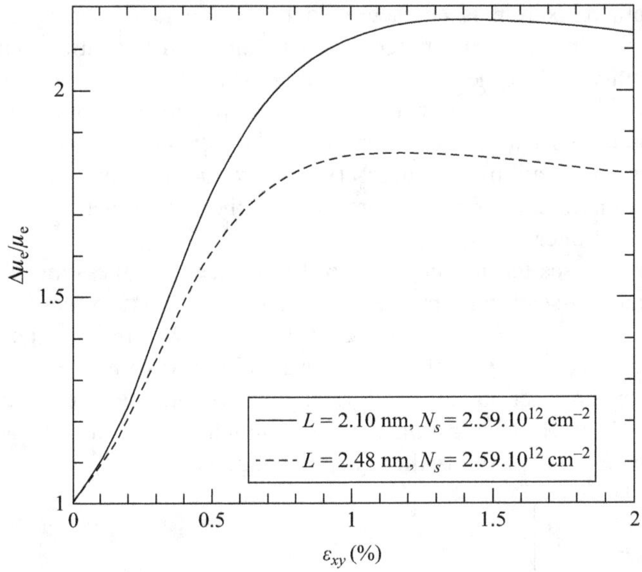

Figure 2.79 Electron mobility enhancement as a function of shear strain for two different film thicknesses at 300 K

thin silicon films at high stress increases by a factor of two as compared to its value in relaxed structures. The increase depends on the electron concentration and the film thickness. A strong mobility increase is observed up to 0.5% shear strain with a further saturation at higher strain value.

The mobility enhancement in silicon was previously explained by the effective mass reduction. However, the effective mass decrease only accounts for roughly one half of the mobility enhancement and cannot explain the two-fold mobility enhancement. Another contribution is the usually ignored dependence of the scattering matrix elements on strain. The electron–phonon scattering matrix elements do not display any substantial dependence on strain. On the other hand, the surface-roughness scattering matrix elements decrease strongly at higher strain values and account for the missing part in the mobility enhancement. As the total mobility is due to a combination of the surface-roughness and electron–phonon scattering limited mobilities, with the surface-roughness contribution becoming more pronounced in narrow structures, the total mobility enhancement is stronger in thinner films, in complete agreement with the simulations in Figure 2.79.

2.3.5.1 Spin lifetime enhancement in strained silicon films

The electron spin is another intrinsic physical characteristic of charge carriers in semiconductors. When injected into a non-magnetic material, the electron spin diffuses away from the spin accumulation region. In contrast to the electron charge, while diffusing the electron spin also relaxes toward its equilibrium value, which is zero. The spin relaxation is characterized by the characteristic spin lifetime. In silicon the spin-flip scattering processes cause the spin relaxation. The small but finite probability to flip the spin after scattering is due to the fact that the wave function of an injected electron with a fixed spin projection is not the eigenfunction of the Hamiltonian because of the presence of the spin–orbit interaction. Therefore, to describe spin relaxation, it is necessary to know the spin-flip scattering rates and the spin–orbit interaction or, to be more specific, the wave functions in the presence of the spin–orbit interaction.

The spin-flip scattering rates due to the surface roughness and the electron–phonon interaction scattering are obtained in a way similar to the relaxation rates (2.82, 2.83) in the previous section [2.346]. The wave functions needed for the evaluation of the scattering matrix elements are obtained within the $\vec{k} \cdot \vec{p}$ formalism [2.364]. For the conduction band valleys along [001] direction relevant to describe the properties in (001) silicon films, the corresponding Hamiltonian written in the vicinity of the X-point at the Brillouin zoned edge is

$$\mathcal{H} = \begin{bmatrix} H_1 & H_3 \\ H_3^\dagger & H_2 \end{bmatrix} \tag{2.84}$$

where H_1, H_2, and H_3 are written as

$$H_1 = \begin{bmatrix} \frac{\hbar^2 k_z^2}{2m_l} + \frac{\hbar^2 \left(k_x^2 + k_y^2\right)}{2m_t} - \frac{\hbar^2 k_0 k_z}{m_l} + U(z) & 0 \\ 0 & \frac{\hbar^2 k_z^2}{2m_l} + \frac{\hbar^2 \left(k_x^2 + k_y^2\right)}{2m_t} - \frac{\hbar^2 k_0 k_z}{m_l} + U(z) \end{bmatrix} \tag{2.85}$$

$$H_2 = \begin{bmatrix} \frac{\hbar^2 k_z^2}{2m_l} + \frac{\hbar^2 \left(k_x^2 + k_y^2\right)}{2m_t} + \frac{\hbar^2 k_0 k_z}{m_l} + U(z) & 0 \\ 0 & \frac{\hbar^2 k_z^2}{2m_l} + \frac{\hbar^2 \left(k_x^2 + k_y^2\right)}{2m_t} + \frac{\hbar^2 k_0 k_z}{m_l} + U(z) \end{bmatrix} \tag{2.86}$$

$$H_3 = \begin{bmatrix} D\varepsilon_{xy} - \frac{\hbar^2 k_x k_y}{M} & (k_y - k_x i)\Delta_{SO} \\ (-k_y - k_x i)\Delta_{SO} & D\varepsilon_{xy} - \frac{\hbar^2 k_x k_y}{M} \end{bmatrix} \tag{2.87}$$

Here, ε_{xy} denotes the shear strain component, $M^{-1} \approx m_t^{-1} - m_0^{-1}$, $D = 14$ eV is the shear strain deformation potential, m_t and m_l are the transversal and the longitudinal silicon effective masses, $k_0 = 0.15X2\pi/a$ is the position of the valley minimum relative to the X-point in unstrained silicon, and $U(z)$ is the confinement potential.

The spin–orbit term $\tau_y \otimes \Delta_{SO}(k_x\sigma_x - k_y\sigma_y)$ with [2.333]

$$\Delta_{SO} = \frac{\hbar^2}{c^2 2m_0^3}\left|\sum_n \frac{\langle X_1|p_j|n\rangle\langle n|[\nabla V \times \vec{p}]_j|X_{2'}\rangle}{E_n - E_X}\right| \tag{2.88}$$

couples the states with the opposite spin projections from the opposite valleys. In the perturbation theory expression for Δ_{SO} E_n is the energy of the nth band at the X-point, E_X is the energy of the two lowest conduction bands X_1 and X_2' degenerate at the X-point, $\vec{p}$ is the momentum operator, V is the bulk crystal potential, σ_x, σ_y, and σ_z are the spin Pauli matrices, τ_y is the y-Pauli matrix in the valley degree of freedom, and c is the speed of light.

Strain and confinement are lifting the four-fold degeneracy of the nth unprimed subband by forming an n^+ and n^- subladder (the valley splitting). The degeneracy of the eigen states with the opposite spin projections $n \pm \Uparrow$ and $n \pm \Downarrow$within each subladder is preserved so that the wave function with an arbitrary spin projection can be constructed. When the spin injection direction is fixed, the degenerate states satisfy

$$\langle \Uparrow n \pm |f|n \pm \Downarrow\rangle = 0 \tag{2.89}$$

with the operators defined as

$$f = cos(\theta)\sigma_z + sin(\theta)\left[cos\,(\varphi)\sigma_x + sin(\varphi)\sigma_y\right] \tag{2.90}$$

where θ is the polar and φ is the azimuthal angle defining the orientation of the injected spin. Due to the spin–orbit interaction, the expectation value of the operator f computed between the wave functions from different subladders is non-zero

$$\bar{\bar{f}} = \langle \Uparrow n \pm |f|n \pm \Downarrow\rangle \neq 0 \tag{2.91}$$

In the two valleys' plus two spin projections' basis the subband wave functions possess four components. These wave functions are written as ($k_x = 0$)

$$|n - \Downarrow\rangle = \begin{pmatrix} \Psi_{1,1} \\ \Psi_{1,2} \\ \Psi^*_{1,1} \\ -\Psi^*_{1,2} \end{pmatrix} |n - \Uparrow\rangle = \begin{pmatrix} -\Psi_{1,2} \\ \Psi_{1,1} \\ \Psi^*_{1,2} \\ \Psi^*_{1,1} \end{pmatrix} |n + \Downarrow\rangle = \begin{pmatrix} \Psi_{2,2} \\ \Psi_{2,1} \\ -\Psi^*_{2,2} \\ \Psi^*_{2,1} \end{pmatrix} |n + \Uparrow\rangle = \begin{pmatrix} -\Psi_{2,1} \\ \Psi_{2,2} \\ -\Psi^*_{2,1} \\ -\Psi^*_{2,2} \end{pmatrix} \tag{2.92}$$

$|n \pm \Downarrow$ and $|n \pm \Uparrow$ are the up- and down-spin wave functions for the first (second) subband. Wave functions with opposite spin in the same valley are orthogonal.

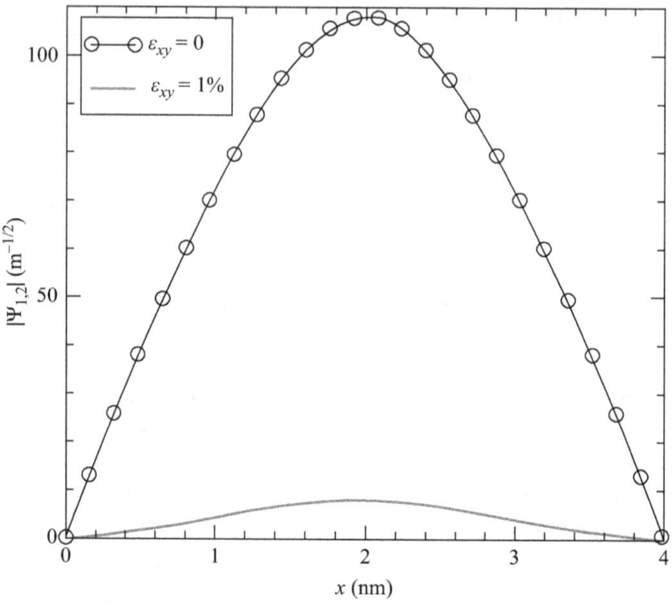

Figure 2.80 Spin–orbit interaction induced components of the wave function in relaxed film and in a film with 1% shear strain. The components are significantly reduced with strain

When the spin is injected along [001] direction, the dominant components are $\Psi_{1,1}$ and $\Psi_{2,2}$ for $|n \pm \Downarrow$ and $|n \pm \Uparrow$, respectively.

The small components of the wave functions are due to the spin–orbit interaction and thus proportional to the spin–orbit interaction strength. Shear strain ε_{xy} considerably suppresses these components as shown in Figure 2.80. Indeed, $\Psi_{1,2}$ for a strain value of 1% has almost vanished. Vanishing values of the small components reduce the spin mixing between the states with opposite spin projections, which results in longer spin lifetime. A significant spin lifetime increase in a silicon film under strain is shown in Figure 2.81. In contrast to the mobility, which is only enhanced by a factor of two, the spin lifetime increases almost exponentially by orders of magnitude. This makes silicon films perfectly suited for spin interconnects as uniaxial stress is now routinely used by the semiconductor industries to boost the MOSFET performance.

2.3.6 *Quantum and quantum-corrected transport models*

Quantum-mechanical effects influence the characteristics of modern semiconductor devices. Due to this fact, purely classical device simulation may not be sufficient to accurately reproduce all the details of transport. Size quantization of carrier motion in the confining potential of an inversion layer is the most investigated quantum-mechanical effect in modern MOSFETs. Because of the size quantization, the energy spectrum becomes discrete in the confinement direction, while it is still continuous

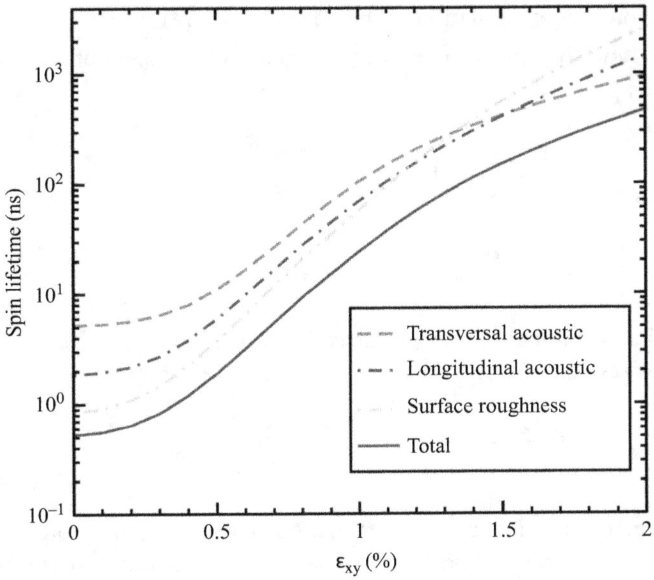

Figure 2.81 Spin lifetime enhancement by shear strain for T = 300 K, the film thickness t = 2.1 nm

along the transport direction. The three-dimensional energy band structure is partitioned into a set of two-dimensional subbands. In order to find the subband energy levels and the corresponding wave function, the Schrödinger equation has to be solved.

The potential entering the Schrödinger equation depends on the charge carrier density in the inversion layer. By knowing the wave functions and occupations of the subbands, the charge carrier concentration in the inversion layer can be obtained. Therefore, the potential has to be determined self-consistently by solving the Schrödinger equation and the Poisson equation simultaneously. This procedure is time-consuming and should be avoided whenever timely results must be obtained. One option is to exploit the well-established semi-classical transport models, while correcting them in such a way that they mimic the quantum-mechanical behavior [2.291, 2.312, 2.324, 2.344, 2.347].

The quantum correction that mimics the local density of states close to the interface can be interpreted as an additional quantum potential [2.294], which has to be added to the classical self-consistent potential in order to describe the decrease of carrier concentration at the interface correctly. Because this additional quantum potential enters into the current relations, it opens an opportunity to introduce quantum-mechanical effects into the drift-diffusion- and higher moments-based transport models. In fact, the appearance of the quantum potential can be easily illustrated, if one just substitutes the wave function represented as the product of the real amplitude and the exponent containing the phase factor

$$\psi(\vec{r},t) = A(\vec{r},t)exp(i\phi(\vec{r},t)) \tag{2.93}$$

into the Schrödinger equation. The density $n(\vec{r},t) = A^2(\vec{r},t)$, the velocity $v = \hbar\nabla\frac{\phi(\vec{r},t)}{m}$, and the current $\vec{j} = n\vec{v}$ are introduced and one obtains the following two equations [2.368]:

$$\frac{\partial n(\vec{r},t)}{\partial t} + \nabla[n(\vec{r},t)\vec{v}] = 0 \tag{2.94}$$

$$\frac{\partial \vec{v}}{\partial t} = \nabla\left(-\frac{v^2}{2} + \frac{q}{m}V(\vec{r},t) + V_{QC}(\vec{r},t)\right) \tag{2.95}$$

Here, the additional quantum correction potential $V_{QC}(\vec{r},t)$ is given by the following expression

$$V_{QC}(\vec{r},t) = \frac{\hbar^2}{2m^2}\frac{\Delta\sqrt{n}}{\sqrt{n}} \tag{2.96}$$

The form of the quantum potential (2.96) is commonly referred to as the density gradient correction and is extensively used in quantum hydrodynamic calculations [2.278, 2.281]. The expression for the current density including the quantum correction reads [2.283]:

$$\vec{J} = q\mu n\vec{E} + qD\nabla n - \mu n\frac{\hbar^2}{2mr}\nabla\frac{\Delta\sqrt{n}}{\sqrt{n}} \tag{2.97}$$

Here an additional parameter $r \geq 1$ is introduced.

Substitution of the current relation (2.97) into the continuity equation (1.41) results in a differential equation for the particle concentration n of fourth order. Such an equation needs two boundary conditions. If one considers the interface between the semiconductor and the dielectric, the first boundary condition is the standard one to set on the normal derivative of the concentration to zero required by the absence of the normal current component at the interface. The second boundary condition allows set the carrier concentration to zero at the interface simultaneously. Thus, the quantum drift-diffusion theory based on (2.97) supplemented with the corresponding boundary conditions automatically reproduces the concentration decrease at the interface, mimicking the quantum-mechanical behavior. A review of quantum hydrodynamic models is given in, e.g., Reference 2.322.

For numerical transport calculations it is convenient to avoid the discretization of the fourth order equation and to include the quantum potential correction term into a generalized electro-chemical potential [2.283]. After a careful calibration of the resulting density gradient model the transport calculations including source-drain tunneling in ultra-scaled MOSFETs becomes possible [2.283]. The density gradient formalism is successfully used for handling discrete charges in drift diffusion "atomistic" simulations [2.301]. It enables the comprehensive statistical investigation of the effects associated with charge trapping on the defects in a wide range of doping values for 20 nm CMOS bulk and 14 nm FinFET transistors

[2.277]. To justify the approach, an accurate calibration to the experimental data is required. A comparison with more sophisticated transport models is also needed.

An approach capable of handling both the quantum coherent propagation and dissipative scattering is based on the Wigner function formalism. The Wigner function is defined as the density matrix in a mixed coordinate/momentum representation [2.330, 2.371]. A practically used approximation to incorporate realistic scattering processes into the Wigner equation is to utilize a properly adapted Boltzmann scattering operator [2.327]. In this way well-established scattering models already calibrated within semi-classical transport approaches can be employed in quantum transport calculations. The inclusion of dissipation through the Boltzmann scattering operator, although intuitively appealing, raises some concerns about the validity of such a procedure. The Boltzmann scattering operator is semi-classical by its nature and represents a good approximation for sufficiently smooth device potentials.

The kinetic equation for the Wigner function is similar to the semi-classical Boltzmann equation, except for a non-local quantum potential term. In the case of a slowly varying potential this non-local term reduces to the local classical force term, and the semi-classical description given by the Boltzmann equation is obtained from the Wigner equation. This semi-classical limit of the Wigner transport equation allows link a semi-classical description of the extended contact regions with the quantum-mechanical description of the active region of a device using the same formalism [2.327].

Implementations of MC methods for solving the Wigner equation have been reported, e.g., [2.342, 2.359]. Recent advances in solving the Wigner transport equation by MC methods are reviewed in Reference 2.357.

To account for scattering more rigorously, spectral information has to be included in the Wigner function formalism, resulting in energy dependence in addition to the momentum dependence [2.338]. The non-equilibrium Green's function method addresses the quantum transport problem in a most consistent and complete way. However, the method is computationally complex. Scattering requires the knowledge of the corresponding self-energies and thus complicates computations significantly [2.364]. The self-consistent Born approximation for the self-energy is an extremely time consuming but necessary step, because it guarantees current continuity. The convergence of the self-consistent iteration is a critical issue, where fine resonances at some energies have to be resolved accurately [2.321, 2.351]. For that purpose an adaptive method for selecting the energy grid is essential [2.321, 2.351].

Nowadays atomistic quantum transport simulators designed for ultra-scaled CMOS and beyond CMOS device simulations are available, e.g., [2.340, 2.353]. They allow not only to compute the electron transport by using the material-dependent atomic parameters and scattering but even to obtain the parameters and currents from the first-principle density-functional calculations with specially designed exchange correlations [2.284]. Being able to accurately study realistic ultra-scaled devices these methods, however, are computationally demanding, as they require using several interacting complex simulation tools.

2.4 Alternative materials and device structures

2.4.1 Introduction

At the beginning of the twenty-first century, the International Technology Roadmap for Semiconductors has been accelerating the introduction of new and diverse technologies to extend the CMOS fabrication technology into nanoscale MOSFET structures [2.373]. Since then, rather than replacing CMOS the new materials and devices have been combined with a CMOS platform to extend microelectronics to new applications domains not accessible at that time to CMOS technology alone [2.374]. However, device cost and performance will continue to be strongly correlated to dimensional and functional scaling of CMOS as information processing technology is driving the semiconductor industry into a broadening spectrum of new applications according to 2013 ITRS [2.375]. Not only the continuation of the Moore's Law is the driver for the search of new materials, in the early 2000s was invented the term More than Moore (MtM) to stress the fact that the value of a packaged system doesn't rely only on the performance of the CMOS technology for the digital information processing, but also on diversified technologies which doesn't necessarily perform better through a dimensional scaling. MtM is the other facet of the microelectronic products complementing the digital part of the integrated systems. More specifically the MtM approach allows for the non-digital functionalities of a product – which do not necessarily scale according to Moore's Law, but provide additional value in different ways – to migrate from the system board-level into the package (SiP) or onto the chip (SoC) [2.376]. As it is pointed in the 2013 ITRS, because the More than Moore domain is multidisciplinary, involving expertise from many different areas, such as electrical and mechanical engineering, materials science, biology and medical science, the search of new and compatible materials with the silicon fabrication technology is now a short-term goal that needs to be fulfilled.

Strained silicon, high-k dielectrics, metal gate, multigate transistors, Ge, SiSe, and II-V semiconductors are now used in IC manufacturing based on the promise of high mobility. All of these triggered the emerging research on completely new transistor operating and new materials. Nanoelectronics is now the common place for the future of the ICs. With a mix of chemistry, physics, biology, and engineering, nanoelectronics may provide a solution to increasing fabrication cost and may allow ICs to be scaled beyond the limits of modern transistors [2.377]. However, the fabrication methods that this approach requires are still waiting for a breakthrough for its implementation in mass production. Nevertheless, the devices that that will be used in the nanoelectronic circuits as well as the new materials to build them are being developed and, among the most popular proposals we can mention the following:

Carbon nanotube (CNT). Are cylindrical carbon molecules that exhibit unique and interesting physical properties including current carrying ability, long ballistic transport length, high thermal conductivity, and mechanical strength [2.378]. The carbon nanotube field-effect transistor (CNFET) is one possible candidate for future high-performance nanoelectronics, because it has near ballistic transport at room temperature, tight electrostatic control on its 1D channel, which provides superior speed and reduced SCEs, as a consequence, it may enable further downscaling [2.379]. By developing noise and manufacturing process variability to the Stanford CNFET compact model in reference the authors predicted that the CNFET devices would outperform Si-CMOS in RF applications in terms of the requirements of the International Technology Roadmap for Semiconductors. However, the current fabrication technology for CNFET are still affected by several shortcomings, it is not possible to provide exact control over CNT diameter and doping, or the removal of tubes with metallic behavior [2.380].

Graphene. Since its discovery in 2004 [2.381], it has been proposed as replacement for silicon in MOSFET for high-frequency applications mainly because its high field effect mobility (close to 10,000 cm^2/Vs at room temperature) [2.382]. Graphene is a two-dimensional material with sp^2 structure that has many excellent physical properties such as extremely high intrinsic carrier mobility, ultra-thin body, long mean-free-path, great thermoelectric property and stability. An excellent summary of the physics, chemistry and engineering of this marvelous material may be found in Reference 2.383. Graphene also suffers from major drawbacks; the most prominent is that is a zero band gap material, which results in small on/off current ratio and no stable saturation region in the output characteristics of transistors made on this material [2.384]. Nevertheless, the search of a graphene FET has not ended but up to this time, they appear with a high off-state leakage current and lack of drive current as a consequence of the gapless band structure. Among all the strategies for producing Graphene, CVD on transition metals substrates seems to be the most promising approach to produce large area and inexpensive materials. In spite of all the developments in the deposition of graphene, there still important challenges to be solved [2.385]: First of all, synthesizing graphene with large and controlled grain size would be very important for various electronic applications. For instance, is it possible to grow single-grain graphene of centimeter or even wafer scale size? Second, controlling the number of layers and stacking order of graphene is also very important, as bilayer and trilayer graphene may offer functions and properties different from monolayer graphene. In addition, growing graphene directly on insulating substrates such as Si/SiO_2 and h-BN would help to overcome the quality degradation caused by the transfer process. Furthermore, low temperature graphene growth will be attractive to reduce the cost and may enable the direct growth on flexible polymer-based substrates. From the aforementioned, one of the possible route for graphene devices is to complementarily integrate them with the CMOS ICs or systems in order to supply more functionality to the silicon technology as an active element in the MtM area. In this respect Huang [2.386] has demonstrated the integration of a graphene Hall sensor with the CMOS amplifier in a silicon chip. So far this is tone of the approaches for incorporating graphene to the silicon mainstream technology and for building system in package.

Nanowires (NW). Semiconductor NWs have attracted considerable attention because its improved electrostatic control in the cylindrical geometry by using wraps gates [2.387], and because they offer the possibility have including heterostructures in transistor design [2.388]. Although many different types of semiconductor NW have been investigated, silicon NWs have become prototypical NWs because they can be readily prepared, the Si/SiO2 interface is chemically stable, and Si NWs are utilized in a number of device demonstrations that have well-known silicon-technology-based counterparts [2.389]. The Si nanowire FET (SiNWFET) has already been demonstrated in continuing with the scaling of the transistor and, seems to be a better alternative than the CNFET because it always is a semiconductor independently of the diameter. When used for FET fabrication they have demonstrated, by comparing it with a SOI FET scaled to the NW dimensions that the SiNWFET has larger on-state current and the average subthreshold slope approaches the theoretical limit and the average transconductance is up to ten times larger [2.390]. In particular epitaxially grown silicon (Si) NWs are considered as promising candidates for post-CMOS logic elements owing to their potential compatibility with existing CMOS technology. One major advantage of vapor-liquid-solid (VLS) grown NWs compared to top-down fabricated devices is that they have well-defined surfaces. This reduces surface scattering, an issue that becomes important for devices on the nanoscale. Moreover, epitaxially grown NWs circumvent the problem of handling and positioning nanometer-sized objects that arise in the conventional pick-and-place approach, where devices are fabricated by manipulating horizontally lying VLS-grown NWs. In this respect, a generic process flow to fabricate silicon NW vertical surround-gate FETs has been demonstrated; the intrinsic advantage of the process developed is that no chemical or mechanical polishing steps, which are difficult to control at this length scale, are needed [2.391].

Because the growth of NWs is a bottom-up self-assembly process, it brings an additional benefit to all of the aforementioned: then freedom in materials design where highly perfect heterostructures like In/InP, InAs/InSb, and Si/Ge may be formed [2.388], combination of materials that cannot be realized by the conventional top down techniques because the constraint of lattice matching. The incorporation of vertically integrated NWFET may enable the implementation of 3D CMOS ICs.

Gallium Nitride (GaN). The excitement generated about GaN stems from its unique material and electronic properties. GaN devices offer five key characteristics: high dielectric strength, high operating temperature, high current density, high speed switching, and low on-resistance. These characteristics are due to the properties of GaN, which, compared to silicon, offers ten times higher electrical breakdown characteristics, three times the bandgap, and exceptional carrier mobility. But this material offers yet more advantages, the high electron mobility and charge density possible in AlGaN/GaN and InAlN/GaN heterostructures has enabled the demonstration of power amplifiers with at least one order of magnitude higher output power density than in their GaAs or Si-based counterparts. However, in spite of this excellent performance, nitride-based devices cannot compete with Si CMOS electronics in terms of cost, scalability, and circuit complexity. The seamless integration of these two semiconductor families would give the circuit and system designer unprecedented flexibility to use the best material and devices for each function [2.392]. A fabrication technique for the manufacturing GaN on Si power devices and the heterogeneous integration with Si devices has been demonstrated for 4” Si wafer [2.392] and for 8” wafer substrates [2.393].

So far, we have dealt with the materials that the main stream of research has considered for increasing the performance of the Si technology, but recently and offering new functionality to this technology the polymers and small molecular thin films have emerged refreshing the materials catalog. The main advantage that makes these materials very attractive is their low deposition temperature. In addition to that, the possibility of tailoring the properties of the films by using multi-components organic semiconductors [2.394] is a new area of research for obtaining new functionalities in the Si technology. For instance, light emission to the Si technology has already been demonstrated [2.395] and is only an example of all the new functionalities that can be added to CMOS chips.

2.4.2 Nanostructured materials, amorphous and SiGe alloys, and its applications

We all are used to the SiGe in the HBT fabrication and as a strained layer in the CMOS technology. HBT fabricated with SiGe heterojunction in the BiCMOS technology show a performance that is superior to its III–V counterparts. The addition of Carbon to SiGe opened new capabilities to the HBT performance and is very important step in the strained-layer epitaxy for this application. An excellent review of these applications can be found in the books edited by John D. Cressler [2.396]. However, the use of amorphous SiGe alloys or nanostructured materials that are obtained from Plasma Enhanced Chemical Vapor Deposition (PECVD) are "structure sensitive" amorphous material because a perplexing diversity of structures and properties can result depending upon the preparative processes and conditions [2.397]. But, what are the advantages or novelty in incorporating amorphous or nanostructured materials obtained from PECVD? In answering the question we will follow the discussion given by Street [2.398]:

> The disorder is the main feature that distinguishes amorphous from crystalline materials. This is of special significance in semiconductors, because periodicity of the atomic structure is central to the theory of crystalline semiconductors. Bloch's theorem is a direct consequence of the periodicity and properly describes the electrons and holes by wave functions, which are extended in space with quantum states defined by the momentum. The theory of lattice vibrations has similar basis in the lattice symmetry. The description of amorphous materials is developed instead from the chemical bonding between atoms, with emphasis on the short range bonding – that is in the bond length and bonding angle – rather than in the long-range order (periodicity). This structural disorder influences the electronic properties in several different ways are summarized in the following text.

Bonding disorder. The disorder represented by deviations in the bond lengths and bond angles broadens the electron distribution of states and causes electron and hole localization as well as strong scattering on the carriers.

Structural defects. Such as broken bonds have corresponding electronic states which lie in the band gap, here the emphasis on the local bonding rather than the long range translational symmetry leads to a strong interaction between the electronic and structural states and causes the phenomenon of metastability.

Electronic properties. The wave functions of the electronic states are the solutions to the Schrödinger equation, the periodic potential of the ordered crystal leads to the familiar Bloch solutions. The wave function has a well-defined momentum that extends through the entire crystal. The energy bands are described by energy-momentum dispersion relations, which, in turn, determine the effective mass, electronic excitations, etc. The aforementioned solutions to Schrödinger equations do not apply to an amorphous semiconductor because the potential is not periodic. A weak disorder potential results in only a small perturbation of the wave function and has the effect of scattering the electron form one Block state to another. The disordering effect of an amorphous semiconductor is strong enough to cause such frequent scattering that the wave function loses phase coherence over a distance of one or two atomic spacing. This strong scattering causes a large uncertainty in the electron momentum trough the uncertainty principle; the uncertainty in the moment is similar to the magnitude of the moment and therefore is not a good quantum number and is not conserved in electronic transitions. The loss of momentum conservation is one of the most important results of disorder and changes much of the basis description of the electronic states. Some consequences of the loss of momentum conservation are listed as follows.

- The dispersion relations do not longer describe the energy bands, but instead also a density of states distribution must redefine the electron and hole effective masses redefined as they are usually expressed as the curvature of the dispersion relations.
- The conservation of momentum selection rules does not apply to optical transitions in amorphous semiconductors. Consequently, the distinction is lost between a direct and an indirect band gap, the later being those transitions which are forbidden by momentum conservation. Instead transitions occur between states that overlap in real space. This distinction is most obvious in Si, which has, and indirect band gap in its crystalline phase but not in the amorphous phase.
- The disorder reduces the carrier mobility because of frequent scattering and causes the much more profound effect of localizing the wave function.

If we try to do a balance of the opportunities offered from amorphous materials, we will be tempted to refuse its incorporation to the CMOS technology. But, looking carefully we find that because the deposition method a perplexing diversity of structures and properties can result depending upon the preparative processes and conditions. We are actually, by changing the depositing conditions, tailoring the density of defects, the band gap and therefore creating new materials from the same components. Additionally, the lack of conservation of the momentum also brings the benefit of not influence of the lattice constant, which results in that all possible ratios among elements of a compound can be achieved.

As an example of the above asseveration, the properties of a-SiGe:H are presented below and we will see the influence of the depositing conditions on the properties of the deposited semiconductor [2.399–2.401]. The samples of amorphous silicon germanium films were prepared by LF PECVD decomposition using a PECVD system from Applied Materials Inc. (Santa Clara, CA), Model 3300. Silane,

SiH_4, and germane, GeH_4, were used as the feed gases and hydrogen and argon as the dilution gases. Two different substrates were used: Corning 1737 glass for the conductivity measurements and crystalline silicon (c-Si) for the measurements of infrared (IR) spectra. The films were deposited at substrate temperature $T_s = 300$ °C. The deposition parameters were as follows: pressure, $P = 0.6$ Torr; power, $W = 350$ W; and frequency $f = 110$ kHz. The total flow of silane and germanium $Q_{SiH4,GeH4}$ was 50 sccm for all depositions. Three types of the samples were fabricated: 20:1 hydrogen dilution (H-dilution), 10:1 argon dilution (Ar-dilution), and undiluted.

Depending on the gas dilutor the growth rate of the films changes for X ranging from 0 (pure Si) to 1 (pure Ge). As is depicted in Figure 2.82, the growth rate increases monotonically with X for dilution, while for dilution at low X values the growth rate suffers changes in opposite direction.

But not only the growth rate depends on the depositing conditions, the spectral dependence of the optical absorption coefficient $\alpha(h\nu)$ is shown in Figure 2.83a for all the samples studied along with that for a-Si for comparison. As can be seen, $\alpha(h\nu)$ curves shift to lower photon energy with increasing Ge content in the feed gas. The total shift is about 0.9 eV when Y changes from 0 to 1. The shift is not proportional to Y; nearly half of the total shift occurs in the low Y region between $Y = 0$ and $Y = 0.42$. The effect of dilution is rather pronounced for $Y = 0.23$ to 0.6; the films deposited with H-dilution are shifted more than the others. Also the

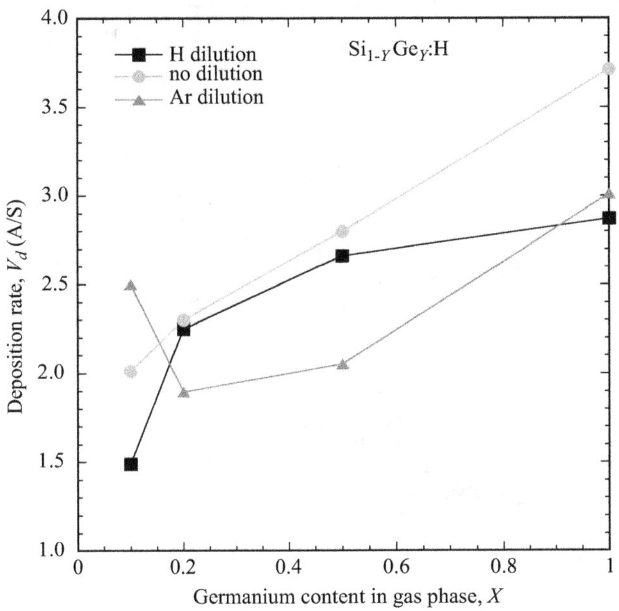

Figure 2.82 Deposition rate as a function of X, the Ge content in the feed gas, defined as the flow gas ratio $X = Q_{GeH4}/Q_{SiH4+GeH4}$

optical gap decreases linearly with Ge content to as small E_g in the range 0.95–0.95 eV for a-Ge as it is shown in Figure 2.83b.

The temperature dependence of the conductivity in the films $\sigma(T)$ is described by an activation dependence

$$\sigma(T) = \sigma_0 \exp\left(-\frac{E_a}{kT}\right) \tag{2.98}$$

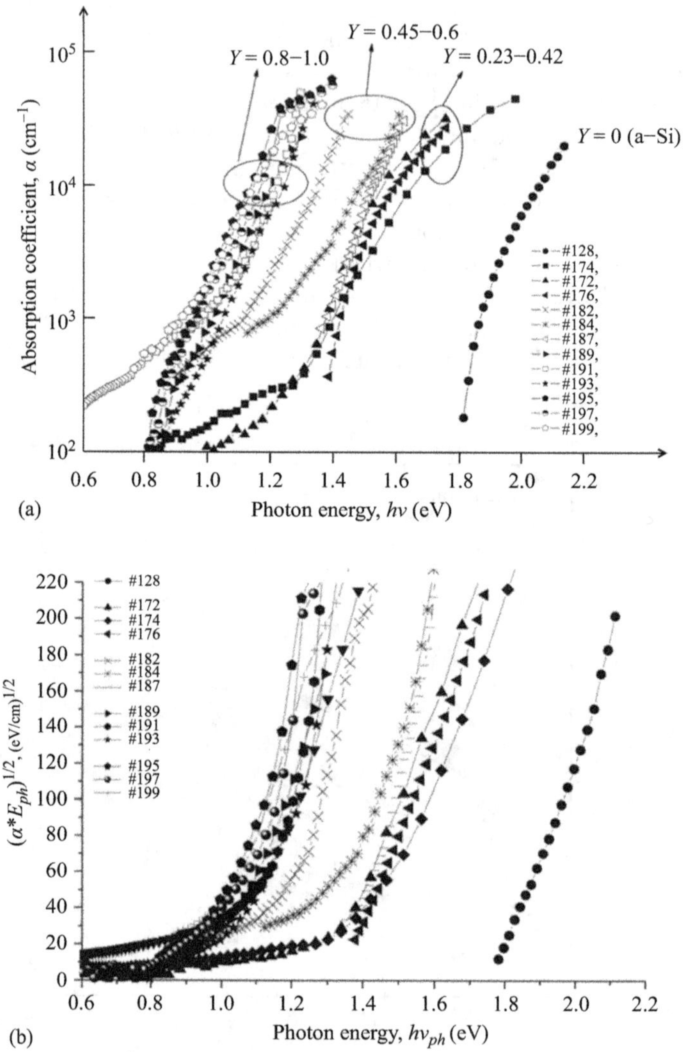

Figure 2.83 Spectral dependence of optical absorption coefficient: (a) α(hν) and (b) Tauc plots of the $Si_{1-Y}Ge_Y$:H films deposited from the gas mixtures with different Ge-content Y and different dilutions

where E_a is the activation energy determined from the slope of the experimental curves logs versus $1/T$. Figure 2.84 shows the measured conductivity of the films; we can see that Ge incorporation significantly changes conductivity of the films. Room temperature (RT) conductivity increases from $s_{RT} = 2 \times 10^{-8}$ at $Y = 0$ to $2 \times 10^{-1}\ \Omega^{-1}\text{cm}^{-1}$ at $Y = 1$, i.e., by seven orders of magnitude while for the same range of change in Y, the E_a changes from 0.60 to 0.22.

Also de deposition parameters influence the surface morphology and grain-like structures are observed trough an Atomic Force Microscope. Average height roughness, $<H>$, height distribution $F(H)$, lateral correlation length, L_c, and kurtosis, γ, were calculated from the AFM images and studied as a function of the Y content of Ge in the films. Figure 2.85 shows the average height $<H>$ for the different gas dilutions. As it can be observed, the average roughness increases from no dilution to hydrogen dilution and the highest roughness is obtained from Ar dilution. By increasing the hydrogen dilution rate the deposited films result in micro- and nano-crystallites [2.374]

By changing the dilution rate in the SiGe film deposition there is also a change in the structure and electronic properties of the deposited material. Defining the Hydrogen dilution parameter as $R = (Q_{H2})/(Q_{SiH4} + Q_{GeH4})$, where the flows $Q_{SiH4} = 25$ sccm and $Q_{GeH4} = 25$. The depositing conditions in addition of maintaining these flows constant are: pressure $P = 0.6$ Torr, RF Power = 300 W, discharge frequency $f = 110$ kHz, and a substrate temperature $T_s = 300$ °C. R was varied from 20 to 80 and in order to characterize optical properties an optical gap E_{04}, a characteristic energy E_{03} determined as the photon energy at which absorption $\alpha(E_{04}) = 10^4\ \text{cm}^{-1}$ and $\alpha(E_{03}) = 10^3\ \text{cm}^{-1}$, respectively, are used, in addition to $\Delta E = E_{04} - E_{03}$ which reflects the density of band tail states, a plot of the optical

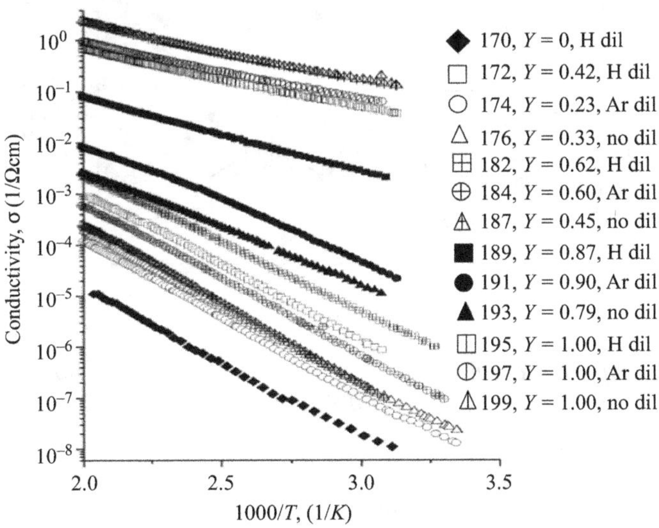

Figure 2.84 Temperature dependence of conductivity $\sigma(T)$ for the SiGe films deposited with the various Ge content Y and dilutions

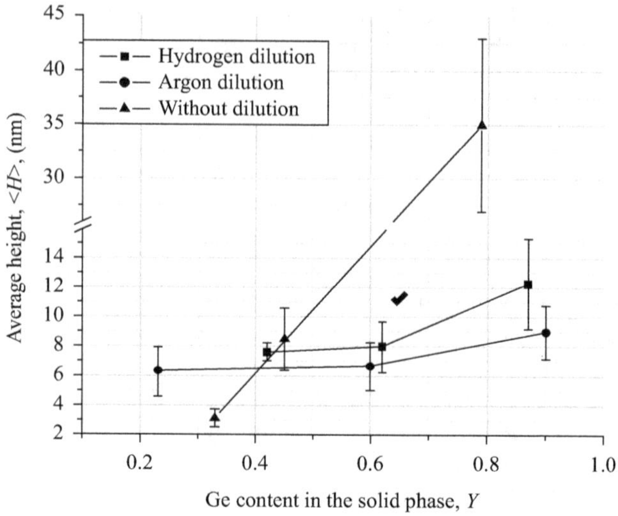

Figure 2.85 *Average height as a function of Ge content, <H> for different gas dilutions. The films were deposited on Si substrates. Solid lines are just guides for the eye*

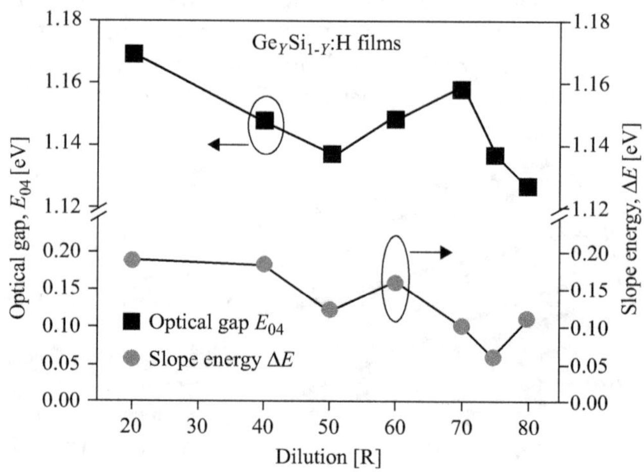

Figure 2.86 *Optical gap, E_{04}, and slope energy, ΔE, as a function of hydrogen dilution in Ge_YSi_{1-Y}:H (b) films*

gap versus dilution rate R is built and we can see this dependence and its corresponding change in ΔE. Figure 2.86 shows this dependence.

So far we have shown how the depositing conditions can produce a great variety of materials, here the case of SiGe:H alloys was used, but still many other aspects in which the same alloy may be deposited by PECVD resulting in a material that gives new and surprising properties [2.402].

2.4.3 Photodetectors and micro-machined bolometers

The major factor limiting the properties of amorphous semiconductors is the density of defects in the energy gap. But we have seen in the precedent paragraph that it is possible by controlling the depositing conditions, to control the density of defects and, therefore, controlling the semiconductor properties. In the following text, we will show how the SiGe:H alloy by just changing the depositing conditions, it can be used in a variety of devices with an excellent advantage, that it can be added by post-processing to the CMOS fabrication technology without affecting the performance of the CMOS ICs.

A planar amorphous SiGe:H SAMAPD [2.403, 2.404]. For optical fiber communications, the Separated Absorption and Multiplication Avalanche Photodetector (SAMAPD) is fabricated on III–V semiconductor because the useful wavelength lies in the range of 0.9–1.55 nm. The quantum efficiency of these devices is determined by the properties of the absorption layer and the hetero-interface, while the avalanche multiplication process in the wide bandgap semiconductor determines the multiplication and excess noise properties. When the SAMAPD idea is exported to silicon, two major facts make this highly attractive: (1) the ionization coefficient ratio in Si is largely different from unity, which will result in low excess noise factor, and (2) the absorption coefficient of amorphous Si is very large compared to its crystalline counterpart. For the absorption layer of the SAMAPD, amorphous silicon (a-Si) and its alloys (a-SiGe) are a good option because they show at least one order of magnitude higher absorption coefficient with respect to crystalline material. Additionally amorphous materials have a very low deposition temperature (when PECVD is used), and do not contain any materials harmful to Si IC's fabrication processing. a-Si and a-SiGe alloys, used as absorption layers in a SAMAPD structure, do not need to have smaller band gap than the crystalline silicon, as in III–V-based APDs. Because a-Si and its alloys have optical absorption coefficients larger than 10^4 cm^{-1} (for energies 0.2 eV larger than the optical gap), for most optoelectronic applications, only a 1 mm thick absorption layer is sufficient which is two to three orders of magnitude thinner than that needed in crystalline silicon. One of the important characteristics of amorphous materials is that materials of arbitrary composition can be obtained.

A cross-section of the a-SiGe SAMAPD, fabricated in an n-type Si wafer (100), $r = 3$–5 Ω-cm of the INAOE 10-μm-CMOS IC's process, is shown in Figure 2.87. Only one extra *p* implantation step is required to prepare the substrate for the SAMAPD fabrication, and the IC's fabrication process is not altered. The p implantation is done with a dose of 2×10^{12} cm^{-2} at 150 keV. With this implantation the p-region of the p–n multiplication-junction is formed. The n$^+$ contact as well as the p region described earlier, are activated with the same heat treatment used in the formation of the source and drain regions. An a-SiGe:H from SiH_4 and GeF_4 as sources and doped with boron was deposited at 250 °C and a pressure of 0.8 Torr. The deposited film resulted a-$Si_{0.67}Ge_{0.33}$:H,F with an optical bandgap = 1.33 eV, a

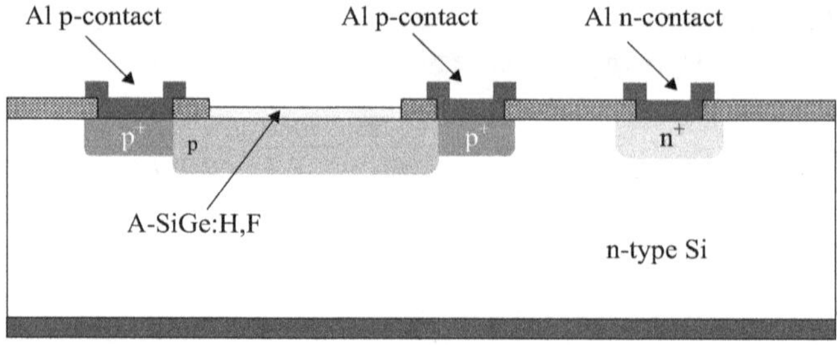

Figure 2.87 Cross-section of the a-SiGe:H,F SAMAPD and the surface n^+ contact to the n-type substrate. The junction depth of the n^+ and p^+ regions is 1.5 mm. The active window of the a-SiGe regions is $100 \times 100\ \mu m^2$

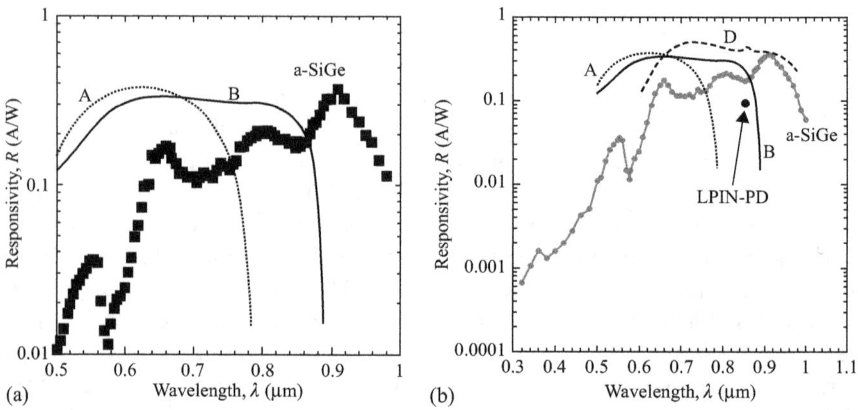

Figure 2.88 (a) Experimental responsivity R of the a-SiGe:H,F SAMAPD versus wavelength at $V_d = 0$ V, with a-SiGe:H,F thickness layer of 0.065 mm. The experimental results of a super lattice GaAs-$Al_{0.4}Ga_{0.6}As$ photo diode are also shown by the continuous lines. Curve A corresponds to 30 GaAs/AlGaAs periods of 46/48 angstroms thick, and curve B is for 30 GaAs/AlGaAs periods of 139/84 angstroms thick. A spectrophotometer is used to sweep the device from 0.3 to 0.9 mm, and a coherent ring laser is used on the 0.9–1.0 μm wavelength range. (b) Device response with AR coating (D)

diffusion length of 0.08 mm and a lifetime of 769.6 μs. Therefore a film thickness of 0.08 will avoid the risk of recombination inside the absorption layer. The addition of an antireflective (AR) coating helps to increase the performance of the device. The AR films Spin on Glass cured at 200°C in order to maintain the properties of the absorber material. Figure 2.88 shows (a) the device responsivity plotted as function of the incident wavelength and (b) with the AR deposited.

So far, an IC-compatible a-SiGe planar SAMAPD, suitable for optical communications when AlGaAs laser source is used, has been demonstrated. Its simple fabrication technology, the gain that can be obtained when operated in avalanche and the ability of tailoring the absorption layer characteristics, makes this device very attractive for Si optoelectronic IC fabrication.

MSM thin film high speed photodetector [2.405]. For the development of low-cost optical receivers for optical communications, the silicon technology has emerged as the best choice because its maturity, very well-developed fabrication process and lower fabrication cost for mass production. In order to obtain the proper absorption coefficient for the second and third fiber spectral windows (1.3–1.55 μm), the use of thin films of materials with the proper characteristics have been used. For operation at the wavelengths of 1.3 and 1.55 μm, it is proposed in this work the use of a-SiGe:H,F. On this material a MSM photo detector is constructed, and this structure was chosen because its inherent lower parasitic capacitance, hence a lower RC delay time and very simple fabrication process, because only a single mask is required. For operation at high speed the use of the transient photocurrent in amorphous hydrogenated semiconductors [2.406] is proposed. The transient photocurrent mechanism can be understood in terms of the large density of localized band tail and midgap states of the amorphous semiconductors. Therefore, the transport of carriers involves the frequent trapping, detrapping, and motion (transport in extended sates). When a photoconductive detector is built on such material, and then is illuminated by a light pulse, a photocurrent is generated. Two factors can reduce the free carrier density and cause photocurrent decay. If recombination happens, then a free carrier is lost, thus the photocurrent will be reduced, which happens when the carrier falls in a deep trap. However, when the carrier falls into a shallow trap it is reemitted soon after capture and a rapid decay is observed in the short period time of picoseconds. Physically this decay corresponds to the initial thermallization of the electrons in shallow states (tail states). Since the density of tails state is high near the band gap, the transient photocurrent decay is very fast. After this initial decay, the photocurrent decays gradually for a time, until deep trapping causes another sharp drop. Therefore, the initial thermallization or the deep trapping decay can be used to achieve a short response time of a device. The response time of the device can be adjusted between 1 and 100 ps for the initial thermallization decay. A material with a high density of midgap sates should be used for a short response time, and this condition can be achieved easily by adjusting the depositing conditions of an amorphous semiconductor. The fabrication process of an MSM is described as follows: As a starting material Si wafers of 2 in diameter were used as a mechanical support. After standard cleaning procedures were performed on the wafers, a SiN_x film was deposited by means of a PECVD system under the following deposition conditions: RF power 250 W, 0.6 Torr of pressure, deposition temperature of 260°C. Gases used are SiH_4, NH_3, and N_2. Thickness measured 105 nm. Without opening the deposition chamber, the temperature of the PECVD system was raised to 300°C to perform the a-SiGe:H deposition. GeF_4, SiH_4, and H_2 were used as the reactive gasses. An RF power of 200W was applied and the pressure of the system was 0.5 Torr.

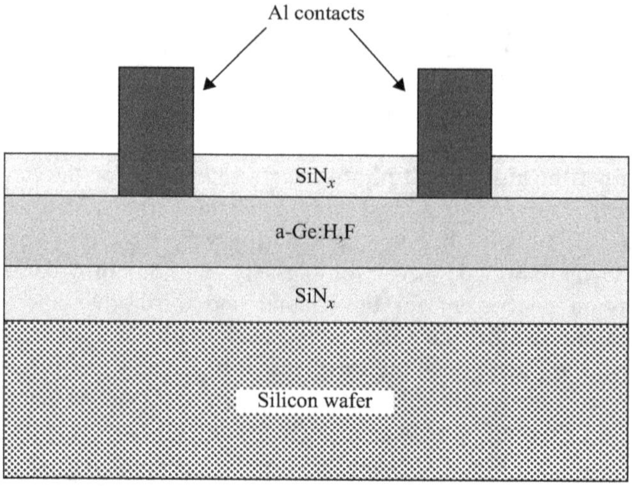

Figure 2.89 Cross-section of the fabricated device

The resulting thickness was 0.3 μm. Then, a second SiN_x layer of the same thickness as the first isolation layer is deposited on top of the a-Ge to act as an anti-reflecting layer. A finger mask containing different sizes and arrays of fingers is then placed on top of the system SiN_x/a-Ge:H,F/SiN_x, using positive photoresist, then the uncovered SiN_x is removed and a layer of Al is evaporated on the photoresist. By removing the photoresist, the finger pattern is placed on top of the a-SiGe. No alloying process for the metal contact to the amorphous layer was performed. A cross-section of the fabricated structure is depicted in Figure 2.89.

The devices were measured under DC bias in the range of ± 3 V under dark and illuminated conditions. The highest dark current measured at 3 V was 2.4×10^{-8} A, for the device with dimensions for the interdigitated area as 260 μm long, separation between fingers 40 μm and finger width 22 μm. When the device is illuminated with different monochromatic sources, it is noticed that as the wavelength increases, the photo generated current also increases. Figure 2.90 shows this behavior. It is seen that for a $\lambda = 0.94$ μm there is almost 2 orders of magnitude in current increase with respect to the dark case, and up to 3 orders of magnitude in the increase of the photo generated current when $\lambda = 1.06$ μm. The observed behavior is in agreement with the transmittance measurements performed on a sample of the system SiN_x/a-Ge:H,F/SiN_x in which no metal deposition was done. Figure 2.91 shows the measured transmittance.

In order to have an idea of the temporal response of the fabricated detector, the device was illuminated by means of a pulsed laser that delivers light pulses of 10 ns at 1060 nm of wavelength. An incident power of 4 mW was used to shine the detector without any bias that is in photovoltaic mode. The generated voltage across the detector was measured by means of an oscilloscope. The load for the probes was 50 W. The time scale that is shown in Figure 2.92 is not related to the pulse duration;

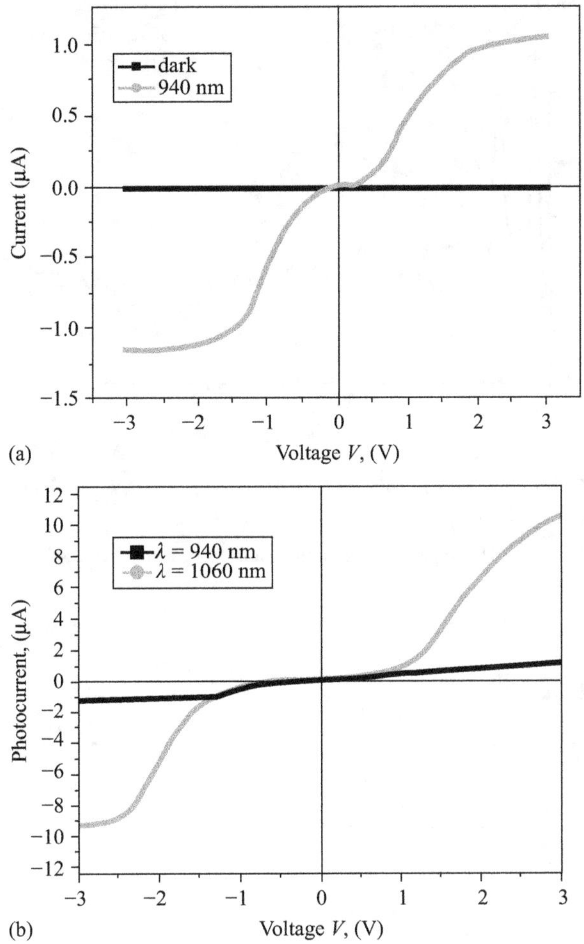

Figure 2.90 Dark and illuminated behavior of the MSM, (a) for illumination at $\lambda = 940$ nm, (b) for illumination at $\lambda = 940$ and 1060 nm

instead it represents the time-scale for the sampling rate (1.54 G samples/s). The measured characteristics under dark and illuminated conditions show that at only 3 V is necessary for the proper performance of the devices. Bias that agrees with the low power consumption requirements of the modern ICs. The device here presented has shown a good response for light pulses of 10 ns of duration.

Microbolometers. For IR imaging the thermal detector that has allowed the fabrication of cameras and video cameras of large format is the microbolometer. The operation of a microbolometer is based on the temperature rise of the thermo-sensing material by the absorption of the incident IR radiation. The change in temperature causes a change on its electrical resistance, which is measured by an external circuit. Microbolometers based on amorphous semiconductors have

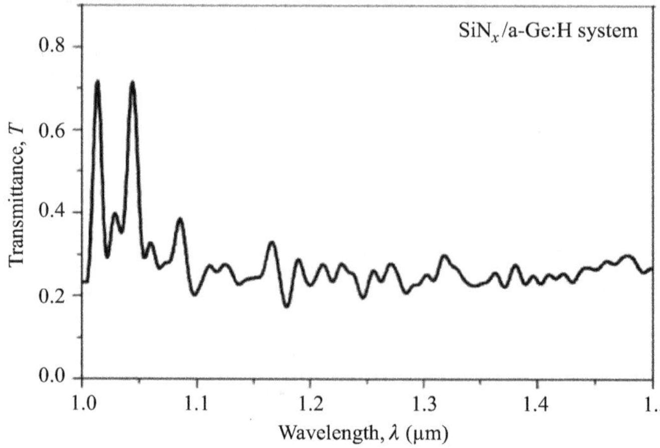

Figure 2.91 Transmittance measured on the system SiN_x/a-Ge:H,F/SiN_x in the range of 1–1.5 mm

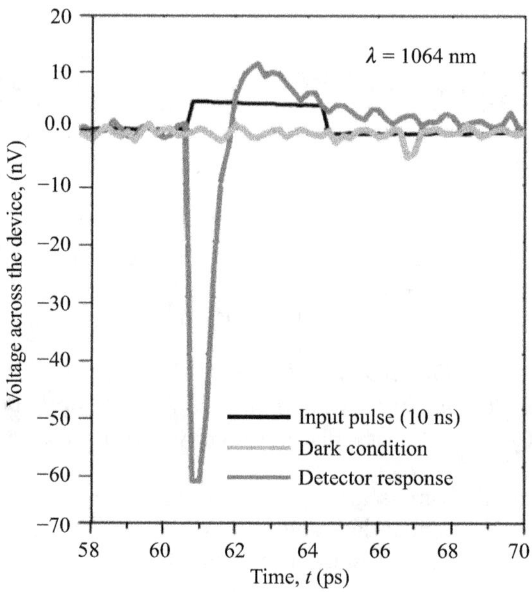

Figure 2.92 Response of the MSM to a 10 ns pulse of $\lambda = 1.064$ mm

advantages over other types of thermal detectors, including microbolometers that use other kind of thermo-sensing materials. The advantages are mainly technological, since these microbolometers are fully compatible with silicon CMOS fabrication technology, and there is no need of additional fabrication equipment in an IC production line as they are relatively of simple fabrication and can be processed at relatively low temperature by PECVD. The latter makes them ideal for a post-process fabrication over a CMOS read-out circuit. At INAOE we have been working on the materials for application of these devices from wavelengths ranging from 10 to 1 mm. But regardless of the wavelength of interest, the key parameter that determines the responsivity is the temperature coefficient of resistance (TCR). The TCR is commonly represented by $\alpha(T)$, which is defined as

$$\alpha(T) = \left(\frac{1}{R}\right)\left(\frac{dR}{dt}\right) \approx \frac{E_a}{kT^2} \tag{2.99}$$

where E_a is the activation energy, k the Boltzmann constant, and T. A large TRC value means that a small temperature change will result in a large change in the resistance of the material. a-Si:H and a-SiGe:H happen to be the set of materials that, by the proper depositing conditions, are able to cover the range of the IR spectra aforementioned.

Uncooled microbolometer. The device is a temperature-dependent resistor working at room temperature, which contains an IR absorbing film deposited on a thermosensing element. Those films are supported by a suspended membrane, which provides thermal isolation. The devices may be fabricated by using bulk micromachinining or through surface micromachining; the former is used for obtaining a better thermal isolation by placing the device on a floating thermal isolator. The later is used when an array of devices is required for getting the larger number of devices for a giver area of silicon, besides that this configuration allows to fabricate under it the conditioning circuitry for every pixel in a focal plane array [2.407]. These two configurations are illustrated in Figure 2.93.

For room temperature operation a-SiGe:H was studied, and resulted in a very attractive thermosensing material. It resulted in a high TCR when deposited from SiH_4 and GeH_4 in a Low Frequency PECVD system at 110 kHz, pressure of 0.6 Torr, and at substrate temperature of 573 K. The TCR measured on the device resulted dependent on the structure fabricated, for the bulk micromachined structure the film showed a TCR = 0.051 K^{-1} and for the surface micromachined structure TCR = 0.037 K^{-1} evidencing the better thermal isolation of the bulk micromachined device [2.407]. Nevertheless, the measured TCR on the fabricated devices resulted much larger than that reported for other devices; for instance Syllaios [2.408] reported a TCR measured on a device of 0.028 K^{-1}. Liang *et al.* [2.409] for poly-SiGe thermosensing film on uncooled microbolometers measured a TCR in the range of 0.014–0.022 K^{-1}.

In order of improving the performance of the surface micromachined bolometers, the position of the electrodes was also varied and two different structures resulted. These devices were named as planar and sandwich structures, in reference to the relative positioning of the electrodes, this is illustrated in Figure 2.94.

As a result the measured TCR on the devices is the same for both with a value of 0.043 K^{-1}, but because electrodes in the sandwich structure resulted in a very high field and a diode like behavior a high current is observed when the bolometer in this structure is illuminated under IR radiation and a gain in the photocurrent is observed

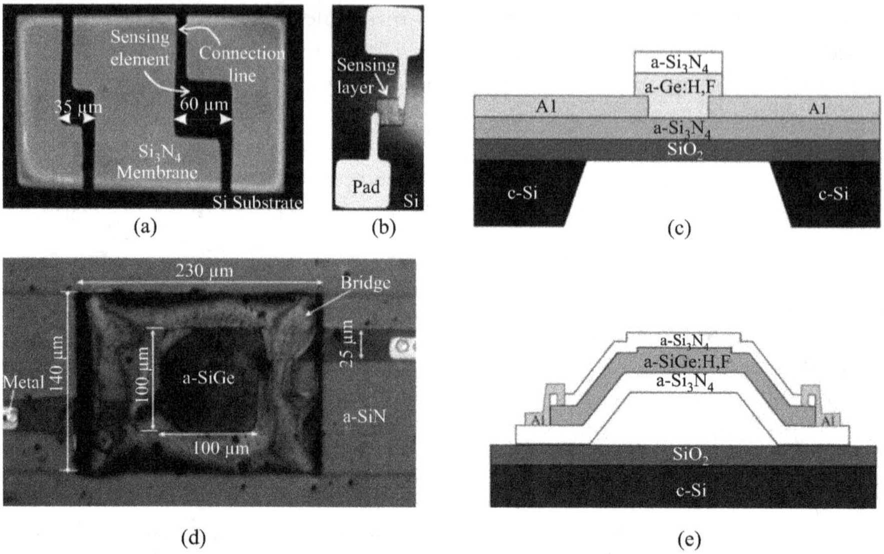

Figure 2.93 Membrane supported bolometer (a) backside view through SiO_2 membrane on a Si substrate. There are two structures of different size on the diaphragm. (b) Top view of a device. (c) Cross-section of the membrane supported microbolometer. (d) Top view of a surface micromachined bolometer. (e) Cross-section of the surface micromachined device

[2.410]. Most of the figures of merit are the same for both structures, the responsivity resulted to be 2×10^{-3} A/W for the planar structure and for the sandwich structure a responsivity dependent on the bias is observed and goes from 0.3 to 14 A/W. Also the thermal constant time for both devices is quite different being 0.1 ms and 125 ms for the planar and sandwich structures respectively [2.410].

Recently, it has been found that by varying the deposition parameters of a-Si:H, it is possible to form nanocrystals (of diameter of ~2–4 nm) distributed along the amorphous matrix. Such a material is commonly referred to as polymorphous silicon (pm-Si:H) [2.411]. The presence of nanocrystals impacts on the properties of the material by reducing the density of states (defects) and improving the transport properties (larger μe) and stability of the films. For a-Si:H thin-film solar cells, where degradation is an issue, pm-Si:H has been used instead of a-Si:H, and it has been demonstrated that this kind of cells suffer less of degradation due to light radiation (light soaking), such improvement is related to the presence of nanocrystals in the films [2.412].

For IR detectors and specifically for microbolometers, polymorphous semiconductors have not been reported, and we believe that its use as IR sensing films may represent a technological improvement, since those materials still preserve the characteristics of their amorphous counterpart, as a direct optical bandgap, large E_a,

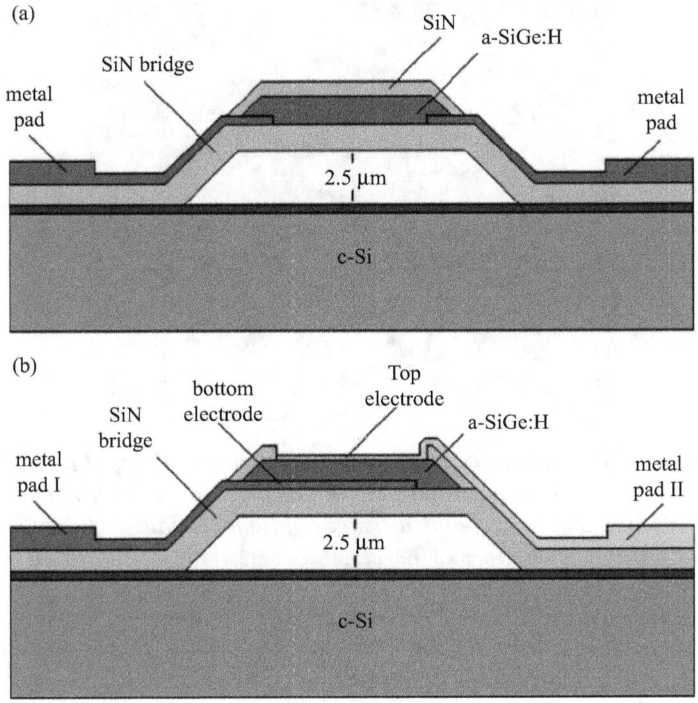

Figure 2.94 Microbolometer structure (a) planar and (b) sandwich contacts

and large TCR. In this aspect, we have studied the IR sensing properties of polymorphous silicon–germanium (pm-Si_xGe_y:H) thin films and have obtained promising results in terms of high E_a and TCR values with an improved σRT [2.413].

The pm-Si_xGe_y:H films were deposited from a SiH_4 (10% in H_2), GeH_4 (10% in H_2), and H_2 gas mixture, at a substrate temperature (T_s) of 200°C, in a capacitively coupled low-frequency PECVD reactor, working at 110 kHz, with an RF power density of 90 mW/cm^2. The films were characterized using high-resolution transmission electron microscopy (HRTEM), Figure 2.95 shows an HRTEM cross-sectional image of a pm-Si_x Ge_y:H film and as one can see, several nanocrystals of diameters in the range of 2–4 nm are distributed in the amorphous film. The presence of nanocrystals reduces the stress in the amorphous matrix, and consequently improves the stability of the Films.

The microbolometer fabricated with pm-Si_xGe_y:H resulted with the following advantages: a large TCR of $-6.6\%\ K^{-1}$, resulting in devices with high-performance characteristics, the voltage Responsivity R_u (9.2×10^5 V/W) and a specific detectivity D^* (2×10^9 cmHz$^{1/2}$/W) with relatively low electrical resistance ($1 \times 10^6\ \Omega$). Coupled with the above, the presence of small nanocrystals in the IR sensing films impacts on a reduction of defects, and improves the stability of the films against radiation [2.414].

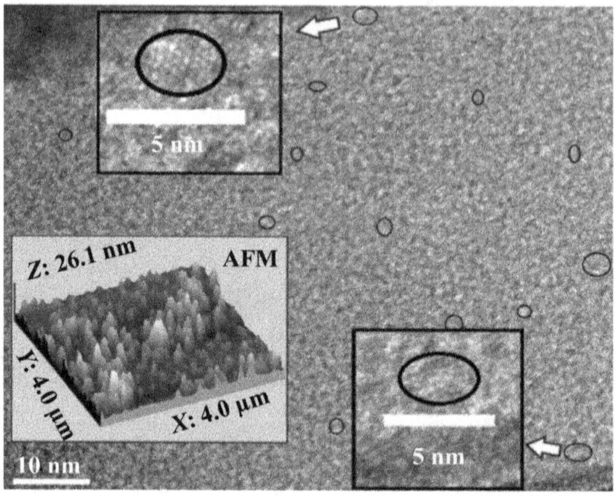

Figure 2.95 HRTEM image of a cross-section of a pm-Si_xGe_y:H film. Inset: 4 µm × 4 µm AFM image of the surface morphology of a pm-Si_xGe_y:H film, and two amplified images of nanocrystals of diameter of 4 nm
© 2015 IEEE. Reprinted with permission of [2.414]

Terahertz and millimeter wavelength microbolometer. Sensing and imaging using pulsed terahertz (THz) radiation have been widely recognized for reconstructing three-dimensional (3D) images of objects [2.415, 2.416]. Recently published works show implemented systems that operate at frequencies below 1 THz. There are two reasons for this: first, humidity interferes with the image due to its high absorption above 1 THz, and second, the ability to find something covered by cloths. THz falls between the RF and IR bands, and is a largely unexplored region of the electromagnetic spectrum with wavelengths ranging from 100 GHz to 10 THz. The technology for this frequency range has been used recently in biology, medicine, and non-destructive control of materials. One key issue for obtaining low-cost detectors using monolithic construction is their easy integration and compatibility with the CMOS technology, therefore we use a-Si-B:H as material sensor. The amorphous silicon boron doped was deposited at a substrate temperature of 250 K by using SiH_4 and B_2H_6 as source gases. The TRC measured at 150K is 0.085 K^{-1}. The current responsivity measured at 77 K is 1.17×10^{-2} A/W at 7 V DC bias, enough for resolution at THz frequencies [2.417]. The same thermosensing material is used in millimeter wavelength microbolometers, here the pixel size is 4 mm because this is the size recommended for avoiding diffraction losses operating in the single mode regime [2.418]. An array of 96 microbolometers was built in a 2 inches Si wafer and is shown in Figure 2.96 mounted in a Cu plate and bonded for testing and waiting for characterization at 4.2 K.

So far we have showed how an amorphous alloy of SiGe is used for very different applications, the depositing conditions were adjusted for obtaining the properties that best fit for the particular application. Additionally the versatility offered for the PECVD system is demonstrated and all the devices are designed for being incorporated to CMOS silicon technology.

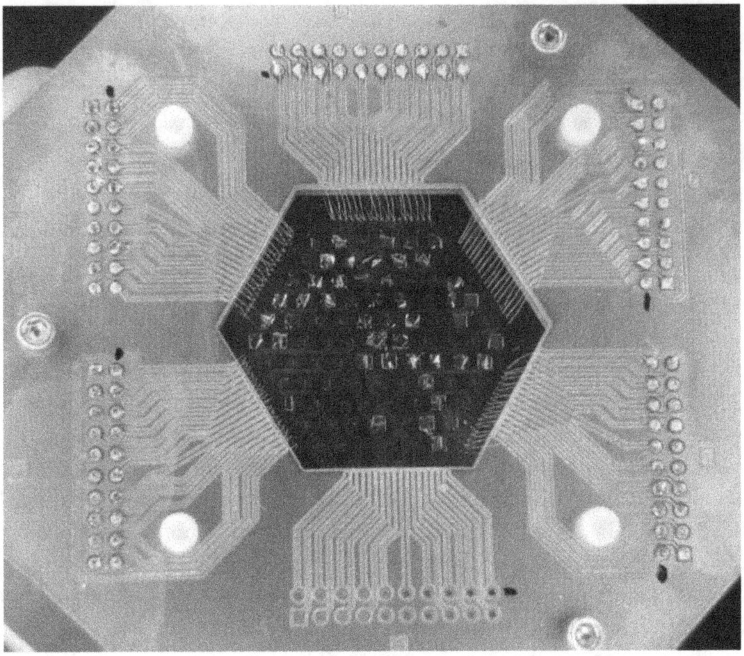

Figure 2.96 Array of 98 microbolometers placed on the Cu plate for testing at 4.2 K

2.4.4 CMOS process-compatible silicon-in-package (SiP)

System integration has shifted from a computational, PC-centric approach to a highly diversified mobile communication approach. The heterogeneous integration of multiple technologies in a limited space (e.g., GPS, phone, tablet, mobile phones) has truly revolutionized the semiconductor industry by shifting the main goal of any design from a performance driven approach to a reduced power driven approach. In few words, in the past performance was the one and only goal; today minimization of power consumption drives IC design.

This is demonstrated by the fact that SOC and SIP products have become the main drivers of the semiconductor industry as total volume of smart phones and tablets has surpassed production volumes of microprocessors in the past few years. The foundation of heterogeneous integration relies on the integration of "More Moore" (MM) devices with "More than Moore" (MtM) elements that add new functionalities (non-CMOS) that do not typically scale or behave according to "Moore's Law" [2.419]. At this point it is necessary to make the distinction between SOC and SIP.

System in Package uses the abbreviation SiP to differentiate itself from the old Single In-line Package (SIP). It is also sometimes called System in a Package. In its most simple definition, a SiP consists of active devices (one or multiple ICs), passive components and discrete devices designed and assembled into a standard or custom package to achieve a modular function previous accomplished by using several separated single chip packages. The SiP forms a functional block or module that can be used for board level manufacturing. The objectives of using SiPs are therefore to [2.420]

- Provide alternate and cheaper solution for SOC, as SOC sometimes take longer to fabricate and reach the market.
- Provide higher levels of integration and better electrical performance.
- Reduce overall assembly size and weight and achieve cost effectiveness.

No doubt SiPs are currently used most widely in portable consumer electronic devices such as digital still camera, music player, video recorders, and mobile phones where size and weight are premium and hence the main drivers for miniaturization.

SoC (IC integration to system-on-a-chip) continues to be the dream of all semiconductor companies. Computer and communication companies have driven this trend for decades through finer lithography, better materials, and larger chips and wafers, all leading to higher clock frequency. Cost was not a factor. But the world has changed, and the primary focus today is more functionality at an affordable cost. Even personal computer companies have moved on to dual – and multi-core processors that perform multiple functions on a single chip – spreadsheet on one quarter of the screen, photograph editing on another, watching a movie on the third, and video-conferencing on the fourth. The key challenges to SoC progress, however, are formidable, and include design and design verification, manufacturability, intellectual property (IP) and legal issues, time-to-market, and cost [2.421].

For achieving the objective of high-level interconnection, size reduction and high packing density, there are four essential enabling technologies that support the progress of SiP. These four technologies are [2.420]: Wafer thinning, Substrate technology, Interconnection, and Embedded Passives. Here the Substrate technology is the topic of discussion, and a very good evaluation of materials for SOP substrates is found in reference [2.422]. In this work, the authors state that the next-generation packaging involves dramatic reduction in size and cost, coupled with higher performance. This can be accomplished with increased integration of passives, RF, and optoelectronics components within the package. For achieving all of this an appropriate choice or design of materials is important, because the major building block for the SOP technology is ultra-high-density wiring within the dielectric layers using single-sided build-up. This high-density interconnection is typically fabricated by a sequential deposition of alternate layers of copper metallization and polymer dielectric on the baseboard. The coefficient of thermal expansion (CTE) mismatch between the dielectric layers and the baseboard, coupled with the curing strain, the sequencing and asymmetry of the building layers etc., induce severe stresses and warpage in the board during or after the process. Results derived from analytical and finite element modeling show that a board with sufficient stiffness is needed to prevent the warpage and hence enable the fineline wiring and microvia build-up process.

The other building block for the SOP concept is a reliable flip-chip technology involving interconnection of unpackaged ICs directly to low-cost boards for increasing the Si efficiency (area of Si within a package) and system level performance further. A chief concern is the CTE mismatch between the silicon die and the board. A silicon die has an approximate CTE of 2–3 ppm/°C, while conventional organic boards such as FR4 have a CTE ranging from 18 to 20 ppm/°C. The expansion mismatch induces plastic strain in the solder joint during operation resulting in low-cycle fatigue. All the materials chosen for the study are composite materials or ceramics, but nevertheless the conclusions of the comparison are challenging and can be summarized as follows:

- Matched CTE with Si die to prevent solder joint cracking without underfill.
- High enough elastic modulus to prevent cracking from cyclic warpage during heat cycle resulting from the CTE mismatch between the board and build-up dielectric layer. High elastic modulus is also necessary to reduce warpage during sequential build up process enabling ultra-high-density wiring of SOP.
- A dielectric material with low CTE, close to that of the board, and low modulus to minimize the dielectric stress and warpage caused by CTE mismatch between the base board and the dielectric. This can be another option to enhance SOP reliability performance from dielectric material design or selection standpoint.

Researchers at INAOE are proposing Teflon as the material for SiP because it offers all the advantages needed for this purpose, is a single component material, and has a very low tangent loss. The latter makes Teflon ideal for high-frequency applications because of its low dielectric constant and the offering of very low dielectric looses. Additionally if the Teflon surface is polished and an adherence to metals like Cu is improved, it also will result in low loss of the metal.

In order to proof our concept, a microstrip is going to be used [2.423]. The ohmic losses for a microstrip depend on DC resistance, AC resistance, and the real part of the impedance, Metal conductivity, frequency, and the average of roughness at the interface between metal and dielectric. In a conventional PCB fabrication process, the roughness is used for promoting adhesion of metals to the substrate. The roughness is an important factor in the ohmic losses especially in low-losses PCBs, it can enhance the AC resistance up to twice its value, and AC resistance is the most contributing part of the ohmic losses. Thus by reducing the roughness it can be reduced the ohmic losses. In our experiment we ended up a final roughness measured is of 30 nm, an argon plasma treatment on PTFE surface for Cu adherence was made and free bonds are created in the Teflon surface. Following 25 mm of electrodeposited Cu, the test of Copper Peel Strength was done with XLW (PC) Auto Tensile Tester and the result was 3.93 lbs/in, which is as good as some commercial PCB's or better. From S-parameters measurements we obtained for the microstrip fabricated an attenuation of only 2 Np/m at 20 GHz. One of the lowest values reported up to date.

Currently, researchers at INAOE are working on the fabrication of passive components as capacitors, inductors, and resistances on up to seven levels of metal using BCB as interdielectric metal. In the near future Teflon may be an optional material for SiP for very high frequency applications.

2.5 Conclusions

The silicon-based Field-Effect-device technology has evolved from the classical bulk approach based on SiO_2 gate dielectric, toward the most advanced and experimental devices, such as junctionless, tunnel-FET, NWs, SOI-FET, and Fin-FETs, where the FinFET being the mature device already in industry production for the most advanced 14 nm microprocessors. This chapter went through a review of these advanced FET-related devices, analyzing quantum effects.

The maturity of the FinFET technology has enabled a drastic reduction of the cost-per-device, or equivalently, the number of transistors that can be bought per dollar, to a number of 20 million transistors per dollar. However, this amazing evolution brought by the transistor size reduction, which increases the current drive capability and reduce the switching times making the device more functional and faster, is not for free. It comes with physical limitations due to the nanoscale quantum nature and fabrication technology limits, such as metal grain granularity, line edge roughness, RDFs. The electrical operation of advanced FET-related devices brought second-order undesirable effects, such as lack of electrostatic channel charge control, leakage currents, self-heating, that ultimately lead to device degradation reducing the performance and lifetime of electronic systems. The most common advanced FET-related device structures, its fabrication process, and operation characteristics have been reviewed in the first two sections of this chapter. The different approaches for fundamental physic-based device modeling and simulation have been introduced in Section 2.3.

Transport modeling for TCAD applications is a mature field of research with vast applications. Models of different complexity, precision, and accuracy are offered and implemented in various commercial and academic TCAD tools. Depending on parameter values and device scales, either a semi-classical or a quantum-mechanical transport description can be adopted. An MC method requires significant CPU resources and is now relatively rarely used for TCAD applications, when timely but perhaps less accurate results are of primary importance. MC methods can easily be extended and generalized to incorporate strain-induced modifications in the band structure and scattering rates, thus providing valuable input about the carrier mobility and other material parameters dependences on technological parameters and driving electric fields. These newly calibrated parameters can be used again in the drift-diffusion-based transport calculations extending their range of applicability. Spin transport in silicon can also be addressed with a diffusion-like equation with the spin relaxation term properly added.

In modern microelectronic devices quantum-mechanical effects become important and sometimes even dominant, which prompts for the development and use of quantum atomistic transport approaches. With shrinking device dimensions, the demand for fully three-dimensional accurate solvers for the coupled transport/ Poisson equations and atomistic based simulations has grown significantly. With the advances in computer architectures, increased computational power and memory capabilities, state-of-the art software, development of fast numerical algorithms and conceptually new generic simulation platforms a fundamental breakthrough in speed, reliability, and accuracy of multi-scale three-dimensional TCAD simulation tools is anticipated. However, with quantum corrections carefully added, the drift-diffusion based "atomistic" simulations provide often sufficiently accurate and timely statistical results for cutting edge ultra-scaled three-dimensional devices with fluctuation parameters including random dopant distribution. This extends the applicability of the drift-diffusion transport approach proving it again to be amazingly efficient even outside of its formal region of validity.

Finally Section 2.4 showed that the introduction of new materials and device structures has extended the CMOS platform to new applications domains, such as metrology for instance, where there is a plethora of radiation sensors that can benefit from amorphous Si and SiGe alloys, such as photodetectors and micro-machined bolometers that can be built into large arrays for high-resolution IR surveillance or detection. Nanostructured materials based on amorphous and SiGe alloys obtained from Plasma Enhanced Chemical Vapor Deposition (PECVD) are "Structure sensitive," which result in some advantages based on their short range bonding, rather than in the long-range order periodicity of crystalline materials. The structural "disorder" of amorphous materials influence the fundamental electronic properties. The strong scattering causes a large uncertainty in the electron momentum. The loss of momentum conservation implies that the E-k dispersion relations do not longer describe the energy bands. There is also no distinction between a direct and indirect band gap transition. These characteristics can be used to built CMOS-compatible high-efficient IR detectors, as the bolometer mentioned earlier.

Other materials, such as carbon for the fabrication of nanotubes have been considered due to its near ballistic transport properties at room temperature, that with a tight electrostatic control of its 1D channel may enable further downscaling beyond 10 nm. Graphene with its high carrier mobility close to 1×10^4 cm^2/Vs may be a good candidate for high frequency applications. However, graphene suffers from a major drawback; its zero bandgap, which results in small on/off current ratio and no stable saturation region in the output transistor characteristics. An option devised for graphene is its integration as a Hall sensor with a CMOS amplifier, for instance. The incorporation of graphene for building system-in-package is also another attractive route for graphene.

Gallium Nitride (GaN) is another attractive material with unique electronic properties, such as: high dielectric strength, high operating temperature, high current density, high speed switching and low on-resistance. These characteristics are due to the properties of GaN, which, compared to silicon, offers ten times higher electrical breakdown characteristics, three times the bandgap, and exceptional carrier mobility. But this material offers yet more advantages, the high electron mobility and charge density possible in AlGaN/GaN and InAlN/GaN heterostructures has enabled the demonstration of power amplifiers with at least one order of magnitude higher output power density than in their GaAs or Si-based counterparts. However, in spite of this excellent performance, nitride-based devices cannot compete with Si CMOS electronics in terms of cost, scalability, and circuit complexity. The seamless integration of these two semiconductor families would give the circuit and system designer unprecedented flexibility to use the best material and devices for each function.

Acknowledgments

Most of the work discussed in Section 2.2 has been performed at the Nanoelectronics Research Group at the Center for Research in Information and Communication Technologies (CITIC) of the University of Granada. Special thanks to our colleagues from Nanoelectronics Research Group: N. Rodriguez, C. Sampedro, L. Donetti, F. Ruiz, A. Godoy, J. L. Padilla, I. Tienda, B. Biel, C. Navarro, E. Gonzalez.

Viktor Sverdlov and Siegfried Selberherr acknowledge that Section 2.3 was supported by the European Research Council through Grant 247056 MOSILSPIN.

References

[2.1] Kahng, D. and Atalla, M. M. "Silicon-silicon dioxide field induced surface devices". In: IRE Solid-State Device Research Conference, Carnegie Institute of Technology, Pittsburgh, PA, USA, 1960.

[2.2] Kahng, D. "Electric field controlled semiconductor device". U.S. Patent No. 3,102,230 (Filed 31 May 31, 1960, issued August 27, 1963).

[2.3] http://www.computerhistory.org/semiconductor/timeline/1964-Commecial.html.

[2.4] Baker, R. J. CMOS: circuit design, layout, and simulation. 3rd ed. Wiley-IEEE, 2010, p. 7.

[2.5] Kilby, J. S. "Invention of the integrated circuit". IEEE Transactions on Electron Devices 1976;ED-23(7):648–654.

[2.6] Wanlass, F. M. and Sah, C. T. "Nanowatt logic using field-effect Metal-oxide semiconductor triodes". In: International Solid-State Circuits Conference, 1963.

[2.7] http://www-elec.inaoep.mx/lnn/estructura/LIMEMS.php.

[2.8] https://www.mosis.com/vendors/view/on-semiconductor/c5.

[2.9] Momose, H. S., Ono, M., Yoshitomi, T., *et al.*, "1.5 nm direct-tunneling gate oxide Si MOSFETs". IEEE Transactions on Electron Devices 1996;43(8):1233–1242.

[2.10] Huang, C.-L., Arora, N. D., Nasr, A. I., and Bell, D. A. "Effect of polysilicon depletion on MOSFET I–V characteristics". Electronics Letters 1993;29:1208–1209.

[2.11] Yang, N., Henson, W. K., and Wortman, J. J. "A comparative study of gate direct tunneling and drain leakage currents in N-MOSFET's with sub-2-nm gate oxides". IEEE Transactions on Electron Devices 2000;47 (8):1636–1644.

[2.12] Yamaguchi, K., Teshima, T., and Mizuta, H. "Numerical analysis of an anomalous current assisted by locally generated deep traps in pn junctions". IEEE Transactions on Electron Devices 1999;46(6):1159–1165.

[2.13] NG, K. K. and Taylor, G. W. "Effects of hot-carrier trapping in n- and p-channel MOSFET's". IEEE Transactions on Electron Devices 1983; ED-30(8):871–876.

[2.14] Gutiérrez-D., E. A. "Electrical performance of submicron CMOS technologies from 300 K down to 4.2", Chapter 2, PhD dissertation, Catholic University of Leuven, Belgium, 1993.

[2.15] Jenkins, K. A. and Rim, K. "Measurement of the effect of self-heating in strained-silicon MOSFETs". IEEE Electron Device Letters 2002;23 (6):360–362.

[2.16] www.globaltcad.com.

[2.17] Raleva, K., Vasileska, D., Hossain, A., Yoo, S.-K., and Goodnick, S. M. "Study of self-heating effects in SOI and conventional MOSFETs with electro-thermal particle-based device simulator". Journal of Computational Electronics 2012;11:106–117.

[2.18] Zhang, H., Hua, C., Ding, D., and Minnich, A. J. "Length dependent thermal conductivity measurements yield phonon mean free path spectra in nanostructures". Scientific Reports 2015; 5:9121, DOI:10.1038/srep09121.

[2.19] Bufler, F. M., Asahi, Y., Yoshimura, H., Zechner, C., Schenk, A., and Fichtner, W. "Monte Carlo simulation and measurement of nanoscale n-MOSFETs". IEEE Transactions on Electron Devices 2003;50(2):418–424.

[2.20] López-Villanueva, J. A., Cartujo-Casinello, P., Banquieri, J., Gámiz, F., and Rodríguez, S. "Effects if the inversion layer centroid on MOSFET behavior". IEEE Transactions on Electron Devices 1997;44(11):1915–1922.

[2.21] Hänsch, W., "The drift diffusion equation and its applications in MOSFET modeling", Springer-Verlag, New York, 1991, Chapter 2, pp. 48–65.

[2.22] Energy balance equations as stated in page 138 of GtsMinimosNT-Manual_01.pdf.

[2.23] Jones, W. and March, N. H. "Theoretical solid state physics". Courier Dover Publications, 1985.

[2.24] Wettstein, A. "Quantum effects in MOS devices". PhD dissertation, ETH, Switzerland, 2000.

[2.25] Ancona, M. G. "Density-gradient theory: a macroscopic approach to quantum confinement and tunneling in semiconductor devices". Journal of Computational Electronics 2011;10:65–97.

[2.26] Troutman, R. R. "VLSI limitations from drain-induced barrier lowering". IEEE Journal of Solid-State Circuits 1979;SC-14(2):383–391.

[2.27] Hsu, F.-C. and Chiu, K.-Y. "Evaluation of LDD MOSFET's based on hot-electron-induced degradation". IEEE Electron Device Letters 1984;EDL-5 (5):162–165.

[2.28] Hu, G. J., Chang, C., and Chia, Y.-T. "Gate-voltage-dependent effective channel length and series resistance of LDD MOSFET's". IEEE Transactions on Electron Devices 1987;ED-34(12):2469–2475.

[2.29] Ogura, S., Tsang, P. J., Walker, W. W., Critchlow, D. L., and Shepard, J. F. "Design and characteristics of the lightly doped drain-source (LDD) insulated gate field-effect transistor". IEEE Journal of Solid-State Circuits 1980;SC-15(4):424–432.

[2.30] Chang, W.-H., Davari, B., Wordeman, M. R., Taur, Y., Hsu, C.-H., and Rodriguez, M. D. "A high-performance 0.25-μm CMOS technology: I-Design and characterization". IEEE Transactions on Electron Devices 1992;39(4):959–966.

[2.31] Taur, Y., Wind, S., Mii, Y. J., *et al.*, "High performance 0.1 μm CMOS devices with 1.5 V power supply". In: Proceedings of the IEDM, 1993, pp. 127–130.

[2.32] De, I. and Osburn, C. M. "Impact of super-steep-retrograde channel doping profiles on the performance of scaled devices". IEEE Transactions on Electron Devices 1999;46(8):1711–1717.

[2.33] Morimoto, T., Momose, H. S., Ozawa, Y., Yamabe, K., and Iwai, H. "Effects of boron penetration and resultant limitations in ultra thin pure-oxide and nitride-oxide gate-films". In: Proceedings of the IEDM, 1990, pp. 429–432.

[2.34] Wong, S. S., Bradbury, D. R., Chen, D. C., and Chiu, K. Y. "Elevated Source/Drain MOSFET". In: Proceedings of the IEDM, 1984, pp. 634–637.

[2.35] Izuka, T., Chiu, K. Y., and Moll, J. L. "Double threshold MOSFET's in bird's beak free structure". In: Proceedings of the IEDM, 1981, pp. 380–383.

[2.36] Gross, M. E., Lingk, C., Brown, W. L., and Drese, R. "Implications of damascene topography for electroplated copper interconnects". Solid State Technology 1999:47–52.

[2.37] Aipperspach, A. G., Allen, D. H., Cox, D. T., Phan, N. V., and Storino, S. "A 0.2-μm, 1.8-V, SOI, 550-MHz, 64-b PowerPC microprocessor with copper interconnects". IEEE Journal of Solid-State Circuits 1999;34 (11):1430–1435.

[2.38] Park, S. B., Kim, Y. W., Ko, Y. G., *et al.*, "A 0.25-μm, 600-MHz, 1.5-V, Fully Depleted SOI CMOS 64-bit Microprocessor". IEEE Journal of Solid-State Circuits 1999;34(11):1436–1444.

[2.39] Colinge, J. P. Silicon-on-insulator technology: materials to VLSI. Kluwer Academic Press, 1997. ISBN 0-7923-8007-X.

[2.40] Wharam, D. A., Thornton, T. J., Newbury, R., *et al.*, "One-dimensional transport and the quantisation of the ballistic resistance". Journal of Physics C: Solis State Physics 1988;21:L209–L214.

[2.41] Sikorski, C. and Merkt, U. "Spectroscopy of electronic states in InSb quantum dots". Physical Review Letters 1989;62(18):2164–2167.

[2.42] Zhu, S., Chen, J., Li, M.-F., *et al.*, "N-type Schottky source/drain MOSFET using Ytterbium silicide". IEEE Electron Device Letters 2004;25(8): 565–567.

[2.43] Kim, K. R., Kim, D. H., Song, K.-W., *et al.*, "Silicon-based field-induced band-to-band tunneling effect transistor". Electron Device Letters 2004;25 (6):439–441.

[2.44] Shang, H., Jain, S., Josse, E., *et al.*, "High performance bulk planar 20 nm CMOS technology for low power mobile applications". In: Symposium on VLSI Technology Digest of Technical Papers, 2012, pp. 129–130.

[2.45] Jan, C.-H., Agostinelli, M., Deshpande, H., *et al.*, "RF CMOS technology scaling in high-k/metal gate era for RF SoC (System-on-Chip) applications". In: Proceedings of the IEDM, 2010, pp. 604–607.

[2.46] Jiang, X., Ramachandran, N. P., Kang, D. W., *et al.*, "Digitally-assisted analog and analog-assisted digital design techniques for a 28 nm mobile System-on-Chip". In: Proceedings of the ESSCIRC, 2014 Conference, pp. 475–478.

[2.47] Jiang, X., Yu, X., Lin, F., *et al.*, "A 28 nm analog and audio mixed-signal front end for 4G/LTE cellular System-on-Chip". In: Proceedings of the ESSCIRC, 2014 Conference, pp. 471–474.

[2.48] Sakuma, K., Andry, P. S., Dang, B., *et al.*, "3D chip stacking technology with low-volume lead-free interconnections". In: Electronic Components and Technology Conference, 2007, pp. 627–632.

[2.49] Han, Y., Lau, B. L., Zhang, H. Z., and Zhang, X. "Package-level Si-based micro-jet impingement cooling solution with multiple drainage micro-trenches". In: IEEE 16th Electronics Packaging Technology Conference (EPTC), 2014, pp. 330–334.

[2.50] Cheng, K., Khakifirooz, A., Kulkarni, P., *et al.*, "Extremely thin SOI (ETSOI) CMOS with record low variability for low power System-on-Chip applications". In: Proceedings of the IEDM, 2009, pp. 49–52.

[2.51] Auth, C., Allen, C., Blattner, A., *et al.*, "A 22 nm high performance and low-power CMOS technology featuring fully-depleted tri-gate transistors, self-aligned contacts and high density MIM capacitors". In: Symposium on VLSI Technology Digest of Technical Papers, 2012, pp. 131–132.

[2.52] Natarajan, S., Agostinelli, M., Akbar, S., *et al.*, "A 14 nm logic technology featuring 2nd – generation FinFet transistors, air-gapped interconnects, self-aligned double patterning and a 0.0588 μm^2 SRAM cell size". In: Proceedings of the IEDM, 2014, pp. 71–73.

[2.53] Lin, C.-H., Greene, B., Narashima, S., *et al.*, "High performance 14 nm SOI FinFet CMOS technology with 0.0174 μm^2 embedded DRAM and 15 levels of Cu metallization". In: Proceedings of the IEDM, 2014, pp. 74–76.

[2.54] Ho, B., Sun, X., Xu, N., *et al.*, "First demonstration of quasi-planar segmented-channel MOSFET design for improved scalability". IEEE Transactions on Electron Devices 2012;59(8): 2273–2276.

[2.55] De Marchi, M., Sacchetto, D., Frache, S., *et al.*, "Polarity control in double-gate, gate-all-around vertically stacked silicon nanowire FETs". In: Proceedings of the IEDM, 2012, pp. 183–186.

[2.56] Park, C.-H., Ko, M.-D., Kim, K.-H., *et al.*, "Electrical characteristics of 20-nm junctionless Si nanowire transistors". Solid-State Electronics 2012;73:7–10.

[2.57] Zhuang, L., Guo, L., and S. Chou, Y. "Silicon single-electron quantum-dot transistor switch operating at room temperature". Applied Physics Letters 1998;72(10):1205–1207.

[2.58] Appelbaum, I., Huang, B., and Monsma, D. J. "Electronic measurement and control of spin transport in silicon". Nature 2007;447:295–298.

[2.59] Windbacher, T., Ghosh, J., Makarov, A., Sverdlov, V., and Selberherr, S. "Modelling of multipurpose spintronic devices". International Journal of Nanotechnology 2015;12(¾):313–331.

[2.60] Buccafurri, E. "Analytical modeling of silicon based resonant tunneling diode for RF oscillator application". PhD dissertation, Institut National Des Sciences Appliquées de Lyon (INSA), France, 2010.

[2.61] Schwierz, F. "Graphene transistors". Nature Nanotechnology 2010; 5:487–496.

[2.62] Novoselov, K. S., Geim, A. K., Morozov, S. V., *et al.*, "Electric field effect in atomically thin carbon films". Science 2004;306:666–669.

[2.63] Moore, G. "Cramming more components into integrated circuits". Electronics 1965;38(8).

[2.64] Sze, S. M. and Lee M. K. Semiconductor devices: physics and technology. 3rd ed. John Wiley & Sons, 2013.

[2.65] Jacoboni, C., Canali, C., Ottaviani, G., and Quaranta, A. A. "A review of some charge transport properties of silicon". Solid-State Electronics 1977;20:77–89.

[2.66] Gamiz, F., Lopez-Villanueva, J., Banqueri, J., Carceller, J. E., and Cartujo, P. "Universality of electron mobility curves in MOSFETs: a Monte Carlo study". IEEE Transactions on Electron Devices 1995;42(2):258–265.

[2.67] Tang, S., Wallace, R. M., Seabaugh, A., and King-Smith, D. "Evaluating the minimum thickness of gate oxide on silicon using first-principles method". Applied Surface Science 1998;135(1):137–142.

[2.68] Huff, H. and Gilmer, D. (Eds.). High dielectric constant materials: VLSI MOSFET applications, Vol. 16. Springer Science & Business Media, 2009.

[2.69] Wong, K., Shiraishi, K., Kakushima, K., and Iwai, H. "High-K gate dielectrics". In: Deleonibus, S. Electronic Devices Architectures for the NANO-CMOS Era. Pan Stanford Publishing, 2009.

[2.70] Andrieu, F., Faynot, O., Rochette, F., *et al.*, "Impact of Mobility Boosters (XsSOI, CESL, TiN gate) on the Performance of "100" or "110" oriented FDSOI cMOSFETs for the 32 nm Node". In: IEEE Symposium on VLSI Technology, 2007, pp. 50–51.

[2.71] Uchida, K., Zednik, R., Lu, C. H., *et al.*, "Experimental study of biaxial and uniaxial strain effects on carrier mobility in bulk and ultrathin-body SOI MOSFETs". In: IEEE International Electron Devices Meeting, 2004. IEDM Technical Digest, pp. 229–232.

[2.72] Chandrakasan, A. P., Sheng, S., and Brodersen, R. W. "Low-power CMOS digital design". IEICE Transactions on Electronics 1992;75 (4):371–382.

[2.73] Roy, K., Mukhopadhyay, S., and Mahmoodi-Meimand, H. "Leakage current mechanisms and leakage reduction techniques in deep-submicrometer CMOS circuits". Proceedings of the IEEE 2003;91(2):305–327.

[2.74] Kawaura, H., Sakamoto, T., and Baba, T. "Observation of source-to-drain direct tunneling current in 8 nm gate electrically variable shallow junction metal–oxide–semiconductor field-effect transistors" Applied Physics Letters 2000;76(25):3810–3812.

[2.75] Liu, D. and Svensson, C. "Power consumption estimation in CMOS VLSI chips". IEEE Journal of Solid-State Circuits 1994;29(6):663–670.

[2.76] Taur, Y. and Ning, T. H., Fundamentals of modern VLSI devices. Cambridge University Press, 2009.

[2.77] Asenov, A. and Cheng, B. "Modeling and simulation of statistical variability in nanometer CMOS technologies". In: Casier, Herman, Steyaert, Michiel, van Roermund, and Arthur H. M. (Eds.), Analog Circuit Design, Robust Design, Sigma Delta Converters, RFID, Springer, 2011.

[2.78] Bernstein, K., Frank, D. J., Gattiker, A. E., *et al.*, "High-performance CMOS variability in the 65-nm regime and beyond". IBM Journal of Research and Development 2006;50(4.5):433–449.

[2.79] Brown, A. R., Roy, G., and Asenov, A. "Poly-Si-gate-related variability in decananometer MOSFETs with conventional architecture" IEEE Transactions on Electron Devices 2007;54(11):3056–3063.

[2.80] Cheng, B., Roy, S., and Asenov, A. "The impact of random doping effects on CMOS SRAM cell". In: Proceeding of the 30th European Solid-State Circuits Conference (ESSCIRC), 2004, pp. 219–222.

[2.81] Agarwal, A., Chopra, K., Blaauw, D., and Zolotov, V. "Circuit optimization using statistical static timing analysis". In: Proceedings of the 42nd Annual Design Automation Conference, 1995, pp. 321–324. ACM.

[2.82] Asenov, A. "Random dopant induced threshold voltage lowering and fluctuations in sub-0.1 μm MOSFET's: A 3-D "atomistic" simulation study" IEEE Transactions on Electron Devices 1998;45(12):2505–2513.

[2.83] Asenov, A., Brown, A. R., Davies, J. H., Kaya, S., and Slavcheva, G. "Simulation of intrinsic parameter fluctuations in decananometer and nanometer-scale MOSFETs." IEEE Transactions on Electron Devices 2003;50(9):1837–1852.

[2.84] Markov, S., Cheng, B., Zain, A. S. M., and Asenov, A. "Understanding variability in complementary metal oxide semiconductor (CMOS) devices manufactured using silicon-on-insulator (SOI) technology" In: Silicon-On-Insulator (SOI) Technology: Manufacture and Applications, Springer, 2014, pp. 212–242.

[2.85] Fossum, J. G. and Trivedi, V. P. Fundamentals of Ultra-thin-body MOSFETs and FinFETs. Cambridge University Press, 2013.

[2.86] Rabaey, J. M., Chandrakasan, A. P., and Nikolic, B. Digital integrated circuits, Vol. 2. Prentice Hall, Englewood Cliffs, 2003.

[2.87] Cristoloveanu, S. and Li, S. Electrical characterization of silicon-on-insulator materials and devices, Vol. 305. Kluwer Academic Press, 1995.

[2.88] Celler, G. and Wolf, M. Smart Cut™ A guide to the technology, the process, the products. Parc Technologique des Fontaines. SOITEC, 2003.

[2.89] Bernstein, K. and Rohrer, N. J. SOI circuit design concepts. Springer Science & Business Media. Springer, 2007.

[2.90] Rodriguez, N., Navarro, C., Andrieu, F., Faynot, O., Gamiz, F., and Cristoloveanu, S. "Self-heating effects in ultrathin FD SOI transistors". In: IEEE International SOI Conference (SOI), 2011, pp. 1–2.

[2.91] Lim, H. K. and Fossum, J. G. "Threshold voltage of thin-film silicon-on-insulator (SOI) MOSFET's." IEEE Transactions on Electron Devices 1983;30(10):1244–1251.

[2.92] Wainwright, S. P., Hall, S., and Flandre, D. "The effect of series resistance on threshold voltage measurement techniques for fully depleted SOI MOSFETs" Solid-State Electronics 1996;39(1):89–94.

[2.93] Colinge, J. P. Silicon-on-Insulator Technology: Materials to VLSI: Materials to VLSI. Kluwer Academic Press, 2004.

[2.94] Okhonin, S., Nagoga, M., Sallese, J. M., and Fazan, P. "A SOI capacitor-less 1T-DRAM concept". In: IEEE International SOI Conference, 2001, pp. 153–154.

[2.95] Bawedin, M., Cristoloveanu, S., Flandre, D., and Udrea, F. "Floating-body memory: concepts, Physics and challenges". In Meeting Abstracts of the Electrochemical Society 2009;23:960–960.

[2.96] Morishita, F., Hayashi, I., Gyohten, T., *et al.*, "A capacitorless twin-transistor random access memory (TTRAM) on SOI". IEICE Transactions on Electronics 2007;90(4):765–771.

[2.97] Waite, A. M., Lloyd, N. S., Ashburn, P., *et al.*, "Raised source/drain (RSD) for 50nm MOSFETs – effect of epitaxy layer thickness on short channel effects", ESSDERC 2003 Conference, September 2003, pp. 223–226.

[2.98] Colinge, J. P., Park, J. T., and Colinge, C. A. "SOI devices for sub-0.1 μm gate lengths". In: IEEE 23rd International Conference on MIEL, 2002, Vol. 1, pp. 109–113.

[2.99] Yoshimi, M., Hazama, H., Takahashi, M., Kambayashi, S., Wada, T., and Tango, H. "Two-dimensional simulation and measurement of high-performance MOSFETs made on a very thin SOI film" IEEE Transactions on Electron Devices 1989;36(3):493–503.

[2.100] Celler, G. K. and Cristoloveanu, S. "Frontiers of silicon-on-insulator". Journal of Applied Physics 2003;93(9):4955–4978.

[2.101] Cauchy, X. and Andrieu, F. "Questions and answers on fully depleted SOI technology for next generation CMOS node". SOI Industry Consortium 2010;1–17.

[2.102] Cauchy, X. "Fully Depleted SOI Designed for low power" SOI Industry Consortium 2010;1–8.

[2.103] Yoshino, A., Kumagai, K., Hamatake, N., Kurosawa, S., and Okumura, K. "Comparison of fully depleted and partially depleted mode transistors for practical high-speed, low-power 0.35 μm CMOS/SIMOX circuits". In: International SOI Conference, 1994, pp. 107–108.

[2.104] Stadele, M., Schmitt-Landsiedel, D., and Risch, L. "Comparison of partially and fully depleted SOI transistors down to the sub-50-nm gate length regime". In: Proceedings of the International Symposium on Silicon-on-Insulator Technology and Devices XI, The Electrochemical Society, 2003, p. 361.

[2.105] Marshall, A. and Natarajan, S. "PD-SOI and FD-SOI: a comparison of circuit performance". In: IEEE 9th International Conference on Electronics, Circuits and Systems, 2002, Vol. 1, pp. 25–28.

[2.106] Bawedin, M., Cristoloveanu, S., and Flandre, D., Novel capacitor-less 1T-DRAM using MSD effect. In: IEEE International SOI Conference, 2006, pp. 109–110.

[2.107] Gamiz, F., Rodriguez, N., Marquez, C., Navarro, C., and Cristoloveanu, S., A2RAM: Low-power 1T-DRAM memory cells compatible with planar and 3D SOI substrates. In: IEEE SOI-3D-Subthreshold Microelectronics Technology Unified Conference (S3S), 2014, pp. 1–2.

[2.108] Gamiz, F., Rodriguez, N., and Cristoloveanu, "Advanced concepts for floating-body memories". International Journal of High Speed Electronics and Systems 2012;21(01):1250002.

[2.109] Ulicki, B. and Reiter, H. "De-myth-tifying the SOI Floating body effect" SOI Industry Consortium 2009:1–7.

[2.110] Chen, C.-D., Matloubian, M., Sundaresan, R., Mao, B.-Y., Wei, C. C., and Pollack, G. P. "Single-transistor latch in SOI MOSFETs". IEEE Electron Device Letters 1988;9(12):636–638.

[2.111] Suh, D. and Fossum, J. G. "Dynamic floating-body instabilities in partially depleted SOI CMOS circuits". In: IEEE International Electron Devices Meeting, 1994, pp. 661–664.

[2.112] Ouisse, T., Ghibaudo, G., Brini, J., Cristoloveanu, S., and Borel, G. "Investigation of floating body effects in silicon-on-insulator metal-oxide-semiconductor field-effect transistors". Journal of Applied Physics 1991;70(7):3912–3919.

[2.113] Wei, A. Sherony, M. J., and Antoniadis, D. A. "Transient behavior of the kink effect in partially-depleted SOI MOSFET's". IEEE Electron Device Letters 1995;16(11):494–496.

[2.114] Tack, M. R., Gao, M., Claeys, C. L., and Declerck, G. J. "The multistable charge-controlled memory effect in SOI MOS transistors at low temperatures". IEEE Transactions on Electron Devices 1990;37 (5):1373–1382.

[2.115] Wann, H. J. and Hu, C. "A capacitorless DRAM cell on SOI substrate". In: IEEE International Electron Devices Meeting, 1993, pp. 635–638.

[2.116] Ioannou, D. E., Cristoloveanu, S., Mukherjee, M., and Mazhari, B. "Characterization of carrier generation in enhancement-mode SOI MOSFET's". IEEE Electron Device Letters 1990;11(9):409–411.

[2.117] Atlas user's manual. Silvaco International Software, Santa Clara, CA, USA, 2015.

[2.118] Zhang, Y., Schroder, D. K., Shin, H., Hong, S., Wetteroth, T., and Wilson, S. R. "Abnormal transconductance and transient effects in partially depleted SOI MOSFETs". Solid-State Electronics 1999;43 (1):51–56.

[2.119] Eaton, S. and Lalevic, B. "The effect of operating frequency on propagation delay in silicon-on-sapphire digital integrated circuits". In: IEEE International Electron Devices Meeting, 1976, Vol. 22, pp. 192–194.

[2.120] Sinitsky, D., Tu, R., Liang, C., Chan, M., Bokor, J., and Hu, C. "AC output conductance of SOI MOSFETs and impact on analog applications". IEEE Electron Device Letters 1997;18(2):36–38.

[2.121] Lim HK and Fossum JG. "Transient drain current and propagation delay in SOI CMOS circuits". IEEE Transactions on Electron Devices 1984;31 (9):1251–1258.

[2.122] Balestra, F., Jomaah, J., Ghibaudo, G., Faynot, O., Auberton-Herve, A. J., and Giffard, B. "Analysis of the latch and breakdown phenomena in N and P channel thin film SOI MOSFET's as a function of temperature". IEEE Transactions on Electron Devices 1994;41(1):109–112.

[2.123] Chen, C. D., Matloubian, M., Sundaresan, R., Mao, B. Y., Wei, C. C., and Pollack, G. P. "Single-transistor latch in SOI MOSFETs. IEEE Electron Device Letters 1998;9(12):636–638.

[2.124] Wan, J., Le Royer, C., Zaslavsky, A., and Cristoloveanu, S. "Gate-induced drain leakage in FD-SOI devices: what the TFET teaches us about the MOSFET". Microelectron Engineering 2011;88(7):1301–1304.

[2.125] Luyken, R., Hartwich, J., Specht, M., *et al.*, "Impact ionization and band-to-band tunneling in ultra-thin body SOI devices with undoped channels". In: IEEE International SOI Conference, 2003, pp. 166–167.

[2.126] Chang, L., Yang, K. J., Yeo, Y.-C., Polishchuk, I., King, T.-J., and Hu, C. M. "Direct-tunneling gate leakage current in double-gate and ultra-thin body MOSFETs". IEEE Transactions on Electron Devices 2002;49 (12):2288–2295.

[2.127] Pretet, J., Matsumoto, T., Poiroux, T., *et al.*, "New mechanism of body charging in partially depleted SOI-MOSFETs with ultra-thin gate oxides," In: Proceedings of the 32nd European Solid-State Device Research Conference, 2002, pp. 515–518.

[2.128] Kuo, J. B. and Lin, S. C. Low-voltage SOI CMOS VLSI devices and circuits. John Wiley & Sons, 2004.

[2.129] Burnett, D. "Partially depleted (PD) silicon-on-insulator (SOI) technology: circuit solutions". In: Kononchuk, O. and Nguyen, B. Y. (Eds.). Silicon-on-insulator (SOI) technology: Manufacture and applications. Elsevier, 2014.

[2.130] Tsuchiya, T., Sato, Y., and Tomizawa, M. "Three mechanisms determining short-channel effects in fully-depleted SOI MOSFETs". IEEE Transactions on Electron Devices 1998;45(5):1116–1121.

[2.131] Trivedi, V. P. and Fossum, J. G. "Scaling fully depleted SOI CMOS". IEEE Transactions on Electron Devices 2003;50(10):2095–2103.

[2.132] Ouisse, T., Cristoloveanu, S., and Borel, G. "Influence of series resistances and interface coupling on the transconductance of fully-depleted silicon-on-insulator MOSFETs". Solid-State Electronics, 1992;35 (2):141–149.

[2.133] Balestra, F., Benachir, M., Brini, J., and Ghibaudo, G. "Analytical models of subthreshold swing and threshold voltage for thin-and ultra-thin-film SOI MOSFETs." IEEE Transactions on Electron Devices 1990;37(11):2303–2311.

[2.134] Omura, Y., Horiguchi, S., Tabe, M., and Kishi, K. "Quantum-mechanical effects on the threshold voltage of ultrathin-SOI nMOSFETs". IEEE Electron Device Letters 1993;14(12):569–57.

[2.135] Cristoloveanu, S. "From SOI Basics to Nano-Size MOSFETs". In: Nanotechnology for Electronic Materials and Devices, Springer, 2007, pp. 67–104.

[2.136] Ernst, T., Tinella, C., Raynaud, C., and Cristoloveanu, S. "Fringing fields in sub-0.1 μm fully depleted SOI MOSFETs: optimization of the device architecture" Solid-State Electronics 2002;46(3):373–378.

[2.137] Cristoloveanu, S., Ernst, T., Munteanu, D., and Ouisse, T. "Ultimate MOSFETs on SOI: ultra thin, single gate, double gate, or ground plane". International Journal of High Speed Electronics and Systems 2000;10 (01):217–230.

[2.138] Berger, M. and Chai, Z. "Estimation of heat transfer in SOI-MOSFETs". IEEE Transactions on Electron Devices 1991;38(4):871–875.

[2.139] Su, L. T., Chung, J. E., Antoniadis, D., Goodson, K. E., and Flik, M. "Measurement and modeling of self-heating in SOI nMOSFET's". IEEE Transactions on Electron Devices 1994;41(1):69–75.

[2.140] Su, L. T., Goodson, K. E., Antoniadis, D. A., Flik, M. I., and Chung, J. E. "Measurement and modeling of self-heating effects in SOI nMOSFETs." In: IEEE International Electron Devices Meeting, 1992, pp. 357–360.

[2.141] Tenbroek, B. M., Lee, M. S. L., Redman-White, W., Edwards, C. F., Bunyan, R. J. T., and Uren, M. J. "Measurement and simulation of self-heating in SOI CMOS analogue circuits". In: IEEE International SOI Conference, 1997, pp. 156–157.

[2.142] Roldan, J. B., Gámiz, F., Lopez-Villanueva, J. A., and Cartujo-Cassinello, P. "Deep submicrometer SOI MOSFET drain current model including series resistance, self-heating and velocity overshoot effects". IEEE Electron Device Letters 2000;21(5):239–241.

[2.143] Ando, T., Fowler, A. B., and Stern, F. "Electronic properties of two-dimensional systems". Reviews of Modern Physics 1982;54(2):437.

[2.144] Gámiz, F., López-villanueva, J., Roldán, J. B., Carceller, J. E., and Cartujo, P. "Monte Carlo simulation of electron transport properties in extremely thin SOI MOSFET's". IEEE Transactions on Electron Devices 1998;45(5):1122–1126.

[2.145] Gamiz, F., Roldán, J. B., Cartujo-Cassinello, P., Carceller, J. E., López-Villanueva, J. A., and Rodriguez, S. "Electron mobility in extremely thin single-gate silicon-on-insulator inversion layers". Journal of Applied Physics 1999;86(11):6269–6275.

[2.146] Gamiz, F., Roldan, J., Lopez-Villanueva, J., Cartujo-Cassinello, P., and Carceller, J. "Surface roughness at the SiSiO2 interfaces in fully depleted silicon-on-insulator inversion layers". Journal of Applied Physics 1999;86(12):6854–6863.

[2.147] Gamiz, F. and Fischetti, M. V. "Monte Carlo simulation of double-gate silicon-on-insulator inversion layers: The role of volume inversion". Journal of Applied Physics 2001;89(10):5478–5487.

[2.148] Price, P. "Two-dimensional electron transport in semiconductor layers. I. Phonon scattering". Annals of Physics 1981;133:217–239.

[2.149] Gamiz, F., Roldan, J., and Godoy, A. "Phonon-limited electron mobility in ultrathin silicon-on-insulator inversion layers". Journal of Applied Physics 1998;83(9):4802–4806.

[2.150] Gamiz, F., Roldan, J., Cartujo-Cassinello, P., Lopez-Villanueva, J., and Cartujo, P. "Role of surface-roughness scattering in double gate silicon-on-insulator inversion layers". Journal of Applied Physics 2001;89 (3):4764–4770.

[2.151] Dennard, R. H., Rideout, V. L., Bassous, E., and Leblanc, A. R. "Design of ion-implanted MOSFET's with very small physical dimensions". IEEE Journal of Solid-State Circuits 1974;9(5):256–268.

[2.152] Colinge, J. P. FinFETs and other multi-gate transistors. Springer Publishing Company, 2007.

[2.153] Gamiz, F., Roldan, J. B., Lopez-Villanueva, J. A., Cartujo-Cassinello, P., and Jimenez-Molinos, F. "Monte Carlo simulation of electron mobility in silicon-on-insulator structures". Solid-State Electronics 2002;46(11): 1715–1721.

[2.154] Cristoloveanu, S. "How Many Gates do we Need in a Transistor?" In Proceedings of International Semiconductor Conference, 2007. (Sinaia, Romania), pp. 3–10.

[2.155] Fenouillet-Beranger, C., Perreau, P., Denorme, S., *et al.*, "Impact of a 10 nm ultra-thin box (UTBOX) and ground plane on FDSOI devices for 32 nm node and below". Solid-State Electronics 2010;54(9):849–854.

[2.156] Trivedi, V. P. and Fossum, J. G. "Nanoscale FD/SOI CMOS: thick or thin box?". IEEE Electron Device Letters 2005;26(1):26–28.

[2.157] Morita, Y., Tsuchiya, R., Ishigaki, T., *et al.*, "Smallest Vth variability achieved by intrinsic silicon on thin BOX (SOTB) CMOS with single metal gate". In: IEEE Symposium on VLSI Technology, 2008, pp. 166–167.

[2.158] Ohtou, T., Saraya, T., and Hiramoto, T. "Variable-body-factor SOI MOSFET with ultrathin buried oxide for adaptive threshold voltage and leakage control". IEEE Transactions on Electron Devices 2008;55 (1):40–47.

[2.159] Sampedro, C., Gámiz, F., Godoy, A., Valín, R., García-Loureiro, A., and Ruiz, F. G. "Multi-Subband Monte Carlo study of device orientation effects in ultra-short channel DGSOI". Solid-State Electronics 2010;54 (2):131–136.

[2.160] Sampedro, C., Gámiz, F., Godoy, A., *et al.*, "Multi-Subband Ensemble Monte Carlo simulation of bulk MOSFETs for the 32 nm-node and beyond". Solid-State Electronics 2011;65:88–93.

[2.161] Sampedro, C., Gamiz, F., Donetti, L., and Godoy, A. "Reaching sub-32 nm nodes: ET-FDSOI and BOX optimization". Solid-State Electronics 2012;70:101–105.

[2.162] International technology roadmap for the semiconductor industry, "http://www.itrs.net/".

[2.163] Cheng, K., Khakifirooz, A., Kulkarni, P., *et al.*, "Extremely thin SOI (ETSOI) CMOS with record low variability for low power system-on-chip applications". In: IEEE International Electron Devices Meeting, 2009, pp. 1–4.

[2.164] Yamashita, T., Yoshida, N., Sakamoto, M., *et al.*, "A 450 MHz 64 b RISC processor using multiple threshold voltage CMOS". In: IEEE International Solid-State Circuits Conference, 2000, pp. 414–415.

[2.165] Kunie, S., Hiraga, T., Tokue, T., Torii, T. T. S., and Ohsawa, T. "Low power architecture and design techniques for mobile handset LSI Medity™ M2" In Asia and South Pacific Design Automation Conference, 2008. pp. 748–753.

[2.166] Zhang, Z. B., Song, S. C., Choi, K., Sim, J. H., Majhi, P., and Lee, B. H. "An integratable dual metal gate/high-k CMOS solution for FD-SOI and MuGFET technologies". In: IEEE International SOI Conference, 2005, pp. 157–158.

[2.167] Pham, D., Luan, H., Mathur, K., *et al.*, "Single metal gate with dual work functions for FD-SOI and UTB double gate technologies". In: IEEEE International SOI Conference, 2006, pp. 25–26.

[2.168] Mistry, K., Allen, C., Auth, C., *et al.*, "A 45 nm logic technology with high-k+ metal gate transistors, strained silicon, 9 Cu interconnect layers, 193 nm dry patterning, and 100% Pb-free packaging". In: IEEE International Electron Devices Meeting, 2007, pp. 247–250.

[2.169] Lai, C. M., Lin, C. T., Cheng, L. W., *et al.*, "A Novel "hybrid" high-k/metal gate process for 28 nm high performance CMOSFETs". In: IEEE International Electron Devices Meeting, 2009, pp. 1–4.

[2.170] Noel, J. P., Thomas, O., Fenouillet-Beranger, C., Jaud, M. A., Scheiblin, P., and Amara, A. "A simple and efficient concept for setting up multi-V T devices in thin box fully-depleted SOI technology". In: Proceedings of the European Solid State Device Research Conference, 2009, pp. 137–140.

[2.171] Fenouillet-Beranger, C., Thomas, O., Perreau, P., *et al.*, "Efficient multi-V T FDSOI technology with UTBOX for low power circuit design". In: IEEE Symposium on VLSI Technology, 2010, pp. 65–66.

[2.172] Noel, J. P., Thomas, O., Jaud, M. A., *et al.*, "Multi-UTBB FDSOI Device Architectures for Low-Power CMOS Circuit". IEEE Transactions on Electron Devices 2011;58(8):2473–2482.

[2.173] Huang, X., Lee, W. C., Kuo, C., *et al.*, "Sub 50-nm FinFET: PMOS". In: IEEE International Electron Devices Meeting, 1999, pp. 67–70.

[2.174] Hisamoto, D., Kaga, T., Kawamoto, Y., and Takeda, E. ''A fully depleted lean-channel transistor (DELTA) – a novel vertical ultra thin SOI MOSFET". In: IEEE International Electron Devices Meeting, 1989, pp. 833–836.

[2.175] Kedzierski, J., Fried, D. M., Nowak, E. J., *et al.*, "High-performance symmetric-gate and CMOS-compatible VTH asymmetric-gate FinFET devices". In: IEEE International Electron Devices Meeting, 2001, pp. 19–22.

[2.176] Park, T. S., Choi, S., Lee, D. H., *et al.*, "Fabrication of body-tied FinFETs (Omega MOSFETs) using bulk Si wafers". In: IEEE Symposium on VLSI Technology, 2003, pp. 135–136.

[2.177] Doyle, B., Boyanov, B., Datta, S., *et al.*, "Tri-gate fully-depleted CMOS transistors: fabrication, design and layout". In: IEEE Symposium on VLSI Technology, 2003, pp. 133–134.

[2.178] Colinge, J. P., Gao, M. H., Romano-Rodriguez, A., Maes, H., and Claeys, C. "Silicon-on-insulator gate-all-around device". In: IEEE International Electron Devices Meeting, 1990, pp. 595–598.

[2.179] Lee, C. W., Afzalian, A., Akhavan, N. D., Yan, R., Ferain, I., and Colinge, J. P. "Junctionless multigate field-effect transistor". Applied Physics Letters 2009;94(5):053511.

[2.180] Sekigawa, T. and Hayashi, Y. "Calculated threshold-voltage characteristics of an XMOS transistor having an additional bottom gate". Solid-State Electronics 1984;27(8):827–828.

[2.181] Allibert, F., Zaslavsky, A., Pretet, J., and Cristoloveanu, S. "Double-gate MOSFETs: is gate alignment mandatory?". In: Proceeding of the 31st European Solid-State Device Research Conference, 2001, pp. 267–270.

[2.182] Widiez, J., Daugé, F., Vinet, M., *et al.*, "Experimental gate misalignment analysis on double gate SOI MOSFETs". In: IEEE International SOI Conference, 2004, pp. 185–186.

[2.183] Cristoloveanu, S. "Silicon on insulator technologies and devices: from present to future". Solid-State Electronics 2001;45(8):1403–1411.

[2.184] Ferain, I., Colinge, C. A., and Colinge, J. P. "Multigate transistors as the future of classical metal-oxide-semiconductor field-effect transistors". Nature 2011;479(7373):310–316.

[2.185] Colinge, J. P. "Multiple-gate SOI MOSFETs". Solid-State Electronics 2004;48(6):897–905.

[2.186] Lee, C. W., Yu, C. G., Park, J. T., and Colinge, J. P. "Device design guidelines for nano-scale MuGFETs. Solid-State Electronics 2007;51 (3):505–510.

[2.187] Singh, N., Lim, F. Y., Fang, W. W., *et al.*, "Ultra-narrow silicon nanowire gate-all-around CMOS devices: impact of diameter, channel-orientation and low temperature on device performance". In: IEEE International Electron Devices Meeting, 2006, pp. 1–4.

[2.188] Xiong, W. W. "Multigate MOSFET technology". In: Colinge, J. P. (Ed.), FinFETs and Other Multi-Gate Transistors. Springer, 2008, pp. 49–111.

[2.189] Choi, Y. K., King, T. J., and Hu, C. "Nanoscale CMOS spacer FinFET for the terabit era". IEEE Electron Device Letters 2002;23(1):25–27.

[2.190] Johnson, C., Ogura, S., Riseman, J., Rovedo, N., and Shepard, J. "Method for making submicron dimensions in structures using sidewall image transfer techniques". IBM Technical Disclosure Bulletin 1984;26(9):4587–4589.

[2.191] Choi, Y. K., Chang, L., Ranade, P., *et al.*, "FinFET process refinements for improved mobility and gate work function engineering". In: IEEE Internationals Electron Devices Meeting, 2002, pp. 259–262.

[2.192] Xiong, W., Gebara, G., Zaman, J., *et al.*, "Improvement of FinFET electrical characteristics by hydrogen annealing". IEEE Electron Device Letters 2004;25(8):541–543.

[2.193] Kedzierski, J., Ieong, M., Nowak, E., *et al.*, "Extension and source/drain design for high-performance FinFET devices". IEEE Transactions on Electron Devices 2003;50(4):952–958.

[2.194] Dixit, A., Kottantharayil, A., Collaert, N., Goodwin, M., Jurczak, M., and Meyer, K. D. "Analysis of the parasitic S/D resistance in multiple-gate FETs". IEEE Transactions on Electron Devices 2005;52(6): 1132–1140.

[2.195] Shang, H., Chang, L., Wang, X., *et al.*, "Investigation of FinFET devices for 32 nm technologies and beyond". In: IEEE Symposium on VLSI Technology, 2006, pp. 54–55.

[2.196] Yan, R. H., Ourmazd. A., and Lee., K. F. "Scaling the Si MOSFET: from Bulk to SOI to Bulk" IEEE Transactions on Electron Devices 1992;39 (7):1704–1710.

[2.197] Auth, C. P. and Plummer, J. D. "Scaling theory for cylindrical, fully-depleted, surrounding-gate MOSFET's". IEEE Electron Device Letters 1997;18(2):74–76.

[2.198] Suzuki, K., Tanaka, T., Tosaka, Y., Horie, H., and Arimoto, Y. "Scaling theory for double-gate SOI MOSFET's". IEEE Transactions on Electron Devices 1993;40(12):2326–2329.

[2.199] Ritzehnthaler, R., Faynot, O., Jahan, C., Kuriyama, A., Deleonibus, S., and Cristoloveanu, S. "Coupling effects in FinFETs and triple-gate FETs". In: Proceedings of the 7th European Workshop on Ultimate Integration of Silicon, 2006, pp. 25–28.

[2.200] Stadele, M., Luyken, R. J., Roosz, M., *et al.*, "A comprehensive study of corner effects in tri-gate transistors". In: Proceeding of the 34th European Solid-State Device Research Conference, 2004, pp. 165–168.

[2.201] Ruiz, F. J. G., Godoy, A., Gámiz, F., Sampedro, C., and Donetti, L. "A comprehensive study of the corner effects in Pi-gate MOSFETs including quantum effects". IEEE Transactions on Electron Devices 2007;54(12):3369–3377.

[2.202] Fossum, J. G., Yang, J. W., and Trivedi, V. P. "Suppression of corner effects in triple-gate MOSFETs". IEEE Electron Device Letters 2003;24 (12):745–747.

[2.203] Colinge, J. P., Park, J. W., and Xiong, W. "Threshold voltage and subthreshold slope of multiple-gate SOI MOSFETs". IEEE Electron Device Letters 2003;24(8):515–517.

[2.204] Park, T. S., Choi, S., Lee, D. H., *et al.*, "Fabrication of body-tied FinFETs (Omega MOSFETs) using bulk Si wafers". In: IEEE Symposium on VLSI Technology, 2003, pp. 135–136.

[2.205] Park, T., Park, D., Chung, J. H., *et al.*, "PMOS body-tied FinFET (Omega MOSFET) characteristics" In: Device Research Conference, 2003, pp. 33–34.

[2.206] Jan, C. H., Bhattacharya, U., Brain, R., *et al.*, "A 22 nm SoC platform technology featuring 3-D tri-gate and high-k/metal gate, optimized for ultra low power, high performance and high density SoC applications". In: IEEE International Electron Devices Meeting, 2012, pp. 3.1–3.4.

[2.207] Balestra, F., Cristoloveanu, S., Benachir, M., Brini, J., and Elewa, T. "Double-gate silicon-on-insulator transistor with volume inversion: a new device with greatly enhanced performance". IEEE Electron Device Letters 1987;8(9):410–412.

[2.208] Frank, D. J., Laux, S. E., and Fischetti, M. V. "Monte Carlo simulation of a 30 nm dual-gate MOSFET: how short can Si go?" In: IEEE International Electron Devices Meeting, 1992, pp. 553–556.

[2.209] Ouisse, T. "Self-consistent quantum-mechanical calculations in ultrathin silicon-on-insulator structures". Journal of Applied Physics 1994;76 (10):5989–5995.

[2.210] Majkusiak, B., Janik, T., and Walczak, J. "Semiconductor Thickness Effects in the Double-Gate SOI MOSFET". IEEE Transactions on Electron Devices 1998;45(5):1127–1134.

[2.211] Janik, T. and Majkusiak, B. "Analysis of the MOS transistor based on the self-consistent solution to the Schrodinger and Poisson equations and on the local mobility model". IEEE Transactions on Electron Devices 1998;45(6):1263–1271.

[2.212] Shoji, M., Omura, Y., and Tomizawa, M. "Physical basis and limitation of universal mobility behavior in fully depleted silicon-on-insulator Si inversion layers". Journal of Applied Physics 1997;81(2):786–794.

[2.213] Shoji, M. and Horiguchi, S. "Electronic structures and phonon-limited electron mobility of double-gate silicon-on-insulator Si inversion layers". Journal of Applied Physics 1999;85(5):2722–2731.

[2.214] Esseni, D., Mastrapasqua, M., Fiegna, C., Celler, G. K., Selmi, L., and Sangiorgi, E. "An experimental study of low field electron mobility in double-gate, ultra-thin SOI MOSFETs". In: IEEE International Electron Devices Meeting, 2001, pp. 19.7.

[2.215] Esseni, D., Abramo, A., Selmi, L., and Sangiorgi, E. "Study of low field electron transport in ultra-thin single and double-gate SOI MOSFETs". In: IEEE International Electron Devices Meeting, 2002, pp. 719–722.

[2.216] Esseni, D., Mastrapasqua, M., Celler, G. K., Fiegna, C., Selmi, L., and Sangiorgi, E. "An experimental study of mobility enhancement in ultra-thin SOI transistors operated in double-gate mode". IEEE Transactions on Electron Devices 2003;50(3):802–808.

[2.217] Gamiz, F. and Godoy A. "Mobility in multigate MOSFETs" In: Colinge, J. P. (Ed.), FinFETs and Other Multi-Gate Transistors. Springer, 2008, pp. 191–256.

[2.218] Majima, H., Ishikuro, H., and Hiramoto, T. "Experimental evidence for quantum mechanical narrow channel effect in ultra-narrow MOSFET's". IEEE Electron Device Letters 2000;21(8):396–398.

[2.219] Wong, H. S. P. "Nanoelectronics: nanotubes, nanowires, molecules, and novel concepts". In: Proceedings of 35th European Solid-State Device Research Conference, 2005, pp. 55–61.

[2.220] Koo, S. M., Fujiwara, A., Han, J. P., Vogel, E. M., Richter, C. A., and Bonevich, J. E. "High inversion current in silicon nanowire field effect transistors". Nano Letters 2004;4(11):2197–2201.

[2.221] Morales, A. M. and Lieber, C. M. "A laser ablation method for the synthesis of crystalline semiconductor nanowires". Science 1998;279 (5348):208–211.

[2.222] Huang, Y., Duan, X., Cui, Y., Lauhon, L. J., Kim, K. H., and Lieber, C. M. "Logic gates and computation from assembled nanowire building blocks". Science 2001;294(5545):1313–1317.

[2.223] Duan, X., Huang, Y., Cui, Y., Wang, J., and Lieber, C. M. "Indium phosphide nanowires as building blocks for nanoscale electronic and optoelectronic devices". Nature 2001;409(6816):66–69.

[2.224] Yang, F. L., Lee, D. H., Chen, H. Y., *et al.*, "5 nm-gate nanowire FinFET". In: IEEE Symposium on VLSI Technology, 2004, pp. 196–197.

[2.225] Singh, N., Agarwal, A., Bera, L. K., *et al.*, "High-performance fully depleted silicon nanowire (diameter/spl les/5 nm) gate-all-around CMOS devices". IEEE Electron Device Letters 2006;27(5):383–386.

[2.226] Wang, J., Rahman, A., Ghosh, A., Klimeck, G., and Lundstrom, M. "On the validity of the parabolic effective-mass approximation for the I–V calculation of silicon nanowire transistors". IEEE Transactions on Electron Devices 2005;52(7):1589–1595.

[2.227] Sanders, G. D., Stanton, C. J., and Chang, Y. C. "Theory of transport in silicon quantum wires". Physical Review B 1993;48(15):11067.

[2.228] Wang, J., Polizzi, E., and Lundstrom, M. "A three-dimensional quantum simulation of silicon nanowire transistors with the effective-mass approximation". Journal of Applied Physics 2004;96(4):2192–2203.

[2.229] Gilbert, M. J., Akis, R., and Ferry, D. K. "Phonon-assisted ballistic to diffusive crossover in silicon nanowire transistors". Journal of Applied Physics 2005;98(9):094303.

[2.230] Jin, S., Park, Y. J., and Min, H. S. "A three-dimensional simulation of quantum transport in silicon nanowire transistor in the presence of electron-phonon interactions". Journal of Applied Physics 2006;99(12):123719.

[2.231] Trellakis, A., Galick, A. T., Pacelli, A., and Ravaioli, U. "Iteration scheme for the solution of the two-dimensional Schrödinger-Poisson equations in quantum structures". Journal of Applied Physics 1997;81 (12):7880–7884.

[2.232] Trellakis, A. and Ravaioli, U. "Lateral scalability limits of silicon conduction channels". Journal of Applied Physics 1999;86(7):3911–3916.

[2.233] Godoy, A., Ruiz-Gallardo, A., Sampedro, C., and Gámiz, F. "Quantum-mechanical effects in multiple-gate MOSFETs". Journal of Computational Electronics 2007;6(1–3):145–148.

[2.234] Kubo, R. "Statistical-mechanical theory of irreversible processes. I. General theory and simple applications to magnetic and conduction problems". Journal of the Physical Society of Japan 1957;12(6):570–586.

[2.235] Greenwood, D. A. "The Boltzmann equation in the theory of electrical conduction in metals". Proceedings of the Physical Society 1958;71 (4):585–596.

[2.236] Keldysh, L. V. "Diagram technique for nonequilibrium processes". Soviet Physics, JETP 1965;20(4):1018–1026.

[2.237] Kadanoff, L. P. and Baym, G., Quantum statistical mechanics, Benjamin, New York, 1962.

[2.238] Ramayya, E. B., Vasileska, D., Goodnick, S. M., and Knezevic, I. "Electron mobility in silicon nanowires". IEEE Transactions on Nanotechnology 2007;6(1):113–117.

[2.239] Godoy, A., Ruiz, F., Sampedro, C., Gámiz, F., and Ravaioli, U. "Calculation of the phonon-limited mobility in silicon Gate All-Around MOSFETs". Solid-State Electronics 2007;51(9):1211–1215.

[2.240] Jacoboni, C. and Reggiani, L. "The Monte Carlo Method for the solution of charge transport in semiconductors with application to covalent materials". Review of Modern Physics 1983;55(3):645–705.

[2.241] Tienda-Luna, I., Ruiz, F. G., Godoy, A., Marín, E. G., and Gamiz F. "Spatial dependence of the phonon-limited mobility in arbitrarily oriented Si-nanowires" In: International Workshop on Computational Electronics, 2012.

[2.242] Marin, E. G., Ruiz, F. G., Godoy, A., Tienda-Luna, I. M., and Gamiz, F. "Mobility and Capacitance Comparison in Scaled InGaAs Versus Si Trigate MOSFETs". IEEE Electron Device Letters 2015;36(2):114–116.

[2.243] Marin, E. G., Ruiz, F. G., Tienda-Luna, I. M., Godoy, A., and Gamiz, F. "Analytical model for the threshold voltage of III–V nanowire transistors including quantum effects". Solid-State Electronics 2014;92:28–34.

[2.244] Marin, E. G., Ruiz, F. G., Godoy, A., Tienda-Luna, I. M., and Gamiz, F. "The unexpected beneficial effect of the L-valley population on the electron mobility of GaAs nanowires. Applied Physics Letters 2015;106 (2):022113.

[2.245] Marin, E. G., Ruiz, F. J. G., Tienda-Luna, I. M., Godoy, A., and Gamiz, F. "Analytical gate capacitance modeling of III–V nanowire transistors. IEEE Transactions on Electron Devices 2013;60(5):1590–1599.

[2.246] Marin, E. G., Ruiz, F. G., Tienda-Luna, I. M., Godoy, A., Sánchez-Moreno, P., and Gamiz, F. "Analytic potential and charge model for III-V surrounding gate metal-oxide-semiconductor field-effect transistors". Journal of Applied Physics 2012;112(8):084512.

[2.247] Cui, Y., Duan, X., Hu, J., and Lieber, C. M. "Doping and electrical transport in silicon nanowires". The Journal of Physical Chemistry B 2000;104(22):5213–5216.

[2.248] Cui, Y., Zhong, Z., Wang, D., Wang, W. U., and Lieber, C. M. "High performance silicon nanowire field effect transistors". Nano Letters 2003;3(2):149–152.

[2.249] Suk, S. D., Lee, S.-Y., Kim, S.-M., *et al.*, "High performance 5 nm radius twin silicon nanowire MOSFET (TSNWFET): fabrication on bilk Si wafer, characteristics and reliability". IEEE IEDM Proceedings, December 2005, pp. 717–720.

[2.250] Colinge, J. P., Quinn, A. J., Floyd, L., *et al.*, "Low-temperature electron mobility in trigate SOI MOSFETs". IEEE Electron Device Letters 2006;27(2):120–122.

[2.251] Jain, S. H., Griffin, P. B., Plummer, J. D., *et al.*, "Low resistance, low-leakage ultrashallow p+-junction formation using millisecond flash anneals". IEEE Transactions on Electron Devices 2005;52(7):1610–1615.

[2.252] Colinge, J. P., Lee, C. W., Afzalian, A., *et al.*, "Nanowire transistors without junctions". Nature Nanotechnology 2010;5(3):225–229.

[2.253] Barraud, S., Berthomé, M., Coquand, R., *et al.*, "Scaling of trigate junctionless nanowire MOSFET with gate length down to 13 nm". IEEE Electron Device Letters 2012;33(9):1225–1227.

[2.254] Ansari, L., Feldman, B., Fagas, G., Colinge, J. P., and Greer, J. C. "Simulation of junctionless Si nanowire transistors with 3 nm gate length". Applied Physics Letters 2010;97(6):062105.

[2.255] Jeon, D. Y., Park, S. J., Mouis, M., *et al.*, "Electrical characterization and revisited parameter extraction methodology in junctionless transistors". In Proceedings of EuroSOI conference, 2012 pp. 109–110.

[2.256] Colinge, J. P. "Silicon-on-insulator (SOI) junctionless transistors". In: Silicon-On-Insulator (SOI) Technology: Manufacture and Applications. Springer, 2014, pp.167–194.

[2.257] Boucart, K. and Ionescu, A. M. "Double-gate tunnel FET with high-κ gate dielectric". IEEE Transactions on Electron Devices 2007;54(7):1725–1733.

[2.258] Esaki, L. "New phenomenon in narrow germanium p-n junctions". Physical Review 1958;109(2):603–604.

[2.259] Quinn, J. J., Kawamoto, G., and McCombe, B. D. "Subband spectroscopy by surface channel tunnelling". Surface Science 1978;73:190–196.

[2.260] Takeda, E., Matsuoka, H., Igura, Y., and Asai, S. "A band to band tunneling MOS device (B2T-MOSFET)-a kind of Si quantum device". In: IEEE International Electron Devices Meeting, 1988, pp. 402–405.

[2.261] Banerjee, S. A. N. J. A. Y., Richardson, W., Coleman, J., and Chatterjee, A. "A new three-terminal tunnel device". IEEE Electron Device Letters 1987;8(8):347–349.

[2.262] Baba, T. "Proposal for surface tunnel transistors". Japanese Journal of Applied Physics 1992;31(4B):L455–L457.

[2.263] Reddick, W. M. and Amaratunga, G. A. "Silicon surface tunnel transistor". Applied Physics Letters 1995;67(4):494–496.

[2.264] Koga, J. and Toriumi, A. "Three-terminal silicon surface junction tunneling device for room temperature operation". IEEE Electron Device Letters 1999;20(10):529–531.

[2.265] Schulze, J., Fink, C., Sulima, T., Eisele, I., and Hansch, W. "Vertical MOS-gated pin-diodes: MOS-gated tunneling transistors in Si (100) and Si (111) ". Thin Solid Films 2000;380(1):154–157.

[2.266] Hansch, W., Borthen, P., Schulze, J., Fink, C., Sulima, T., and Eisele, I. "Performance improvement in vertical surface tunneling transistors by a Boron surface phase". Japanese Journal of Applied Physics 2001;40 (5R):3131–3136.

[2.267] Sedlmaier, S., Schulze, J., Sulima, T., *et al.*, "Phonon assisted tunneling in gated pin diodes". Materials Science and Engineering B 2002;89 (1):116–119.

[2.268] Aydin, C., Zaslavsky, A., Luryi, S., *et al.*, "Lateral interband tunneling transistor in silicon-on-insulator". Applied Physics Letters 2004;84 (10):1780–1782.

[2.269] Appenzeller, J., Lin, Y. M., Knoch, J., and Avouris, P. "Band-to-band tunneling in carbon nanotube field-effect transistors". Physical Review Letters 2004;93(19):196805.

[2.270] Appenzeller, J., Lin, Y. M., Knoch, J., Chen, Z., and Avouris, P. "Comparing carbon nanotube transistors-the ideal choice: a novel tunneling device design". IEEE Transactions on Electron Devices 2005;52 (12):2568–2576.

[2.271] Choi, W. Y., Park, B. G., Lee, J. D., and Liu, T. J. K. "Tunneling field-effect transistors (TFETs) with subthreshold swing (SS) less than 60 mV/dec". IEEE Electron Device Letters 2007;28(8):743–745.

[2.272] Björk, M. T., Knoch, J., Schmid, H., Riel, H., and Riess, W. "Silicon nanowire tunneling field-effect transistors". Applied Physics Letters 2008;92(19):3504–3506.

[2.273] Chynoweth, A. G., Feldmann, W. L., and Logan, R. A. "Excess tunnel current in silicon Esaki junctions". Physical Review 1961;121(3):684–694.

[2.274] Datta, S.. Electronic transport in mesoscopic systems. Cambridge University Press, Cambridge, 1995.

[2.275] Sze, S. Physics of semiconductor devices. Wiley, 2007.

[2.276] International Technology Roadmap for Semiconductors: 2013 Edition, 2014. http://www.itrs.net.

[2.277] Amoroso, S. M., Adamu-Lema, F., Brown, A. R., and Asenov, A. "A mobility correction approach for overcoming artifacts in atomistic drift-diffusion simulations of nano-MOSFETs". IEEE Transactions on Electron Devices 2015;62:2056–2060.

[2.278] Ancona, M. G. "Macroscopic description of quantum-mechanical tunneling". Physical Re-view B 1990;42(2):1222–1233.

[2.279] Ancona, M. G. and Tiersten, H. F. "Quantum correction to the equation of state of an electron gas in a semiconductor". Physical Review B 1989;39(13):9536–9540.

[2.280] Ancona, M. G., Yu, Z., Dutton, R. W., Voorde, P. J. V., Cao, M., and Vook, D. "Density-gradient analysis of tunneling in MOS structures with ultra-thin oxides". In: Proceedings of the International Conference on Simulation of Semiconductor Processes and Devices, 1999, pp. 235–238.

[2.281] Ancona, M. G., Yu, Z., Dutton, R. W., Voorde, P. J. V., Cao, M., and Vook, D. "Density-gradient analysis of MOS tunneling". IEEE Transactions on Electron Devices 2000;47(12):2310–2319.

[2.282] Asenov, A., and Saini, S. "Suppression of random dopant-induced threshold voltage fluctuations in sub-0.1-μm MOSFET's with epitaxial and ∂-doped channels". IIEEE Transactions on Electron Devices 1999;46(8):1718–1724.

[2.283] Asenov, A., Brown, A. R., and Watling, J. R. "Quantum corrections in the simulation of deca-nano MOSFETs". Solid-State Electron 2003;47 (7):1141–1145.

[2.284] Blom, A., Pozzoni, U. M., Markussen, T., and Stokbro, K. "First-principles simulations of nanoscale transistors". In: Proceedings of the International Conference on Simulation of Semiconductor Processes and Devices, 2015, pp. pp. 52–55.

[2.285] Blotekjaer, K. "Transport equations for electrons in two-valley semiconductors". IEEE Transactions on Electron Devices 1970;17(1):38–47.

[2.286] Canali, C., Jacoboni, C., Nava, F., Ottaviani, G., and Quaranta, A. "Electron drift velocity in silicon". Physical Review B 1975;12 (4):2265–2284.

[2.287] Datta, S., Quantum transport: atom to transistor. Cambridge University Press, 2005.

[2.288] DeMari, A. "An accurate numerical steady-state one-dimensional solution of the p-n junction". Solid-State Electron 1968;11:33–58.

[2.289] Dhar, S., Kosina, H., Palankovski, V., Ungersboeck, E., and Selberherr, S. "Electron mobility model for strained-Si devices". IEEE Transactions on Electron Devices 2005;52(4):527–533.

[2.290] Doris, B., Ieong, M., Kanarsky, T., *et al.*, "Extreme scaling with ultra-thin Si channel MOSFETs". In: Proceedings of the International Electron Devices Meeting, 2002, pp. 267–270.

[2.291] van Dort, M. J., Woerlee, P. H., and Walker, A. J. "A simple model for quantization effects in heavily-doped silicon MOSFETs at inversion conditions". Solid-State Electron 1994;37(3):411–414.

[2.292] Fawcett, W., Boardman, A., and Swain, S. "Monte Carlo determination of electron transport properties in gallium arsenide". Journal of Physics and Chemistry Solids 1970;31:1963–1990.

[2.293] Fawcett, W. and Paige, E. "Negative differential mobility of electrons in germanium: a Monte Carlo calculation of the distribution function, drift velocity and carrier population in the "111" and "100" minima". Journal of Physics C: Solid State Physics 1971;4:1801–1821.

[2.294] Ferry, D., Akis, R., and Vasileska, D. "Quantum effects in MOSFETs: Use of an effective potential in 3D Monte Carlo simulations in ultra-short channel devices". In: Proceedings of the International Electron Devices Meeting, 2000, pp. 287–290.

[2.295] Fischetti, M. and Laux, S. "Monte Carlo simulation of electron transport in Si: the first 20 years". In: Proceedings of the European Solid State Device Research Conference, 1996, pp. 813–820.

[2.296] Fischetti, M. V., Gämiz, F., and Hänsch, W. "On the enhanced electron mobility in strained-silicon inversion layers". Journal of Applied Physics 2002;92(12):7320–7324.

[2.297] Fischetti, M. V. and Laux, S. E. "Monte Carlo analysis of electron transport in small semi-conductor devices including band-structure and space-charge effects". Physical Review B 1988;38(14):9721–9745.

[2.298] Fischetti, M. V., Ren, Z., Solomon, P. M., Yang, M., and Rim, K. "Six-band k·p calculation of the hole mobility in silicon inversion layers: Dependence on surface orientation, strain, and silicon thickness". Journal of Applied Physics 2003;94(2):1079–1095.

[2.299] Gehring, A., Grasser, T., Kosina, H., and Selberherr, S. "Simulation of hot-electron oxide tunneling current based on a non-Maxwellian electron energy distribution function". Journal of Applied Physics 2002;92 (10):6019–6027.

[2.300] Gilbert, M., Akis, R., and Ferry, D. "Phonon-assisted ballistic to diffusive crossover in silicon nanowire transistors". Journal of Applied Physics 2005;98(9):094, pp. 303–1–8.

[2.301] Gold Standard Simulations Ltd. GARAND. Available: http://www.goldstandardsimulations.com/products/garand/

[2.302] Grasser, T., Jungemann, C., Kosina, H., Meinerzhagen, B., and Selberherr, S. "Advanced transport models for sub-micrometer devices"' In: Proceedings of the International Conference on Simulation of Semiconductor Processes and Devices, 2004, pp. 1–8.

[2.303] Grasser, T., Kosik, R., Jungemann, C., Kosina, H., and Selberherr, S. "Nonparabolic macroscopic transport models for device simulation based on bulk Monte Carlo data". Journal of Applied Physics 2005;97(9):0937, pp. 101–09371, 012.

[2.304] Grasser, T., Kosina, H., Gritsch, M., and Selberherr, S. "Using six moments of Boltzmann's transport equation for device simulation". Journal of Applied Physics 2001;90(5):2389–2396.

[2.305] Grasser, T., Kosina, H., Heitzinger, C., and Selberherr, S. "Characterization of the hot electron distribution function using six moments". Journal of Applied Physics 2002;91(6):3869–3879.

[2.306] Grasser, T., Kosina, H., and Selberherr, S. "An impact ionization model including non-Maxwellian and non-parabolicity effects". In: Proceedings of the International Conference on Simulation of Semi-conductor Processes and Devices, 2001, pp. 46–49.

[2.307] Grasser, T., Kosina, H., and Selberherr, S. "Hot carrier effects within macroscopic transport models". International Journal of High Speed Electron 2003;13(3):873–901.

[2.308] Gritsch, M. "Numerical modeling of SOI MOSFETs". PhD Dissertation, Technische Universität Wien, 2002. http://www.iue.tuwien.ac.at/phd/gritsch.

[2.309] Gritsch, M., Kosina, H., Grasser, T., and Selberherr, S. "Influence of generation/recombination effects in simulations of partially depleted SOI MOSFETs". Solid-State Electron 2001;45(4):621–627.

[2.310] Gritsch, M., Kosina, H., Grasser, T., and Selberherr, S. "Revision of the standard hydrodynamic transport model for SOI simulation". IEEE Transactions on Electron Devices 2002;49(10):1814–1820.

[2.311] Gummel, H. "A self-consistent iterative scheme for one-dimensional steady state transistor calculations". IEEE Transactions on Electron Devices 1964;11:455–465.

[2.312] Hänsch, W., Vogelsang, T., Kircher, R., and Orlowski, M. "Carrier transport near the Si/SiO_2 interface of a MOSFET". Solid-State Electron 1989;32(10):839–849.

[2.313] Herring, C. and Vogt, E. "Transport and deformation-potential theory for many-valley semi-conductors with anisotropic scattering". Physical Review 1956;101(3):944–961.

[2.314] Hockney, R. and Eastwood, J. W., Computer simulation using particles. Adam Hilger, Bristol and Philadelphia, 1988.

[2.315] Hruska, J. "Intel confirms 10 nm delayed to 2017, will introduce Kaby Lake at 14 nm to fill gap". Online: http://www.extremetech.com/computing/210050-intel-confirms-10nm-delayed-to-2017-will-introduce-kaby-lake-at-14nm-to-fill-gap, 2015.

[2.316] Institut für Mikroelektronik: Vienna SHE: A multi-dimensional deterministic Boltzmann equation solver, http://www.iue.tuwien.ac.at/index.php?id=viennashe. Technische Universität Wien, Austria, 2015.

[2.317] Jacoboni, C. "A new approach to Monte Carlo simulation". In: Proceedings of the International Electron Devices Meeting, 1989, pp. 469–472.

[2.318] Jacoboni, C., Minder, R., and Majni, G. "Effects of band non-parabolicity on electron drift velocity in silicon above room temperature". Journal of Physics and Chemistry Solids 1975;36:1129–1133.

[2.319] Jacoboni, C., Poli, P., and Rota, L. "A new Monte Carlo technique for the solution of the Boltzmann transport equation". Solid-State Electron 1988;31(3/4):523–526.

[2.320] Jacoboni, C. and Lugli, P. "The Monte-Carlo method for semiconductor device simulation", Chapter 3, Springer, Wien, 1989, pp. 218–261.

[2.321] John, D. L., Castro, L. C., Pereira, P. J. S., and Pulfrey, D. L. "A Schrödinger-Poisson solver for modeling carbon nanotube FETs". In: Proceedings of the Nanotech, 2004, pp. 1–4.

[2.322] Jungel, A. (Ed.), Quasi-hydrodynamic semiconductor equations, progress in nonlinear differential equations and their applications, Vol. 41. A Birkhauser Press, 2001, p. 191.

[2.323] Jungemann, C. and Meinerzhagen, B. (Eds.), Hierarchical device simulation. The Monte Carlo perspective. Springer, 2003, pp. 162–172.

[2.324] Jungemann, C., Nguyen, C. D., Neinhüs, B., Decker, S., and Meinerzhagen, B. "Improved modified local density approximation for modeling of size quantization in nMOSFETs". In: Proceedings of the International Conference on Modeling and Simulation of Microsystems, 2001, pp. 458–461.

[2.325] Kennedy, D. "On the ambipolar diffusion of impurities into silicon". Proceedings of the IEEE 1969;54(6):1202–1203.

[2.326] Kosina, H., Nedjalkov, M., and Selberherr, S. Theory of the Monte Carlo method for semicon-ductor device simulation. IEEE Transactions on Electron Devices 2000;47(10):1899–1908.

[2.327] Kosina, H., Nedjalkov, M., and Selberherr, S. "A Monte Carlo method seamlessly linking classical and quantum transport calculations". Journal of Computational Electronics 2003;2(2–4):147–151.

[2.328] Kosina, H., Nedjalkov, M., and Selberherr, S. "A stable backward Monte Carlo method for the solution of the Boltzmann equation". Lecture Notes in Computer Science 2003;2907:170–177.

[2.329] Kosina, H. and Selberherr, S. "Device simulation demands of upcoming microelectronics devices". International Journal of High Speed Electron 2006;16(1):115–136.

[2.330] Kosina H. and Nedjalkov M. "Wigner function based device modeling". In: Rieth, M. and Schommers, W. (eds.), Handbook of Theoretical and Computational Nanotechnology. American Scientific Publishers Editions Frontiers, 2006, pp. 731–763.

[2.331] Kunikiyo, T., Takenaka, M., Kamakura, Y., *et al.*, "A Monte Carlo simulation of anisotropic electron transport in silicon including full band structure and anisotropic impact-ionization model". Journal of Applied Physics 1994;75(1):297–312.

[2.332] Kurosawa, T. "Monte Carlo calculation of hot electron problems". In: Proceedings of the International Conference on Physics of Semiconductors, 1966, pp. 424–426.

[2.333] Li, P. and Dery, H. "Spin-orbit symmetries of conduction electrons in silicon". Physical Review Letters 2011;107:107203.

[2.334] Lin, C. H., Greene, B., Narasimha, S., *et al.*, "High performance 14 nm SOI FinF ET CMOS technology with 0.0174 μm^2 embedded DRAM and 15 levels of Cu metallization". In: Proceedings of the International Electron Devices Meeting, 2014, pp. 3.8.1–3.8.3.

[2.335] Loeb, H., Andrew, R., and Love, W. "Application of 2-dimensional solutions of the Shockley-Poisson equation to inversion-layer M.O.S.T. devices". Electronics Letters 1968;4:352–354.

[2.336] Lundstrom, M., Fundamentals of carrier transport. Cambridge University Press, Cambridge, 2000, Chapter 2.

[2.337] Lundstrom, M. "Drift-diffusion and computational electronics – Still going strong after 40 years!". In: Proceedings of the International Conference on Simulation of Semiconductor Processes and Devices, 2015, pp. 1–4.

[2.338] Mahan, G. "Many-particle physics". Premium Press, New York, 1990, Chapter 7.

[2.339] Moglestue, C. "Monte Carlo particle modelling of small semiconductor devices". Computer Methods in Applied Mechanics and Engineering 1982;30:173–208.

[2.340] Nanoelectron Modeling Group: OMEN, https://engineering.purdue.edu/gekcogrp/software-projects/omen/. Purdue University, USA, 2015.

[2.341] Natarajan, S., Agostinelli, M., Akbar, S., *et al.*, "A 14 nm logic technology featuring 2nd-generation FinFET, air-gapped interconnects, self-aligned double patterning and a 0.0588 μm^2 SRAM cell size". In: Proceedings of the International Electron Devices Meeting, 2014, pp. 3.7.1–3.7.3.

[2.342] Nedjalkov, M., Kosik, R., Kosina, H., and Selberherr, S. "Wigner transport through tunneling structures – Scattering interpretation of the potential operator". In: Proceedings of the International Conference on Simulation of Semiconductor Processes and Devices, 2002, pp. 187–190.

[2.343] Nedjalkov, M. and Vitanov, P. "Iteration approach for solving the Boltzmann equation with the Monte Carlo method". Solid-State Electronics 1989;32(10):893–896.

[2.344] Nguyen, C. D., Jungemann, C., and Meinerzhagen, B. "Modeling of size quantization in strained Si-nMOSFETs with the improved modified local density approximation". In: Proceedings of the Nanotech, 2005, Vol. 3, pp. 33–36.

[2.345] Online Simulations and More: http://www.nanohub.org, 2015.

[2.346] Osintsev, D., Sverdlov, V., and Selberherr, S. "Electron mobility and spin lifetime enhancement in strained ultra-thin silicon films". Solid-State Electronics 2015;112:46–50.

[2.347] Paasch, G. and Ubensee, H. "Carrier density near the semiconductor-insulator interface – local density approximation for non-isotropic effective mass". Physica Status Solidi (b) 1983;118(1):255–266.

[2.348] Palestri, P., Esseni, D., Eminente, S., Fiegna, C., Sangiorgi, E., and Selmi, L. "Understanding quasi-ballistic transport in nano-MOSFETs: Part I – Scattering in the channel, and in the drain". IEEE Transactions on Electron Devices 2005;52(12):2727–2735.

[2.349] Pham, A., Jungemann, C., and Meinerzhagen, B. "Deterministic multi-subband device simulations for strained double gate PMOSFETs including magnetotransport". In: Proceedings of the International Electron Devices Meeting, 2008, pp. 895–898.

[2.350] Pourfath, M., The non-equilibrium Green's function method for nanoscale device simulation. Springer, 2014, Chapter 6.

[2.351] Pourfath, M. and Kosina, H. "Fast convergent Schrödinger-Poisson solver for the static and dynamic analysis of carbon nanotube field effect transistors". Lecture Notes in Computer Science 2006;3743:578–585.

[2.352] Price, P. J. "Monte Carlo calculation of electron transport in solids". Semiconductors and Semimetals 1979;14:249–308.

[2.353] Quantum Wise Ltd. http://quantumwise.com/

[2.354] Scharfetter, D. and Gummel, H. "Large-signal analysis of a silicon read diode oscillator". IEEE Transactions on Electron Devices 1969;16 (1):64–77.

[2.355] Schroeder, J. and Muller, R. "IGFET analysis through numerical solution of Poisson's equation". IEEE Transactions on Electron Devices 1968;15 (12):954–961.

[2.356] Selberherr, S., Analysis and simulation of semiconductor devices. Springer, 1984. Chapter 7.

[2.357] Sellier, J. M., Nedjalkov, M., and Dimov, I. "An introduction to applied quantum mechanics in the Wigner Monte Carlo formalism". Physics Reports 2015;577:1–34.

[2.358] Shichijo, H. and Hess, K. "Band-structure-dependent transport and impact ionization in GaAs". Physical Review B 1981;23(8):4197–4207.

[2.359] Shifren, L. and Ferry, D. K. "A Wigner function based ensemble Monte Carlo approach for accurate incorporation of quantum effects in device simulation". Journal of Computational Electronics 2002;1:55–58.

[2.360] Silvaco, Santa Clara, CA: ATLAS User's Manual, 2010.

[2.361] Slotboom, J. "Iterative scheme for 1- and 2-dimensional d.c.-transistor simulation". Electronics Letters 1969;5:677–678.

[2.362] Stratton, R. "Diffusion of hot and cold electrons in semiconductor barriers". Physical Review 1962;126(6):2002–2014.

[2.363] Sverdlov, V. "Strain-Induced Effects in Advanced MOSFETs". Springer Wien New York, 2011.

[2.364] Sverdlov, V. and Selberherr, S. "Silicon spintronics: progress and challenges". Physics Reports 2015;585:1–40.

[2.365] Svizhenko, A. and Anantram, M. P. "Role of scattering in nanotransistors". IEEE Transactions on Electron Devices 2003;50: 1459–1466.

[2.366] Synopsys, Mountain View, CA: Sentaurus Device User's Manual, 2015.

[2.367] Tavernier, C., Pereira, F. G., Nier, O., *et al.*, "TCAD Modeling Challenges for 14 nm fully depleted SOI technology performance assessments". In: Proceedings of the International Conference on Simulation of Semiconductor Processes and Devices, 2015, pp. 4–7.

[2.368] Trellakis, A., Zibold, T., Andalauer, T., Smith, S. B. A. K., Morschal, R., and Vogl, P. "The 3D nanometer device project nextnano3: concepts, methods, results". In: Proceedings of the Workshop on Computational Electroncis, Wien, 2006, pp. 173–174.

[2.369] Vasileska, D., Ferry, D., and Goodnick, S, "Computational nanoelectronics" In Rieth, M. and Schommers, W. (Eds.) Handbook of theoretical and computational nanotechnology. American Scientific Publishers Editions Frontiers, LA, 2006, pp. 1–135.

[2.370] VMC2.0: Vienna Monte Carlo 2.0 User's Guide. Institut für Mikroelektronik, http://www.iue.tuwien.ac.at/software, Technische Universität Wien, Austria, 2006.

[2.371] Wigner, E. "On the quantum correction for thermodynamic equilibrium". Physical Review 1932;40:749–759.

[2.372] Yoder, P., Higman, J., Bude, J., and Hess, K. "Monte Carlo simulation of hot electron transport in Si using a unified pseudopotential description of the crystal". Semiconductor Science and Technology 1992;7 (3B):357–359.

[2.373] Semiconductor Industry Association (SIA), International Technology Roadmap for Semiconductors 2001 edition. International SEMATECH, Austin, TX, 2001. Available: http://public.itrs.net.

[2.374] Hutchby, J. A., Bourianoff, G. I., Zhirnov, V. V., and Brewer, J. E. "Extending the Road Beyond CMOS". IEEE Circuits and Device Magazine 2002;18(2):28–41.

[2.375] Semiconductor Industry Association (SIA), International Technology Roadmap for Semiconductors 2013 edition. International SEMATECH, Austin, TX, 2013. Available: http://public.itrs.net.

[2.376] Brillouët, M. Towards a "More than Moore" roadmap. CATRENE Scientific Committee, 2011.

[2.377] Haselman, M. and Hauck, S. "The future of integrated circuits: a survey of nanoelectronics". Proceedings of the IEEE 2010;8(1):11–38.

[2.378] Awano, Y., Sato, S., Nihei, M., Sakay, T., Ohno, Y., and Mizutani, T. "Carbon nanotubes for VLSI: interconnect and transistor applications". Proceedings of the IEEE 2010;98(12):2015–2031.

[2.379] Landauer, G. M. and Gonzáles, J. L. "Radio-Frequency of Carbon Nanotube-Based Devices and Circuits Considering Noise and

Process Variation". IEEE Transactions on Nanotechnology 2014;13 (2):228–237.

[2.380] Zhang, J., Patil, N., Wong, H. P., and Mitra, S. "Overcoming carbon nanotube variations through co-optimized technology and circuit design". Proceedings of the IEEE International Electron Device Meeting, Washington, DC, USA, 2011. pp. 4.6.1–4.6.4.

[2.381] Baughman, R. H., Zakhidov, A. A., and de Heer, W. A. "Carbon nanotubes—the route toward applications". Science 2002;297(5582): 787–792.

[2.382] Banerjee, S. K., Register, L. F., Tutuc, E., *et al.*, "Graphene for CMOS and Beyond CMOS Applications". Proceedings of the IEEE 2010;98 (12):2032–2046.

[2.383] Pati, S. K., Enoki, T., and Rao, C. N. R. (Eds.) Graphene and its fascinating attributes. World Scientific Publishing, London, 2011.

[2.384] Xu, D., Ivan, S., Anthony, B., and Eva, Y. A. "Approaching ballistic transport in suspended graphene". Nature Nanotechnology 2008. 3:491–495.

[2.385] Zhang, Y., Zhang, L., and Zhou, C. "Review of chemical vapor deposition of graphene and related applications". Accounts of Chemical Research 2013;46(10):2329–2339.

[2.386] Huang, L., Xu, H., Zhang, Z., *et al.*, "Graphene/Si CMOS hybrid hall integrated circuits". Scientific Reports 2014;4:5548; DOI:10.1038/ srep05548.

[2.387] Auth C. P. and Plummer J. D. "Scaling for theory for cylindrical, fully-depleted, surrounding gate MOSFETs". IEEE Electron Device Letters 1997;18(2):74–76.

[2.388] Wernersson, L.-E., Thelander, C., Lind, E., and Samuelson, L. "III–V nanowires-Extending a narrowing road". Proceedings of the IEEE 2010;98(12):2047–2059.

[2.389] Yu, J. Y., Chung, S. W., and Heath, J. R. Silicon nanowires: preparation, device fabrication, and transport properties. Journal of Physical Chemistry B 2000;104:11864–11870.

[2.390] Cui, Y., Zhong, Z., Wang, D., and Lieber, C. M. "High performance silicon nanowire field effect transistors". Nano Letters 2003;3(2):149–152.

[2.391] Schmidt, V., Riel, H., Senz, S., Karg, S., Riess, W., and Gösele, U. "Realization of a silicon nanowire vertical surround-gate field-effect transistor". Small 2006;2(1):85–88.

[2.392] Lee, H. -S., Ryu, K., Ghung, J., and Palacios, T. "The best material for the function: seamless on-wafer integration of GaN and Si devices". In: CS MANTECH Conference. Palm Springs, California, 2011.

[2.393] Van Hove, M., Boulay, S., Stoffels, S., *et al.*, "CMOS process-compatible high-power low- leakage AlGaN/GaN MISHEMT on silicon". Abstract of ESA/ESTEC contract 20713/NL/07/SF, 2013.

[2.394] Cavallini, M., D'Angelo, P., Criado, V. V., *et al.*, "Ambipolar multi-stripe organic field-effect transistors". Advanced Materials 2011;23: 1173–1181.

[2.395] Zulkarnaen, S., Piliego, C., Gao, J., and Loi, M. A. "Outlook and emerging semiconducting materials for ambipolar transistors". Advanced Materials 2014;26:1176–1199.

[2.396] Cressler, J. D. (Ed.). Fabrication of SiGe HBT BiCMOS technology: Cressler John D (Ed.). SiGe and Si strained-layer epitaxy for silicon heterostructure devices. CRC Press, 2008.

[2.397] Tanaka, K., Maruyama, E., Shimada, T., and Okamoto, H. Amorphous silicon. John Wiley & Sons LTD, 1999. p. ix–x.

[2.398] Street, R. A., Hydrogenated amorphous silicon. Cambridge University Press, 1991, pp. 1–17.

[2.399] Kosarev, A., Torres, A., Hernandez, Y., Ambrosio, R., and Zúñiga, C. "Silicon-germanium films deposited by low-frequency plasma-enhanced chemical vapor deposition: Effect of H_2 and Ar dilution". Journal of Materials Research 2006;21(1):88–104.

[2.400] Kosarev, A., Sánchez, L., Torres, A., *et al.*, "Effect of hydrogen dilution on structure and electronic properties of Ge:H and GeYSi1-Y films deposited by low frequency plasma". Materials Research Society Symposia Proceedings 2006;910:0910-A07-02.

[2.401] Sánchez, L., Kosarev, A., Torres, A., Felter, T., and Ilinskii, A. "AFM morphology study of Si_{1-Y} Ge_Y:H films deposited by LF PE CVD from silane- Germane with different dilution". Materials Research Society Symposia Proceedings 2005;862:A18.5.1–A18.5.6.

[2.402] Poulsen, P. R., Wang, M., Xu, J., *et al.*, "Role of hydrogen surface coverage during anodic plasma deposition of hydrogenated nanocrystalline germanium". Journal of Applied Physics 1998;84(6):3386.

[2.403] Torres-J., A. and Gutiérrez, E. A. "A planar amorphous $Si_{1-x}Ge_x$ separated-absorption-multiplication avalanche photo diode". IEEE Electron Device Letters 1997;18(11):568–570.

[2.404] Torres J. A., Munguía C. A., and Zúñiga I. C. "Spin on glass as an antireflection layer on amorphous absorption layer photodetectors". In: Third IEEE International Caracas Conference on Devices, Circuits and Systems, 2000, pp. 70–73.

[2.405] Castillo, A. M., Jacome, A. T., Heredia, A., and Ramos, R. "An a-SiGe: H,F based thin film MSM high-speed photodetector". In: Proceedings of the 6th International Caribbean Conference on Devices, Circuits and Systems, 2006, pp. 91–94.

[2.406] Shen D. S. and Wagner S. "Transient photocurrent in hydrogenated amporphous Silicon and implications for photodetector devices". Journal of Applied Physics 1996;79(6):794–801.

[2.407] García, M., Ambrosio, R., Torres, A., and Kosarev, A. "IR bolometers based on amorphous silicon germanium alloys". Journal of Non-Crystalline Solids 2004;338–340:744–748.

[2.408] Syllaios, A. J., Schimert, T. R., Gooch, R. W., McCardel, W. L., Ritchey, B. A., and Tregilgas, J. H. "Amorphous silicon microbolometer technology". Materials Research Society Symposia Proceedings 2000;609:A14.4.1.

[2.409] Liang, D., Rui-Feng, Y., and Li-Tian, L. "Characterization of uncooled microbolometer for infrared detection". Chinese Physics Letters 2003;20 (5):770–773.

[2.410] Moreno, M., Kosarev, A., Torres, A., and Ambrosio, M. "Fabrication and performance comparison of planar and sandwich structures of micro-bolometers with Ge thermo-sensing layer". Thin Solid Films 2007;515:7607–7610.

[2.411] Bronner, W., Kleider, J. P., Brüggemann, R., Cabarrocas, P. R. I., Mencaraglia, D., and Mehring, M. "Comparison of transport and defects properties in hydrogenated polymorphous and amorphous silicon". Journal of Non-Crystalline Solids 2002;299–302:551–555.

[2.412] Kim, K.-H., Johnson, E. V., and Cabarrocas, P. R. I. "Irreversible light-induced degradation and stabilization of hydrogenated polymorphous silicon solar cells". Solar Energy Materials and Solar Cells 2012 105:208–212.

[2.413] Moreno M., Torres A., Calleja C., *et al.*, "Exploring the infrared-sensing properties of polymorphous silicon-germanium (pm-Si_xGe_y:H) thin Films". Canadian Journal of Physics 2014;92(7–8):565–569.

[2.414] Moreno, M., Jimenéz, R., Torres, A., and Ambrosio, R. "Micro-bolometers based on amorphous silicon–germanium films with embedded nanocrystals". IEEE Transactions on Electron Devices 2015;62(7): pp. 2120–2127.

[2.415] Liu, H. B. Chen, Y., Bastians, G. J., and Zhang, X. C. "Detection and identification of explosive RDX by THz diffuse reflection spectroscopy". Optics Express 2006;14(1):415–423.

[2.416] Zhang, L., Zhong, H., Deng, C., Zhang, C., and Zhao, Y. "Terahertz wave reference-free phase imaging for identification of explosives". Applied Physics Letters 2008;92(9):091117.

[2.417] Orduña-D., A., Castillo-D., E., Torres-J., A., De la Hidalga-W., F. J., and Treviño-P., C. G. "Design and fabrication of a bidimensional microbolometer array for Terahertz detection characterized at different temperatures". Journal of Physics: Conference Series. 2011;274:012112.

[2.418] Chuss, D. T., Wollack, E. J., Moseley, S. H., Withington, S., and Saklatvala, G. "Diffraction considerations for planar detectors in the few-mode limit." The Publications of the Astronomical Society of the Pacific 2008;120:430–438.

[2.419] Semiconductor Industry Association (SIA). International technology roadmap for semiconductors 2013 edition. Executive summary. International SEMATECH, Austin, TX, 2013. Available: http://public.itrs.net.

[2.420] Koh, W. "System in packaged (SiP) technology applications". In: 6th International Conference on Electronic Packaging Technology, 2005, pp. 1–6.

[2.421] Tummala, R., SoC vs. MCM vs SiP vs. SoP. 2006. Available from: http://electroiq.com/blog/2006/07/soc-vs-mcm-vs-sip-vs-sop/. [accessed 17.10.15].

[2.422] Shinotani, K.-I., Raj, P. M., Seo, M., *et al.*, "Evaluation of alternative materials for system-on-package (SOP) substrates". IEEE Transactions On Components and Packaging Technologies 2004;27(4):694–701.

[2.423] Perez, A., Torres, A., and Torres, R. "Novel PCB fabrication process roughness free for high frequency applications". In: To be presented at The 12th International Conference on Electrical Engineering, Computing Science and Automatic Control (CCE 2015), to be held from October 26 to 30 in Mexico City.

Chapter 3
Advanced device characterization techniques

E.A. Gutiérrez-D.[1], *Stewart E. Rauch*[2], *J. Molina*[1] *and E.A. Miranda*[3]

In this chapter the conventional electrical characterization techniques are reviewed. This includes the extraction of basic electrical parameters, such as threshold voltage, parasitic components, effective electrical dimensions, and measurement of second-order effects, such as self-heating and parasitic tunneling currents. The device degradation and reliability is reviewed based on different techniques, such as BTI, for instance. Other advanced mechanisms related to soft- and hard-breakdown (HBD) are reviewed, including very specific phenomena like percolation and filamentary conduction, which may also serve as a reference for potential nanowire-like transport. As an alternative characterization technique approach, the use of a magnetic field combined with conventional electrical characterization techniques is introduced. This magneto-electrical characterization technique allows the study of local effects across small sections in the order of a few nanometers. This technique is very useful for mapping the conductance properties at the atomistic level, and could be very useful in optimizing the design of futuristic atomic-scale devices.

3.1 Conventional electrical characterization and modeling techniques

The proper modeling and characterization has been a key for the appropriate evolution of the FET-based device technology. The electrical performance of semiconductor devices is intimately linked to the device physics and technology, so doing the right electrical testing for extracting the right technology-related model parameters is crucial for the evolution of semiconductor devices, as well as for their circuit application. In this section the standard testing techniques for extracting the basic electrical intrinsic and parasitic device parameters are introduced.

[1]National Institute for Astrophysics, Optics and Electronics, INAOE, Puebla, Mexico
[2]GlobalFoundries, Inc., East Fishkill, NY, USA
[3]Universitat Autònoma de Barcelona, Spain

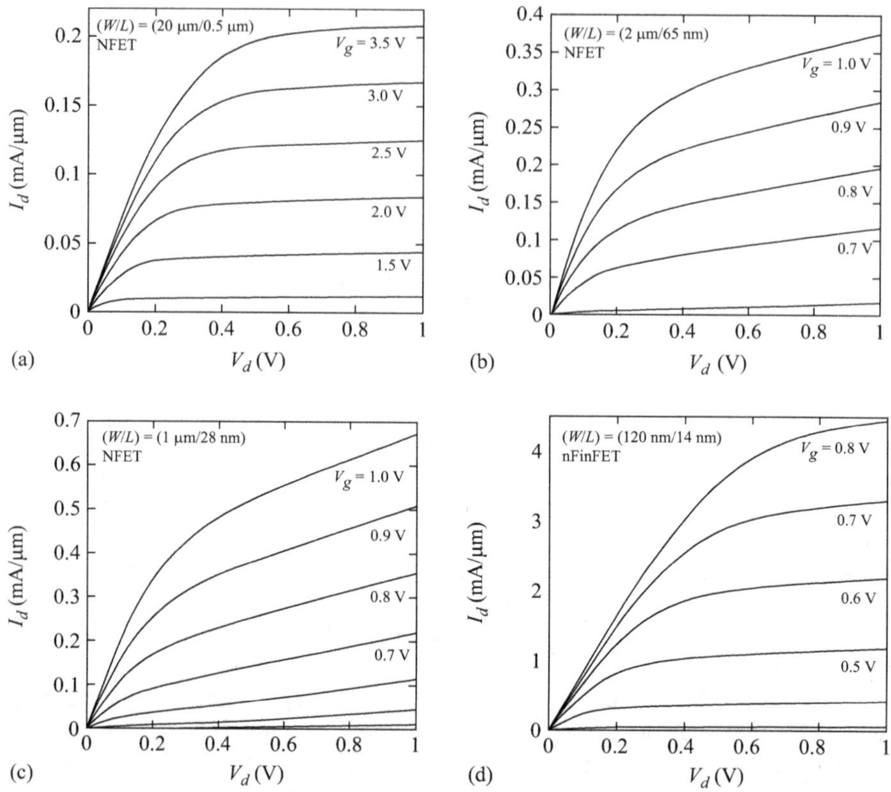

Figure 3.1 I_d–V_d *characteristics of four different bulk technologies, (a) 0.5 μm, (b) 65 nm, (c) 28 nm, and (d) a 14 nm SOI FinFET*

3.1.1 Measurement and extraction of basic electrical parameters

The basic electrical parameters of an FET device are extracted from conventional current versus voltage *I–V* and capacitance versus voltage *C–V* characteristics. The I_d–V_d characteristics of four different technology generations, 0.5 μm, 65 nm, 28 nm, and 14 nm SOI FinFET, are shown in Figure 3.1 as an example.

The drain voltage V_d axis is scaled to 3.3, 1.2, 1.0, and 0.8 V, for the 0.5 μm (a), 65 nm (b), 28 nm (c), and 14 nm (d), respectively.

It is worth to mention that the 0.5 μm technology shows an I_d–V_d profile closer to the ideal case with an output resistance $R_o = 3.0 \times 10^4$ Ω at the highest V_g voltage. For the 65 nm technology $R_o = 5 \times 10^3$ Ω, for the 28 nm technology $R_o = 3.8 \times 10^3$ Ω, while for the 14 nm $R_o = 9.2 \times 10^3$ Ω. The better electrostatic control of the FinFET technology brings back a slight improvement on the output resistance. On the other hand, the current drive capability for the highest V_d and V_g voltage conditions has a positive evolution that goes from 0.22 mA/μm for 0.5 μm, 0.37 mA/μm for 65 nm, 0.68 mA/μm for 28 nm, and 4.6 mA/μm for 14 nm. The dramatic enhancement of the electrical performance for a SOI-based FinFET technology is also seen from the I_d–V_g characteristics shown in Figure 3.2.

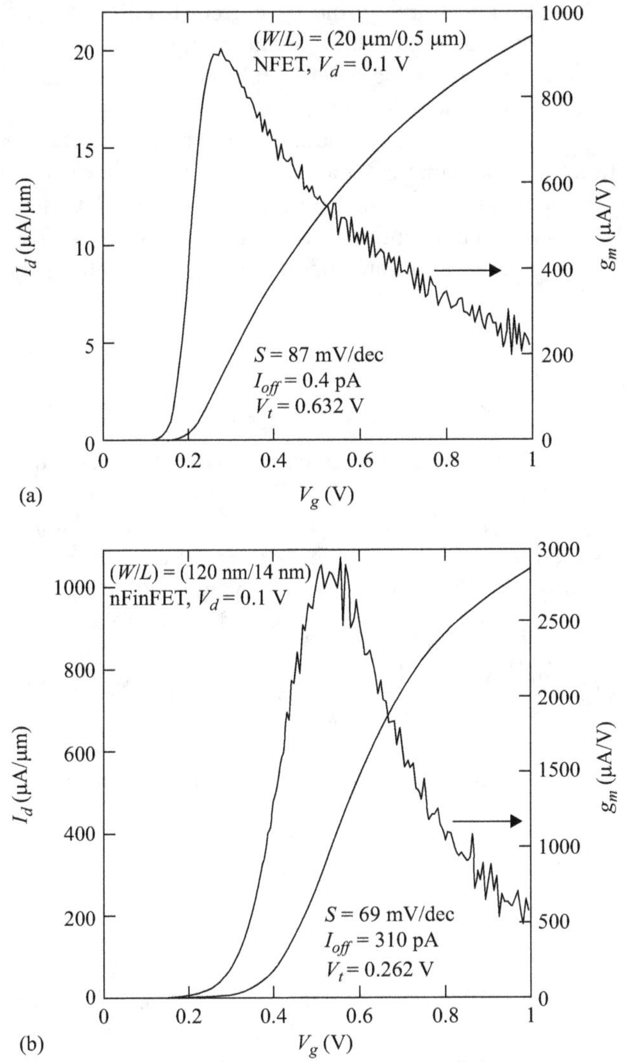

Figure 3.2 Measured I_d–V_g and g_m–V_g curves at $V_d = 0.1$ V for (a) 0.5 μm technology and (b) for a 14 nm SOI FinFET technology. The V_g axis is normalized to 3.3 for the 0.5 μm, and to 0.8 for the 14 nm technologies

The transconductance $g_m = (\partial I_d / \partial V_g)$ is 2.8×10^3 μA/V for a 14 nm technology, while for the 0.5 μm it is 9.0×10^2 μA/V. The threshold voltage V_t is 0.632 V for the 0.5 μm technology, while for the 14 nm it is only 0.262 V. The subthreshold slope S is 87 mV/dec for 0.5 μm and 69 mV/dec for the 14 nm technologies. This makes the 14 nm technology much more efficient as a logic switch with a superb current drive capability for high-performance computing applications. However, there is a parameter, which is on the background, and we need to pay special attention to it. It is the channel leakage current I_{off} at $V_g = 0.0$ V. For the conventional 0.5 μm bulk technology the I_{off} current is only 0.4 pA, while for the SOI FinFET technology is 310 pA. The I_{off} value is high if we consider that a high-performance microprocessor has the order of billions of transistors. Other parasitic leakage current components need to be accounted for, such as that of the gate oxide I_g, and bulk I_b. We will pay special attention to these parasitic currents in the following sections. A mathematical model in some extent can characterize the electrical performance. In principle such a model is the Berkeley Short-channel I_{gfet} Model (BSIM) [3.1]. For a conventional technology one can start from the BSIM3 approach [3.2], which models the three operation regions, subthreshold, linear, and saturation, as follows:

$$I_{ds} = \mu n \frac{W_{eff}}{L_{eff}} \sqrt{\frac{q\varepsilon_{Si} N_{CH}}{4\phi_B}} v_{th}^{2} \left(1 - e^{\left(\frac{-V_{ds}}{nv_{th}}\right)}\right) e^{\left(\frac{V_{gs}-V_t-V_{OFF}}{nv_{th}}\right)} \tag{3.1}$$

where V_{OFF} is a voltage determined from experimental *I–V* characteristics, and has a value that ranges from -0.06 to -1.2 V [3.3]. The thermal voltage is given by V_{th}, n is a fitting factor, μ is the carrier mobility, N_{CH} is the channel doping concentration, V_t is the threshold voltage. ϕ_B is the potential difference between the Fermi level at energy band bending and at flat-band conditions.

The linear I_{ds}–V_{ds} characteristics are modeled by the following equation.

$$I_{ds} = \mu_{eff} C_{ox} \left(\frac{W_{eff}}{L_{eff}}\right) \frac{1}{1 + \left(\frac{V_{ds}}{E_{sat}}\right) L_{eff}} \left(V_{gs} - V_t - \left(\frac{A_{bulk} V_{ds}}{2}\right)\right) V_{ds} \tag{3.2}$$

The parasitic source and drain resistance introduce a modification to previous equation.

$$I'_{ds} = \frac{I_{ds}}{1 + \left(\frac{R_{ds}}{I_{ds}}\right) V_{ds}} \tag{3.3}$$

where R_{ds} accounts for both source and drain parasitic series resistance components. E_{sat} is the electric field at the saturation V_{dsat} voltage. The carrier mobility is given by μ. C_{ox} is the gate oxide capacitance, and W_{eff} and L_{eff} are the effective width and length, respectively. A_{bulk} accounts for the bulk charge effect due to the drain-source voltage V_{ds}. The description and equation that models this and other BSIM3 parameters can be found in Reference 3.2. E_{sat} and V_{dsat} are also described in detail in Reference 3.2. V_{dsat} describes the V_{ds} voltage at the transition from the linear to the saturation region in the I_{ds}–V_{ds} characteristics. The channel current in the saturation region is given as

$$I_{ds} = W_{eff} v_{sat} C_{ox} \left(V_{gst} - A_{bulk} V_{dsat}\right) \left[1 + \frac{V_{ds} - V_{dsat}}{V_A}\right] \tag{3.4}$$

$$V_A = W_{eff} v_{sat} C_{ox} \left(V_{gst} - A_{bulk} V_{dsat}\right) \frac{1}{\left(\frac{\partial I_{ds}}{V_{ds}}\right)} \tag{3.5}$$

v_{sat} is the saturation velocity, which varies between 6 and 10 × 10^7 cm/s for electrons, and is between 4 and 8 × 10^7 cm/s for holes [3.4]. $V_{gst} = (V_{gs} - V_t)$, E_{sat} is then given by $E_{sat} = v_{sat}/\mu_{eff}$, where μ_{eff} is the effective mobility. The V_A is the early voltage, which accounts for the output resistance of the MOSFET, and has contributions from different mechanisms, such as the channel length modulation (V_{ACLM}) and the drain-induced barrier lowering (V_{ADIBL}) [3.5]. Other mechanisms, which also affect the output resistance, are the parasitic bulk currents that modify the bulk potential and threshold voltage and thus the output resistance. One of those parasitic bulk currents is the one originated by hot carriers [3.6]. The other parasitic bulk current comes from band-to-band direct tunneling (DT) at the reverse-biased drain-bulk junction [3.7]. These effects give rise to an additional modification for V_A (V_{ASCBE}). With all these mechanisms accounted for the drain current in the saturation region is given by

$$I_{ds} = W_{eff} v_{sat} C_{ox} \left(V_{gst} - A_{bulk} V_{dsat} \right) \left(1 + \frac{V_{ds} - V_{dsat}}{V'_A} \right) \left(1 + \frac{V_{ds} - V_{dsat}}{V_{ASCBE}} \right) \tag{3.6}$$

$$V'_A = V_{Asat} + \left(1 + \frac{P_{VAG} V_{gsteff}}{E_{sat} L_{eff}} \right) \frac{1}{\left(\frac{1}{V_{ACLM}} + \frac{1}{V_{ADIBL}} \right)} \tag{3.7}$$

V_{Asat} is a voltage parameter that accounts for the parasitic resistances, the effective width W_{eff} and length L_{eff} of the transistor, and the saturation voltage. P_{VAG} is a parameter that accounts for the influence of the gate voltage. A description in detail of the voltage parameter V_{Asat} and P_{VAG} can be found in Reference 3.2.

Finally the BSIM3 electrical model that describes the current in a single expression can be written as

$$I_{ds} = \frac{I_{ds0}}{1 + \frac{R_{ds} I_{ds0}}{V_{dseff}}} \left(1 + \frac{V_{ds} - V_{dseff}}{V'_A} \right) \left(1 + \frac{V_{ds} - V_{dseff}}{V_{ASCBE}} \right) \tag{3.8}$$

$$I_{ds0} = \frac{W_{eff} C_{ox} \mu_{eff} V_{gsteff} V_{dseff} \left(1 - \frac{V_{dseff}}{2 V_b} \right)}{L_{eff} \left(1 + \frac{V_{dseff}}{E_{sat} L_{eff}} \right)} \tag{3.9}$$

The effective drain-to-source voltage V_{dseff} as well as V_{gsteff} is a function introduced to guarantee continuity of I_{ds} through the different operation regimes, subthreshold, linear, and saturation with the use of a single equation (3.8). A complete description of these effective voltages is also found in Reference 3.2. The model BSIMs tested with the spice simulator [3.8] works fine as shown in Figure 3.3, where a (W/L) = (10 μm/0.5 μm) nMOSFET of a 0.5 μm technology is used as sample. The simulation results are compared with experimental values taken at $T = 300$ K. The one-equation BSIM3 model is able to reproduce the experimental results with continuity between the different operation regimes, subthreshold, linear, and saturation.

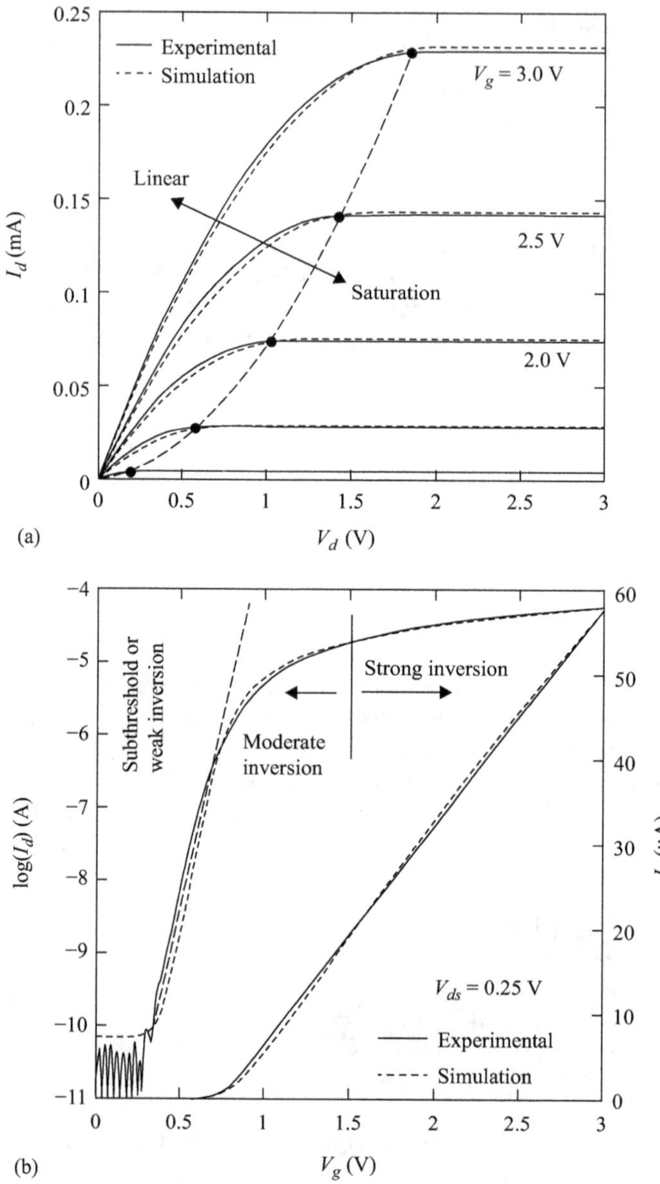

Figure 3.3 Simulated results of (W/L) = (10 μm/0.5 μm) nMOSFET using spice with BSIM3 as a model. The long dashed line is a parabolic fit to the dot points that represent the V_{dsat} voltage on (a). The long dashed line on (b) is a fit to the exponential subthreshold region

The number of parameters required for a basic DC simulation of a 0.35 μm MOS technology is above 100 [3.9]. The threshold voltage has 16 parameters, the carrier mobility has 4, the subthreshold region has 4, the saturation region requires 14, 14 parameters related to geometry modulation are also required, the temperature effects require 9, the overlap capacitances uses 11, the parasitic resistance-related parameters are 7, the process-related parameters are 3, and the substrate-related parameters are 2. Other parameters related to noise and common extrinsic components are in the order of 22.

The silicon foundries that offer the fabrication capacity as a commercial product provide that large amount of parameters. However, at the research environment, when studying new experimental device structures, one has to rely on parameter extraction programs such as the ICCAP offered by Keysight (former Agilent) [3.10]. The ICCAP program works in conjunction with the Agilent B1500A Semiconductor Device Analyzer, which performs the electrical characterization. The ICCAP as well as other parameter extraction programs, such as UTMOST III from Silvaco [3.11], requires *I–V* and *C–V* measurements on both types of MOSFETs, p- and n-type with wide and long, wide and short, and narrow and short devices over a temperature range from 25 to 125°C and under a wide variety of bias conditions.

The BSIM model is now in its version 6, which provides coverage for bulk analog and radio frequency (RF) applications [3.12]. There is also a BSIMSOI version that models SOI devices based on the BSIM3 version, BSIMCMG for common multi-gate FETs, and BSIMIMG for independent multi-gate ultra-thin-body (UTB) SOI transistors.

Besides BSIM, there are other options for MOSFET modeling for circuit applications. In 1995, for instance, C. Enz, F. Krummenacher, and E. Vittoz published their model [3.13], which is known as the EKV model. It is a fully analytical MOS transistor model dedicated to the design and analysis of low-voltage, low-current analog circuits. In its first version it had only nine physical parameters, three fine-tuning fitting coefficients, and two additional temperature parameters. This model is valid in all modes of the MOSFET operation. There is a recent list of designs that makes use of the EKV3 version model for RF applications, such as Reference 3.14 where the design methodology for a 5.5 GHz LNA is introduced for a 120 nm CMOS technology. Vertical MOSFETs (VMOSFETs) from a 0.35 μm technology have also been characterized and modeled, with a measured f_T and f_{MAX} can reach 30.5 GHz and 41 GHz, respectively [3.15]. Another paper reports on a reliable and portable design of an LNA validated for a 2.4 GHz LNA in a 28 nm CMOS technology [3.16]. The EKV3 version for DC operation can be modeled with 42 parameters for a 0.12 μm CMOS technology [3.17]. RF noise parameters for a 90 nm CMOS technology have been characterized and modeled, up to 24 GHz using the EKV3 model [3.18].

As in the case of BSIM there is also a parameter extraction methodology available [3.19]. The parameter extraction procedure is accurate for both submicron CMOS and fully depleted SOI technologies. It is capable to extract the 13 main EKV intrinsic model parameters for first- and second-order effects.

The third option for MOSFET modeling is the so-called PSP or surface-potential-based MOSFET model [3.20]. The PSP model incorporates a wide variety of mechanisms, such as polysilicon depletion, deviation from the "universal" dependence of the effective channel mobility, trap-assisted tunneling (TAT) through the drain-bulk reverse-biased junction, degradation of the output resistance due to the incorporation of Halo doping implants, RF noise sources, gate tunneling current, and gate-induced drain and source leakage (GIDL and GISL) currents. The PSP model is based on the calculation of the surface potential at the source and drain sides of the channel, the intrinsic drain channel current, and the terminal charges. The extrinsic part of the model includes the overlap charges, gate tunneling current, substrate impact ionization current, the series resistance, and spectral densities of the noise. The model has been tested for 180, 130, 90, and 65 nm technologies. The PSP model has approximately 70 parameters, which can be extracted by measuring *C–V* and *I–V* curves. This data is processed with a special particle swarm optimization (PSO) algorithm reported in Reference 3.21, which allows the extraction of 34 DC parameters of a 65 nm technology. As published by Fujitsu [3.22] the PSP model is far more accurate than BSIM in the subthreshold region for high-speed applications. This fact is also verified by the ability of the PSP model to reproduce the slopes of the first to fourth harmonics and a fifth-order harmonic that follows the gradient of the correct slope at significant power levels for a 90 nm nMOSFET [3.23]. The PSP model then can achieve much more accurate distortion results.

There is a fourth option for a MOSFET model, the MOS9 by Philips [3.24], which is available as level 50 in Star-HSPICE [3.25] and has around 100 parameters. The MOS9 model is intended for analog applications, which incorporates nodal currents and charges, noise-power spectral densities, and weak-avalanche currents. The model is based on the gradual-channel approximation with a number of first-order corrections to account for small-size devices. The updated MOS11 model is also suitable for digital and analog RF applications.

The fifth option for RF MOSFET modeling for advanced technologies, such as 28 nm, is the Hiroshima University model called HiSIM [3.26]. The HiSIM2 model has been used for designing low-power RF-transceiver circuit for the K-band in a wireless localization system [3.27]. The design methodology based on HiSIM2 has been proven for a 130 nm CMOS technology 24 GHz wireless localization system.

The compact model council (CMC) [3.28] has selected the BSIM, PSP, and HiSIM as standard models for the design of integrated circuits.

In general there is a trend to use three fundamental compact models for advanced technologies of 28 nm and below as shown in Table 3.1. The BSIM advanced models are based on the iterative calculation of the inversion channel

Table 3.1 MOSFET technologies and corresponding compact models

Technology	Models
0.5 μm	MOS9, MOS11, BSIM2, BSIM3
90, 65, 45 nm	MOS11, BSIM4, BSIM5, EKV, PSP
28 nm and below	BSIM6, BSIMSOI, BSIMCMG, BSIMIMG, PSP, HiSIM2

charge Q_{inv} with the use of smooth functions to make a continuous transition between different operation regimes. The HiSIM, MOS9, MOS11, and PSP models are based on the calculation of the surface potential along the different operation regimes. These models do not require smooth functions to make a continuous transition between different operation regimes. The EKV is a charge-based continuous model combined with the surface potential.

The Silicon Integration Initiative (Si2) and the CMC [3.27], which is the largest organization of industry-leading semiconductor, systems, EDA and manufacturing companies, have accomplished three industry standard models: (1) BSIM, (2) PSP, and (3) HiSIM.

Detailed information about the BSIM, PSP, and HiSIM models is found in References 3.12, 3.28, and 3.29, respectively. These three advanced compact models have a number of model parameters that overpass 100, which require special parameter extraction techniques, such as that reported in Reference 3.30 for BSIM, Reference 3.21 for PSP, and Reference 3.31 for HiSIM.

A comparison of these three compact models has been done in Reference 3.32. In such a study DC small signal and large signal analysis have been compared to experimental results. For low overdrive voltages the BSIM4 model is accurate for small-signal performance, which makes it a good option for LNAs, mixers, and other low-power circuits. For drain voltages closer to zero, the PSP model seems to be more reliable for correct linearity estimation. The HiSIM seems to be a good choice for large signal analysis, but its scalability is not good.

From the three different MOSFET operation regimes, subthreshold, linear, and saturation, several basic parameters can be extracted using regular or conventional techniques based on *I–V* or *C–V* characteristics, such as the subthreshold slope S; the threshold voltage V_t; the drain saturation voltage V_{dsat}; the carrier mobility μ; the effective channel width and length W_{eff} and L_{eff}; the output resistance $R_o = (\partial V_{ds}/\partial I_{ds})$; the parasitic series source/drain resistances R_{ss} and R_{ds}; and the gate-to-source C_{gs}, gate-to-drain C_{gd}, gate-to-bulk C_{gb}, and the other related capacitances.

3.1.2 *Parasitic effects and de-embedding under nonconventional bias conditions*

The RF characterization techniques, which are usually used for extracting the RF MOSFET model parameters, can also be used to extract DC parameters, or to study second-order effects. This is the case of the series resistance for instance [3.33, 3.34], where the use of *S*-parameters serves to extract the series resistance from a single device, instead of a channel length array as done with DC extraction techniques.

The series resistance at both the source and the drain side is conventionally extracted using DC measurement techniques [3.35–3.37]. A DC measurement technique that allows the extraction of the gate-voltage dependence of the series resistance is given in Reference 3.38. All these DC extraction techniques require data taken from devices of different channel length. Therefore, in Reference 3.39 a *S*-parameters-based extraction technique is introduced, which allows the series resistance to be extracted from a single device and as a function of the gate

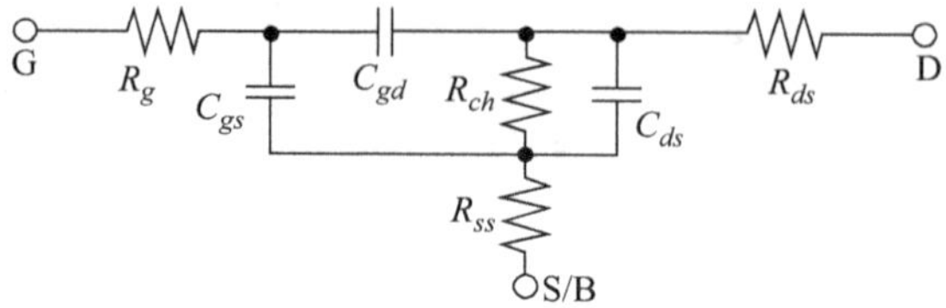

Figure 3.4 A first-approach model for a MOSFET biased at $V_{ds} = 0.0$ V and $V_g > V_t$ at low frequencies

voltage [3.40]. However, this *S*-parameter-based extraction technique does not allow the identification of the bias-dependent components from the bias-independent ones. In Reference 3.33 an *S*-parameter-based measurement technique is proposed to extract the bias-dependent and bias-independent series resistance components from a single device. In order to eliminate the influence of pad parasitics and the large influence of gate resistance R_g, the on-wafer measurements are de-embedded from pad parasitics using two multi-fingered common-source/bulk $(W/L) = (10\ \mu m/0.18\ \mu m)$ n-type MOSFET, one with two and the other one with 20 fingers. Under these conditions, the total series resistance R_{st} is considered as the sum of a bias-independent and bias-dependent term.

$$R_{st} = R_{ss\_cons} + R_{ss\_bias} + R_{ds\_cons} + R_{ds\_bias} \tag{3.10}$$

The low-frequency and simplified MOSFET model for $V_{ds} = 0.0$ V and gate voltage V_g larger than the threshold voltage V_t is shown in Figure 3.4. The R_{ss_cons} and R_{ds_cons} refer to the bias-independent series resistance components, while R_{ss_bias} and R_{ds_bias} refer to the bias-dependent series resistance components. The constant term accounts for the contact and sheet resistance, and the bias-dependent term accounts for the gate-voltage-dependent components. The bias-dependent term follows the model [3.36]:

$$R_{ss\_bias} = \frac{R_{ss0}}{V_g - V_t} \tag{3.11}$$

where R_{ss0} is a fitting parameter. A MOSFET biased at $V_{ds} = 0.0$ V and high V_g has a negligible transconductance gm and a channel resistance R_{ch} very small compared to the substrate impedance. Therefore, under this bias condition, the substrate effects are negligible at relatively low frequencies, which validate the equivalent circuit shown in Figure 3.5. Then R_{ss} and R_{ds} can be modeled as [3.40]

$$R_{ss} = Re(Z_{12}) - \frac{1}{2} \cdot \frac{R_{ch}}{1 + (\omega C_x R_{ch})^2} \tag{3.12}$$

$$R_{ds} = Re(Z_{22}) - R_{ss} - \frac{R_{ch}}{1 + (\omega C_x R_{ch})^2} \tag{3.13}$$

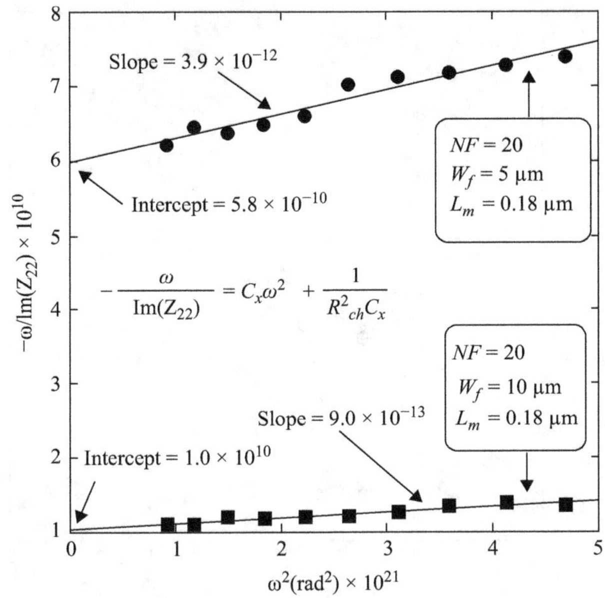

Figure 3.5 Linear regression used to extract R_{ch} and C_x from the experimental data of transistors at $V_{ds} = 0.0$ V and $V_{gs} = 0.9$ V

where

$$C_x = C_{ds} + \frac{C_{gs}}{(C_{gs} + C_{gd})} \tag{3.14}$$

R_{ss} and R_{ds} can then be evaluated from (3.12) and (3.13), provided C_x and R_{ch} are obtained. From the equivalent circuit in Figure 3.5, the imaginary part of Z_{22} can be expressed as

$$Im(Z_{22}) = -\frac{\omega C_x R_{ch}^2}{1 + \omega^2 C_x^2 R_{ch}^2} \tag{3.15}$$

Therefore, C_x and R_{ch} are calculated from the slope and the intercept of the linear regression from Figure 3.5. The inset in Figure 3.5 shows the rearrangement of (3.15) used for the linear regression on the experimental data. There is a good linearity for the narrow transistor ($NF = 2$) up to 10 GHz, which is the frequency limit where the substrate effects are negligible. On the other hand, data dispersion is observed for the wide transistor ($NF = 20$). The value of C_x, in the case of wide transistors, dominates over R_{ch}, which implies the parameter Im(Z_{22}) is not represented by (3.15) any longer. Therefore narrow transistors are preferred over wide transistors to extract R_{ss} and R_{ds}. This procedure is repeated for different values of V_g, so R_{ss} and R_{ds} can be plotted as a function of $(V_{gs} - V_t)^{-1}$ as shown in Figure 3.6. The constant components of the series resistance, R_{ss_cons} and R_{ds_cons}, are obtained from the extrapolation to

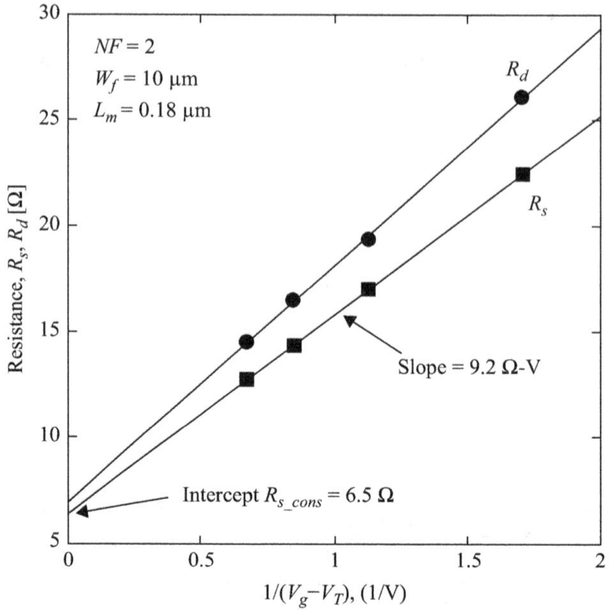

Figure 3.6 Extracted series resistance R_{ss} and R_{ds} from the linear regression on experimental data of transistor with NF = 2

$(V_g - V_t)^{-1} = 0$. The R_{ss0} slope, which determines the rate of change of the parasitic resistance with respect to the V_g voltage and which is a fundamental term of R_{ss_bias} and R_{ds_bias}, is also extracted from Figure 3.6. One important aspect to remark here is the difference between R_{ss} and R_{ds}. The differences in the slope R_{ss0} and intercept (R_{ss_const}, R_{ds_const}) are attributed to the transistor asymmetry [3.41].

The constant and bias-dependent values for the series resistance, within the frequency range of 10 GHz and with the equivalent circuit proposed in Figure 3.4, have been validated through HSPICE simulation. The simulated results are shown in Figure 3.7.

The HSPICE simulation shows a deviation beyond 10 GHz since substrate parasitics were not considered in the equivalent circuit. However, for frequencies up to 10 GHz the extraction procedure and model works fine. For frequencies above 10 GHz the device substrate elements should be extracted and added to the equivalent circuit. Since the accuracy of this extraction procedure depends on the value of the channel resistance, narrow transistors are preferred over wide devices.

From a general perspective, if this method wants to be extended beyond 10 GHz, one has to account for a complete equivalent circuit model for the shielded pad design as shown in Figure 3.8 [3.34]. An equivalent circuit, for this particular test structure, with contact at the third metal level, is shown in Figure 3.8b. In this model R_i and R_o represent purely resistive elements since they are inherent to the sheet resistance of the pad metal layer and contact resistance. Y_i and Y_o are not purely capacitive, but a series connection with capacitors. Other elements in the

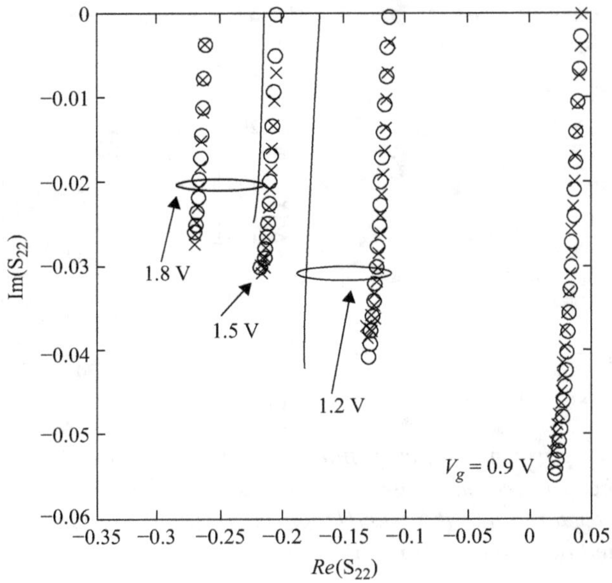

Figure 3.7 *Measured and simulated S22 up to 10 GHz for different V_g voltages. $(W_f/L_m) = (10\ \mu m/0.18\ \mu m)$ MOSFET with NF = 2. The solid lines correspond to simulated results (V_g = 1.2, 1.5, and 1.8 V), while the symbols correspond to measured results. The solid lines are for the constant values of R_{ss} and R_{ds}. Circles = experiment, crosses = simulation*

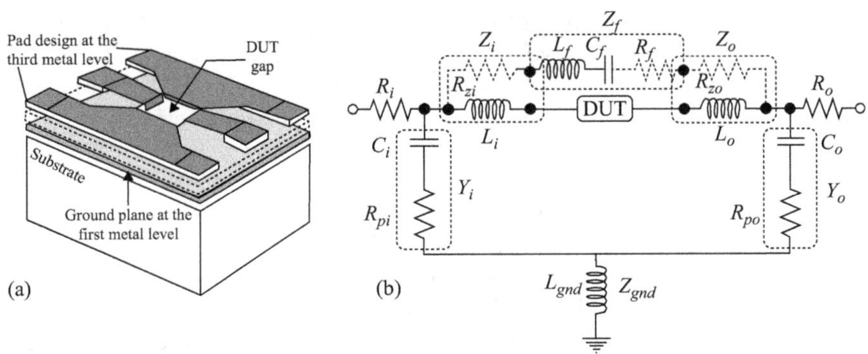

Figure 3.8 *(a) Conventional shielded test structure pad design with a third metal level, and (b) equivalent circuit model for the shielded pad. The solid lines represent the model for conventional structures, while the dotted lines are the components that need to be added to improve the model*

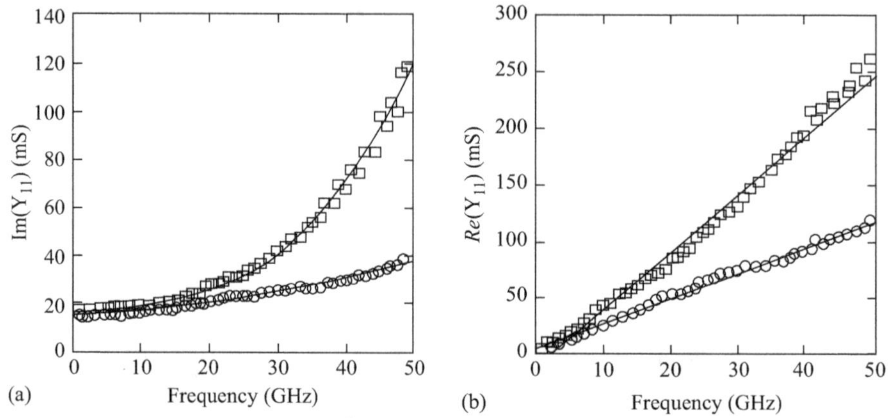

Figure 3.9 Comparison of experimental and simulated results (continuous line) of the real and imaginary terms of parameter Y_{11} for the open pad structure up to the 50 GHz. Open circles correspond to metal 3 level and open squares to metal 2 level

equivalent circuit have inductance behavior as those related to the output, input, and ground impedances. A complete and detailed description of an analytical parameter extraction methodology can be found in Reference 3.34.

As an example of the accuracy of the equivalent circuit and model, a comparison between the simulated and experimental data of the real and imaginary parts of parameter Y_{11}, for the open dummy structure, is shown up to 50 GHz in Figure 3.9.

The proposed equivalent circuit and model agrees well with the experimental data. Using higher-level metals to reduce the input/output shunt capacitance is an advantage, but at the cost of introducing new resistive elements in the equivalent circuit. These new resistive elements come from the metal thickness differences in each layer, and the use of deep vias to connect the device under test (DUT) to the pad.

Besides the parasitic equivalent circuit of the pad, and those related to the intrinsic device parasitics, such as resistances, capacitors, and inductors, at high frequency one needs to care about device intrinsic second-order effects, which are notorious at DC. Examples of these effects are the gate oxide tunneling current [3.42] and the band-to-band tunneling (BTBT) from the reverse-biased drain-bulk junction [3.43], just to mention a couple of these second-order effects. The gate oxide leakage I_g at DC, for advanced FET technologies, may be in the range of hundreds of pico-Amperes (pA) for a single transistor. Therefore, at high frequency such a high level of parasitic current, coupled with a relatively large parasitic gate resistance R_g, leads to a considerable degradation of the transistor electrical performance at high frequency [3.44]. Therefore, the use of gate-interdigitated electrode is a must to reduce R_g [3.45].

For nanometric devices the use of high doping levels at both the bulk and drain/source terminals is a need for the electrical performance optimization. However, such a high-doping levels leads to BTBT at the drain/bulk junction [3.43]. The BTBT charge leakage generates a parasitic current in the intrinsic

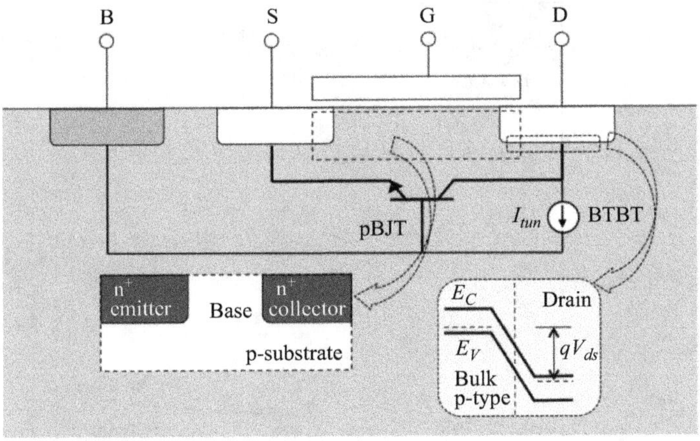

Figure 3.10 MOSFET model biased at $V_g = 0.0$ V and $V_d > 0.0$ V showing the BTBT I_{tun} current and associated parasitic bipolar transistor

MOSFET bulk. The magnitude of this current is negligible when compared to the channel current operating in the strong inversion regime. However, the magnitude of this parasitic current is high enough to turn on the parasitic bipolar transistor (pBJT) formed between the intrinsic source, bulk, and drain MOSFET nodes as shown in Figure 3.10 [3.46].

Because the current gain β of the bipolar effect, schematically shown as a pBJT device in Figure 3.10, is a strong function of the channel length L, the pBJT will turn on much stronger at shorter L_s. This effect is shown by the measured data on n-MOSFETs of three different technologies, 90 nm, 65 nm, and 28 nm, biased at $V_g = 0.0$ V.

As shown in Figure 3.11, the shorter the transistor, the larger is the channel leakage current. The increase of the leakage current at very short channel lengths is either related to BTBT at the drain-bulk junction, or to tunneling from source-to-drain, or to a combination of both together with the bipolar transistor action.

The BTBT effect is shown to happen at the drain-bulk junction when the drain is biased at its maximum value as shown in Figure 3.12. Due to the heavy doping effects at both sides, the drain and bulk, the band bending is quite pronounced, which leads to direct BTBT or trap-assisted band tunneling [3.47]. This mechanism is responsible for the parasitic tunneling current shown in Figure 3.10.

When the transistor shortens, an additional tunneling current may happen from source-to-drain as shown in Figure 3.13. This is the case when the length of the potential barrier L_b is reduced by the effect of the drain voltage. The other effect is the DIBL that decreases the potential barrier height E_b.

As shown in Figure 3.13, the potential barrier length L_b reduces as the drain voltage V_d increases. This mechanism as a function of V_d is shown in Figure 3.14.

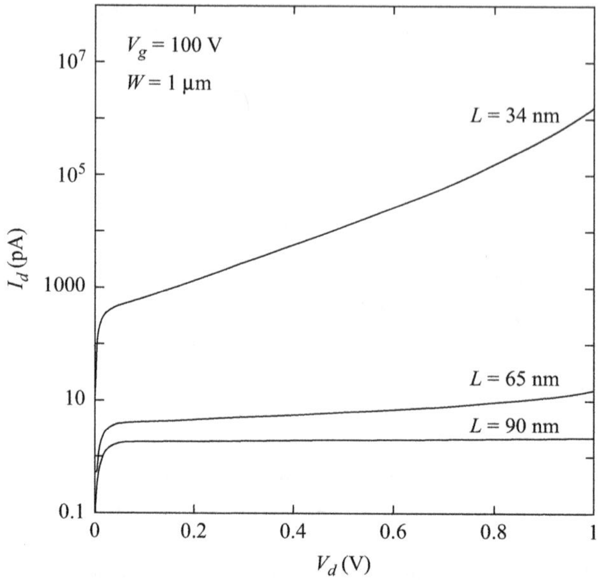

Figure 3.11 Measured leakage channel current versus V_d for $V_g = 0.0$ V for transistors of three different technologies (90, 65, and 28 nm)

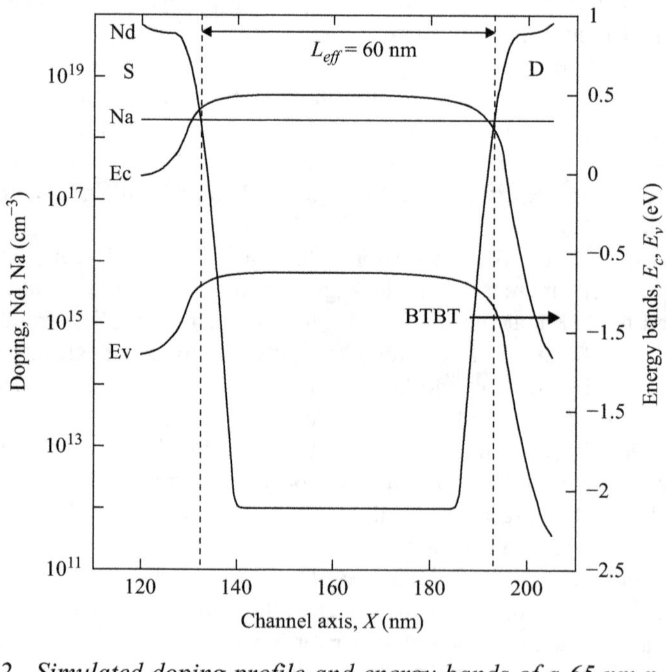

Figure 3.12 Simulated doping profile and energy bands of a 65 nm n-type MOSFET biased at $V_g = 0.0$ V and $V_d = 1.2$ V

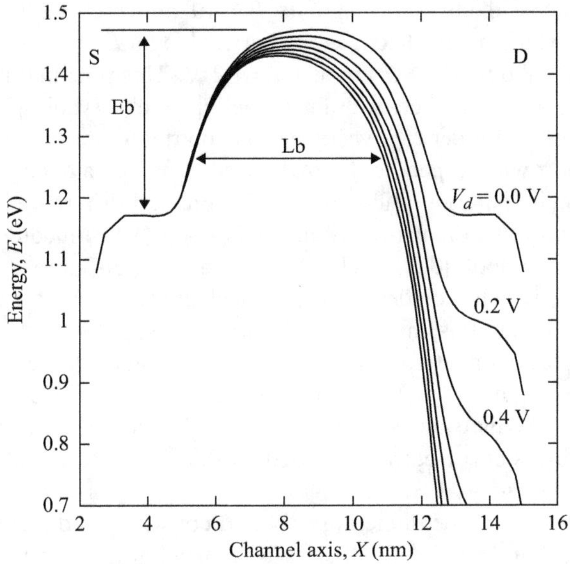

Figure 3.13 Simulated potential barrier for a (W/L) = (1 μm/10 nm) n-type MOSFET at $V_g = 0.0$ V, and V_d varying from 0.0 to 1.2 V

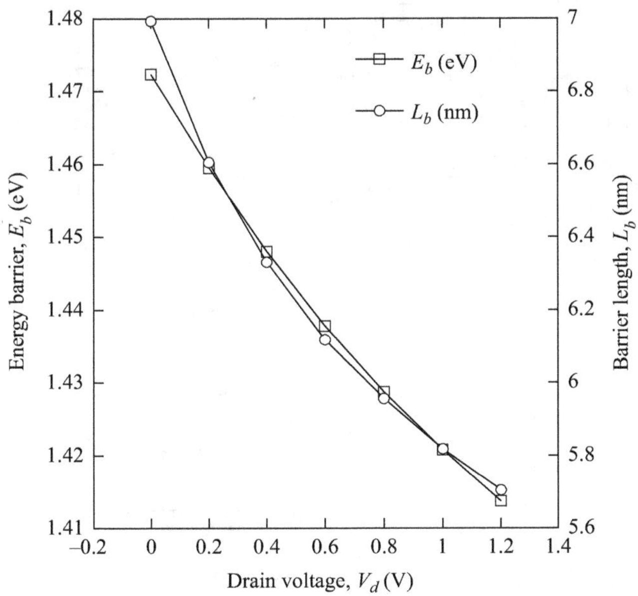

Figure 3.14 Simulated results of E_b and L_b as a function of V_d for $V_g = 0.0$ V

Although the physical channel length is $L = 10$ nm, the effective or electrical channel length, which here is taken as the barrier length L_b, is shorter. It goes from 7 nm at $V_d = 0.0$ V down to 5.7 nm at $V_d = 1.2$ V. This potential barrier length is within the range where source-to-drain tunneling is observable [3.48, 3.49], i.e., shorter than 10 nm. In order to investigate the contribution of impact of source-to-drain DT transport with respect to thermal-based transport, a double gate MOSFET structure with a silicon body thickness t_{Si} of 5 nm and different channel length L was simulated with the Vienna-Schrödinger-Poisson (VSP) module of the gtsFramework simulation tool [3.50]. The VSP is a general-purpose semiconductor device simulator, which includes quantum mechanical solvers for both open- and closed-boundary conditions, and outputs information such as the wavefunction and tunneling transport parameters, for instance. The implemented device for simulation is shown in Figure 3.15. Two main charge transport mechanisms lead to the channel current: thermal and source-to-drain DT. The first process is basically dominated by diffusion of carriers excited in the source side and moving to the drain side over the energy barrier.

The second process, the tunneling transport consists of a direct source-to-drain tunneling mechanism [3.51] and a thermally-assisted-at-the-source source-to-drain tunneling [3.52]. In a first approximation, as shown in Reference 3.49, the energy barrier can be seen as a rectangular-shaped potential with a barrier height E_b and a barrier length L_b. The direct source-to-drain tunneling may be, in principle, assumed as a temperature-independent transport mechanism, while the thermal component is diffusive in nature and thus has a positive temperature coefficient.

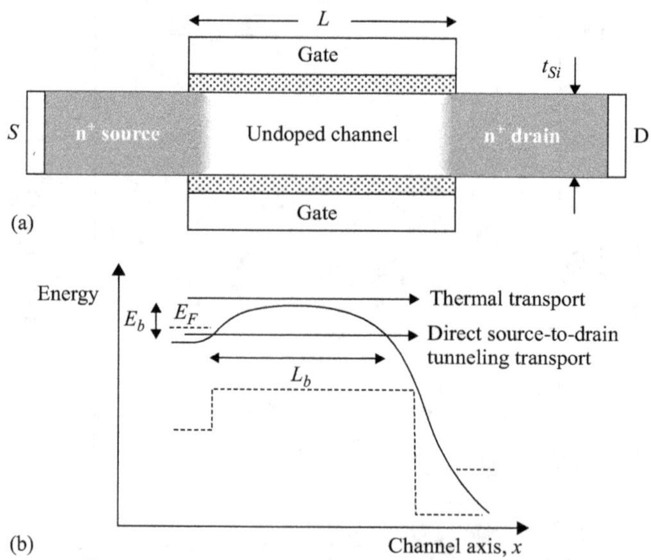

Figure 3.15 (a) Schematic representation of the double-gate simulated MOSFET device, and (b) energy potential barrier profile at $V_g = 0.0$ V and $V_d = 1.2$ V

The thermal-assisted tunneling component is a mixture of thermal and DT transport. The VSP simulation of the double-gate n-type MOSFET, biased at $V_g = 0.0$ V and $V_d = 1.2$ V, i.e., the off-state channel leakage current, shows a rapid increase as the channel length shrinks from 80 nm down below 10 nm as shown in Figure 3.16. The off-state leakage channel current I_{dn} has been normalized to its value at $L = 80$ nm. From 80 nm down to 40 nm an exponential increase of the leakage current is observed, which is attributed to DIBL-based thermal and source-to-drain tunneling. The pure direct source-to-drain leakage component is extracted from the total leakage current and plotted as a percent on the right axis. The percent tunneling component starts to show up at $L = 15$ nm with a 3.6 per cent of the total leakage current. At $L = 10$ nm the tunneling contributions amounts to 38 per cent of the total leakage current, and reaches 96 per cent for $L = 6$ nm. The percentage component of tunneling mechanism on the leakage current is also a function of technology parameters, such as doping, gate oxide thickness, and gate material. In general we consider that for gate lengths shorter than 10 nm, the source-to-drain DT component amounts for the 50 per cent of the total leakage current.

There are three tunneling current mechanisms in a nano-scaled MOSFET: the tunneling through the gate oxide, the tunneling from source-to-drain, and the tunneling through the reverse-biased drain-bulk junction. All these mechanisms contribute to parasitic leakage currents under DC conditions.

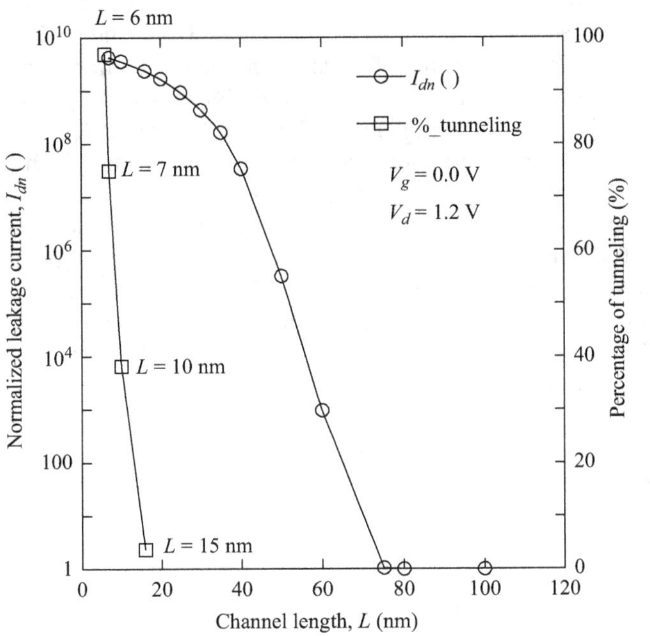

Figure 3.16 Simulated normalized channel current I_{dn}, and percentage contribution of DT to the total channel leakage current

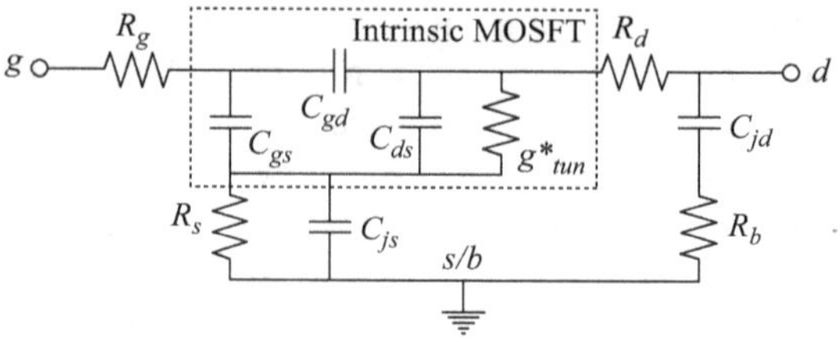

Figure 3.17 Equivalent circuit model for the RF MOSFET biased at $V_g = 0.0$ V and $V_d > 0.0$ V

Now we analyze the impact of one of these tunneling currents, the reverse-biased drain-bulk tunneling current, on the microwave performance of a MOSFET up to 40 GHz.

The drain-bulk tunneling current I_{tun} is placed at the drain terminal, which should impact the output impedance as shown in Figure 3.10. For showing the case we use the same device as that described in Figure 3.7.

By measuring the S-parameters of the MOSFET biased at $V_g = V_d = 0.0$ V, we observe that an additional admittance should be included in the RF MOSFET equivalent circuit model if the BTBT effect is to be accounted for. Such an admittance accounts for the parasitic current introduced by the BTBT current I_{tun}. The measured RF MOSFET has a mask length $L = 100$ nm and two gate fingers ($NF = 2$), with finger width $W_f = 5$, 10, 15, and 20 μm, respectively. The S-parameters are measured at $V_g = 0.0$ V and $V_d = 0.3$, 0.55, 0.8, and 1.05 V. At the $V_g =$ 0.0 V condition there is no channel formation, so there is no impact of any parasitic current at all. The equivalent circuit of the MOSFET biased at $V_g = 0.0$ V including the intrinsic and extrinsic components is shown in Figure 3.17.

The parasitic source, gate, drain, and bulk resistances are given by R_s, R_g, R_d, and R_b, respectively. The other two extrinsic parasitic components are the source- and drain-bulk junction capacitances, C_{js} and C_{jd}, respectively. The intrinsic part of the MOSFET is represented by the three capacitances, C_{gs}, C_{gd}, and C_{ds}, and a complex admittance g^*_{tun}, associated with the BTBT I_{tun} current. As $V_g = 0$ V all the extrinsic parameters are considered as drain bias-independent components [3.53]. Because of the high doping of the p–n junctions, the drain and source junction capacitances, as well as the substrate resistance, are also considered as bias-independent [3.54]. The equivalent circuit of Figure 3.17 is then used to extract the substrate parameters. Under the cold-FET condition, i.e., $V_d = 0.0$ V, the effect of g^*_{tun} is neglected, which allows to rearrange the equation for the real part of Z_{22} in the following form:

$$\frac{\omega^2}{Re\left(\frac{1}{Z_{22}}\right)} \approx R_b\omega^2 + \frac{1}{C_{jd}^2 R_b} \tag{3.16}$$

where ω is the angular frequency in radians. From this equation R_b and C_{jd} are readily obtained from the slope and the intercept with the abscises of a linear regression of the measured $\omega^2/Re(1/Z_{22})$ versus ω^2 data. The effect of these extracted parameters is removed from experimental data by applying the following equation:

$$Y_A = Y_{meas} - \begin{bmatrix} 0 & 0 \\ 0 & \dfrac{\omega^2 C_{jd}^2 R_b}{1+\omega^2 C_{jd}^2 R_b^2} + \dfrac{j\omega C_{jd}}{1+\omega^2 C_{jd}^2 R_b^2} \end{bmatrix} \tag{3.17}$$

where Y_{meas} is the Y-parameter matrix obtained from the direct transformation of measured S-parameters into Y-parameter data, whereas Y_A is the Y-parameter matrix after removal of R_b and C_{jd} effects. The bias-dependent source and series resistances are then obtained by using the method proposed in Reference 3.33, and their corresponding effect de-embedded from experimental data by applying the following equation:

$$Z^* = Z_A - \begin{bmatrix} R_g & X - j\omega C_{js} R_s X \\ X - j\omega C_{js} R_s X & R_d \end{bmatrix} \tag{3.18}$$

where

$$X = \frac{R_s}{1+\omega^2 C_{js}^2 R_s^2} \tag{3.19}$$

Z_A is the Z-parameter transformation of Y_A and Z^* is the corresponding Z-parameter matrix after removal of the extrinsic resistances effect. Due to the device symmetry $C_{jd} \approx C_{js}$. Now that the effect of the extrinsic parameters has been removed from experimental data, the extraction of parameters is focused on the intrinsic MOSFET. The intrinsic MOSFET behavior is considered as pure capacitive even for high V_d values, which makes the real part of the intrinsic Y-parameters approximately equal to zero. However, the measured real part of the intrinsic Y_{22}^* parameter shows a frequency dependence as shown in Figure 3.18.

A sinusoidal behavior of $Re(Y_{22}^*)$ is fitted to the measured data, which suggests an additional current is flowing from the intrinsic drain toward the source, in parallel with C_{ds}. Then, this current needs to be added to the equivalent circuit model. As demonstrated in Reference 3.46, such a current corresponds to the BTBT tunneling I_{tun} current.

According to the equivalent circuit in Figure 3.17, when g_{tun}^* is connected in parallel with C_{ds}, the admittance Y_{22}^* can be expressed as

$$Y_{22}^* = g_{tun}^* + j\omega\left(C_{ds} + C_{gd}\right) \tag{3.20}$$

where g_{tun}^* is the admittance associated to the BTBT I_{tun} current given by [3.55]

$$g_{tun}^* = g_{tun} e^{-j\left(\omega\tau_0 - \frac{\pi}{2}\right)} \tag{3.21}$$

where g_{tun} and τ_0 are, respectively, the magnitude and the phase delay introduced by the BTBT effect. Y_{22}^* has real and imaginary parts, which allows the BTBT admittance to be modeled as

$$g_{tun}^* = g_{tun}\left[cos\left(\omega\tau_0 - \frac{\pi}{2}\right) - jsin\left(\omega\tau_0 - \frac{\pi}{2}\right)\right] = g_{tun}[sin(\omega\tau_0) + jcos(\omega\tau_0)] \tag{3.22}$$

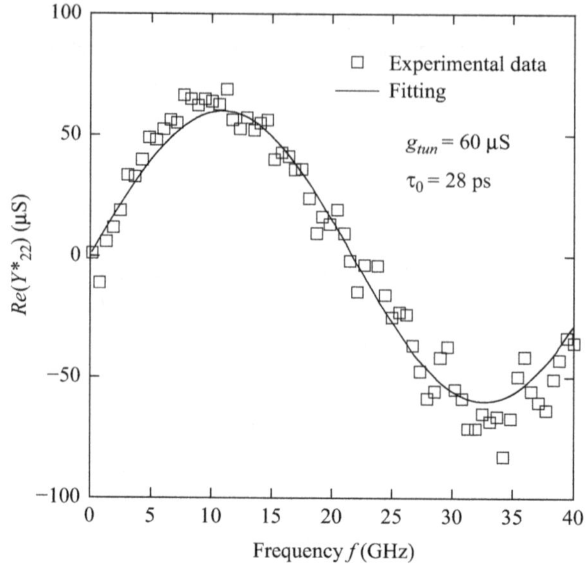

Figure 3.18 Measured $Re(Y^*_{22})$ as a function of frequency. A sinusoidal function, represented by the solid line, is fitted to the experimental data

From (3.20) and (3.21) the real part of Y^*_{22} is indeed represented by a sinusoidal function

$$Re\left(Y^*_{22}\right) = g_{tun} sin(\omega \tau_0) \tag{3.23}$$

It is now evident that both g_{tun} and τ_0 are directly determined from the magnitude and phase velocity of a sinusoidal fitting of the measured Y^*_{22} as shown in Figure 3.15, where $g_{tun} = 60$ μS and $\tau_0 = 28$ ps. With these extracted values, the intrinsic capacitances are directly obtained from the imaginary part of the intrinsic Y-parameters.

The BTBT current density J_{BTBT} or J_{tun} generated at the reverse-biased drain-bulk junction is given by [3.56]

$$J_{tun} = \frac{\sqrt{2m^*}q^3\sqrt{\frac{2qN_a(V_{ds}+\psi_{bi})}{\epsilon_{Si}}}V_d}{4\pi^3 h\sqrt{E_g}} exp\left(-\frac{4\sqrt{2m^*}E_g^{1.5}}{3qh\sqrt{\frac{2qN_a(V_{ds}+\psi_{bi})}{\epsilon_{Si}}}}\right) \tag{3.24}$$

where N_a is the bulk doping concentration, ψ_{bi} is the built-in potential of the drain-bulk junction, ε_{Si} is the silicon relative permittivity, q is the fundamental charge, m^* is the electron effective mass, h is the Planck constant, and E_g is the energy bandgap of silicon.

In order to verify BTBT is occurring at the drain–bulk junction and that the model of (3.24) applies, the different currents flowing through the circuit model of

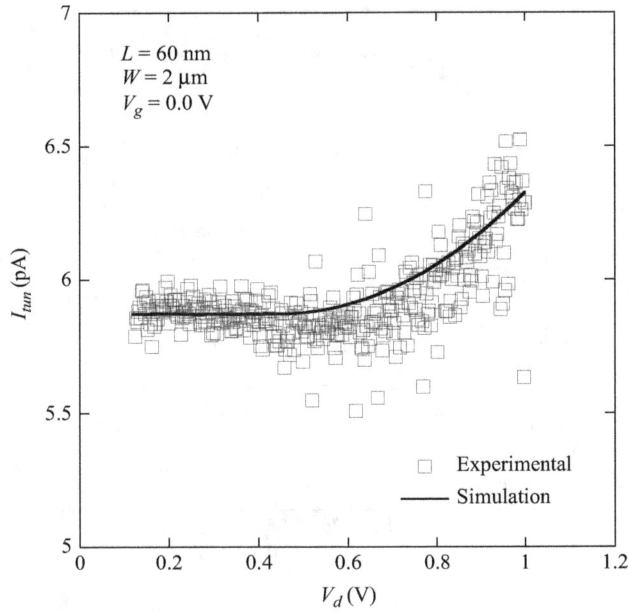

Figure 3.19 Measured (open squares) and simulated (solid thick line) results of the I_{tun} current for a $(W/L) = (2\ \mu m/60\ nm)$ n-type MOSFET

Figure 3.10, i.e., I_{source}, I_{bulk}, and I_{drain}, were measured at $V_g = 0.0$ V and V_d varying from 0.0 to 1.0 V. Under this condition there is a lack of channel formation, but then the pBJT activates when the I_{tun} current is fed through its base (p-type substrate). The I_{tun} current is then amplified by the transistor β gain factor, which is then measured at the source and drain terminals. Thus the collector (drain) parasitic bipolar current I_c can be modeled by including the β gain factor into (3.23). The drain and source currents are then defined as $I_{drain} = I_c + I_{tun}$ and $I_{source} = I_e$, with I_e as the emitter current. For $\beta >> 1$, $I_c \approx I_e$, which results in $I_{tun} = I_{drain} - I_{source}$. Using the procedure to extract the pBJT current gain factor introduced in Reference 3.34, the $\beta = 16.2$ for $L = 100$ nm, while it reaches a value of 31 for $L = 70$ nm, and a value of 130 for $L = 60$ nm. The bipolar β gain factor increases as the base width reduces. The base width happens to be the MOSFET gate length L; therefore, the shorter the MOSFET gate length L, the larger is the parasitic β gain factor.

The extracted I_{tun} current, when compared with the calculated value using the model of (3.24), shows an excellent correlation as depicted in Figure 3.19.

Comparing the measured real and imaginary parts of the MOSFET output impedance with the simulated data further validates the parameter extraction procedure, when using S-parameters, to detect the presence of BTBT. The simulations were performed with HSPICE. The results are shown in Figure 3.20.

Above 10 GHz the inclusion of BTBT considerably improves the matching between experimental and simulation results for the real part of Z_{22}. When no

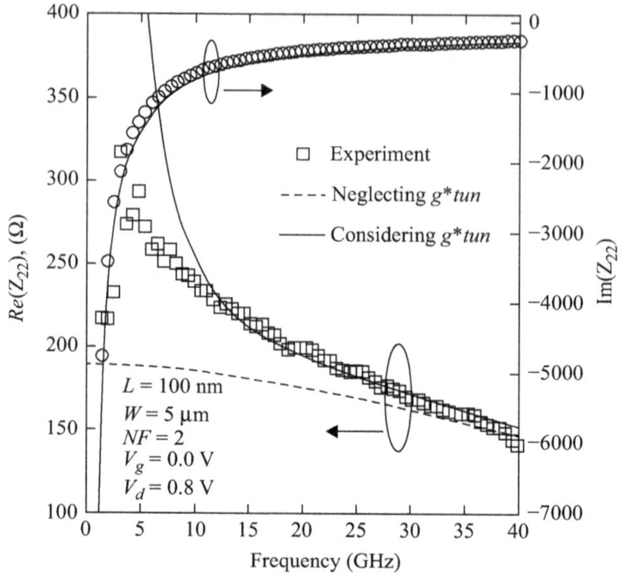

Figure 3.20 Measured and simulated output impedance of a MOSFET with and without considering BTBT at the reverse-biased drain-bulk junction

BTBT is considered an almost frequency-independent curve is obtained. In contrast, below 10 GHz the MOSFET intrinsic part shows a higher impedance when biased at $V_{ds} > 0.0$ V, which is not accounted for in conventional models yielding an underestimation of this parameter. Although the incorporation of the BTBT effect improves the simulation results, the inclusion of additional components may be necessary to improve the model for frequencies below 10 GHz.

So far the use of high-frequency, or RF, measurement methods have been shown as an alternative or nonconventional way to DC-based methods to extract parasitic components, such as the series resistances or capacitances, or even the BTBT effect. In these examples only a single gate length MOSFET is required to extract parasitic components in contrast to DC-based methods that require an L-array MOSFET structure. Now a very particular or nonconventional DC-based method, where the drain-bulk junction is forward-biased is introduced.

A negative resistance device (NRD) is recreated by using a MOSFET with the voltage at the drain-bulk junction swept in the forward direction, the voltage at the bulk is grounded, the voltage at the source-bulk junction is set in the reverse direction, while the key condition is to leave the gate electrode floating or connected to a zero-current source, which acts as a voltmeter. Under these conditions the I_d–V_d characteristics of a high-k oxide metal gate n-type MOSFET with a 2-nm-thick oxide and a (W/L) relation of (1 μm/34 nm) are measured and shown in Figure 3.21.

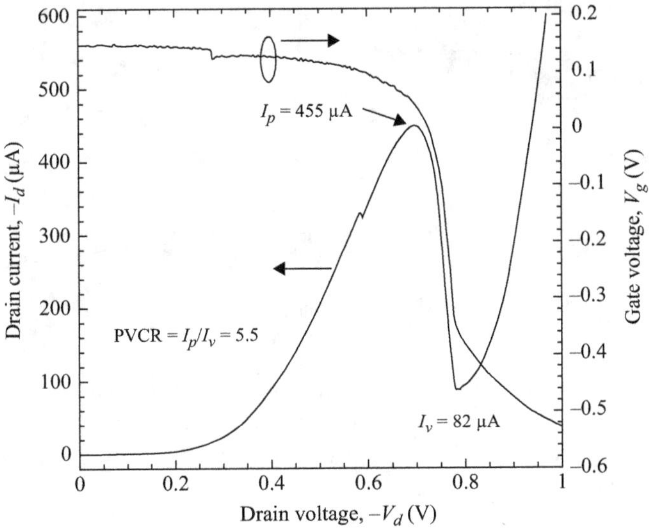

Figure 3.21 Measured I_d current versus $-V_d$ voltage for $V_s = 1.0$ V

Notice that at $-V_d = 0.7$ V, $I_d = I_p = 455$ μA, and rolls down to $I_d = I_v = 82$ μA, which gives a peak-to-valley current ratio (PVCR) = 5.5. The measured gate floating voltage V_g has a positive value until I_{ds} rolls down. The shape of the V_g–V_d curve follows that of the electrons pumped from the forward-biased drain-bulk junction into the surface channel, which forces charges to diffuse from drain to source, where they are pulled out to the source contact by the action of the reverse-biased electric field in the source-bulk junction. As the gate is floating, the charge pumped into the channel modulates the gate voltage, and thus we define this as a charge-effect transistor action, in contraposition to the conventional field-effect transistor action. In the low V_d charge-pumping regime, 0 to 0.4 V approximately, the channel operates in the weak to moderate inversion region. For higher V_{db} voltages the channel slightly enters into the moderate-to-strong inversion region, but the whole drain-bulk junction starts entering into the forward regime, which pulls charges from the surface channel down to a sub-surface conduction path as shown in Figure 3.22. The gtsFramework simulation tool is also used to calculate the I_d–V_d characteristic curve at $V_s = 1.0$ V as shown in Figure 3.23. The simulated device current I_d (circles in Figure 3.23) does not agree in magnitude but matches the shape of the experimental result. However, what is most relevant is the electron distribution as shown in Figure 3.22, which confirms an electron flow distributed across two paths. There is a surface path, in this simulation case located at $y = 0$ nm, which is basically controlled by a field-effect transistor action. There is a sub-surface path located at and below $y = 5$ nm, where most of the bipolar action of the forward D-B diode takes place.

In a real-life device operation, these two current paths do not exist as parallel connected paths, but as a distributed current all along the Si–Gate Oxide interface

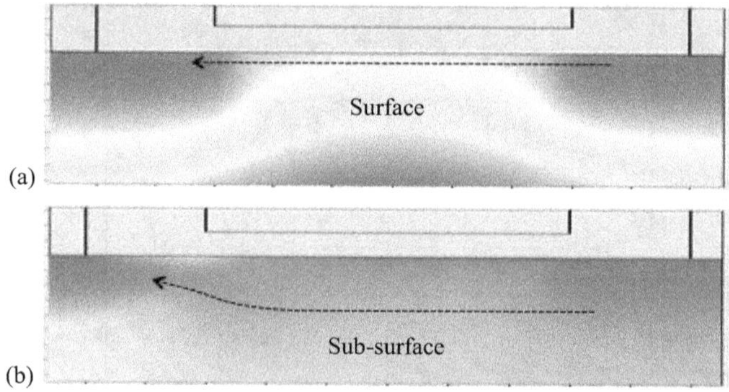

Figure 3.22 Numerical simulation of the electron distribution n for (a) $-V_d = 0.2$ V and (b) $-V_d = -1.0$ V. For case (a) electrons are confined in a surface channel, while for case (b) the electrons spread out in a sub-surface path

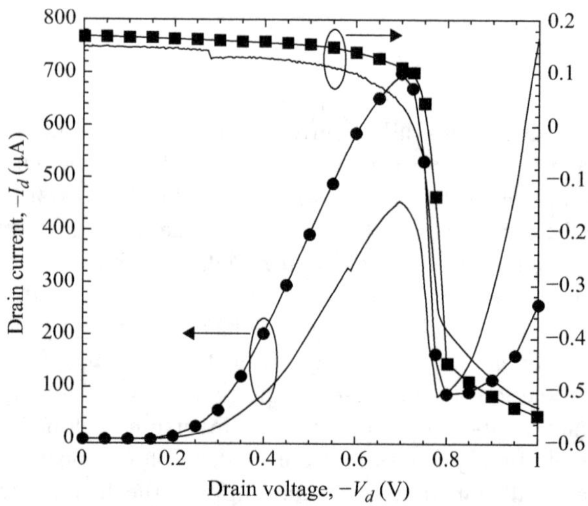

Figure 3.23 Simulated results of the $-I_d$ (circles) current and floating gate voltage V_g (squares) as a function of $-V_d$. As a comparison the continuous lines without symbols show the experimental results

down to the bottom plate of the source and drain junctions. Therefore, it can be readily observed that the negative resistance effect comes from the composition of a close-to-the-surface current path, which gives rise to the current bump, and a diode current component below the surface. At low $-V_d$ voltages the surface path activates resulting in a surface current controlled by the Field-Effect conduction mechanism. As $-V_d$ approaches the forward-biasing point of the Drain-Bulk junction, the sub-surface path starts conducting, which pulls electrons from the

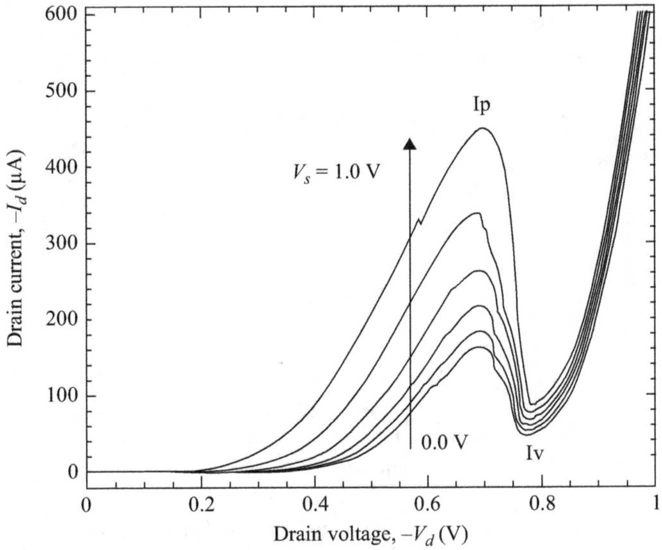

Figure 3.24 Measured $-I_d$–V_d curves for $V_s = 0.0$, 0.2, 0.4, 0.6, 0.8, and 1.0 V

surface turning off the surface path, and turning on the sub-surface path. The sub-surface path is controlled by a bipolar conduction mechanism. Due to the forward-biased D-B junction, the electrons are withdrawn from the surface down to the sub-surface path, which reflects into an excursion of the floating gate voltage from positive to negative values [3.57].

The reverse-biased S-B junction electric field pulls the electrons injected by the drain. Then the higher the V_s voltage, the higher the collected I_d current as shown by experimental results in Figure 3.24. As seen from Figure 3.24 the PVCR can be modulated from about 3.0 to 5.5 by the action of the source voltage V_s. If both transport mechanisms, diffusion and drift, are active players, then one should expect that a rising of the local temperature will increase the diffusion and will decrease the drift component [3.58]. The other way around will happen if the local temperature decreases.

Figure 3.25 shows the experimental results of $-I_d$ for $T = 325$ K, 300 K, and 275 K [3.59]. In the 0.0–0.8 V $-V_d$ range $-I_d$ increases with T, which is a characteristic of the diffusion process. This temperature-dependent mechanism is shown through $\Delta I_{ds} = (I_{ds@T\neq 300\mathrm{K}} - I_{ds@T=300\mathrm{K}})/I_{ds@T\neq 300\mathrm{K}}$ shown at the right axis of Figure 3.25. Around the peak to valley current transition, the ΔI_d factor shows a change of sign indicating the presence of a drift mechanism, which corresponds to the strong inversion of a MOSFET device. At $-V_d$ above 0.9 V the ΔI_d switches back to a similar trend as in the first region (0.0–0.8 V), which is an indication of a prevalence of diffusion over drift. Around $-V_d = 0.7$ V where $\Delta I_d = 0$ both diffusion and drift compensates each other. The simulated results

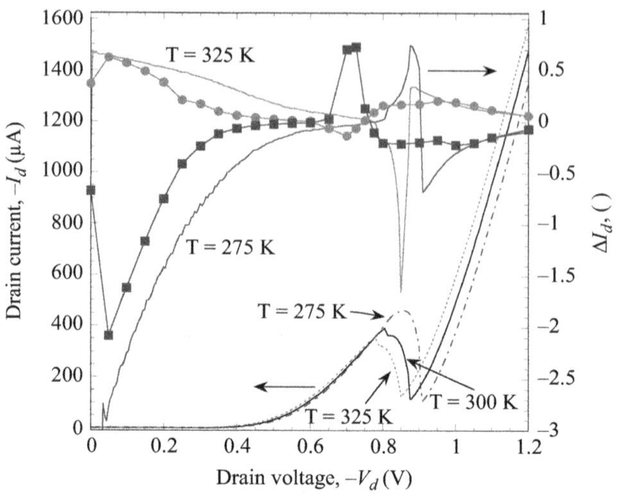

Figure 3.25 Measured and simulated results for $-I_d$ and ΔI_d for $T = 325$, 300, and 275 K. The symbols, circles and squares, correspond to the simulated ΔI_d for 325 and 275 K, respectively. The continuous lines are experimental results. The measured $-I_d$ are represented by a dotted line for $T = 325$ K, a continuous line for $T = 300$ K, and a dash-dotted line for $T = 275$ K

(symbols in Figure 3.25) do not match quantitatively the experimental data, but they qualitatively probe the combination of drift and diffusion transport mechanisms.

The qualitative match between experiment and simulations is further tested with the experiment shown in Figure 3.26, where the two conduction paths, a surface and a sub-surface path, are demonstrated. While measuring the I_d current, the sub-surface path is blocked by putting the bulk in high impedance condition ($I_b = 0$ A). This leaves only the surface path conduction active (dash-dotted line in Figure 3.26). This reproduces the activation of the surface channel with a peak current. By putting the source in high impedance condition ($I_s = 0$ A), the exponential drain-bulk sub-surface current path is then activated, while the surface path gets blocked. The summation of both reproduces the negative resistance effect.

This alternative or nonconventional bias condition allows extract information from devices, that other way is not possible. This is the case of the identification of surface and sub-surface conduction paths in the MOSFET. Other interesting finding is the recreation of an NRD implemented with a single MOSFET without the use of quantum tunneling mechanism as required for this effect [3.60]. From the electrical modeling point of view, this NRD can be viewed as the parallel connection of a surface MOSFET and a sub-surface drain-bulk junction (bipolar conduction).

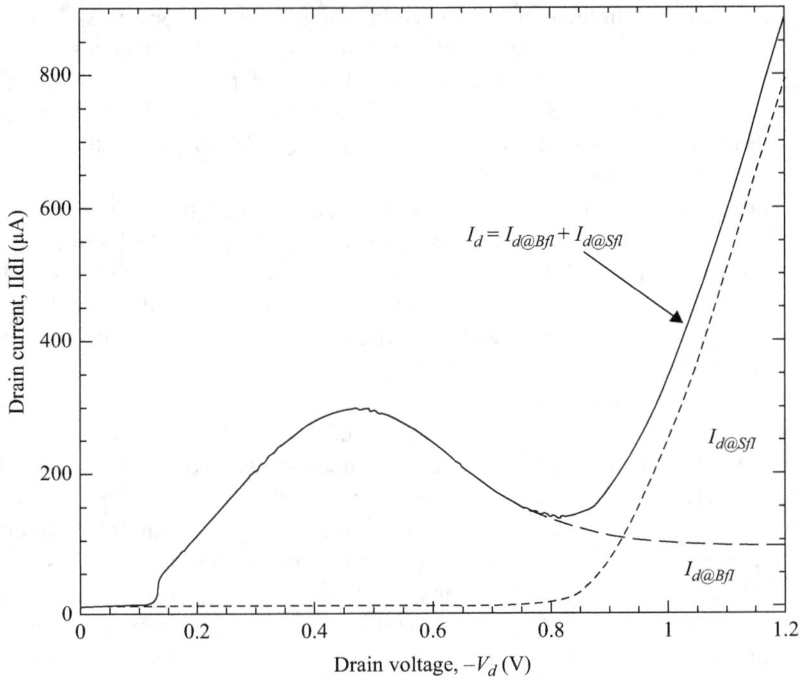

Figure 3.26 Measured drain current with source floating ($I_{d@Sfl}$) at $V_b = 0.0$ V, with bulk floating ($I_{d@Bfl}$) at $V_s = 1.0$ V, and the summation of both currents I_d

3.2 Reliability, degradation, lifetime prediction, and failure

3.2.1 Bias temperature instability of high-k materials beyond HfO_2

One of the major reliability concerns regarding the long-term performance of modern N- and P-type MOSFET devices is bias temperature instability (BTI), which predicts the ability of these devices to withstand any change or shift in the threshold voltage V_{th}, mobility, drain current I_d, transconductance, etc., of the transistor under high bias and temperature conditions over time. Even though the last decades have seen a lot of research effort devoted to the final fabrication of MOSFET devices with outstanding electrical performance, the long-term operation of transistors shows that their original performance is highly degraded thus affecting circuit performance and, therefore, its reliability. This takes special importance if we consider the aggressive scaling down of the MOSFET geometry (down to nanometer dimensions), the replacement of conventional materials for metal gate and high-dielectric constant materials (decreasing the physical quality of a standard SiO_2/Si interface), increased operation temperatures and also the introduction of complex three-dimensional geometries (FinFET structures with multiple gates for the sub 22 nm technology nodes) in which more than one crystalline plane of the semiconductor substrate is able to contribute specific densities of charge during FET inversion. All of this has resulted in an increase of the electric fields developed throughout the biased regions of the MOSFET devices and, therefore, making BTI characterization and important tool in order to predict the long-term operation of the FET under consideration.

BTI characterization uses positive or negative biasing depending on which type of transistor is characterized: positive BTI (PBTI) for N-type MOSFET and negative BTI (NBTI) for P-type MOSFET. This way, both PBTI and NBTI deal with the continuous degradation of the transistor in inversion mode and with an oxide electric field which is usually lower than the required for hot carrier degradation. In this section, we will mostly show the effect of PBTI and NBTI characterization with respect to the shift in threshold voltage V_{th} and another important parameter known as density of states D_{it}. Correlating BTI to V_{th} and D_{it} is quite important since we will be able to examine some features of the continuous stressing of the gate oxide (BTI) to the degradation in the MOSFET operation (V_{th} shift) and also to the resulting quality of the gate oxide interface with the semiconductor substrate (D_{it}). Given that the gate oxide of modern transistors is now based on alternative materials (mostly binary oxides with high dielectric constant, better known as *high-k*), the reduced physical and electronic quality of a high-k/Si interface is then prone to an increase in the degradation of the MOSFET device after BTI characterization and we will show some of these phenomena in the following paragraphs. This way, BTI could be simultaneously related to important operative and physical parameters of the MOSFET device during continuous stressing by inverted carriers (coming through the gate oxide/semiconductor interface). This continuous degradation could be low enough to make the MOSFET device still useful when embedded to particular circuit applications until failure occurs. In this sense, the lifetime prediction of the MOSFET parameters after BTI is also important and we will also show some of these important lifetime predictions, useful to guarantee the continuous operation of the transistor to a certain extent or condition.

The basic biasing configuration for BTI characterization in MOSFET devices is shown in Figure 3.27, in which the gate terminal of a MOSFET device is positively (PBTI) or negatively (NBTI) biased in order to form an inversion channel connecting source and drain. With the exception of the gate, all the other terminals are grounded. Generally, BTI characterization is done under direct current *dc* conditions (constant gate voltage stress) but alternate current *ac* conditions are also used (along with specific waveforms and frequencies) in order to characterize the dynamics of charge trapping and its influence on MOSFET degradation.

By continuously stressing a MOSFET transistor with a positive or negative bias in inversion mode, a progressive shift in the threshold voltage V_{th} of this device will be observed, whereas the degree of displacement for this parameter (and many others) is greatly influenced by temperature. It is the type of charge in the inverted channel (electrons or holes), its density (related to a gate area of the transistor, gate voltage $V_g > V_{th}$), and the duration of stress that dictates the initial direction and amount of V_{th} shift during BTI. By looking at the PBTI configuration during stressing of N-type MOSFET devices, we notice that are mostly inverted electrons

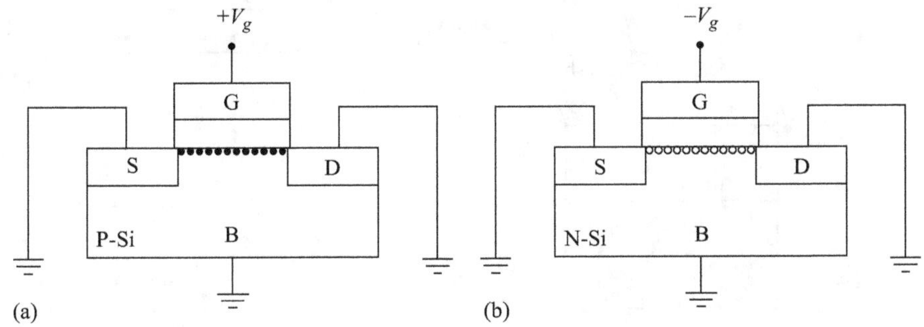

Figure 3.27 (a) PBTI biasing scheme for characterization of N-type MOSFET; we notice an inverted channel formed of electrons. (b) NBTI biasing scheme for characterization of P-type MOSFET; we notice an inverted channel formed of holes. In both cases, all other terminals are grounded

the charge being attracted and trapped in the gate oxide (whether at the interface with silicon or in the bulk of the oxide). Trapping of a high enough density of negative charge in the gate oxide will produce a positive shift of the MOSFET's threshold voltage which in turn, decreases its transconductance (g_m), increases power consumption, increases the original density of states D_{it}, and in general, it lowers the reliability of the device, see Figure 3.28a. We can see that PBTI in N-type MOSFET devices effectively induces a positive shift in V_{th} which in turn will greatly reduce the transconductance g_m of the transistor when compared to the original (fresh) condition. On the other hand, if only holes were the major charge being trapped in the gate oxide (mostly from the metal gate being positively biased), a negative shift of V_{th} is expected which in turn would increase the transconductance of the MOSFET device (see Figure 3.28b).

It is, however, the continuous trapping of inverted electrons or inverted holes (negative and positive charge) that causes a lot of electrical instabilities during BTI and this directly affects the performance of MOSFET devices not only in terms of a shifted V_{th} or transconductance but also on the creation of more defects that are directly related to the quality of the gate oxide material and its interface with the semiconductor substrate. Figure 3.29 clearly shows the exponential trend in V_{th} and g_m degradation of a P-type MOSFET device (based on a dry-thermal SiO_2 as gate oxide) when subjected to continuous stressing during NBTI conditions [3.61, 3.62]. This general trend is seen in most of the reported BTI results obtained from conventional or advanced MOS structures (using SiO_2/Si or the most recent high-k/IL/Si structures) and it could greatly reduce the desired lifetime of the device when considering deeply scaled down transistor geometries along with advanced materials.

From a physical point of view, degradation in MOSFET devices by BTI initially occurs by the continuous trapping of inverted charge (which in turn, locally modifies the energy gap of the gate oxide), creation of interface traps, increase in the density of a fixed oxide charge and, if we look at the atomistic origin of BTI degradation, by the continuous breaking of initially passivated Si–H chemical

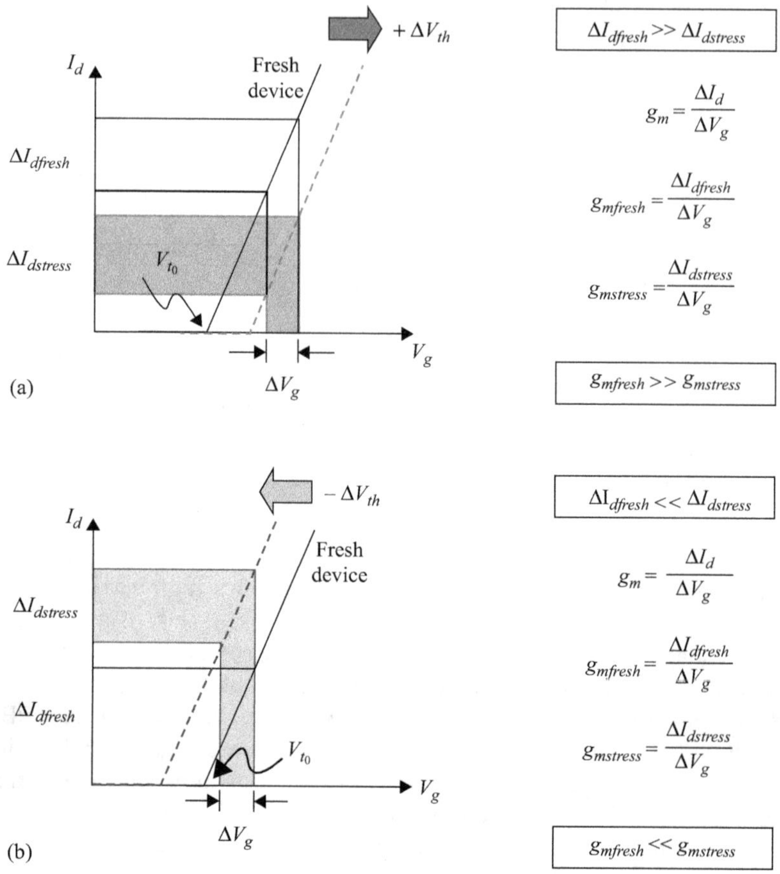

Figure 3.28 (a) Simplified drain current versus gate voltage (I_d–V_g) characteristic for an N-type MOSFET device in which a fixed amount of trapped negative charge shifts the threshold voltage to the right of the curve (positive shift). (b) I_d–V_g characteristic for an N-type MOSFET device in which a fixed amount of trapped positive charge shifts the threshold voltage to the left of the curve (negative shift). For both conditions, BTI would greatly affect a MOSFET device in the subthreshold mode of operation because of the exponential dependence of drain current to gate voltage

bonds at the gate oxide/silicon interface, breaking of additional chemical bonds at the high-k/interfacial layer (IL) interface, among other phenomena [3.63–3.68]. Given the vast amount of experimental conditions dealing with the proposed physical mechanisms related to degradation of MOSFET devices by BTI characterization (from atomistic/chemical to modeling/simulation approaches), we strongly recommend reading those excellent reviews that are already published in literature. Here, we will focus on the discussion of BTI applied to long-channel MOSFET

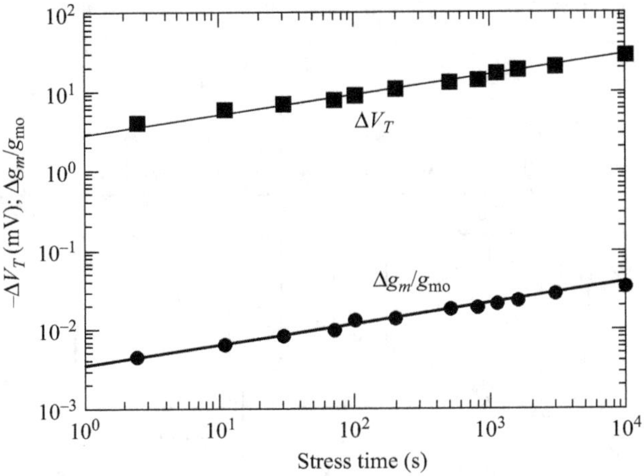

Figure 3.29 *V_{th} shift and g_m degradation induced by NBTI over stressing time ($V_g = -2.3$ V, $T = 100°C$, $L = 0.1$ μm, $T_{ox} = 2.2$ nm, dry-thermal SiO_2)*

devices, which make use of advanced metal gate/high-k materials that are well beyond HfO_2-based configurations, and whose emphasis is in the close relationship between both V_{th} and D_{it} during BTI stressing. With this in mind, a brief introduction to high-k materials is provided.

The use of high-dielectric constant materials (mostly HfO_2) as the replacement for ultra-thin SiO_2 in the most advanced MOSFET devices (from planar MOSFET to FinFET architectures) has based their success in the reduction of an exponentially increased gate leakage current (I_g) by means of increasing their physical thickness (*Thk*) while maintaining a high enough gate capacitance for proper formation of an inverted channel at relatively low gate voltages ("off" to "on" transition). A metal gate is also used on top of the high-k material in order to prevent the effect of polysilicon depletion and in general, is this metal gate/high-k combination, the actual structure that is found in the most advanced transistor devices since their commercial introduction back in 2007 and up to now [3.69–3.71]. Nevertheless, the replacement of HfO_2-based materials as the gate oxide in the most advanced MOSFET devices is now foreseeable given the aggressive scaling down of its physical thickness (less than 4 nm), and therefore, the unavoidable appearance of higher levels of gate leakage current I_g that were initially reduced by this material. Since a higher level of I_g directly increases BTI-related degradation of MOSFETs (shifts in V_{th}, g_m, mobility, D_{it}, etc), it is important to explore other high-k materials with enhanced physical parameters able to reduce BTI-related degradation.

La_2O_3 has one of the largest bandgap values from the rare earth oxides ($E_g \sim 5$ eV), while also having the lowest lattice energy and a relatively high dielectric constant, $\varepsilon = 27$ [3.72, 3.73]. The use of La_2O_3-gated MOSFET devices will significantly reduce gate leakage-current density (especially true for N-type MOSFET devices), this is

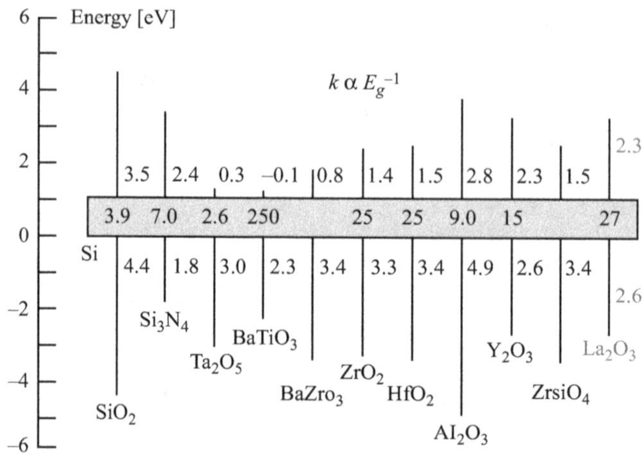

Figure 3.30 Band offsets of several high-k materials with respect to the conduction and valence bands of silicon ($E_g = 1.12$ eV). The numbers within the silicon bandgap are the average dielectric constant values for all dielectrics shown. Even though HfO_2 and La_2O_3 have similar dielectric constants (25 and 27, respectively), a higher band offset of La_2O_3 to the conduction band of silicon could help to greatly minimize PBTI

because of its larger band offset for electrons in the conduction band of silicon as compared to HfO_2, see Figure 3.30. Figure 3.31 clearly shows that La_2O_3 poses the great advantage of reduced leakage current for the same equivalent-oxide thickness (EOT) when compared to other high-k materials like Hf-based oxides [3.74–3.78]. Even for *EOT* below 1 nm (thus having a well controlled thermal stability in order to minimize the IL between the high-k material and silicon), La_2O_3 still conserves this advantage. The SiO_2-equivalent oxide thickness *EOT* can be expressed in terms of the physical thickness for IL and La_2O_3 as:

$$EOT = \left(\frac{\varepsilon_{ox}}{\varepsilon_{IL}}\right) t_{IL} + \left(\frac{\varepsilon_{ox}}{\varepsilon_{hk}}\right) t_{hk} \tag{3.25}$$

Because of the larger band offset of La_2O_3 to the conduction band of silicon, a reduction of PBTI-related degradation in N-type MOSFET devices should be expected as compared to HfO_2-gated transistors [3.79] when they are stressed under similar conditions and while having similar EOTs. This is because of the larger energy barrier seen by electrons in the inverted silicon channel of the La_2O_3-gated transistor and which are drawn by a positive bias during PBTI (substrate injection).

Figure 3.32 shows a high-resolution transmission-electron microscopy (HRTEM) image of an $Al/La_2O_3/Si$ stacked structure in which sharp interfaces between the La_2O_3 and its surroundings are noticed. We are able to distinguish the crystalline nature of the silicon substrate, the amorphous state for La_2O_3, and some polycrystalline faces of the aluminum electrode. It is important to consider that in

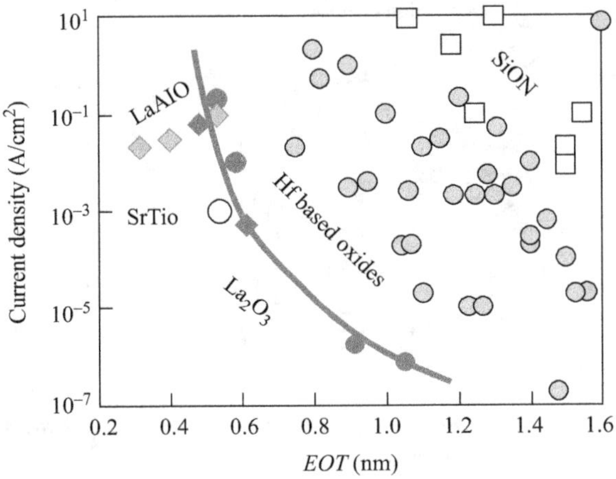

Figure 3.31 *Gate leakage current density of La_2O_3-based oxides compared to Hf-based oxides [3.74–3.78]. Even for EOT < 1 nm, a greater reduction in the gate current density still holds because of the larger band offset for La_2O_3 to the conduction band of silicon and this characteristic should also be able to minimize its PBTI*

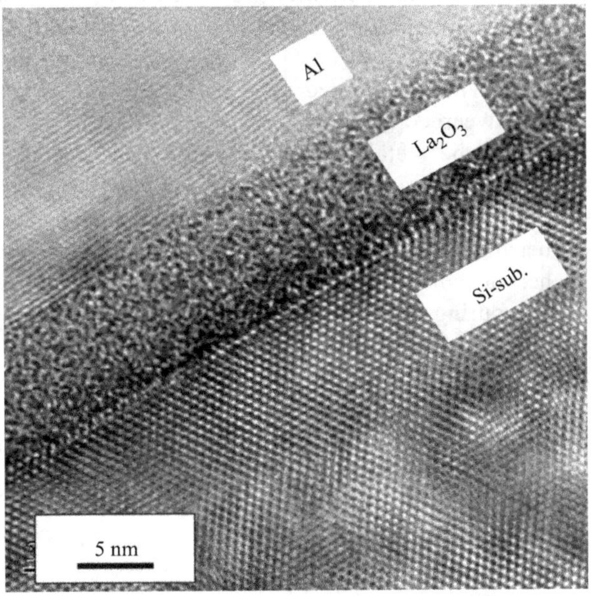

Figure 3.32 *TEM image of an Al/La_2O_3/Si stacked structure in which La_2O_3 shows sharp interfaces with the silicon substrate and the aluminum electrode. The amorphous nature of La_2O_3 is observed, whereas some polycrystalline faces of aluminum are also noticed. In order to minimize a low-k IL of the metallic electrode with La_2O_3, inert electrode materials should be used atop this high-k layer*

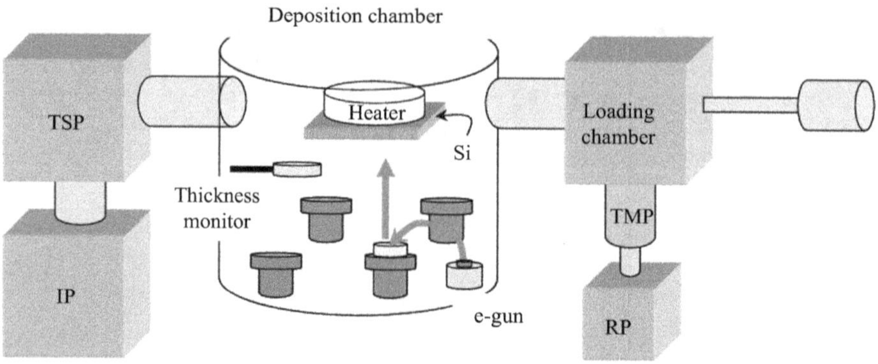

Figure 3.33 Schematic of the La_2O_3 deposition system using an MBE chamber and UHV conditions

order to minimize a low-k IL of the metallic electrode with La_2O_3 (which could prevent a desired reduction in *EOT* for deeply scaled-down devices) inert electrode materials should be used atop this high-k layer and this issue will be further discussed in the following text.

In order to obtain high-quality La_2O_3 films, they are deposited by E-beam evaporation in ultra-high vacuum conditions (10^{-8} Pa as base pressure), as shown in Figure 3.33. There are four compartments to allocate same/different solid sources at the bottom of the chamber. La_2O_3 is then bombarded and subsequently heated by the E-beam, which is located near the source and its beam being controlled by a magnetic sweep controller.

On the other hand, the metal's work function (ϕ_M) is also important when considering NBTI-related degradation because of the variable energy barrier (ϕ_B) seen by electrons coming from the metal electrode and being injected into the high-k layer by a negative bias (gate injection condition). This is clearly observed in Figure 3.34, where a reduction in the gate leakage current density J_g (I_g normalized to gate area) during gate injection is observed by just replacing the gate electrode from a low to a high work function material (aluminum to ruthenium) even though the same physical thickness for La_2O_3 is used (measured by spectroscopic ellipsometry). It is interesting to notice that not only lower J_g are obtained during gate injection when using metals with higher ϕ_M (thus minimizing NBTI); even *EOT* is enhanced by going well below 1 nm, which is an indirect evidence that any IL (with low-k) developed at the metal/high-k interface could be minimized. In this sense, the use of inert metal electrodes for high-k gated MOSFET devices has a great potential to further scale down the *EOT* of these devices.

When selecting a metal electrode having any desired work function, not only a high energy barrier ϕ_B is important in order to minimize NBTI degradation. As stated earlier, the chemical stability of these metals with the underlying high-k oxide is of the utmost importance in order to obtain reproducible behavior in terms of flat-band voltage (V_{fb}) of capacitor structures, and therefore, stable threshold voltages V_{th} in MOS transistors. Figure 3.35 shows the capacitance–voltage

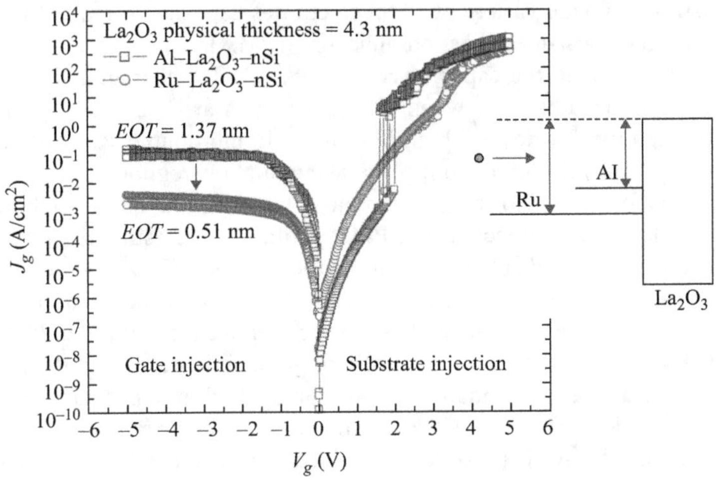

Figure 3.34 *J_g–V_g characteristics for Al and Ru-gated La_2O_3 (4.3 nm in thickness) MIS capacitors after post-deposition annealing in N_2 at 400°C. Metals with a higher work function lead to a reduced J_g during gate injection. The inset shows the difference in ϕ_B for electrons during gate injection using both metals*

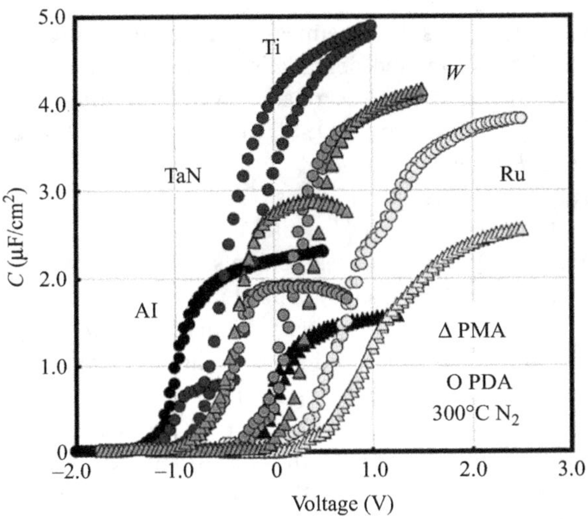

Figure 3.35 *C_g–V_g characteristics of Metal/La_2O_3/n-Si capacitors after PDA/PMA in N_2. Even though lower EOT can be achieved with other metals, W gets the more stable and reproducible characteristics. Here, La_2O_3 with different physical thicknesses were used so that these data is presented only for comparison purposes*

characteristics of a wide variety of metal electrodes deposited on top of La_2O_3 after post-deposition or post-metallization annealing (PMA).

Large variations in the capacitance–voltage (C_g–V_g) characteristics of La_2O_3-based capacitors are observed when different metals are used as gate electrodes: positive or negative V_{fb}, low or high hysteresis during double V_g sweeps, slow or fast transitions from the inversion to the accumulation regimes (an indirect indication of surface states D_{it}), etc. Importantly, the thermodynamic stability of the metal electrodes when subjected to a PMA could also be estimated by the observation of any possible displacement from the original C_g–V_g curve (with PDA only). From all these metals, we can see that tungsten is the more stable in terms of producing similar C_g–V_g characteristics when the capacitors follow a PDA or PMA treatment. By using tungsten, a material highly valued because of its highly inert properties (resistance to oxidation even during high temperature conditions), obtaining MOS devices with $EOT < 1$ nm can be also guaranteed because of its ability to minimize any IL (with lower dielectric constant) when in contact to the gate oxide. Also, tungsten is considered as a mid-gap-metal gate electrode able to produce "normally-off" N- and P-type channel MOSFETs with low enough threshold voltages V_{th}. However, from the viewpoint of process integration, the use of inert materials as the gate electrode for high-k-gated devices imposes a big challenge related to the etching of this metal for the precise definition of gate area patterns having deeply scaled-down areas. This analysis, however, is beyond the scope of this chapter but it is a research topic that needs to be further investigated because it will also impact MOSFET degradation during NBTI and PBTI characterization.

With all the former physical, chemical, and electronic ideas in mind, we will now show the PBTI/NBTI related degradation of La_2O_3-gated MOSFETs in order to obtain any experimental ΔV_{th}–ΔD_{it} relationship. First, Figure 3.36a–c shows the shift in V_{th} for La_2O_3-gated N-type MOSFET devices after PBTI (as a displacement of the drain current I_d vs gate voltage V_g) at room temperature and by using three different constant voltage stress conditions (CVS = +1, +1.5, and +2 V, substrate injection mode). Initial $V_{th0} = 0.5$ V while $V_d = 0$ or 100 mV were used for all samples during stressing or V_{th} measurements, respectively. Please notice that injected charge density Q_{inj} (in C/cm^2) is used instead of stressing time in order to minimize any variation in ΔV_{th} by extrinsic fluctuations in J_g.[†] Figure 3.36d shows a summary of the ΔV_{th} trends observed in (a–c) during different positive constant-voltage stress CVS and where interesting results are found.

We can see that at the lowest CVS = +1 V, the V_{th} shifts toward negative direction in proportion to the duration of stress (or injected charge density Q_{inj}). When a CVS = +1.5 V is applied to the gate of the transistor, the shift in V_{th} continues to increase in the negative direction but is not as pronounced as in the

[†]By integrating the injected gate current density J_g with stressing time, we are able to quantize the effective density of charge being injected at the gate oxide. This way, we would be able to measure the intrinsic capacity of the whole oxide stack for BTI characterization.

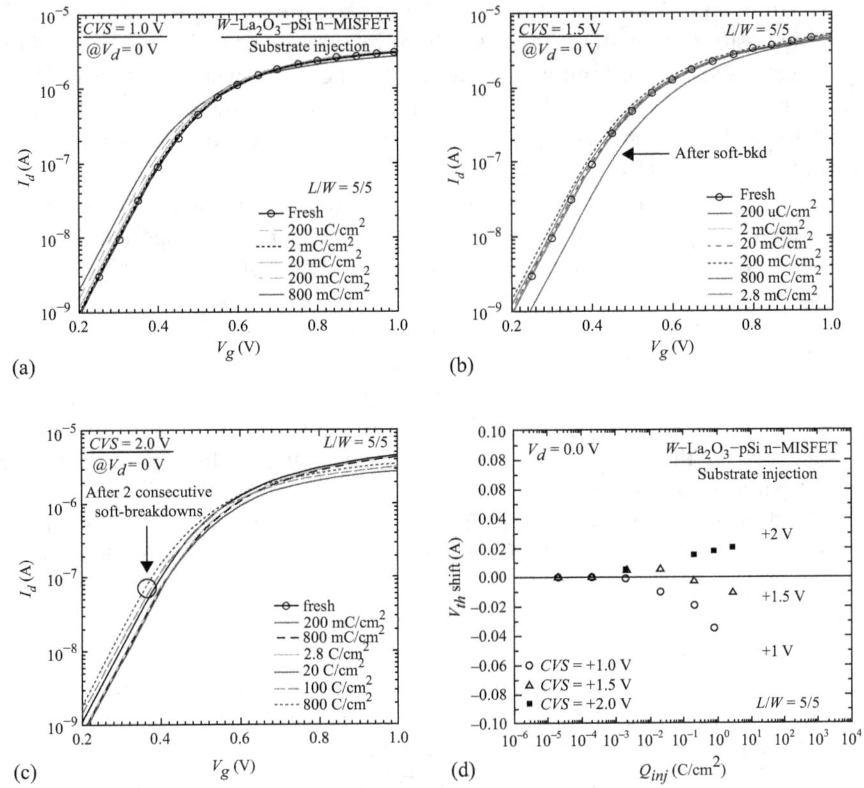

Figure 3.36 (a) ΔV_{th} during PBTI of W/La$_2$O$_3$/pSi based N-type MOSFET devices at room temperature and with CVS = V_g = +1 V. (b) CVS = +1.5 V. (c) CVS = +2 V. (d) Summary of the ΔV_{th} trends observed in (a–c) during different positive CVS. Injected charge density Q_{inj} is used instead of stressing time

former case. Interestingly, when a soft-breakdown (SBD) event occurs during the stressing of the device, the V_{th} shifts to the opposite (yet expected) direction and the final V_{th} is drastically increased. Any SBD event is regarded as an initial failure mode (which was already described earlier along with other breakdown modes) and yet, for this particular case of positive stressing of a La_2O_3-gated MOSFET, the device continues to operate as a transistor with no appreciable decrease in the subthreshold slope. When the highest voltage stress conditions are applied (CVS = +2 V), the V_{th} shifts to the expected positive direction (because of trapping of electrons from the inverted channel into La_2O_3 and its interfaces). It should be noted that the subthreshold slope (SS) of all I_d–V_g curves did not change before or after stress for any voltage condition, thus suggesting a higher endurance during BTI degradation even with the appearance of some SBD events. These results are plotted altogether in (d) but without showing the effect of SBD events on V_{th} for the sake of simplicity. This plot represents the general trend that V_{th} follows as the voltage intensity during stress increases from low to higher CVS conditions. Plotting the results shown in Figure 3.36d using $|\Delta V_{th}|$ gives a clear lineal trend of this parameter with respect to Q_{inj} (only one stressing condition) as shown in Figure 3.37a. For comparison purposes, $|\Delta V_{th}|$ for P-type MOSFETs (same W/L ratios) after NBTI is also shown in (b).

Figure 3.37b summarizes the shift in $|V_{th}|$ (also using absolute value) for La_2O_3-gated P-type MOSFET devices after NBTI at room temperature and by using three different constant voltage stress conditions (CVS $= -1.5$, -2, and -3 V, gate injection mode). Drain voltages $V_d = 0$ or 100 mV were used for all samples during stressing or V_{th} measurements, respectively. Here, we have also used injected charge density Q_{inj} (in C/cm^2), instead of stressing time. Even though not shown, all I_d–V_g curves were shifted to the negative side as expected, thus indicating the continuous trapping of holes from the inverted channel (these holes were detected as the majority carriers within the total gate leakage current after carrier-separation measurements). We notice that independently of the magnitude of CVS, all measured samples immediately produced a large shift in V_{th} even after the first injection of charge, and all subsequent shifts were always into the negative direction. An initially large $\Delta V_{th} \sim 10$ mV was kept almost constant for the samples stressed with low CVS even after injecting large charge densities in the gate injection condition (up to 20 C/cm^2). For the higher stressing condition however, a slight increase in V_{th} in proportion to injected charge is observed and this is because higher gate currents are now injected by this particular biasing (up to 200 C/cm^2) which in turn, activate conduction mechanisms that could promote additional trapping of holes and therefore, further degradation in V_{th}. It is important to notice that even with the largest stressing conditions (able to produce significant shifts in V_{th} for both MOSFET devices), the final V_{th} measured after the largest injection of charge was kept always below 100 mV in both transistors. This is important considering that the La_2O_3 films are intended to be applied in devices with gate voltage polarizations of $V_g \leq 1$ V, so that even lower shifts in V_{th} can be expected thus increasing their lifetimes before any preset condition of failure.

Given that for N-type and P-type MOSFET devices, BTI involves both the trapping of charge from an inverted channel (minority carriers) and the injection of charge from the gate electrode, the major contribution for charge trapping (electrons or holes) into the high-k/IL layers is quite important. By using a MOSFET structure, the total gate leakage current I_g can be separated into its electron and hole components (I_e and I_h) using the carrier separation method [3.80] and, therefore, correlate the injection of a specific type of charge during BTI with the observed shift in V_{th} as was shown in Figure 3.37a,b. A schematic illustration of the carrier separation measurement is shown in Figure 3.38a,b for N- and P-type MOSFET structures, respectively, where the applied gate voltage V_g promotes substrate and gate injection of electrons (using positive and negative V_g, respectively).

During substrate injection of electrons, a positive voltage is applied to the gate of an N-type MOSFET, whereas in gate injection of electrons, a negative voltage is applied to the gate of a P-type MOSFET. This way, the inversion condition can be

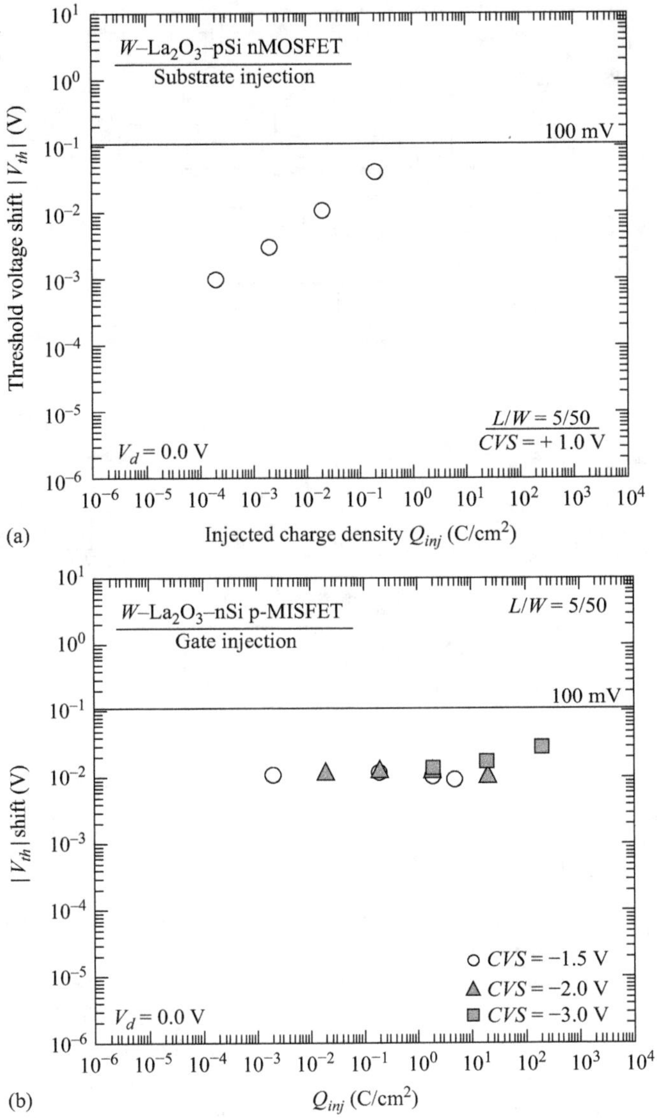

Figure 3.37 *(a) V_{th} shift for a W/La_2O_3–IL/pSi structure in a N-type MOSFET after substrate injection. A progressively linear shift in V_{th} after stress is found. (b) V_{th} shift for a W/La_2O_3–IL/nSi structure in a P-type MOSFET after gate injection. V_{th} hardly shifts after the first injection of charge. All I_d–V_g curves (not shown) shifted to the negative side, thus indicating continuous hole trapping from the inverted channel*

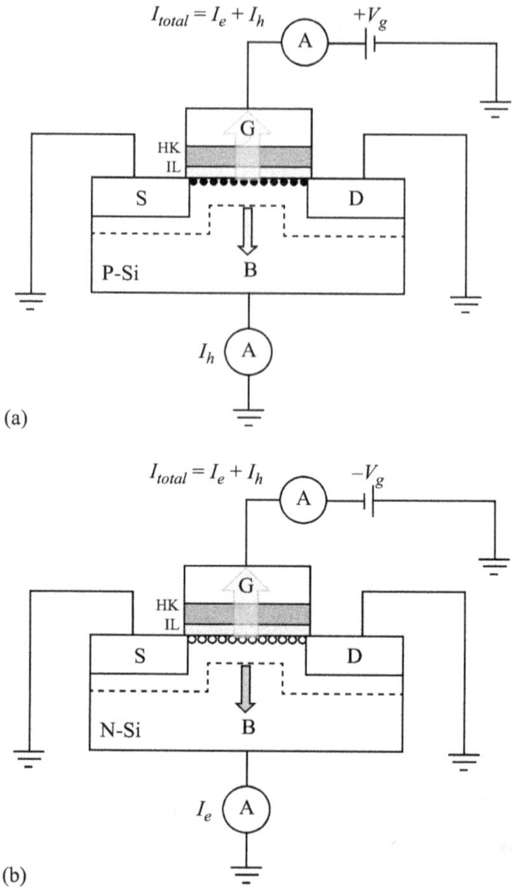

Figure 3.38 (a) Carrier-separation measurement for N-type MOSFET in inversion mode (PBTI condition). Both electron and hole current components are detected at the gate and substrate respectively. (b) Carrier-separation measurement for P-type MOSFET in inversion mode (NBTI condition). Although both electrons and holes tunnel simultaneously through the gate oxide, only the hole current component tunneling through the oxide is shown for clarity purposes. With this scheme, the major carrier component for I_g is detected

reached for both transistors in order to properly separate the contribution of electron and hole components on the total gate leakage current. The total tunneling gate leakage current is then

$$I_{total} = I_g = I_e + I_h \tag{3.26}$$

It is important to also consider a low-k IL that develops in between the La_2O_3 and silicon. Such an IL is the product of a chemical reduction of La_2O_3 into a lanthanum-based silicate following thermal annealing at relatively low temperatures [3.81]. This IL is detected after X-ray photoelectron spectroscopy (XPS) measurements and it plays an important role considering that because of this IL, a direct compromise between EOT and better reliability properties of MOSFET devices (like BTI characterization) could be observed. In the end, annealing these types of structures in forming gas atmospheres (containing H_2) is often necessary in order to passivate some of the dangling bonds at the newly formed La_2O_3–IL and IL–Si interfaces and therefore, increase their reliability at the cost of an even lower *EOT*. Figure 3.39a,b shows the electron and hole current components (tunneling at the gate of an N-type MOSFET device) after carrier-separation measurements in the (a) fresh condition and (b) after stress conditions using a positive $V_g = +1$ V (substrate injection), such that 180 mC/cm^2 were injected into the La_2O_3/IL oxide stack. For both cases, a clear contribution of holes is seen as the major component of the total I_g for $V_g \leq 0.5$ V (as seen by the arrow), which in fact, is the threshold voltage V_{th} for these devices. For $V_g \geq 0.5$ V, electrons are now the major component of the total I_g for both conditions. This clear separation of the dominant carrier component in I_g at different V_g regimes is the origin of the observed negative and positive shift in V_{th} for N-type MOSFET devices when using low or high CVS conditions, see Figure 3.36d. On the other hand, the electron and hole current components (tunneling at the gate of a P-type MOSFET device) are shown Figure 3.40a,b. These carrier-separation measurements are also shown in the (a) fresh condition and (b) after stress conditions using a negative $V_g = -1.5$ V (gate injection), such that 5 C/cm^2 were injected in to the same La_2O_3/IL oxide stack.

From Figure 3.40a,b, we can see that holes are the major carrier component tunneling into the La_2O_3–IL stacked oxide of a P-type MOSFET before and after stress for almost the complete range of applied gate voltage (gate injection mode). This in turn, shifts the V_{th} to the negative side of all the I_d–V_g curves (not shown) during NBTI while this trend is kept almost constant as compared to PBTI conditions for the N-type transistors, see Figures 3.37a,b. We need to remember that, independently of the magnitude of CVS during NBTI conditions, all P-type MOSFET devices immediately produced a large shift in V_{th} even after the first injection of charge, and all subsequent shifts were always into the negative direction. Because holes are the major carrier contributors to leakage current during gate injection and since they have a heavier mass than electrons during conduction, it is thought that the density of available sites for hole trapping (Nt) during stress is rapidly filled and saturated after the first injection and trapping of these carriers. In this sense, identifying the dominant current component (electrons or holes) in the total gate leakage current I_g, is a powerful tool in order to correlate precise conduction mechanisms during BTI stressing of the device and therefore, enabling a better estimation of lifetime predictions for advanced MOSFET devices. On the other hand, since substrate injection of electrons (from a strongly inverted channel into the oxide stack) can severely degrade the MOSFET's V_{th} characteristics (as compared to gate injection, see Figures 3.37a,b) because of a linear increase in ΔV_{th} with stressing time or Q_{inj}, further in-depth studies correlating both PBTI and NBTI for N- and P-type MOSFET devices are required when considering that both

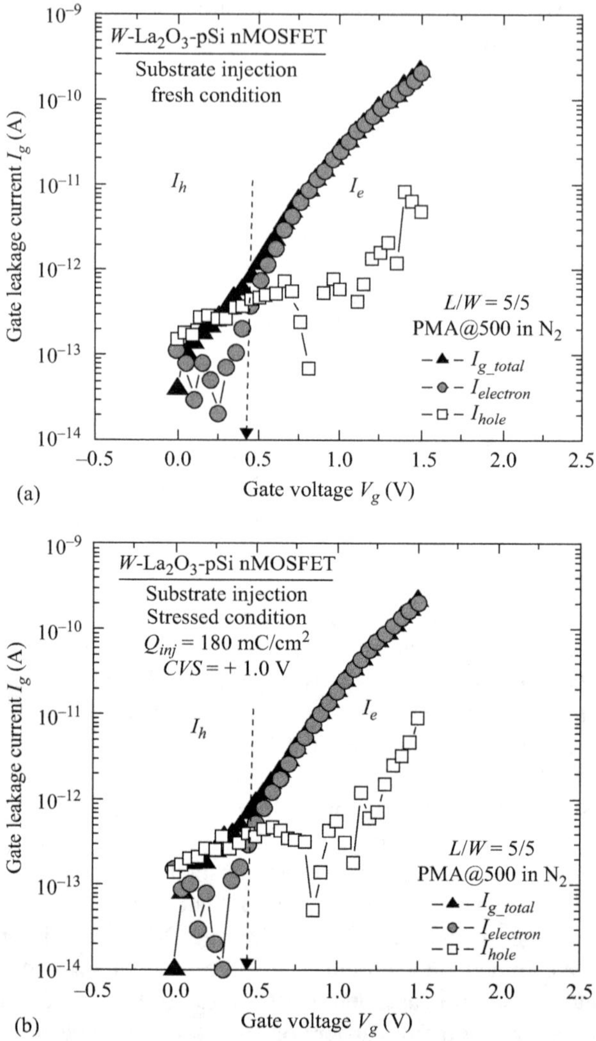

Figure 3.39 *(a) Separation of the total gate leakage current I_g into its hole and electron carrier components (I_h, I_e) for an N-type MOSFET before constant voltage stress. Holes are the major tunneling carriers for $V_g \leq 0.5$ V whereas electrons are the major contributors to gate leakage current for $V_g \geq 0.5$ V. (b) The same situation occurs even after positive CVS (substrate injection) where 180 mC/cm^2 are injected into the gate oxide stack*

types of transistors will share the same electric fields within a single CMOS fabrication process (same magnitude of V_{dd}). This is important since their ΔV_{th} distribution over time will be quite different, thus highly affecting the functionality of an integrated circuit containing mostly N-type MOSFET devices.

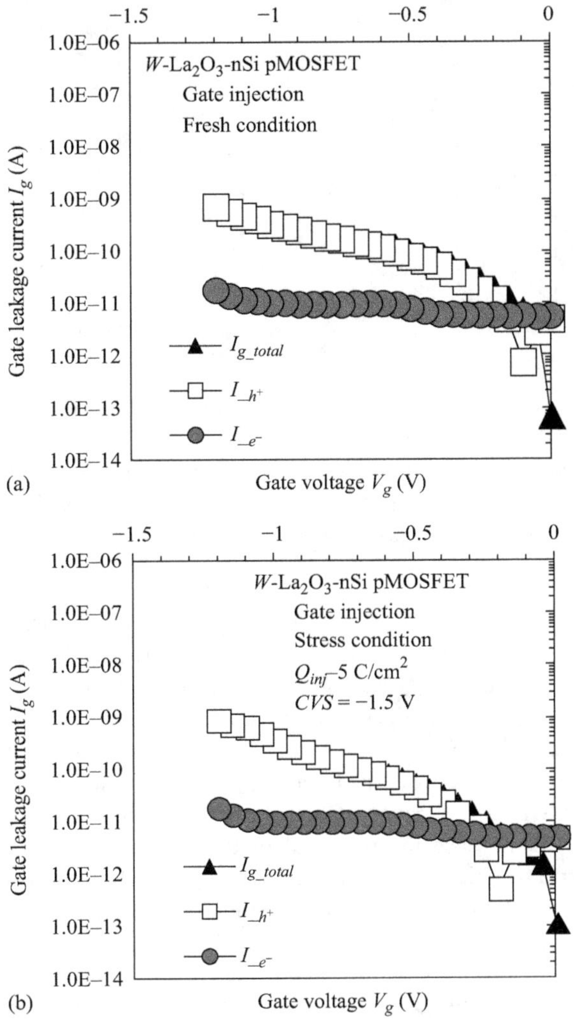

Figure 3.40 Separation of total I_g into its hole and electron carrier components (I_h, I_e) for an P-type MOSFET (a) before and (b) after constant voltage stress. In both cases, holes are the major tunneling carriers for almost the complete range of applied gate voltage so that ΔV_{th} is always in the negative direction (hole trapping)

In order to get a physical picture of carrier injection, tunneling, and degradation mechanisms involved during PBTI or NBTI of La_2O_3–IL/Si stacked structures, simplified energy band diagrams for substrate and gate injection conditions are shown in Figure 3.41a,b. During substrate injection (PBTI for N-type MOSFET), a positive voltage is applied to the gate and both current components I_e and I_h are detected as shown previously. From Figures 3.27a,b, we recall that holes are the

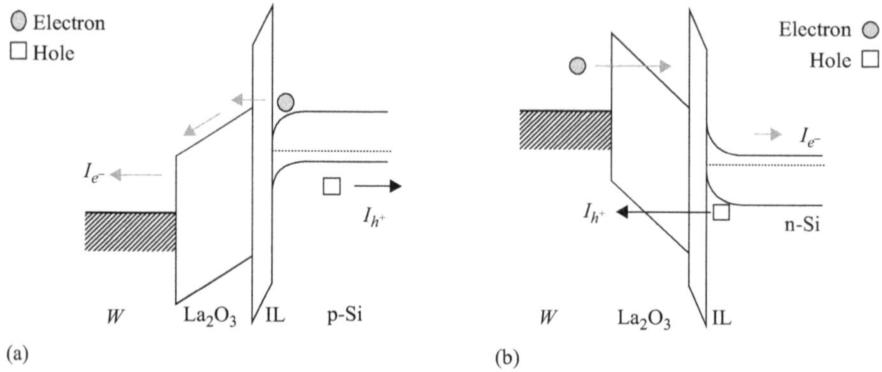

Figure 3.41 (a) Simplified energy band diagram for W/La_2O_3–IL/pSi (N-type MOSFET) under substrate injection conditions. (b) Simplified energy band diagram for the same W/La_2O_3–IL/nSi structure (P-type MOSFET) under gate injection conditions. In both diagrams, the major electron and hole currents are shown

major tunneling carriers for $V_g \leq 0.5$ V, whereas electrons are the major contributors to gate leakage current for $V_g \geq 0.5$ V. Now, by looking at the corresponding energy diagram in Figure 3.41a, we notice that at an initially low positive potential $+V_g$, the injection of electrons from an inverted silicon surface should be at a minimum because this current component I_e, "sees" a double oxide stack whose total thickness is composed of the IL and the La_2O_3 whose conduction band is lightly bended downward because of the low $+V_g$ applied. Once this $+V_g$ is high enough in order to increase the potential of the electrons at the inverted silicon surface, they could be able to tunnel directly through the IL and into the conduction band of La_2O_3 (which should be strongly bended downward because of a greater $+V_g$) by means of a Fowler-Nordheim, Poole-Frenkel, or even TAT mechanism (given the defective nature of the La_2O_3–IL interface). At the largest $+V_g$ applied, the electrons become the major current component of I_g and this condition seems to apply when $V_g \geq V_{th} \geq 0.5$ V.

The origin of I_e then comes from the tunneling of electrons through the La_2O_3–IL stack. The source for these electrons is the S/D regions that are connected once the channel is formed at sufficiently positive V_g. Determining the origin of the hole current I_h, is not so straightforward. Because of the metallic tungsten gate, the possibility of cold or hot-hole injection from the gate is eliminated. The possible mechanisms that could explain the origin of I_h are (1) anode–hole injection (AHI), (2) electron–hole (e–h) pair generation in the oxides' bulk/interface via trap levels (TL), (3) tunneling of valence-band electrons from the substrate directly into the La_2O_3 conduction band (leaving holes behind) or via TL, and (4) generation-recombination currents after impact ionization of electrons. It is thought that hole creation via (1) AHI (injected electrons from the substrate generate holes at the anode that can tunnel back into La_2O_3) and consequent (2) e–h pair generation via

TL within the bulk/interfaces of La_2O_3 are the more plausible mechanisms for the origin of I_h because of the highly defective nature of La_2O_3 after its deposition [3.82], which in turn creates a high density of trap levels. With the application of a continuous electrical stress, more traps or defects within the insulator are created [3.83] which can be filled with the corresponding carrier component depending on the magnitude of $+V_g$.

During gate injection (NBTI for P-type MOSFET), a negative voltage is applied to the gate and (according to Figure 3.41b), even though some electrons can tunnel into the conduction band of La_2O_3 via Fowler-Nordheim mechanism, the major component of the total I_g is the hole current I_h, which has been detected for almost all range of applied $-V_g$, see Figures 3.40a,b. A heavier effective mass for holes (m_h^*) during conduction would enable a rapid filling/saturation of the available sites for hole trapping during stress after the first injection of these carriers. This would create a large shift of V_{th} during the initial stages of injection and keep ΔV_{th} almost constant with any subsequent injection. In fact, this has been observed in our samples where a $\Delta V_{th} \sim 10$ mV was kept almost constant even with higher degrees of injected charge Q_{inj}, see Figure 3.37b. The origin of I_h also comes from the tunneling of holes through the La_2O_3–IL stack and then into the valence band of La_2O_3. The source for these holes is the S/D regions that are connected once the channel is formed at sufficiently negative V_g. Given a larger band offset of La_2O_3 with the valence band of silicon (with respect to its band offset in the conduction band), the hole current component I_h, should "see" the double oxide stack (composed of the IL and the La_2O_3) for the complete range of $-V_g$ applied. Because of the continuous trapping of holes at the La_2O_3 bulk and/or its interface with the IL, a localized modification of the energy band diagram at the trapping location or trap energy level E_t could be expected. This is important since significant variations in the total gate leakage current I_g (when holes or electrons are the major tunneling components) could increase or decrease the power consumption in the long term along with a higher/lower degree of V_{th} instabilities.

By using MOSFET structures during continuous electrical characterization (like PBTI and NBTI) we are able to extract many device parameters and correlate them together in order to highlight specific physical mechanisms leading to degradation. Density of interface states D_{it}, which is a parameter related to the density of dangling bonds at the insulator/semiconductor interface, could be then extracted and correlated to the shift in V_{th} after continuous stressing of the transistors during BTI. This correlation between ΔV_{th} and ΔD_{it} is quite important since it can give us an idea with respect to where is the damage more concentrated during BTI, at the La_2O_3–IL or IL–Si interface.

Figure 3.42a,b shows the correlation between ΔV_{th} and ΔD_{it} for N-type MOSFET device after PBTI conditions (similar transistors were measured for V_{th} and D_{it}). As shown earlier, Figure 3.42a shows a linear relationship between ΔV_{th} and the injected charge density Q_{inj}. On the other hand, D_{it} shows no change when using the same log-log plot and the same levels of Qinj. From these results, it is thought that the origin of ΔV_{th} for N-type MOSFET devices is then related to a progressive trapping of carriers (holes for low $+V_g$ and electrons for higher $+V_g$ applied, as shown by carrier-separation measurements) at the bulk of La_2O_3 and/or

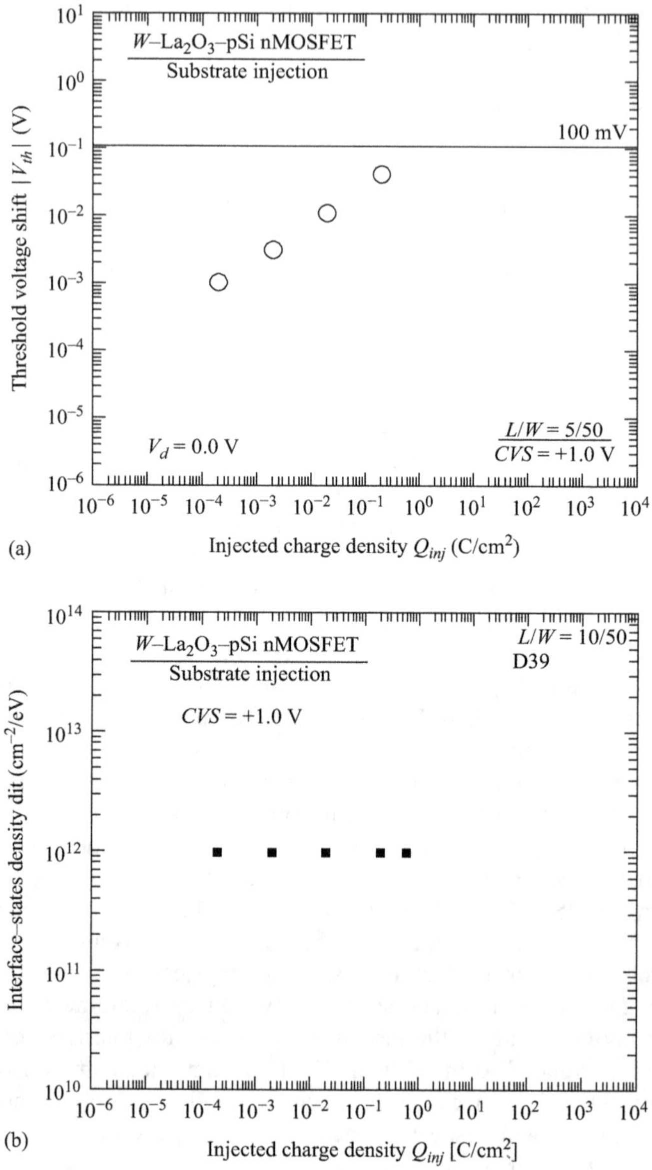

Figure 3.42 (a) V_{th} shift for a W/La_2O_3–IL/pSi structure in a N-type MOSFET after substrate injection. A linear ΔV_{th}–Q_{inj} relationship after stress is observed. (b) D_{it} shift with Q_{inj} for a similar transistor. From these data, no correlation between ΔV_{th} and ΔD_{it} is observed for the same densities of injected charge

at the La_2O_3–IL interface. With substrate injection during PBTI, the defect processes generated during stress are therefore affecting regions far from the silicon surface and its interface with the initial IL. However, even though D_{it} does not change with Q_{inj}, its density is close to ~10^{12} cm^2/eV, a large value for this parameter. One of the reasons for these high D_{it} densities could be the generation and saturation of interface traps created by release of hydrogen at the IL/Si interface and its subsequent diffusion and trapping into the La_2O_3/IL interface during carrier injection (thus shifting V_{th} to the negative side of an I_d–V_g plot as was already discussed). This is similar to the *Reaction-Diffusion* (R-D) model [3.64, [3.82–3.84] which is the most prevalent mechanism regarding NBTI evolution. Since carrier-separation measurements show that at low $+V_g$, holes are the main component of gate leakage current, an R-D model would make sense in terms of hole injection, H–Si bond breaking, hydrogen release at the IL/Si interface, a subsequent diffusion and trapping into the La_2O_3/IL interface and finally, a negative shift in V_{th} with a constant but large initial density of D_{it}.

Figure 3.43a,b shows the correlation between ΔV_{th} and ΔD_{it} for P-type MOSFET device after NBTI conditions (similar transistors were measured for V_{th} and D_{it}). Here, we have a different scenario regarding the correlation between ΔV_{th} and ΔD_{it}, where a similar trend is observed between these important parameters following stressing for the whole range of injected charge (in gate injection). While both V_{th}–D_{it} are kept constant after low stressing conditions, higher $-V_g$ conditions slightly increases them both. These results indicate that after the injection and continuous trapping of holes (the major component of I_g after carrier-separation measurements) into the La_2O_3–IL interface, the shift of V_{th} with stress is preceded by the damage produced by the breaking of H–Si bonds after injection of these holes through the initial IL–Si interface (thus increasing the density of interface-states D_{it}). Although D_{it} of P-type MOSFETs under gate injection is half that of N-type MOSFET devices, its density is close to ~5×10^{11} cm^2/eV, still a large value for this parameter so that an adequate processing for enhancing the IL/Si interface is required.

PMA of already metalized transistor structures in forming gas atmospheres (containing H_2) is important for the enhancement of the electrical and reliability characteristics of La_2O_3-based MOSFETs because of its ability to passivate some of the broken H–Si bonds. Up to now, we only have used dry-N_2-based PMA in our transistors and the results in terms of ΔD_{it} during BTI conditions showed constant and yet high densities of surface-states D_{it} for both N- and P-type MOSFETs. In order to reduce the initial density of Si dangling bonds at the IL/Si interface, P-type MOSFET devices were annealed in forming gas (a mixture composed of 97 per cent N_2 and 3 per cent H_2) and then stressed under similar NBTI conditions so that both ΔV_{th} and ΔD_{it} were measured and correlated as a function of injected charge density Q_{inj}. The results are shown in Figure 3.44a,b, where lower ΔV_{th} and ΔD_{it} after NBTI conditions are obtained using forming-gas based PMA at the same temperature. As before, identical trends in V_{th} and D_{it} shifts after $-V_g$ stress have been found but the initial levels of both ΔV_{th} and ΔD_{it} are lower than those found in N_2-based PMA devices. In the case of surface states, ΔD_{it} is kept constant at low densities of $\Delta D_{it} \leq 2 \times 10^{11}$ cm^2/eV, which is half of the ΔD_{it} for N_2-annealed MOSFETs.

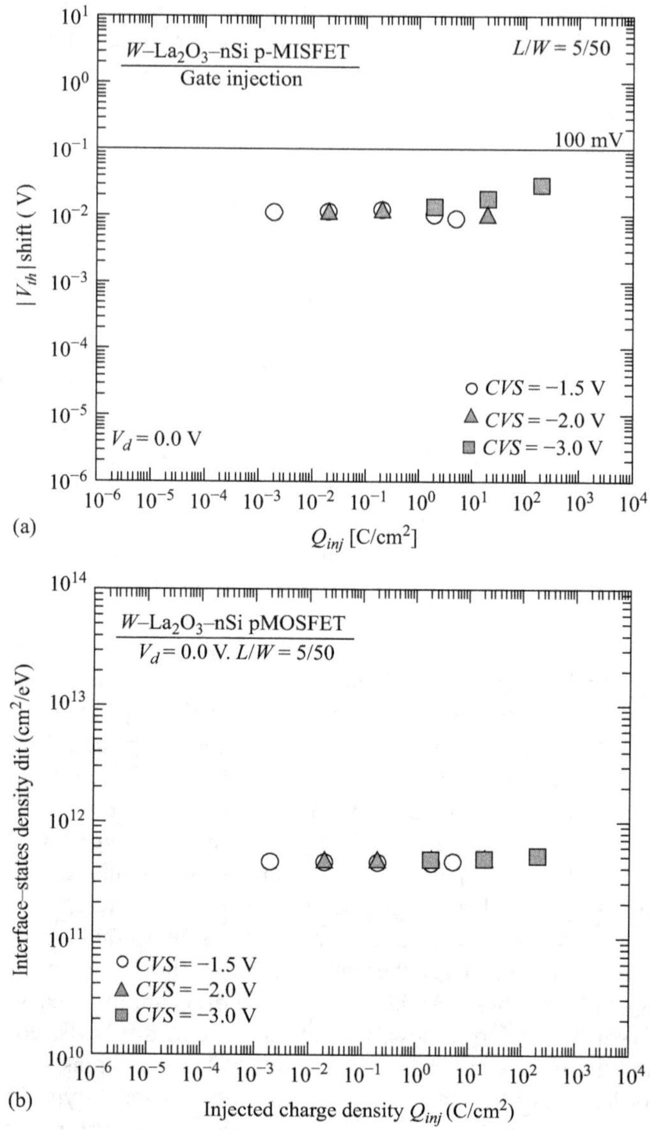

Figure 3.43 *(a) V_{th} shift for a W/La_2O_3–IL/nSi structure in a P-type MOSFET after gate injection. A slight dependence of V_{th} on Q_{inj} is observed at the higher stressing conditions. (b) D_{it} shift with Q_{inj} for a similar transistor. From here, a slight correlation between ΔV_{th} and ΔD_{it} is observed for the same densities of Q_{inj}*

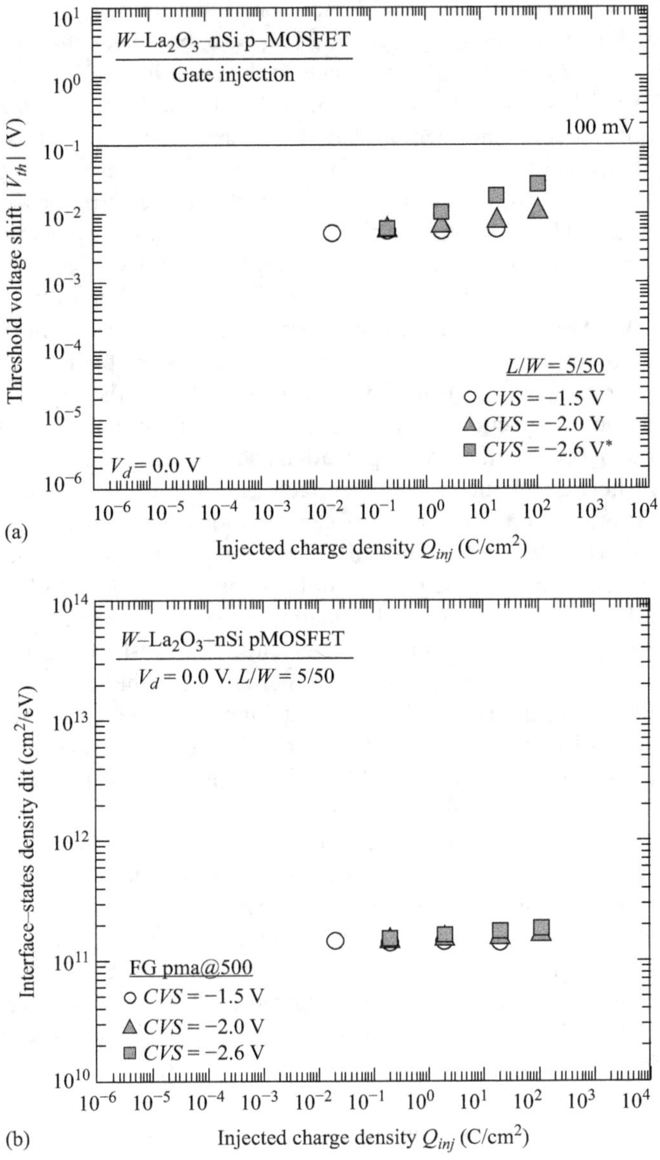

Figure 3.44 (a) V_{th} shift for a W/La_2O_3–IL/nSi structure in P-type MOSFET after gate injection and forming gas annealing. As before, identical ΔV_{th} (Q_{inj}) behavior is obtained for these transistors but now, the initial levels of ΔV_{th} are slightly lower than those found in N_2-based PMA devices. Also, a slight linear dependence of V_{th} on Q_{inj} is observed at the higher stressing conditions. (b) D_{it} shift with Q_{inj} for a similar transistor also annealed in forming gas. We notice a clear reduction in ΔD_{it} as compared to N_2-based PMA devices for the same densities of Q_{inj}. After forming gas annealing, ΔD_{it} is kept constant at $\Delta D_{it} \leq 2 \times 10^{11}$ cm^2/eV

With the former results, we have demonstrated that forming-gas based PMA is able to reduce the initial density of dangling bonds so that lower changes in D_{it} with stress are obtained. Additional improvements in the reliability of MOS devices have been found using deuterium (D_2, an isotope of H_2) in the forming gas atmosphere. Compared with H_2-based annealed samples, D_2-annealed gate dielectrics exhibit less charge trapping, less generation of interface state density, a larger charge for hard-dielectric breakdown (Q_{bd}), and longer time-dependent dielectric-breakdown characteristics under BTI conditions [3.85]. The improved reliability is attributed to a higher strength of the D–Si bond, so that a D_2-based PMA of high-k/IL stacked dielectrics on silicon has high potential for integration into ultra large-scale-integration (ULSI) processes with advanced gate dielectric stacks. This is important when considering that all transistors operate under dynamic conditions, where both gate and drain voltages (V_g, V_d) are simultaneously applied to the MOSFET structure so that an additional injection mechanism for carriers (additional source for oxide degradation) is involved in the total degradation and early failure of these devices; see Figure 3.45a where an additional lateral current component I_d is observed for N-type MOSFETs. In other words, the flow of carriers along an inverted silicon surface toward the drain (by means of an applied V_d) will induce an additional damage at the IL/Si interface which in turn, increase D_{it} or the density of more sites for trapping of carriers with this combined stress. If the magnitude of V_d is high enough, this turns into severe gate oxide degradation by hot electrons or hot-holes in what is better known as hot carrier injection (HCI); see Figure 3.45b, where HCI occurs for an N-type MOSFET under high V_d bias so that highly energetic electrons generate additional carriers by impact ionization. Since holes are cooler than electrons, hot-carrier degradation is higher in N-type MOSFET devices. Damage is usually located at the drain because of larger electric fields.

During HCI, as V_d increases, the electric field in the space charge region at the drain increases. At high electric field, additional electron–hole pairs can be generated in the space charge region by impact ionization of highly energetic electrons (N-type MOSFET). The generated electrons tend to be swept to the drain and generated holes swept into the substrate for an N-type MOSFET. Some of the electrons generated in the space charge region are attracted to the La_2O_3–IL/Si region due to the electric field induced by a positive V_g; this effect is shown as a red arrow in 3.45b. These generated electrons have energies far greater than the thermal-equilibrium value and are called hot electrons. With high enough energies, these electrons are able to tunnel into the oxide stack thus causing serious instabilities in the long-term performance of transistors. A fraction of the electrons traveling through the oxide may be trapped, producing negative charge in the oxide and thus a positive shift in V_{th}. The probability of electron trapping is usually less than that of hole trapping; but a hot electron induced gate current may exist over a long period of time, therefore the negative charging effect may build up. The energetic electrons, as they cross the first IL/Si interface, can also generate additional interface states due to a breaking of H–Si bonds. Charge trapping at these interface states causes additional shift in V_{th}, additional surface scattering, and reduced mobility. Since hot electron charging effects are continuous processes, the device continues to degrade with

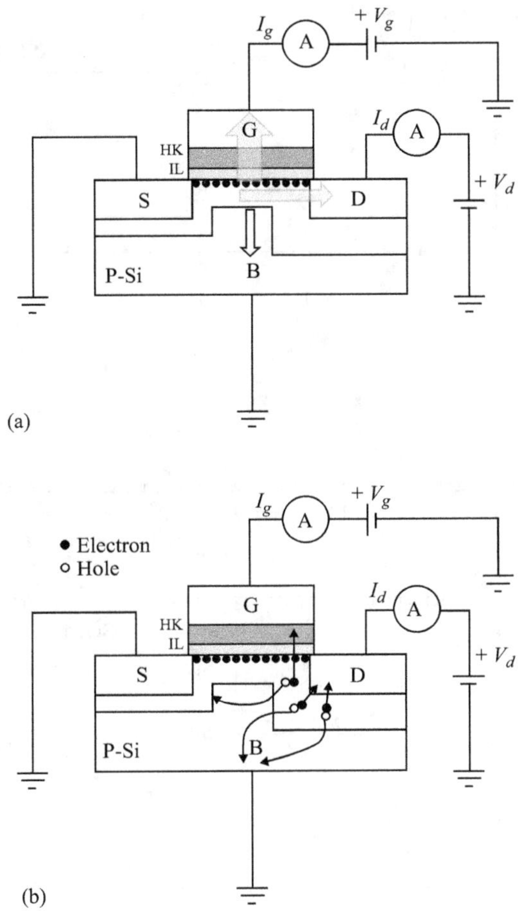

Figure 3.45 *(a) Schematic of an N-type MOSFET with both V_g and V_d polarization during stress. The total gate leakage current I_g is now reduced in proportion to I_d (V_d). This lateral current I_d, will induce additional damage at the IL/Si interface thus increasing D_{it}. (b) Hot carrier mechanism occurring at the drain side of the transistor (region with highest electric fields) where additional carriers are generated by impact ionization. Some of these carriers are injected in the La_2O_3–IL/Si structure thus increasing oxide charge and D_{it}*

operation time, which reduces the useful lifetime of the device. Additional charge trapping at the second La_2O_3–IL interface, could lead to the breakdown of the IL, additional shift in V_{th}, additional surface scattering and reduction in mobility and thus, a still lower operational lifetime of the device.

Figure 3.46a,b shows the result of stressing W/La_2O_3–IL/pSi N-type MOSFET devices with both V_g and V_d applied during PBTI characterization. Because of a

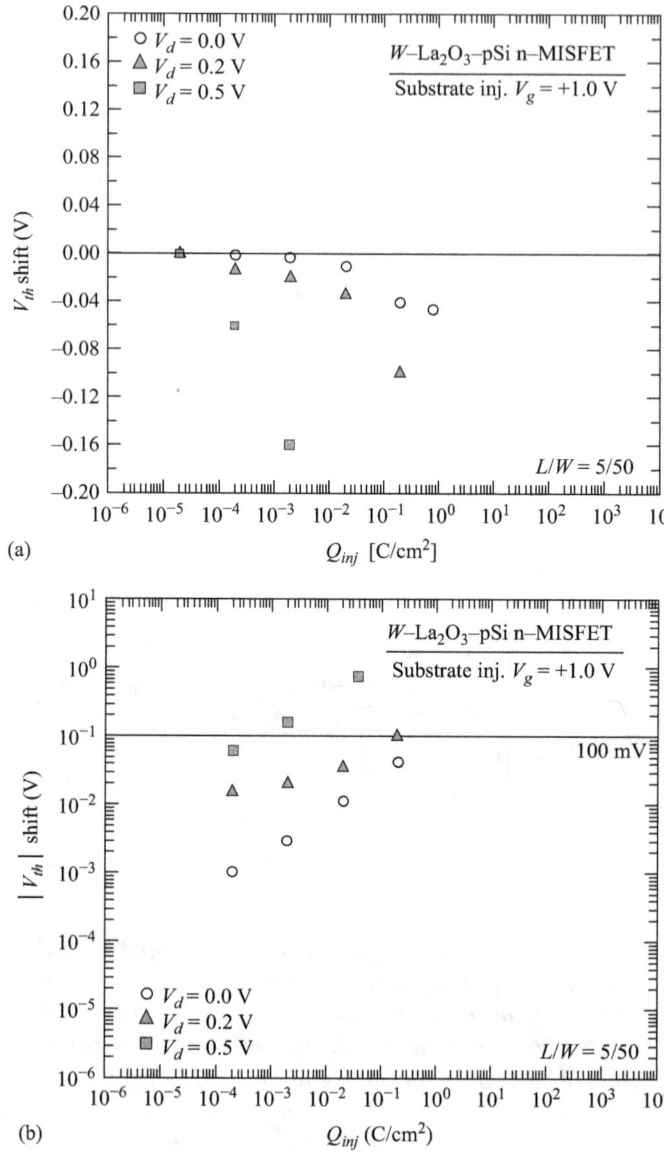

Figure 3.46 *(a) ΔV_{th} versus inject charge density Q_{inj} for N-type MOSFET device when both V_g and V_d are applied. By using even small V_d biases, ΔV_{th} largely increases. The negative shifts in V_{th} are due to hole injection which was shown after carrier-separation measurements. (b) Same data as in (a) but using $|\Delta V_{th}|$. A linear dependence of ΔV_{th} with injected charge is observed while this dependence increases with V_d*

low $+V_g$ used during stressing the devices ($V_g = \text{CVS} = +1.0$ V), V_{th} shifts in the negative direction for all charge injection conditions (as shown in Figure 3.36a,d).

We can see that once V_d is also applied during the stressing of these devices, the additional damage induced by V_d bias increases the shift in V_{th} (even though the total gate leakage current is reduced by I_d). With a $V_d = +0.5$ V, the largest shift in V_{th} is obtained even for the smallest densities of injected charge. The high sensitivity of ΔV_{th} (and ΔD_{it} as well) with respect to a lateral current I_d during BTI conditions, illustrates the need for a more realistic characterization of degradation in these devices. Therefore, it is important to include as much degradation factors as possible when stressing these devices in order to get a more realistic projection of their operational lifetimes. Once all factors that contribute to electrical degradation are determined, they can be used to obtain more complete models for lifetime projections and even propose enhancement techniques to increase reliability.

Finally, a brief comparison between some performance and reliability properties of La_2O_3-gated MOS devices with those of HfO_2 can be found in References 3.81, 3.86–3.90 in order to have more details with respect to the high potential offered by La_2O_3 in order to replace HfO_2 for nodes below 14 nm (logic technology). This is possible given the ability of La-based and La-silicate IL (sometimes combined with HfO_2) high-k materials to aggressively scale down *EOT* below the 0.6 nm regime. However, additional processing of P-type MOSFET devices and the extraction of their performance and reliability when metal gates with high work functions are used (suitable for P-type transistors) are required in order to consider La_2O_3 introduction into any CMOS processes. Also, a precise control in the thickness and stoichiometry of any IL developed at the La_2O_3/Si interface is important because of its role in order to enhance some performance and reliability characteristics at the cost of a reduced *EOT*. The general characteristics of this IL will also depend on the semiconductor material, which important when considering Ge or even III–V semiconductors.

3.2.2 Hot carriers in nanoscale devices

What can be said about the reliability challenges of future nanoelectronic technologies? For the near future (10–20 years), nanoelectronic logic devices are expected to remain FET based. Based on experience of past CMOS technologies, quite a bit can be said about expectations for the reliability of these devices. Contrary to early projections, the hot carrier mechanism has proven to be very persistent as technology scales and power supply voltages are reduced. That hot carrier degradation will continue to evolve into future FET-based nanotechnology is a reasonable prediction. Whereas power supply voltage scaling will reduce hot carrier effects, scaling will also lead to an increase of ballistic and quasi-ballistic transport and carrier–carrier scattering (CCS), which may exacerbate the effects of hot carriers. But, as we shall see, many of these trends have already made impacts to CMOS reliability.

3.2.2.1 Device scaling into the nanotechnology regime

Table 3.2 is adapted from the 2013 International Technology Roadmap for Semiconductors Emerging Research Devices (ERD) document [3.91]:

Table 3.2 Nanoelectronic logic device roadmap. Adapted from Reference 3.91

State variable ↑	Structure/Materials →	Conventional	Novel
	Non-charge		Spin wave logic Nanomagnetic FET Excitonic FET BISFET STMG All spin logic
	Charge	Si FET NW FET TFET n Ge p III-V Graphene FET CNT FET	Spin FET NEMS Atomic switch Mott FET Neg-C_g FET

Table 3.3 Conceptual CMOS device roadmap

Technology node (nm)	Device
45	SiO_xN_y poly-Si gate FET ("Conventional" FET)
28/32	HKMG FET
20/22	HKMG FET/FinFET
10/14	FinFET
5/7	III-V (n) and Ge or SiGe (p) FinFET
<5	Si nanowire/CNT/graphene

The devices expected to be used in near end technologies (developed over the next decade or more) are those in the conventional/charge box. The CMOS device evolution (and plausible extension) into the near end nanoelectronic era suggested by the 2013 roadmap is [3.92] (Table 3.3).

All of these devices operate by an electrically isolated gate electrode, which controls conduction into (or out of) a semiconductor – the basic field effect principle will still reign for the next few decades!

3.2.2.2 Hot carriers

Hot carrier degradation is an FET mechanism resulting in charge buildup – gate insulator bulk charge and interface states – at the insulator/Si interface under the influence of applied drain voltage. The damage is caused by carriers (electrons or holes) that gain high kinetic energy from the steep potential drop (high electric field) in the drain region of an MOS device. If carriers gain enough energy to be injected into the gate oxide or to cause interfacial damage, they will introduce instabilities in the electrical characteristics of a MOSFET device. The damage is localized near the drain junction (Figure 3.47).

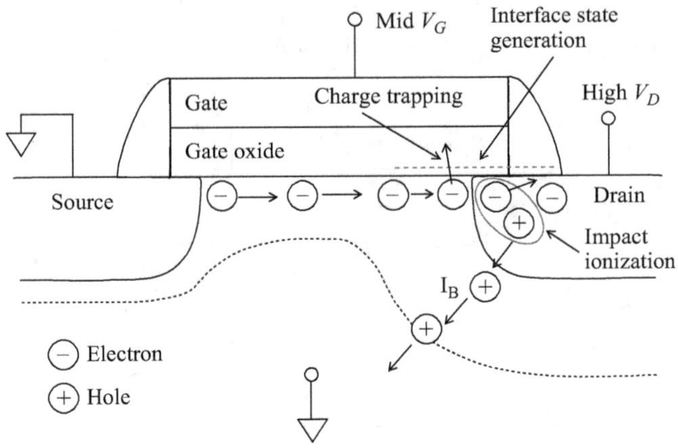

Figure 3.47 Hot carrier processes in an NFET

The history of hot carrier degradation in CMOS goes back to the 1970s. For devices of that era, there was no significant quasi-ballistic transport. Carriers (electrons or holes) gain kinetic energy from transit through regions of high electric field in the drain region of the CMOS device. If the mean carrier energy is larger than the thermal energy at the lattice temperature, they are called "hot," because historically the carrier kinetic energy was assumed to be distributed with a thermal-like distribution ("Maxwellian") at an effective temperature higher than that of the lattice. This distribution is in a steady state with the local electric field, and its effective temperature is dependent on this field. To cause damage, carriers require sufficient energy. The damage rate is thus dependent on the lateral electric field. This is the basis of the popular "Lucky Electron Model" (LEM) [3.93].

However, with scaling, quasi-ballistic transport, at least in the region near the drain, becomes more substantial, and a new picture has emerged, such as the "Energy-Driven Model" [3.94]. The assumption of a Maxwellian-like energy distribution in steady state with the local electric field increasingly breaks down as the size of the high field region is scaled below 100 nm or so (roughly 0.25 μm or less channel length technology), and technology power supply voltages scale down. Quasi-ballistic transport in the high field region of MOSFETs at modern dimensions has been shown to result in a rather shallow carrier energy distribution function (EDF) up to the total energy available in the high field region, but then a steeper tail at higher energies. This sharp downward inflection or "knee" leads to a hot carrier damage rate that is dependent on the total available energy, not the peak electric field (Figure 3.48).

3.2.2.3 The lucky electron model

The LEM of C.M. Hu *et al.* [3.93] has been the dominant guiding principle of most industry standard hot carrier models and projection methodologies, and is often misapplied to current technology nodes. The fundamental concept as applied to silicon can be traced back to Shockley [3.95] as an analogy to the theory of *gas*

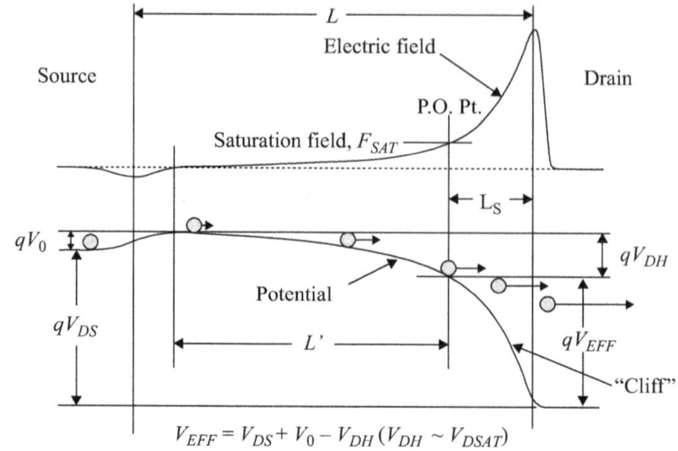

Figure 3.48 Quasi-ballistic transport in a modern FET

discharge developed in the early 1900s. [3.96] In a gas discharge, a free electron travels ballistically in the electric field until it collides with a gas atom, which may ionize the atom, leading to two free electrons. This process is referred to as "avalanche breakdown". This is why the phrase "impact ionization" is used. The probability of an electron's traveling a distance at least d before suffering a collision is,

$$P(d) = e^{\frac{-d}{\lambda}} \tag{3.27}$$

where λ is called the "mean free path". The electron is assumed to lose all of its kinetic energy in the collision. Since the energy, E, gained by the electron from the electric field, F, is $E = q\mathrm{d}F$, the electron energy distribution is given by

$$f(E) = P(E) = e^{\frac{-E}{q\lambda F}} \tag{3.28}$$

This is the basis of the LEM. This distribution has very similar energy dependence to a thermal, or "Maxwellian" energy distribution with an effective temperature, T_{EFF}, of

$$T_{EFF} = \frac{q\lambda F}{k} \tag{3.29}$$

This coincidental resemblance to an energy distribution in thermal equilibrium is the historical reason for the designations "hot electron" and "hot carrier".

Note: The electrons do not actually behave in a thermal way – a thermal velocity distribution would be isotropic, and the LEM actually assumes purely ballistic behavior – velocity is strictly along the field direction.

Next, to model either the impact ionization rate, or the hot carrier damage rate, very sharp energy thresholds are assumed. It is recognized that the electric field is not spatially constant, so the quantity F is replaced by F_m, the maximum field, since this is where the rates should peak. Under these assumptions, the ratio of substrate current, I_{sub}, to drain current I_D (approximately the impact ionization ratio), is given by

$$\frac{I_{sub}}{I_D} = Ae^{\frac{-\phi_i}{q\lambda F_m}} \tag{3.30}$$

where ϕ_i is the threshold energy for impact ionization. And the hot carrier rate, defined as the inverse of the hot carrier lifetime, τ, divided by drain current I_D, is then

$$\frac{1}{\tau I_D} = Be^{\frac{-\phi_{it}}{q\lambda F_m}} \tag{3.31}$$

and ϕ_{it} is the threshold energy for hot carrier damage.

By comparison with photon-induced emission rates, the values of ϕ_{it} and ϕ_i were "inferred" to be about 3.7 eV and 1.3 eV, respectively. Since the peak lateral electric field is proportional to the drain bias to channel pinch-off drop $V_{DS} - V_{DSAT}$, where V_{DSAT} is the potential at the "pinch-off" or "saturation" point in the channel, the following equation is often used for the impact ionization rate:

$$\frac{I_{sub}}{I_D} = Ae^{-\left[\frac{b}{(V_{DS}-V_{DSAT})}\right]} \tag{3.32}$$

A second cruder form of hot carrier voltage acceleration is loosely related:

$$\tau \approx De^{-\frac{V_0}{V_{DS}}} \tag{3.33}$$

which is also highly used for lifetime extrapolation.

Nonuniform electric fields, limited potential drops, and quasi-ballistic transport

The lateral electric field in MOSFETs past the saturation point in the channel is highly nonuniform [3.97], being approximately an exponential function of lateral distance [3.98]:

$$F \propto e^{\frac{y}{l}} \tag{3.34}$$

where l is a scale factor that is roughly on the order of 10 per cent of "L_{nom}" (the minimum design channel length for a given MOSFET device design). The length of the high field region (after the pinch-off or saturation point in the channel) is somewhat larger than l, but even for devices at the 90 nm node, it is less than ~30 nm, which is short enough for significant quasi-ballistic effects. In addition, the drain bias is becoming increasingly limited by the scaling down of the power supply voltage. Many authors under these conditions have simulated the EDF and the general conclusion is that the EDF develops a significant knee near the maximum energy available from the steep potential drop at the drain [3.99–3.107]. This is approximately the potential from the drain to the channel "pinch-off" point [3.101] (called "V_{EFF}" here). Above this knee, there is a "thermal" tail, which is due to the interactions of quasi-ballistic carriers with phonons. As CMOS device dimensions

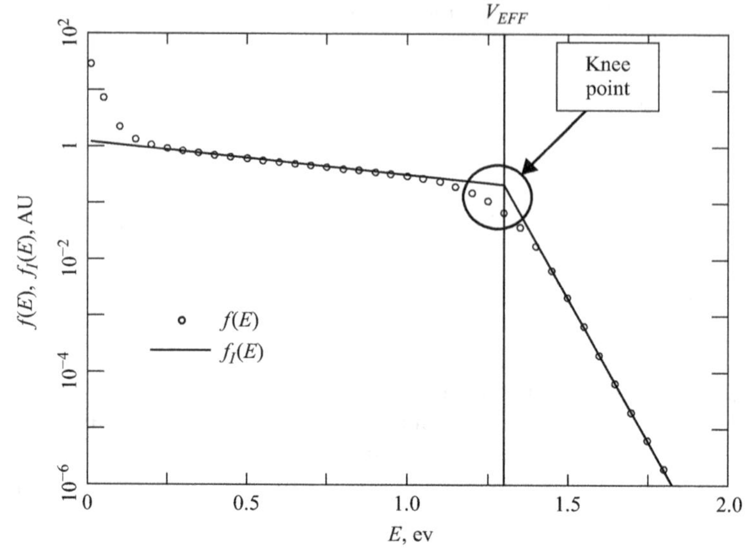

Figure 3.49 Generalized quasi-ballistic EDF [3.101], f(E), and idealized EDF, $f_I(E)$, ($\chi \sim 2$)

are made smaller and the supply voltage decreases, the EDF becomes shallower for $E < qV_{EFF}$, and the knee near $E = qV_{EFF}$ strengthens. The characteristics of these EDFs (at about $L_{nom} = 100$ nm) are shown in Figure 3.49 for one value of V_{EFF} (cf. Reference 3.101). If the position is within the neutral drain region, then there will be a population of cold drain carriers, which can be seen here below 0.2 eV. These will not contribute to hot carrier effects. The EDF can be fit to an idealized distribution that is "LEM-like" for $E < qV_{EFF}$, but is truncated by a thermal tail for higher energies:

$$\begin{aligned} f_1(E) &\propto e^{(-\chi E/qV_{EFF})}, \quad E \leq qV_{EFF} \\ &\propto e^{-\chi} e^{[(qV_{EFF}-E)/nkT]}, \quad E > qV_{EFF} \end{aligned} \tag{3.35}$$

Typical values of χ derived from the literature are dependent on L_{nom}: $\chi \sim 0$ for 25 nm, 1 for 50 nm, 2 for 100 nm, and 4 for 250 nm.

3.2.2.4 The energy-driven model

The basic assumption for energy-mediated processes, such as impact ionization or interface state generation (ISG), is that the rates are approximately determined by an integral of the following form:

$$Rate = \int f(E)S(E)dE \tag{3.36}$$

where f is the EDF, and S = interaction cross-section or scattering rate. The density of states is not explicitly included here, and can be considered to be part of $f(E)$, or else neglected. The integrand of this rate equation will generally peak at one or more points, which are referred to as "dominant energies" because carriers near these energies dominate the respective hot carrier rate.

This occurs when

$$\frac{\mathrm{d}\ln(f)}{\mathrm{d}E} = \frac{\mathrm{d}\ln(S)}{\mathrm{d}E} \tag{3.37}$$

Mathematically, the dominant energy can be controlled by "knee" points (points of high curvature) of either ln(f) or ln(S). While the LEM implicitly assumes that the knee points of the ln(S) functions drive the dominant energies, the energy-driven model is based on the idea that the knee points of ln(f) drive the dominant energies [3.94].

The LEM represents the large (long high field region) device and high voltage limit. However, we know that as the power supply voltage is scaled down, the EDF becomes increasingly limited at energies of importance to hot carrier effects (generally less than 5 eV or so). The LEM assumes that the EDF has no significant knee and the dominant energies for impact ionization (II) and ISG are determined by knee points in the respective cross-sections (which were thought to be very close to energy thresholds ϕ_i and ϕ_{it}). Figure 3.50 is a conceptual schematic of this concept for impact ionization. In keeping with the thinking of the time, the Keldysh model ($\propto (E - E_G)^2$) [3.108] is used for S_{ii}.

The II and ISG rates are then "field driven." The dominant energies are weak functions of bias conditions, and the hot carrier voltage dependence is due almost solely to the changes in the EDF slope with field.

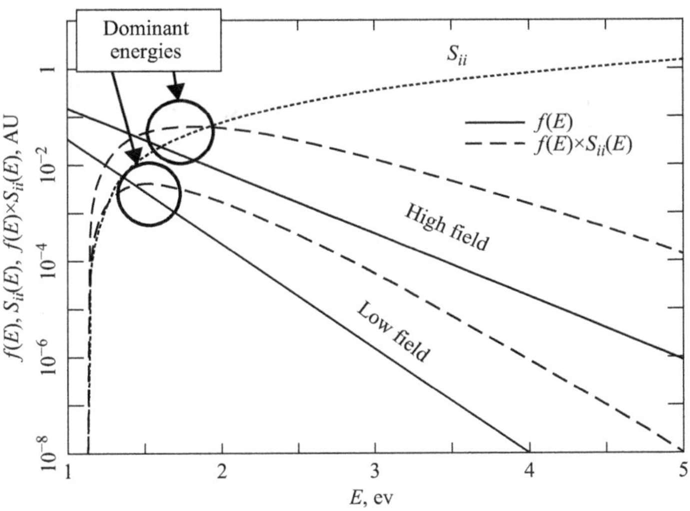

Figure 3.50 The field-driven hot carrier paradigm applied to impact ionization

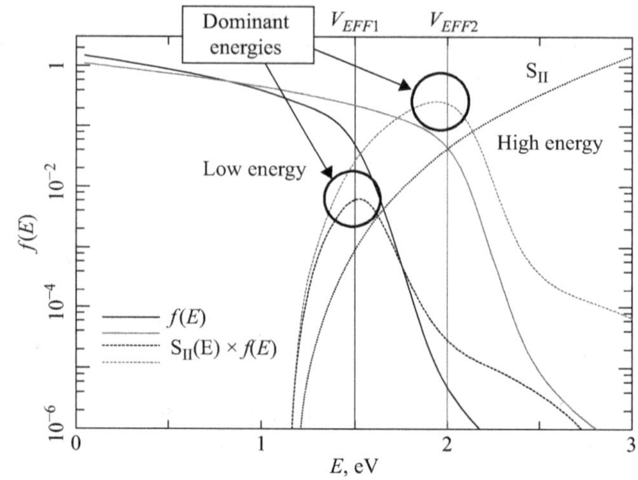

Figure 3.51 The energy-driven hot carrier concept applied to impact ionization

For the case of quasi-ballistic transport, the EDF knee determines the dominant energies, and the II and ISG rates are "energy driven"; the dominant energies track with bias. The hot carrier voltage dependencies are due primarily to the energy dependence of the S functions, through the bias dependencies of V_{EFF}. The field dependence of the electron energy distribution function (EEDF) is secondary as far as voltage dependence is concerned. Figure 3.51 is a conceptual schematic of the energy-driven concept for impact ionization. In this figure the S_{II} used is the more modern model of Kamakura *et al.* ($\propto(E - E_G)^{4.6}$) [3.104], which is "softer" (less curvature near threshold) than the older II models.

Conditions for energy-driven hot carrier rates

To illustrate the conditions under which the hot carrier behavior is energy driven, we use the idealized EEDF, $f_1(E)$, which collapses the knee to a single point. Recall,

$$\begin{aligned} f_1(E) &\propto e^{(-\chi E/qV_{EFF})}, \quad E \leq qV_{EFF} \\ &\propto e^{-\chi} e^{[(qV_{EFF}-E)/nkT]}, \quad E > qV_{EFF} \end{aligned} \tag{3.38}$$

A scattering rate of the following form is used: $S = A(E - E_{TH})^p$. In this idealized case, the energy-driven regime can be defined as when the dominant energy $= qV_{EFF}$. Using (3.38), this can easily be shown to be

$$E_{TH} + pnkT \leq qV_{EFF} \leq \frac{E_{TH}}{1 - \frac{p}{\chi}}, \quad \chi > p \tag{3.39}$$

The field-driven regime is when V_{EFF} is above this region. If $\chi \leq p$, there is no field-driven regime. For V_{EFF} below this region, the dominant energy is in the thermal tail. This might be referred to as the "thermal tail-driven" regime [3.98].

To give some approximate numbers as examples, let S = impact ionization rate for electrons (S_{ii}), $E_{TH} = E_G = 1.12$ eV, $p \sim 4.6$ for electron-induced impact ionization [3.109], and using $n = 1.66$ [3.116], $T = 300$ K,

$$1.317\ eV \leq qV_{EFF} \leq \frac{1.12\ eV}{1 - \frac{4.6}{\chi}}, \quad \chi > 4.6 \tag{3.40}$$

Since the typical values of $\chi < 4.6$ for $L_{nom} < 0.25$ mm, for any NFET device of a quarter micron or smaller technology, the impact ionization will be energy or thermal tail-driven for any V_{EFF}. The dominant energy is equal to V_{EFF} for these values.

In the energy-driven regime, the impact ionization rate is approximately proportional to the scattering rate,

$$r_{ii} = \frac{I_{sub}}{I_D} = \int f_I(E) S_{ii}(E) dE \approx A_{ii} e^{-\chi} S_{ii}(qV_{EFF}) \tag{3.41}$$

Calculation of V_{EFF} in the energy-driven model is performed in the following way:

V_{EFF} = effective potential drop from channel to drain:

$$V_{EFF} = V_0 + V_{DS} - V_{DSAT} \tag{3.42}$$

where V_0 = added potential due to halo [3.110, 3.111] and/or "source function" (total expected to be on the order of several hundred mV), and V_{DSAT} = pinch-off or saturation voltage. One approximate equation for V_{DSAT} is given by Taur and Ning [3.112],

$$V_{DSAT} = \frac{\frac{2(V_{GS} - V_T)}{m}}{1 + \sqrt{1 + \frac{2(V_{GS} - V_T)}{mF_C(L - L_S)}}} \tag{3.43}$$

where F_C = critical field for velocity saturation (~ratio of carrier mobility to saturation velocity μ/v_{sat}), L = channel length, L_S = length of velocity saturated region, and m = body effect coefficient.

3.2.2.5 Short-range carrier–carrier scattering effects

Short-range, or coulombic, CCS is a mechanism whereby a carrier can gain even more energy than qV_{EFF}. There is a small probability that two high energy carriers undergo a coulombic scattering process so that one gains much of the total kinetic energy, leading to a small electron population of carriers up to about twice qV_{EFF}. Many authors [3.113–3.117], employing Monte Carlo or other simulation techniques to nMOSFETs, have predicted that at drain voltages below about 3 V, electrons "heated" by e–e scattering (EES) should dominate the high energy tail of the EEDF. CCS induces a second, weaker knee at just less than $2qV_{EFF}$. (Above this energy, the higher-order knees at somewhat below 3,4 V_{EFF}, etc., are smeared out into a quasi-thermal tail, whose effective temperature increases with carrier concentration.) Adding CCS effects to the base EDF and an idealized EDF is demonstrated in Figure 3.52.

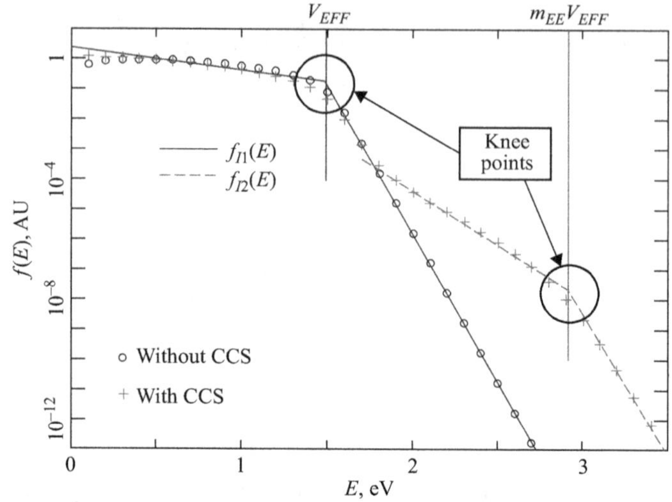

Figure 3.52 EDF with and without CCS tail and idealized EDF

The idealized EDF is

$$f_I(E) = f_{I1}(E) + f_{I2}(E) \tag{3.44a}$$

$$\begin{aligned} f_{I1} &\propto e^{(-\chi_1 E/qV_{EFF})}, \quad E \le qV_{EFF} \\ &\propto e^{-\chi_1} e^{[(qV_{EFF}-E)/nkT]}, \quad E > qV_{EFF} \end{aligned} \tag{3.44b}$$

$$\begin{aligned} f_{I2} &= 0, \ E \le qV_{EFF} \\ &= a_{cc} V_{EFF}^{-3/2} e^{(-\chi_2 E/qV_{EFF})}, \quad qV_{EFF} < E < 2qV_{EFF} \\ &= a_{cc} e^{-2\chi_2} V_{EFF}^{-3/2} e^{[(2qV_{EFF}-E)/nkT]}, \quad E > 2qV_{EFF} \end{aligned} \tag{3.44c}$$

The ${V_{EFF}}^{-3/2}$ term is due to the energy dependence of the CCS cross-section. Since the CCS rate per carrier is approximately proportional to the carrier density in the energy range between V_{EFF} and $2V_{EFF}$, the relative level of ϕ_{I2}, a_{cc}, has a linear I_D dependence. For this example, $V_{EFF} = 1.76\text{V}$, $\chi_1 = 2.5$, $\chi_2 = 11.5$. The value of χ_2 depends on χ_1 and the energy dependence of the CCS cross-section. This is an approximate expression for χ_2:

$$\chi_2 \approx 9 + \frac{\chi_1}{2} \tag{3.45}$$

The relative impact of the knee of the CCS tail to impact ionization can be bounded in the following way. Assuming a reasonable upper limit for the ratio of the tail population at its knee just below $2V_{EFF}$ to the base population at V_{EFF} of about 10^{-5}, the ratio of peak $f(E) \times S_{ii}(E)$ at the tail knee to that at the base knee is

$$ratio < 10^{-5} \frac{S_{ii}(2qV_{EFF})}{S_{ii}(qV_{EFF})} = 10^{-5} \left(\frac{2qV_{EFF} - E_G}{qV_{EFF} - E_G} \right)^p \tag{3.46}$$

For electrons ($p = 4.6$), this ratio will exceed unity only for $V_{EFF} < 1.23$ V, which is inside the thermal tail-driven regime. It appears that EES is too weak at reasonable carrier concentrations to contribute substantial impact ionization in the energy-driven regime (at the tail knee), although lower energy parts of the EES tail will contribute at around bandgap or sub-bandgap V_{EFF} (<1.3–1.4 V) [3.117]. Even for holes ($p = 7$), unity ratio is at $V_{EFF} < 1.47$ V, just barely above the critical point for holes (1.42 V). This will appear as a CCS-induced enhancement of the thermal tail contribution, and will not be modeled correctly by this energy-driven approximation.

For NFETs, there is experimental evidence that the hot carrier damage rate is quadratic in I_D over much of the V_{GS} range (for a given energy) [3.118–3.120] implying that the knee of the EES tail does drive the rate. In this case, the hot carrier damage rate can be written by this (energy-driven) approximation,

$$R_{ISG}(V_{EFF}) \approx A_1 I_D S_{it}(qV_{EFF}) + A_2 {I_D}^2 S_{it}(qm_{EE}V_{EFF}) \tag{3.47}$$

S_{it} is the ISG scattering rate, and the parameter m_{EE} represents the ratio of the dominant energy (for ISG) due to the CCS tail to that due to the base distribution. Because of the relative weakness of the CCS knee, this may tend to be somewhat less than two. Simulations suggest values of 1.7–1.95 (depending on technology) are reasonable.

The "universal" nature of the energy-driven model for NFETs in the quadratic (mid- V_G) regime is demonstrated by Figure 3.53 down to a minimum dominant energy of ~2.6 eV (V_{EFF} ~ 1.3 V).

The solid curve shown has the following empirical form:

$$\begin{aligned} S_{it} &\propto e^{(aE)}, \quad E \le \phi_{it} + p/a \\ &\propto (E - \phi_{it})^p, \quad E > \phi_{it} + p/a \end{aligned} \tag{3.48}$$

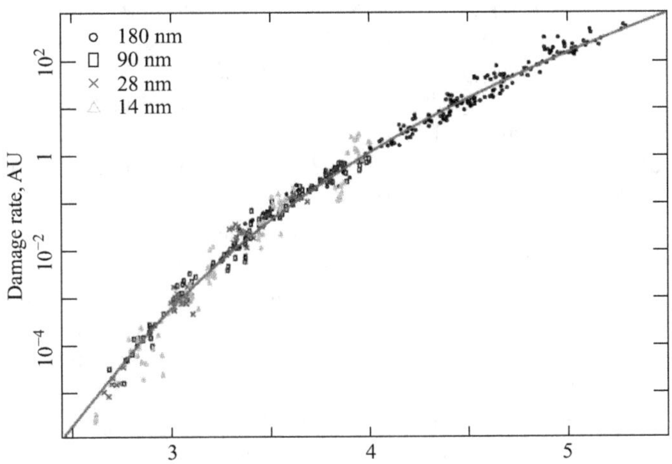

Figure 3.53 Experimental data for a wide range of technologies in the quadratic regime

with the following parameter values:

$$\phi_{it} = 1.75eV, \quad p = 13.5, \quad a = 13eV^{-1}.$$

3.2.2.6 Justification of energy-driven model for ISG damage rates

Let us revisit Section "Conditions for energy driven hot carrier rates" for this ISG scattering rate. Recall the upper limit for an energy-driven dominant energy:

$$qV_{EFF} \leq \frac{E_{TH}}{1 - \frac{p}{\chi}} \tag{3.49}$$

Now, for our experimental S_{it} function, $E_{TH} = 1.75$ eV, $p \sim 13.5$. This implies that a typical quarter micron device ($\chi_2 \sim 11$), or below, will be energy limited. For technologies of greater L_{nom} (say, $V_{DD} = 3.3$ V or more), ISG rates may be field driven at sufficiently high voltages.

3.2.2.7 Temperature dependence

Historically, the major temperature effect to hot carrier has been ascribed to the change of λ, the mean free path, with temperature. This expression was given by Crowell and Sze in 1966 [3.121].

$$\lambda(T) = \lambda_0 tanh\left(\frac{E_R}{2kT}\right) \tag{3.50}$$

where $E_R = 63$ meV is the optical phonon energy in silicon. Around room temperature λ is a weak function of temperature; its value at $T = 400$ K is 0.86 of its value at 300 K. However, the LEM predicts that hot carrier rates are exponentially dependent on λ. In the energy-driven model, the effect can be captured by the parameter χ of the base distribution(χ_1) [3.2.32]. By (3.41), (3.44), and (3.45), the impact ionization rate and the ISG rates in both the linear and quadratic regimes will all have the same χ_1 dependence, $e^{-\chi_1}$. Assuming $\chi_1 \propto 1/\lambda$, and using (3.50), the net temperature dependence of this term from 200 to 400 K can be approximated as a power of absolute temperature T^{-nt}, where $nt \approx 0.4\chi_1(300\text{ K})$. This implies that the effect of mean free path temperature dependence decreases with scaling, and becomes small below an $L_{nom} = 50$ nm technology. For nanoscale devices, $nt \to 0$, and the hot carrier rate temperature dependence will be driven by other parameters. Basically, as the carrier transport becomes increasingly ballistic, the dominance of phonon scattering disappears. Experimentally, the negative temperature dependencies of IIS and EED observed for older technologies have evolved to positive ones with scaling [3.106]. This has been explained as an increase of the V_0 parameter with temperature in the energy-driven model. Examples from NFETs of 1.2 and 2.5 V technologies are shown in Figure 3.54. The hot carrier-induced current shifts at a typical mid-Vg stress bias condition, but three temperatures, for these two device types are shown in Figure 3.54.

The left side of Figure 3.54 corresponds to a 90nm (1.2 V) NFET, $t = 3000$ s, Stress: $V_{DS} = 1.9$ V, $V_{GS} = 1.0$ V. The right side corresponds to a 0.25 μm (2.5 V) NFET, $t = 1000$ s, Stress: $V_{DS} = 3.3$ V, $V_{GS} = 1.65$ V.

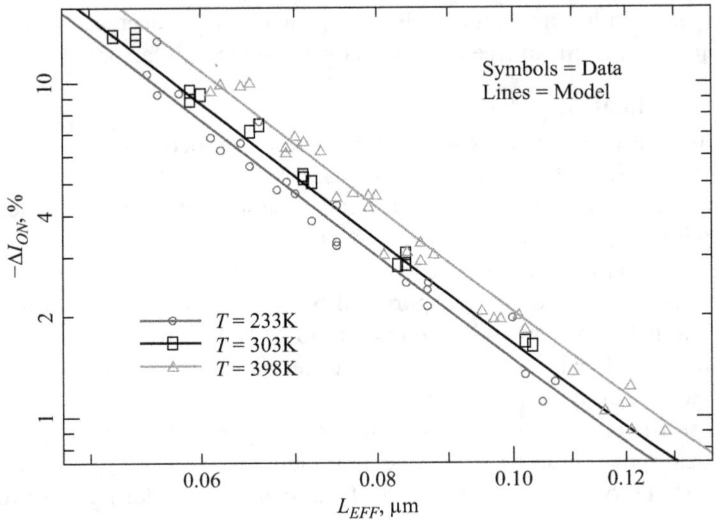

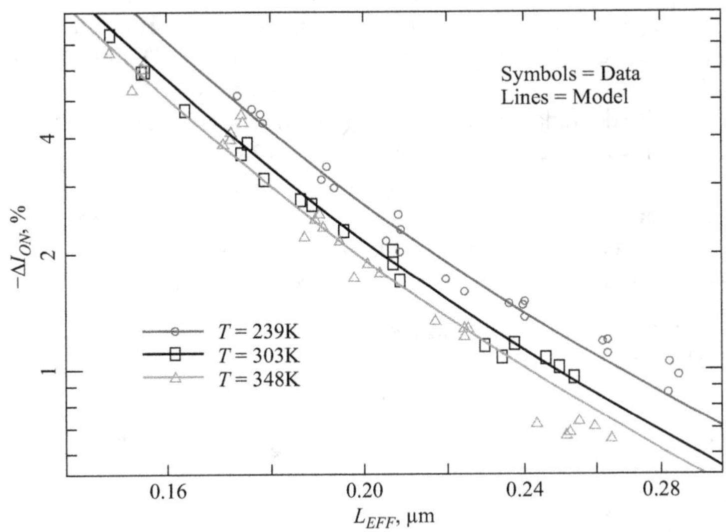

Figure 3.54 Hot carriers-induced current shift, %ΔI_{on}, versus channel length, L, for three ambient temperatures, T, of 1.2 and 2.5 V NFETs

There are opposite temperature dependencies for these two device types, whereas the quarter micron NFET displays the classic behavior of negative temperature coefficient, the hot carrier shift of the 90 nm NFET increases with temperature.

The scaling effects are mainly due to only two of these parameters – V_0 and prefactor. As technology scales down, the reduction in supply voltage and the possible increase in V_0 significantly increase the positive impact of a modest available energy increase with temperature. Also, as channel length decreases, the negative temperature effect of mean free path (as reflected by the prefactor) diminishes.

3.2.2.8 Self-heating effects

As active device volumes decrease, and become more thermally isolated from the substrate, localized self-heating will increase. And, at scales <~20–30nm, especially in the presence of energetic carriers, a simple heat diffusion description breaks down [3.122].

Heat flow over large scales is dominated by the diffusion of acoustic phonons. But, over nanoscales, phonon transport will be ballistic. Optical phonons are generated by energetic carriers and have low group velocity – they tend to accumulate locally. Since the local phonon EDF is not thermal, the concept of "lattice temperature" no longer applies [3.123].

Local self-heating is not a reliable mechanism per se. However, we know that elevated local temperatures (or large local populations of energetic phonons) will exacerbate HCD. As long as local power is dissipated only during short transients with a low duty cycle, there will be minimal impact in a logical switching environment. But, for a device in static on condition (perhaps an analog application), there may be a serious reliability impact. In addition, hot carrier DC stress is normally done in static on conditions with elevated voltages. The result is excessive local self-heating, leading to an unknown acceleration of HC, and potential activation of the BTI degradation mechanism (which tends to mask the HC). This has already been shown to occur in planar PFETs [3.124].

These considerations may necessitate methodology changes for HC estimation and qualification. Data from DC stress conditions need to be corrected for self-heating, which may not be a trivial task to do accurately. DC methods should be augmented by AC HC stress using on-chip pulse generators or ring oscillators.

3.2.2.9 Progression of hot carrier effects in the nanoscale FET regime

- *Evolutionary trends (general scaling)*

  Table 3.4 lists the major hot carrier scaling trends.

- *Revolutionary trends*

  Silicon FinFETs represent a deviation from the traditional scaling of planar, single gate, FETs. A dual/triple gate structure with low fin width has allowed very low or no body doping concentration. This results in higher carrier mobilities in the channel due to reduced impurity scattering. Also, the combined use of a (011) channel orientation causes a particularly strong increase in hole mobility. There are several ramifications to hot carrier degradation: strengthened quasi-ballistic behavior (→ higher EDF at the knee), and a reduction in the critical field, F_c (→ higher V_{EFF}). Both of these will increase hot carrier degradation (HCD).

  Geometrical confinement due to the 3D structure and a small fin width will have several effects: (1) There will be an appreciable flux of energetic carriers on each gate side, which are generated on the opposite side and top. In a planar

Table 3.4 Major hot carrier scaling trends

Scaling parameter	Model parameter	HC (%ΔI_{dsat}) trend
Supply voltage, V_{DD}	V_{EFF}	~$V_{DD}{}^{-p}$, p ~ 12
Channel length, L	V_{EFF}	~L^{-m}, m ~ 0–2 (m decreasing with scaling)
	Sensitivity ($\Delta I_{dsat}/\Delta N_{it}$)	~L^{-1}
Length of high field region, L_s	c	~$e^{-c/2}$, c ~ L_s/l (mean free path)
Drain current density, I_D, & carrier concentration	I_D	~$I_D{}^1$
Effective oxide thickness, T_{ox}	Sensitivity ($\Delta I_{dsat}/\Delta N_{it}$)	~T_{ox}

device, these are injected into the substrate and do not contribute to HCD. (2) Quantum effects introduce sub-bands in the channel, which increases V_{EFF} (through the parameter V_0).

New semiconductor materials, such as Ge, III-V compounds, and graphene or CNT, coupled with ever decreasing channel length, are expected to finally result in true ballistic transport. In this case, the effective "V_{DSAT}" will approach zero.

3.2.2.10 Intrinsic variation considerations

No matter what FET device or dielectric will be used – one thing is common – device size decreases. The ramification is the rise of intrinsic variation effects. Intrinsic process variation (IPV) is due to the inherent atomistic nature of matter in small dimensions, for example, lattice spacing and the discrete charge of the electron. IPV, or random local fluctuation (RLF) between devices, does not disappear at small separation in the manner that extrinsic variation does. Technology scaling makes this gets worse. This is true for devices as manufactured (T_0), and for reliability degradation effects. This will be an increasing challenge for nanoelectronic device reliability. SRAM "V_{min}" failure induced by the BTI mechanism is already driven by intrinsic variation [3.125]. If yield and reliability are maintained by redundancy and/or error correction, it becomes paramount to be able to estimate the reliability failures that occur, to ensure that the error correction systems are robust.

Hot carrier degradation is due to ISG and the accumulation of bulk oxide charge. Because HC degradation is caused by discrete charges, intrinsic fluctuations are induced. This is somewhat analogous to the case of random dopants.

3.2.3 Filamentary conduction in thin dielectric films

3.2.3.1 Introduction

Although filamentary-type conduction in thin dielectric films has been studied for more than about five decades mainly in connection with reliability issues for MOS capacitors and transistors [3.126, 3.127], nowadays there is a renewed interest in this topic because of its close relationship with the working principle of nonvolatile memory devices such as the resistive RAMs (ReRAMs) [3.128, 3.130]. The kind of filamentary conduction we are going to review along this section is that corresponding to a breakdown or forming event of a thin oxide layer, i.e., the final outcome of degradation. As it is well known, in MOSFET devices, the dielectric breakdown of the gate insulator may cause the partial or total loss of the transistor action as well as an increase in standby power consumption, and these effects can have unpredictable consequences for the overall functionality of a circuit application [3.131]. Understandably, most of the research effort until the end of the 90s was primarily focused on investigating the driven forces behind the degradation and breakdown of SiO_2 [3.132–3.136]. However, in recent years, the focus of attention shifted to alternative gate insulators with high dielectric permittivity (high-K) and on seeking feasible ways of taking advantage of the localize nature of conduction as an essential feature of the operational principle of devices [3.130, 3.137]. The breakdown of a thin oxide layer has been related to the formation of a percolation pathway spanning the dielectric film [3.126]. Since this can be related to the formation of a chain of traps, defects, vacancies, metal atoms, etc., the idea of filamentary conduction, i.e., the electron flow through a low dimensional structure attached to two charge reservoirs, becomes essential.

A wide variety of models connecting the degradation process of the oxide layer with the stress conditions, expressed in terms of either the oxide field or the applied voltage, have been proposed, but, despite the huge amount of data and publications, there is still room for discussion. However, beyond trying to understand what triggers ultimately the appearance of the percolative pathway, around 20 years ago, a number of research groups begun to systematically investigate the electron transport characteristics of oxide films after the detection of a failure event. It is clear that although much more information can be gained in that aspect from MOS transistors (because of the control of the substrate voltage and channel potential drop), a vast majority of the analytical approaches proposed to deal with post-breakdown exclusively refer to MOS capacitors. Most of the studies concerning the dielectric degradation and breakdown of MOS devices are performed under constant electrical stress conditions (current or voltage), but dynamical electrical stress [3.138] as well as radiation-induced stress have also been explored [3.139]. From the microscopic viewpoint, the oxide wear-out process has been associated with several mechanisms including molecular dipole flipping [3.140], anode hole injection [3.141], and hydrogen release at the anode interface [3.142]. In connection with the type of defect created, neutral electron traps, interface states, hole traps, and slow states are well-known results of electrical stress that have been linked to the origin of dielectric breakdown [3.143, 3.144]. Nowadays, it is widely accepted that an oxide layer breaks down as a consequence of the accumulation of defects generated during stress (see Figure 3.55a). The current runaway and the energy dissipation dynamics occurring in the damaged region determine the severity of the event and consequently the magnitude of the localize leakage current. In addition, the relationship between breakdown statistics and defect creation has been successfully explained by the so-called percolative models [3.126, 3.145, 3.146]. These models consider the random distribution of defects inside the oxide layer and have shed light on the area and thickness dependence of the failure distribution in terms of the injected charge or the time-to-breakdown. They have also provided a rough estimation of the defect size involved in the formation of the breakdown path (~0.5–1.5 nm). As illustrated in Figure 3.55b, in some cases, the

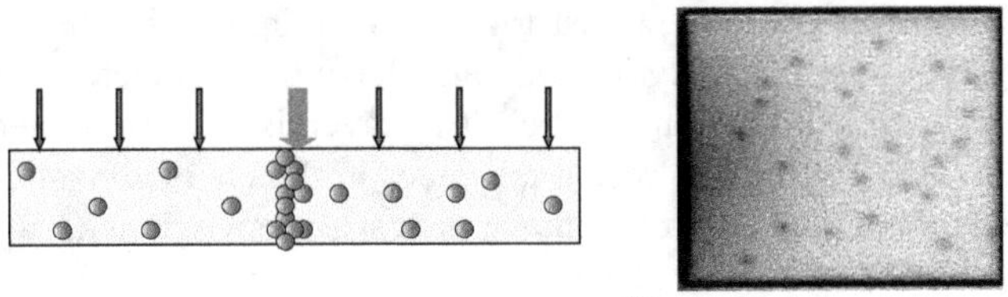

(a) (b)

Figure 3.55 (a) Formation of a percolation pathway spanning the dielectric film caused by the accumulation of traps or defects. The arrows indicate the magnitude of the leakage current. (b) Top view of a MOS device with multiple breakdown spots distributed over the top metal electrode ($60 \times 60\ \mu m^2$)

breakdown spots become visible through an optical microscope. Electron transport during the wear-out stage has been extensively investigated as well. While in thick oxides ($t_{ox} > 10$ nm) charge trapping is the most important factor affecting the Fowler-Nordheim (FN) tunneling current [3.147, 3.148], in thin oxides ($t_{ox} < 7$ nm) stress-induced-leakage-current (SILC) plays a fundamental role [3.149, 3.150]. For intermediate thicknesses both mechanisms affect the device stability [3.151]. The evolution of such conduction mechanisms as monitors of the underlying degradation process and the triggering of dielectric breakdown have also been investigated in the past [3.152–3.154].

On the other hand, until recent years, the physics of post-breakdown conduction has received much less attention. This certainly has happened because of the random nature of the phenomenon as well as its stress- and sample-dependent features. A clear evidence of this fact is the wide gamut of models that were proposed in the past to explain the charge transport mechanism in broken down SiO_2 and later in high-K films, many of them still requiring confirmation. More recently, a second generation of filamentary conduction models, with roots in the memristive device concept, has been developed to deal with the reversible dielectric breakdown phenomenon [3.155]. As it will be shown in this section, the research activity in this field has not decreased its pace along the years. On the contrary, upcoming device applications based on single or multifilamentary conduction (such as ReRAMs, electronic synapses, neuromorphic circuits) demonstrates that the dielectric breakdown field has gained relevance within the framework of functional oxide electronics [3.128].

This section is organized as follows: in Subsection 3.2.3.2, a number of introductory issues such as the terminology used to characterize dielectric breakdown, the classification of conduction modes, the detection problem, and the origin and relationship between failure modes in MOS devices are presented. Subsection 3.2.3.3 focuses on the functional forms of the post-breakdown I–V curves and the different physical approaches that have been proposed to explain the associated electron transport mechanism. An overview of the influence of the dielectric breakdown phenomenon on the transistor action and how it has been treated in the past is provided. A simple analytic model for the output characteristics of a broken down transistor with an advanced dielectric layer is also presented. Finally, in Subsection 3.2.3.4, we refer to recent advances in the characterization and modeling of the reversible dielectric breakdown or resistive switching (RS) effect in thin binary oxides.

3.2.3.2 Oxide breakdown phenomenology

In order to understand filamentary conduction in thin films, it is necessary first to introduce several definitions that, though they are not strictly linked to the modeling of the post-breakdown *I–V* characteristics, constitute the general framework under which the phenomenon should be examined. In this subsection, the different post-breakdown conduction modes that have been identified so far will be described. Then we will focus the attention on the detection methodology and will discuss the origin of the bimodal failure distribution. For historical reasons most of the data reported in this section concerns with SiO_2 but many of the concepts can be straightforwardly applied to other dielectrics. Even though part of the terminology presented here is no longer used for high-*K* dielectrics, we believe that a brief introduction to some of the acronyms that can be found in literature is necessary for understanding the evolution of the ideas in this research area.

Breakdown modes

The dielectric breakdown of the oxide layer in a MOS structure can be defined as the local increase of the system's conductance. This change can be abrupt or gradual depending fundamentally on the oxide thickness, device area, and stress condition and it is often accompanied by a noisy behavior [3.156]. Notice that the early concept of regarding dielectric breakdown as the loss of the insulating capability is no longer valid for ultra-thin ($t_{ox} < 5$ nm) oxides since the distinction between insulating and conducting is not well defined for this thickness range [3.157]. First post-breakdown studies in SiO_2 date back to the 60s and were performed by Klein *et al.* [3.158–3.160], who investigated the switching characteristics of what they called the "high-conduction" state of the oxide. Later on, Shatzkes *et al.* [3.161] further explored this issue and concluded that electron transport under these circumstances was filamentary and governed mainly by the voltage across the structure. It was also pointed out that the high-conduction state was weakly dependent on the oxide thickness and electrode area, features that characterize dielectric breakdown even for current technologies. Almost in the same period, Hickmott [3.162] and Dearnaley *et al.* [3.163] investigated filamentary conduction in a wide variety of materials such as SiO_x, TiO_x, LiF, CaF_2, and Ta_20_5, among others. As an alternative to breakdown, the term forming was introduced at that time. Since those experiments exclusively relied on thick ($t_{ox} \approx$ 20–100 nm) films, the investigated devices presented important thermal damage due to local heating effects. For many years, the breakdown event in a MOS device was associated with the appearance of a resistance-like conducting pathway running between the electrodes [3.164, 3.165]. Perhaps because of this seemingly simple behavior, post-breakdown issues remained practically unexplored until the nineties, when it was claimed that the use of thinner oxides ($t_{ox} \approx$ 5–7 nm), in which thermal effects are absent or under control, would allow examining the signature of the breakdown mechanism [3.166, 3.167]. In 1992, Fukuda and coworkers [3.168] detected a new breakdown mode in ultra-thin SiO_2 films, the associated current being orders of magnitude lower than the mode discussed previously. This led to name this failure mode as B-SILC [3.169], quasi [3.170], partial [3.171], or SBD [3.172], identified with the low-conduction state, and the final, catastrophic or HBD, identified with the high-conduction state. Figure 3.56a shows typical *I–V* curves corresponding to SBD and HBD events in a 4.3-nm-thick SiO_2 layer. For completeness, the FN and SILC current components are also displayed. Although SBD has been mostly reported to occur in sub-5-nm oxides, some authors have reported it in thicker oxides [3.173].

At least six different types of breakdown modes (three SBD and three HBD modes) can be found in literature. SBD has been classified into analog SBD (A-SBD), characterized by random noise, and digital SBD (D-SBD) characterized by a random telegraph noise [3.174]. The third type of SBD, called micro-breakdown (MB), was reported in Reference 3.175. This mode is characterized by a lower current than that corresponding to a typical SBD event and appears to be undetectable in large area devices. In addition, HBD has been associated with and without thermal effects, i.e., with and without lateral propagation of damage [3.176, 3.177]. The third HBD mode has been referred to as nonlinear HBD (NL-HBD) [3.127]. In this case, the *I–V* curve is similar to that corresponding to the high-HBD but exhibits a reduction of its slope for voltages exceeding some

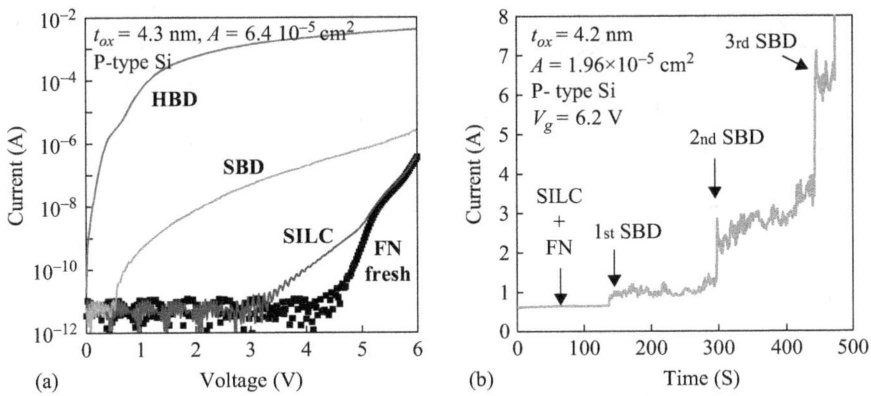

Figure 3.56 *(a) Typical stages in the I–V characteristic of an ultra-thin gate oxide as the degradation proceeds. FN refers to Fowler-Nordheim, SILC to stress-induced leakage current, SBD to soft-breakdown, and HBD to hard-breakdown. (b) Multiple SBD events during a constant voltage stress. Each current jump is associated with the appearance of a new filamentary path*

threshold. Importantly, the failure event can gradually progress with time, be it SBD [3.178, 3.179] or HBD [3.180, 3.181]. This process has been referred to as progressive breakdown (PBD) and is a major issue for present scaling technologies because it affects oxide reliability criteria [3.182]. PBD is mainly observed in hyper-thin oxides and, as revealed by optical studies, it is directly related to the increase of the damaged area of the device. For many years, breakdown modes in SiO_2 have been simply identified as SBD or HBD and the choice of whether a breakdown event belongs to one or another mode has been a rather ambiguous practice. Nevertheless, some criteria were proposed mainly based on the shape and/or magnitude of the *I–V* characteristic, the post-breakdown resistance [3.183] or the normalized differential conductance [3.184–3.186]. Because of the stochastic nature of the phenomenon, SBDs and HBDs corresponding to first events in different devices lead to a bimodal failure distribution whose spread is also influenced by a number of factors such as the geometry of the devices and stress conditions. Moreover, depending on the considered wear-out test, single or multiple breakdown events of either type can occur in the same sample (see Figure 3.56b). Each jump in the *I–V* characteristic is associated with the triggering of a new breakdown pathway across the oxide layer.

Detection window

While in thick oxides a breakdown event can be easily detected as a sudden increase of the system's conductance, in thin oxides, the difficulty arises from the large background tunneling current that can mask the localized current flow. In the case of HBD, this is especially important for hyper-thin oxides ($t_{ox} < 2$ nm) in which the current flowing through the damaged area of the device can be of the same order of magnitude than the area-distributed tunneling current over a wide bias operating range. Concerning the SBD mode, it was shown that the experimental detection window depends not only on the severity of the event but also on the oxide thickness, device area, and substrate type of the DUT [3.187–3.189]. As a general rule, it was found that the detection of SBD is harder for larger areas and thinner oxides. Figure 3.57 compares a typical band of SBD curves (shaded area) with fresh tunneling *I–V* characteristics of MOS capacitors with four different oxide thickness (2.1 nm, 3.0 nm, 3.8 nm, and 4.9 nm) and four different areas ranging from 6.4×10^{-5} cm^2 to 9.3×10^{-3} cm^2. While the complete band of SBD *I–V* curves can be detected for the 4.9-nm-thick oxide in a large voltage window, the same set of curves are completely masked by the DT component in the 2.1-nm-thick oxide. Moreover, the substrate type also affects the experimental detection window because of the different oxide field that governs the tunneling currents, similar SBD events being better resolved in samples with p-type substrates. To overcome these problems, a variety of methods were implemented to improve the detection of SBD occurrence: substrate current increase [3.187], two-level stress test [3.189, 3.190], noise increment criterion [3.191–3.193], and relative current increase [3.194]. Having reliable automatic SBD detection techniques is particularly important during accelerated stress tests for this can have serious impact on the breakdown statistics [3.195–3.197].

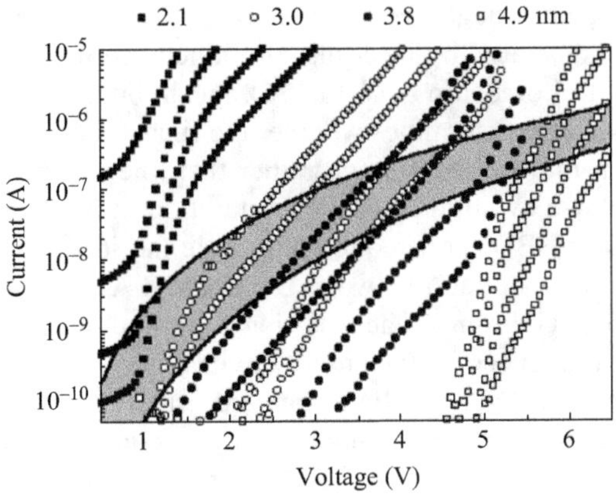

Figure 3.57 *Symbols: fresh current–voltage characteristics measured on p-type substrate samples with different oxide thickness and gate area. For each thickness and from right to left, the curves correspond to gate areas of* 6.4×10^{-5} *cm*^{2}, 3.2×10^{-4} *cm*^{2}, 2.3×10^{-3} *cm*^{2}, *and* 9.6×10^{-3} *cm*^{2}. *The shaded area corresponds to a typical band of SBD curves*

Bimodal failure distribution

Beyond the classification and detection issues discussed in the previous subsections, it has been clearly established that the difference between HBD and SBD is straightforwardly linked to the size of the breakdown spot area. This is supported by combined electrical and optical experiments [3.198, 3.199]. In turn, the severity of the event was shown to depend on the details of the breakdown current runaway and on the final microscopic arrangement of defects that form the percolation path. Several works have pointed out in this direction, claiming that the key factor leading to one mode or the other during stress resides in the energy delivered through the spot at the very first moment of the formation of the current pathway [3.196, 3.197, 3.200–3.203]. In this regard, the prevalence ratio between SBD and HBD can be modified by setting different compliance limits during stress [3.204, 3.205] so that the available energy during the discharge transient through the damaged area may be somehow limited. The same concept is currently applied to RS devices in order to control the damage caused to the sample during the set process [3.155]. By means of equivalent circuit modeling, the appropriate conditions for which SBD turns into HBD and therefore the origin of the bimodal failure distribution were identified [3.206, 3.207]. According to such reports, if the dissipated power exceeds a certain threshold (about 0.1 to 10 μW) for irreversible thermal damage, HBD follows; otherwise the resistance path remains intact resulting in SBD. The assumption that SBD is a previous stage of HBD has also been invoked several times [3.180, 3.208]. It has been claimed that the boundary between SBD and HBD is also determined by how fast the stored energy in the capacitor is discharged through the original conductive filament and this has been experimentally demonstrated by inserting an inductance in series with the device [3.190]. According to Lombardo *et al.* [3.176], the amount of electrons available at the cathode interface also determines whether the final result is HBD or SBD. They have suggested that the applied voltage rather than the energy is at the end what controls the breakdown evolution [3.209]. In addition, it has been claimed that the occurrence of SBD or HBD is related to the sample fabrication process [3.210] and that the specific prevalence ratio between D-SBD and A-SBD also depends on the stress conditions [3.211]. From a practical viewpoint, the severity of the breakdown event is ultimately influenced by the circuit environment of the device [3.126]. For reliability projections, SBD and HBD events are usually treated on equal grounds, thus implicitly assuming that both modes share a common root. Irrespective of whether the first event in a sample set is SBD or HBD, the separate statistical distributions coincide when measured on identical devices stressed under the same conditions [3.212–3.214]. This clearly indicates that SBD and HBD correspond to the same kind of failure, i.e., they are related to the generation of identical defects and they are locally triggered by the same physical mechanism. This issue was explored in depth by analyzing the area dependence of the time-to-breakdown statistics both in transistors and capacitors and it was found a universal behavior over eight orders of magnitude in area [3.200]. This observation justifies merging SBD and HBD data for reliability projections. In addition, when analyzed in MOS transistors, SBD and HBD events exhibit identical location distribution along the device channel [3.215–3.217]. Despite all these common features, some authors have claimed that both breakdown modes should be treated as competing failure mechanisms [3.195] and this is in part supported by the fact that SBD and HBD were reported to have different temperature and electric field acceleration factors [3.218].

3.2.3.3 Modeling of filamentary conduction

As a first step toward the modeling of the conduction characteristic of any electron device, and in particular of a broken down capacitor or transistor, a preliminary set of measurements revealing the fundamental features of the involved charge transport mechanism is required. For many years, post-breakdown conduction in thin SiO_2 films was an almost unexplored field though the phenomenon was indeed matter of intense research in many other materials mainly for its relevance in the field of memory devices. In this subsection we will review first some fundamental observations on the *I–V* curves of broken down oxide films and, second, we will use this information to conform a physical picture of post-breakdown conduction. In what follows, a number of compact models proposed to this aim are presented and discussed.

Functional form of the post-breakdown I–V curves

Although it may sound surprisingly, the details of the *I–V* characteristic after the occurrence of a HBD event were not investigated in depth until recent years. This was motivated perhaps by its simple resistor-like behavior, which in combination with the random nature of the phenomenon, led to think that there was nothing or very little to investigate. Quite the opposite, the HBD *I–V* characteristic is far from

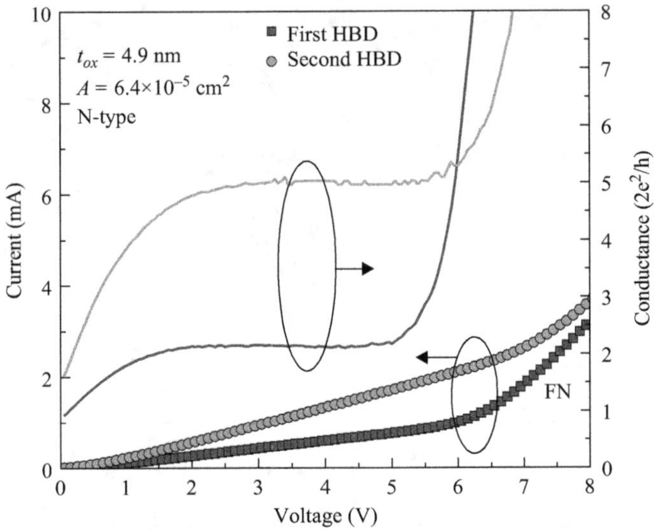

Figure 3.58 Post-breakdown I–V curves and their conductances in units of the quantum conductance unit for two consecutive breakdown events

being linear in the whole bias range and exhibits interesting features, both at low and high biases. Particularly noteworthy is the fact that in many cases the HBD conductance exhibit values typical of atom-sized constrictions [3.127]. In Figure 3.58, two HBD events successively opened in the same sample together with the associated conductance-voltage (*G–V*) curves are illustrated. Three qualitatively different regions can be observed: from 0 V to ~1.5–2 V there is a rapid current increase followed by a linear region with a differential conductance (d*I*/d*V*) of the order of the quantum conductance unit $G_0 = 2e^2/h$, *e* and *h* being the electron charge and the Planck's constant, respectively. For higher biases, it is observed a new increase of the currents and this behavior is associated with the dominant background FN current. On the contrary, the functional form of the SBD *I–V* characteristic has been explored in depth. While many groups claimed that the SBD current could be well fitted by a power-law model [3.211, 3.186, 3.219–3.221], others claimed that an exponential function of the applied bias was more appropriate [3.185, 3.222, 3.223]. In this regard, it can be shown that the SBD *I–V* characteristic in SiO_2 exhibit both trends depending on the voltage range investigated. From 0 to approximately 3 V, a power-law model is a good option, whereas for higher voltages, an exponential model yields better results. Figure 3.59a,b illustrates post-breakdown *I–V* curves corresponding to a high-*K* film analyzed according to both approaches.

In addition, when represented in log-linear axis, the slopes of the SBD *I–V* characteristics of a large number of devices are statistically correlated with the current level [3.184]: the higher the current at a given voltage is, the lower the associated normalized differential conductance is. The empirical law that links

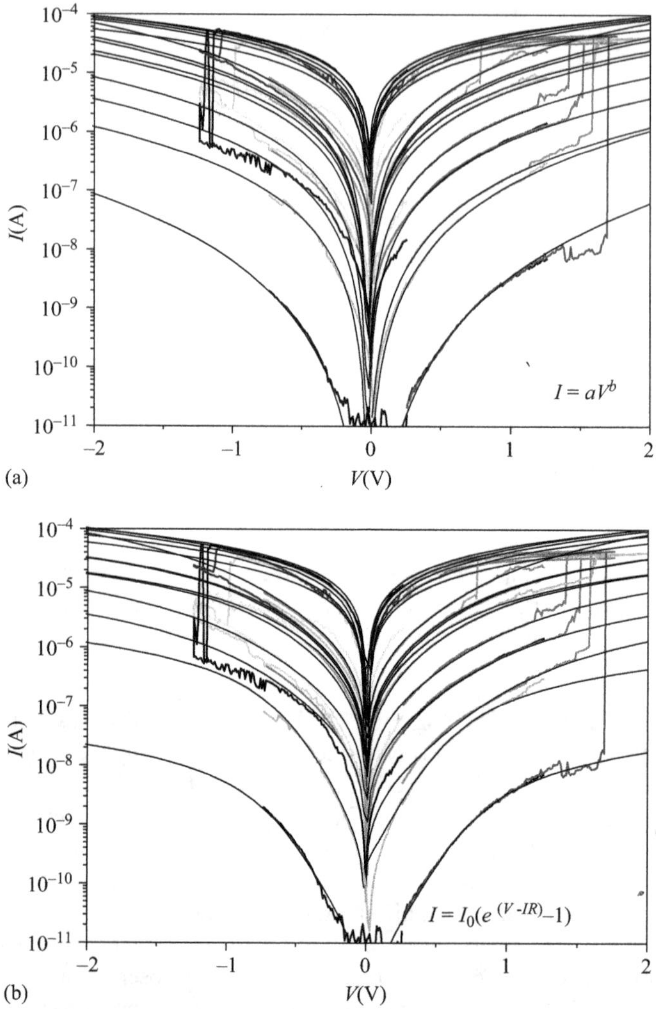

Figure 3.59 Experimental and fitting results using (a) the power-law model and (b) the diode-like (exponential) model for post-breakdown I–V curves in $HoTiO_x$

these magnitudes was also verified by Bruyère [3.224] for electrical stressed samples and by Cester *et al.* [3.225] and Ceschia *et al.* [3.226] in the case of radiation-induced damage. A recent paper has explored this correlation in connection with high-*K* dielectrics and the results are illustrated in Figure 3.60 [3.227].

Another important issue, which has originated a great controversy, is the dependence of the SBD current on the oxide thickness and device area [3.228]. Many authors [3.222, 3.224, 3.229] have shown that the SBD conduction is essentially independent of such factors as expected for filamentary conduction. In

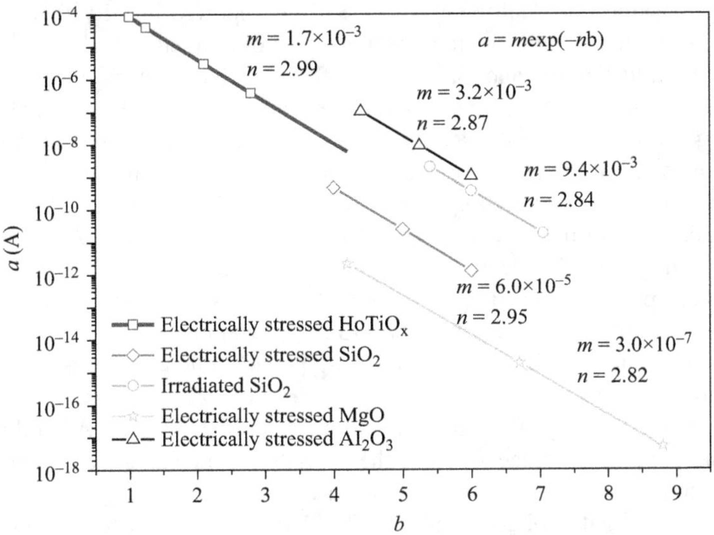

Figure 3.60 Correlation between the parameters a and b associated with the power-law model $I = aV^b$

addition, the fact that the SBD current does not depend essentially on the oxide thickness and injecting electrode indicates that the current is controlled by a highly localized mechanism (interfacial or bulk). Notice that this is quite different from what occurs in FN conduction through fresh oxides, where the oxide field and the injecting electrode play fundamental roles.

Filamentary conduction models in SiO_2

In this subsection, a brief description of the different approaches proposed to deal with the post-breakdown conduction characteristics in MOS devices is presented. The considered models are in general extensions of well-known electron transport mechanisms adapted to the case under study with little or no modifications. Among others, junction-like, hopping, percolation, and tunneling conduction are the most invoked mechanisms. Within the latter mechanism we can find direct, Fowler-Nordheim, trap-assisted, resonant, inelastic quantum tunneling models, and quantum point contact (QPC) conduction. The majority of these models are aimed at explaining exclusively the SBD mode completely disregarding HBD. Notice that they will not be reported in chronological order but according to the underlying physical mechanism involved.

One of the earliest attempts to analytically describe the filamentary current in SiO_2-based MOS devices was based on the diode equation with series resistance [3.167] (see Figure 3.59b). This model is supported by the fact that a plot of log(I) vs. V_G for p-type substrate devices can be fitted by a straight line of slope e/nk_BT, where k_B is the Boltzmann constant, T the temperature, and n the ideality factor which was found to be ~1.2 for poly-Si, and ~1.7 for aluminum and chromium gates. In the same line, Halimaoui *et al.* [3.222] reported ideality factors ranging from 1.8 to 2 for the same voltage range. Later on, Umeda *et al.* [3.230] further studied this failure mode and claimed that the *I–V* characteristic looked like very similar to the current of a forward-biased pn-junction diode. They concluded that the leakage current was controlled by the spreading resistance effect at large biases and that the sizes of the HBD spots were in the order of 10 nm. The idea that HBD resembles diode-like conduction was also supported by Bearda *et al.* [3.231]. By means of comparing finite-element simulations with experimental *I–V* curves, they established that the HBD leakage mechanism is diode-like in n^+/p MOS devices and classical point-contact-like in n^+/n MOS devices. Diode-like conduction for HBD was also reported by Avellán *et al.* [3.232], who investigated the temperature dependence of the *I–V* characteristics for n-type substrate capacitors.

Based on the analysis of the temperature dependence, Okada *et al.* proposed the variable-range hopping (VRH) mechanism to model SBD [3.220]. According to this model, the conduction process across the insulator is mediated by localized states, including traps and interface states. The idea comes from the fact that the SBD current at a fixed gate voltage exhibits a $T^{-1/4}$ dependence for temperatures above 125 K. This particular behavior arises from Mott's prediction about the conductivity of amorphous semiconductors and dielectrics [3.233]. On the contrary, Cester *et al.* [3.234] demonstrated that the VRH temperature-law couldn't fit the radiation-induced SBD current. Moreover, Okada's data were shown to be consistent with an exponential dependence in a wider temperature range [3.235]. Concerning the conduction characteristic, VRH predicts that the current can be expressed as [3.233]

$$I = I_0 \sinh\left(\frac{eR}{t_{ox}k_B T}V\right) \tag{3.51}$$

where I_0 and R are constants and t_{ox} the oxide thickness. The effect of applying a voltage in VRH is to modify the energy of a hop between adjacent sites. Notice that VRH strongly depends on the sample thickness, which is at variance with the experimental findings for SBD. Houssa and coworkers based their SBD model on the percolation theory of nonlinear conductor networks with a distribution of percolation thresholds [3.236, 3.237]. According to this model, during degradation, traps randomly occupy the sites of the SiO_2 lattice eventually forming a backbone that spans between the electrodes. In subsequent publications, the authors extended the proposed model to explain both the SBD temperature dependence as well as its noise characteristics. The temperature dependence was modeled in terms of the scattering of electrons within the percolation path [3.236], while the fluctuations in terms of Lévy flights caused by switchings in the percolative network [3.238, 3.239].

In 1994, Lee *et al.* [3.170] claimed that the SBD current followed a DT model, i.e., tunneling through a trapezoidal-type potential barrier and they proposed that SBD is a consequence of the progressive degradation of a localized region close to the anode interface. According to this model, the ballistic impact of the injected electrons on the silicon surface causes a physical damage, which reduces the effective oxide thickness. Hirose and coworkers reported a variation of this tunneling-based model [3.240] assuming that the oxide layer near the SiO_2/Si interface locally deteriorates during the electrical stress, becoming, at some stage, highly conductive. The thinned barrier model was also investigated by Houssa *et al.* but it was found that good fit to the experimental data could only be achieved assuming unrealistic barrier height values [3.222]. Tunneling through thinned SiO_2 potential barriers has also been used to explain PBD [3.178] as well as HBD [3.241]. However, instead of a thinned barrier, some authors claimed that SBD arises from a lowering of the oxide barrier height [3.168, 3.242]. According to these authors, during stress, hole trapping takes place close to the anode interface thus narrowing the insulator bandgap and inducing a loss of the insulating capability. The idea that SBD corresponds to a lowering of the barrier height rather than to a reduction of its width has also been investigated by means numerical simulations [3.216]. TAT and resonant (RT) tunneling has also been suggested as possible candidates for post-breakdown conduction in SiO_2 [3.172, 3.243]. Cellere *et al.* [3.175] considered TAT to simulate the leakage current through the so-called MBs. The underlying idea is that charges strategically located inside the oxide can reduce the effective cathode barrier, thus increasing the local current through the spot. Cheung *et al.* [3.244] proposed a qualitative model both for SBD and HBD based on RT. According to these authors, during degradation, energy funnels located at the trap sites are created inside the oxide that eventually form a defect's band. HBD would take place when this band connects the electrodes. Based on the temperature dependence of a negative differential resistance region observed in the I–V characteristic after D-SBD, Uno *et al.* suggested phonon-assisted RT as a plausible mechanism for this breakdown mode [3.245]. However, the most comprehensive approach for post-breakdown conduction based on the RT mechanism was presented by Ting [3.246]. The author considered the crowding of the electron wavefunctions into nanoscale wires embedded in the oxide layer connecting partially (SBD) or totally (HBD) the opposite electrodes. In the same line, Nigam *et al.* [3.247, 3.248] proposed that a single electron tunneling (SET) mechanism could explain both the voltage and temperature dependences of the gate current after SBD. In this model the conducting sites forming the breakdown path behave as isolated islands embedded in the oxide layer. To travel between electrodes the electrons must go through such islands by means of successive uncorrelated tunnel events, these transitions being determined by the electrostatic potentials originated by the charges captured by the islands. The I–V characteristic corresponding to such multijunctions can be expressed as [3.249]:

$$I(V) = \sum_{k=0}^{N-1} A_k T^{2k} V^{2(N-k)-1} \qquad (3.52)$$

where N is the number of junctions in series, T is the absolute temperature, and A_k are constants related to the junction resistances.

In 1998, a QPC conduction mechanism was proposed to explain the SBD and HBD I–V characteristics of SiO_2-based MOS capacitors within a unified framework [3.250]. In the QPC model, the breakdown path spanning the oxide film is treated as

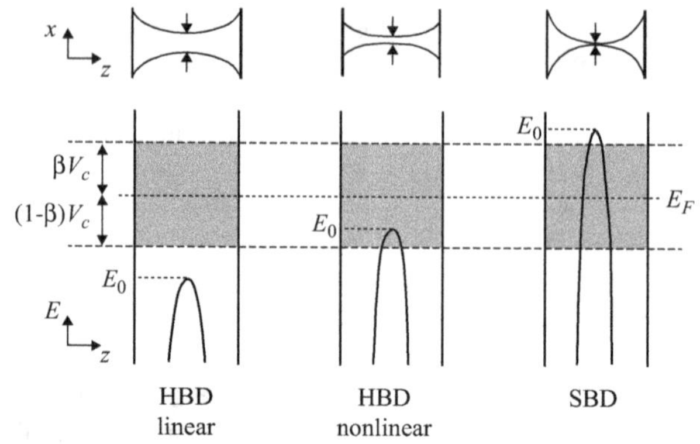

Figure 3.61 Schematic energy diagram of the constriction and its associated potential barrier profile according to the QPC model

an atom-sized constriction [3.250–3.252]. Because of the lateral confinement of the electron wavefunction, the transverse momentum of the particle becomes quantized. This gives rise to conduction sub-bands that act as potential barriers for the incoming electrons, the top of the barriers being located at the bottleneck of the filamentary structure. Even though the electron transport across these sub-bands is ballistic, there is always a contact resistance associated with impedance-mismatching effects at the two ends of the filament [3.253]. According to the QPC model, the confinement barrier determines whether SBD, HBD, or NL-HBD is observed (see Figure 3.61). Starting from the Landauer approach for electron transport in mesoscopic conductors and assuming and inverted parabolic barrier for the longitudinal confinement potential, the current that flows through the filamentary path reads

$$I(V) = \frac{2e}{h}\left\{e(V - V_0) + \frac{1}{\alpha}\ln\left[\frac{1 + \exp\{\alpha[\Phi - \beta e(V - V_0)]\}}{1 + \exp\{\alpha[\Phi + (1-\beta)e(V - V_0)]\}}\right]\right\} \tag{3.53}$$

where Φ is the barrier height, α a constant related to the longitudinal shape of the constriction, β a measure of the potential drop at the injecting electrode, and V_0 the potential drop in the electrodes. Some experimental and model results for SBD and HBD are illustrated in Figure 3.62a,b. The QPC model was further investigated by Cester *et al.* [3.254], who was able to obtain, by means of inverse modeling, direct information on the barrier potential profile associated with SBD. In addition, the model was also invoked to explain the heavy ion-induced SBD conduction [3.255], the PBD *I–V* characteristics in hyper-thin oxides [3.182] the current noise after HBD [3.256], and the temperature dependence of the localized leakage current [3.257].

More recently, the QPC model has been used to explain the experimental SBD and HBD *I–V* characteristics of HfO_2-based MIM devices as well as their temperature dependence [3.258, 3.259]. Degraeve *et al.* proposed an extension of the model to account for the switching dynamics in HfO_2 as well [3.260]. It is worth pointing out that several recent experiments support the idea of conductance

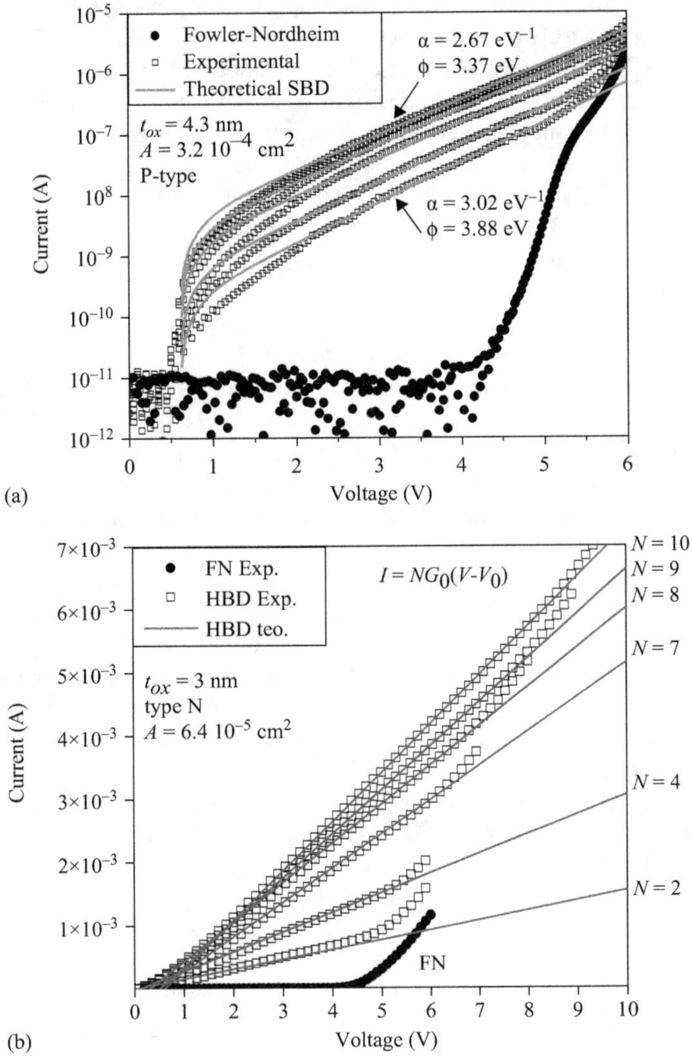

Figure 3.62 (a) SBD I–V characteristics corresponding to first events measured on different MOS devices. Φ is the barrier height associated with the first sub-band level and α a parameter related to the shape of the barrier. (b) HBD I–V characteristics generated by the creation of successive breakdown paths. N is the number of parallel filaments. The deviation at the largest biases corresponds to the background FN tunneling component

quantization in broken down oxides. Observation of discrete conductance steps of the order of G_0 in Ta_2O_5 layers has been reported by Tsuruoka *et al.* [3.261] and by Chen *et al.* [3.262]. Discrete changes of conductance of the order of G_0 were also reported during the unipolar reset transitions of Pt/HfO_2/Pt structures by Long *et al.* [3.263]. The conductance changes were interpreted as the signature of atomic-size variations of the filament nanostructure. Zhu *et al.* have reported conductance quantization effects in Nb/ZnO/Pt and ITO/ZnO/ITO structures as well [3.264]. Mehonic *et al.* [3.265] showed similar effects in silicon-rich silica (SiO_x) resistive switches. Nonlinear quantization effects in CeO_x/SiO_2-based RS devices have been reported by Miranda *et al.* [3.266].

Effects on the transistor output characteristics

Understanding how the different breakdown modes affect the functional capacity of MOSFET devices has also attracted a great deal of interest. In 1994, it was claimed that SBD would have a large impact on device applications because of the large leakage current at operating biases [3.267]. However, some years later, Weir *et al.* [66,77] demonstrated that an increase in the gate current noise was the only detectable effect after SBD on transistors with ultra-thin oxides. No significant variations in the threshold voltage and transconductance were detected. According to these early observations, SBD was not considered as a device failure, at least, for some applications. The issue was then revisited and, on the contrary, several authors reported indeed measurable effects on the transistors' output characteristics [3.135, 3.268–3.271]. Even though there is consensus that SBD is the prevailing breakdown mode for low-voltage operation in SiO_2-based devices [3.197, 3.272], the final consequences upon the device behavior are rather complex and strongly depend on a number of factors such as the location of the breakdown spot along the device channel [3.216, 3.268, 3.272, 3.273] and the dimensions of the damaged area compared to length and width of the DUT [3.200, 3.271]. Briefly, it was found that a breakdown event near the source/drain regions produce more severe consequences than a breakdown in the channel region. In other words, gate-to-substrate leakage paths seem to exhibit higher resistances than those occurring in the source/drain regions. It was also observed that the drain saturation current and transconductance collapses after SBD are linked to the formation of an oxide defective region around the breakdown spot whose effective area is much larger than the conductive path itself. A crucial problem that arises investigating this issue is whether a transistor remains functional or not after the occurrence of a gate oxide breakdown. The question is not so simple and direct because, as pointed out by Wu *et al.* [3.135], even if the device is able to survive after SBD, there may still be some impact of subsequent stress on the already broken down device. Kaczer *et al.* [3.274] demonstrated that a digital CMOS circuit could tolerate many HBDs without affecting its logical functionality. This means that depending on the particular application, reliability criteria may be, in some extent, relaxed. The most comprehensive and complete study about the influence of a localized failure on the transistors' output characteristics was carried out by the IMEC group [3.215]. Representing the breakdown path as a narrow inclusion of highly doped n-type silicon across the oxide layer, the authors were able to reproduce, using a finite-element device simulator, the post-breakdown characteristics of a transistor, including the substrate current, both for gate-to-substrate and gate-to-source/drain breakdowns. They also compared experimental results with the predictions of a discrete-device equivalent circuit formed by passive and active elements. Their model circuit was originally aimed at explaining the characteristics of transistors after HBD but it was later extended to cover the SBD case by replacing the resistance that mimics the breakdown path by a nonlinear resistor [3.216].

As we have seen so far, the effects of a gate oxide breakdown in a MOSFET have been several times investigated [3.275] and although the idea of representing the electrical behavior of the device by means of an equivalent circuit model is not new, little efforts have been devoted to describe the problem in terms of a simple compact model. Most of the proposed approaches basically consist in a combination of reduced area MOS and bipolar transistors, diodes and resistances, which in general rule out any possible analytic solution [3.274–3.279]. The complexity mainly arises because models attempt to account for in a single framework the different conduction modes occurring in thin dielectric films such as SBD and HBD, which, as we have already seen exhibit very dissimilar *I–V* characteristics. In addition, these models often involve a large number of parameters so that a systematic parameter extraction strategy does not result evident. In what follows, the so-called linear response regime will be analyzed. In this regime the transistor action is totally lost becoming essentially dominated by parasitic elements. Although a physical explanation is out of the scope of this subsection, this particular behavior may arise as a consequence of a major obstruction of the channel region [3.280] or by the rearrangement of the internal potential distribution caused by the large gate leakage current [3.277]. In a recent paper, Nicollian *et al.* analyzed in detail the case of a gate-to-drain BD event in a medium-K dielectric material such as SiON using a circuit model [3.279]. Here, a simple model for a broken down transistor with an advanced dielectric like La_2O_3 ($K \sim 27$–30) is presented [3.281]. The central point of this approach is to illustrate how the localization and magnitude of an HBD event affect the output characteristics of the device for conventional and unconventional biases (as required in circuit simulators). Typical *I–V* characteristics both for negative and positive drain voltages (V_{DS}) are illustrated in Figure 3.63a. Notice that the gate current (I_G) is negligible and the important role played by the bulk current (I_B) for $V_{DS} < -0.5$ V. This latest current component arises from the forward-biased drain-bulk junction and affects the drain current (I_D) for $V_{DS} < 0$ as well. In order to induce the oxide BD, a constant voltage stress at $V_{GS} = 4$ V was applied to the gate until the detection of a sudden jump in I_G. The stress was performed with source (S), drain (D), and bulk (B) terminals grounded. The electrical characterization was carried out using combined voltage sweeps (-1 V $\leq V_{DS} \leq 1$ V and -2 V $\leq V_{GS} \leq 2$ V) and a common-source configuration was considered.

The proposed approach is a consequence of the "potentiometer" model reported in References 3.277 and 3.282. This model has long been used, implicitly or explicitly, to calculate the location of the oxide breakdown path along the channel region. In order to conform to the simplest port description of a MOS

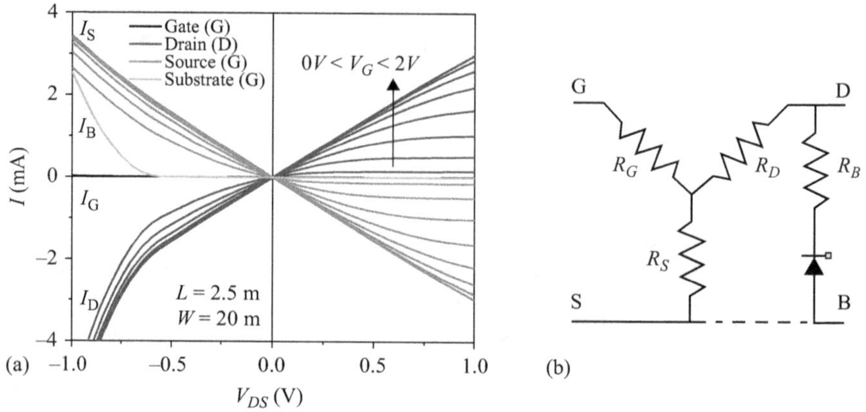

Figure 3.63 (a) Experimental I–V characteristics for positive and negative biases. (b) Model for the MOSFET device after the BD breakdown event. R_G, R_D, and R_S represents the post-breakdown resistances. The dashed line indicates a common reference potential

transistor, a *Y*-type network with resistances R_G, R_D, and R_S for the broken down device is adopted (see Figure 3.63b). The diode with the series resistance R_B represents the D-B current. From the linear resistance network, the *I–V* characteristics can be written using a matrix model as

$$\begin{bmatrix} I_G \\ I_D \end{bmatrix} = \begin{bmatrix} g_{11} & g_{12} \\ g_{21} & g_{22} \end{bmatrix} \begin{bmatrix} V_{GS} \\ V_{DS} \end{bmatrix} - \begin{bmatrix} 0 \\ I_B \end{bmatrix} \tag{3.54}$$

I_B can be expressed, using the Lambert-*W* function [3.283, 3.284], as

$$I_B = \frac{nV_T}{R_B} W\left\{\frac{I_0 R_B}{nV_T}\exp\left[\frac{(I_0 R_B - V_{DS})}{nV_T}\right]\right\} - I_0 \tag{3.55}$$

where I_0 is the diode inversion saturation current, n the ideality factor and $V_T = kT/q = 26$ mV the thermal voltage. Using the *Y*-Δ network transformation, it can be demonstrated that

$$\begin{aligned} g_{11} &= (R_D + R_S)/(R_G R_D + R_G R_S + R_D R_S) \\ g_{12} = g_{21} &= -R_S/(R_G R_D + R_G R_S + R_D R_S) \\ g_{22} &= (R_G + R_S)/(R_G R_D + R_G R_S + R_D R_S) \end{aligned} \tag{3.56}$$

I_S is calculated from Kirchhoff's current law as

$$I_S = -(I_G + I_D + I_B) \tag{3.57}$$

The matrix elements in (3.54) can be found from the experimental *I–V* curves using a port analysis setting first $V_{DS} = 0$ V and then $V_{GS} = 0$ V. As an example, for

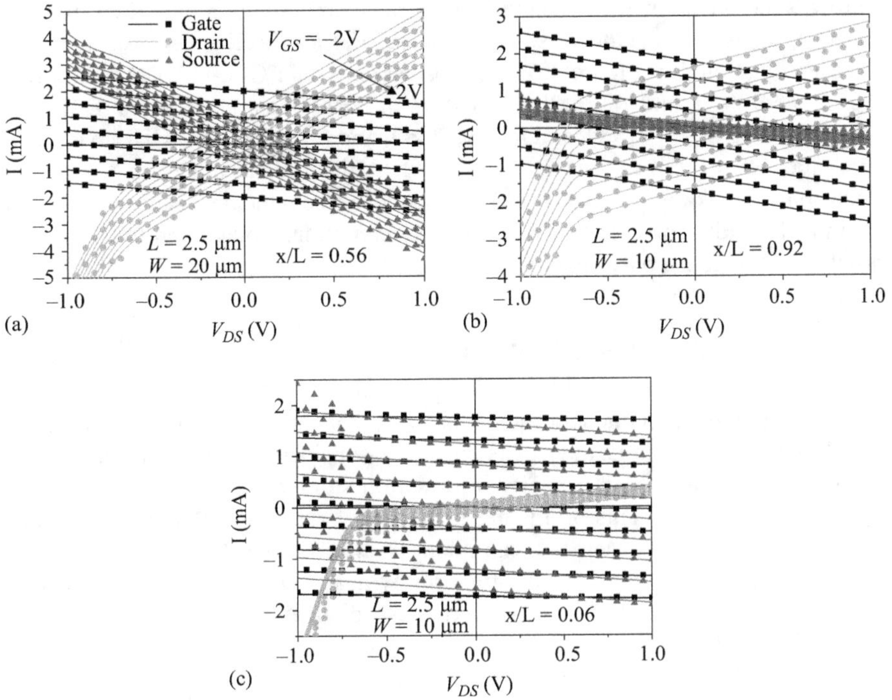

Figure 3.64 Experimental (symbols) and model (lines) results for the post-breakdown output characteristics of an advanced MOSFET. (a) BD site located in the middle of the channel, (b) BD site located close to the D contact, and (c) BD site located close to the S contact

the curves shown in Figure 3.64a, $g_{11} = 1$ mS, $g_{21} = g_{12} = -0.56$ mS, and $g_{22} =$ 3.7 mS were obtained. n and I_0 were extracted from the $\ln(I_B)$ vs. V_{DS} curve in the range $-0.5\text{ V} < V_{DS} < -0.2$ V and R_B was calculated from the linear part of the same curve. Figure 3.64 shows model results for the I–V characteristics using (3.54), (3.55), and (3.57) for three particular cases. In order to achieve further insight into the circuit model under consideration, the connection between the matrix elements in (3.54) and the fundamental parameters that characterize a breakdown event can be established as follows:

$$\frac{x}{L} = \frac{I_D}{I_G}\bigg|_{VDS=0} = abs\left(\frac{g_{12}}{g_{11}}\right) \tag{3.58}$$

$$R_G = \frac{V_{GS} - V_{DS}}{I_G}\bigg|_{ID=0} = \frac{g_{12} + g_{22}}{g_{11}g_{22} - g_{12}} \tag{3.59}$$

$$R_C = R_S + R_D = \frac{V_{DS}}{I_D}\bigg|_{IG=0} \frac{g_{11}}{g_{11}g_{22} - g_{12}^2} \tag{3.60}$$

where x is the breakdown site location along the channel region measured from the S contact, R_G is the breakdown spot resistance, and R_C the channel resistance. For the cases illustrated in Figure 3.64: (a) corresponds to a G-to-C HBD with $x/L = 0.56$, $R_G = 929\ \Omega$, and $R_C = 286\ \Omega$, (b) to an HBD event close to the D contact with $x/L = 0.92$, $R_G = 975\ \Omega$, and $R_C = 2.2\ \text{k}\Omega$, and (c) to a G-to-S HBD event with $x/L = 0.06$, $R_G = 952\ \Omega$, and $R_C = 3.3\text{k}\ \Omega$. Although the model only covers a catastrophic breakdown event, notice the good agreement between experimental and model results for conventional and unconventional bias conditions. To our knowledge, this is the only analytic model for the output characteristics of a broken down MOS transistor.

3.2.3.4 Reversible dielectric breakdown

Resistive switching and memristive approach

The use of filamentary conduction in metal-insulator-metal (MIM) structures as the central element in memory devices is currently considered a viable alternative for the implementation of high capacity information storage systems and logic applications [3.285–3.288]. This promising technology, called ReRAM for Resistance change Random Access Memory, can adopt the form of stackable crossbar arrays and is the subject of extensive investigation by academia and industry worldwide [3.289–3.291]. The working principle of these devices relies on their ability to withstand the alternate formation and rupture of a single or multiple filamentary defect or metal pathways spanning the dielectric layer for a large number of read-write cycles ($>10^{10}$ [3.292]). The physical phenomenon is known as the RS effect and it has been observed in a large number of binary transition metal oxides (TMO) and multinary oxides such as NiO, TiO_2, HfO_2, Ta_2O_5, ZnO, SiO_x, CeO_x, Al_2O_3, SnO_2, ZrO_2, MgO, Nb_2O_5, Yb_2O_3, $SrTiO_3$, $Pr_{1-x}Ca_xMnO_3$, and $La_{1-x}Ca_xMnO_3$, among others. The simplicity, low cost, high endurance, fast switching time, scalability, multi-bit storage capability, controllable programming current, and in some cases compatibility with conventional CMOS technology make ReRAM devices serious contenders to memories based on charge transfer mechanisms such as Flash [3.293–3.295]. However, in spite of these promising news, variability and reliability are still major concerns for ReRAMs [3.296–3.298]. The most accepted picture for RS attributes the alternate formation and dissolution of the filamentary pathways to an electrochemical reduction-oxidation (REDOX) process occurring within the insulating matrix [3.285]. Modifications of the oxide layer at the interfaces with the electrodes caused by the accumulation of mobile charge may also be involved [3.299, 3.300]. Memories based on this principle are referred to as Valence Change Memories (VCM). As the result of this electron–ion interplay, the oxide becomes more or less permeable to the electron flow with the consequent resistance change between a low (LRS) and a high (HRS) resistance state. It is widely recognized the correspondence of the HRS and LRS with the SBD and HBD conduction modes occurring in thin dielectric films [3.301, 3.302].

A similar phenomenology has also been reported in the case of the formation and rupture of metallic-like filamentary paths [3.303, 3.304]. This is the operational principle of Conducting Bridge RAMs (CBRAM), also called Electrochemical

Metallization Memories (ECM) or Programmable Metallization Cells (PMC), which are based on the relocation of metal ions within a solid electrolyte.

The transitions HRS↔LRS can be abrupt (digital RS) or gradual (analog RS) indicating the sudden or progressive opening/closing of multiple parallel leakage paths or the narrowing/widening of the cross-section area of a single filament [3.305–3.308]. This latest property (tunable resistance) has been suggested for multi-level or multi-bit storage systems [3.309–3.312]. In general, the switching occurs after reaching certain threshold voltages with the same (unipolar RS) or with opposite (bipolar RS) polarities and they are called the SET (HRS→LRS) and RESET (LRS→HRS) voltages (see Figure 3.65). The particular features of the switching processes seem to be related not only to the properties of the dielectric material but also to the metal electrodes and forming conditions. Current compliances (CC) are often applied during the SET process in order to limit the thermal effects caused by the current runaway [3.313, 3.314]. On the other hand, Joule heating at the bottleneck of the filament has been implicated in the dissolution mechanism of the conducting bridges, mainly for unipolar RS [3.290]. Within the field of RS, the first oxide breakdown event upon the application of electrical stress is called electroforming. From the statistical viewpoint, the electroforming event is fully consistent with the percolation theory of dielectric breakdown [3.315]. After this, the application of increasing and decreasing voltage sweeps leads to a pinched hysteretic behavior of the *I–V* characteristics, which has been often interpreted in terms of the memristor theory. This theory, which links electric charge and flux linkage in a device, was originally proposed by Leon Chua in 1971 [3.316], later reformulated for memristive systems by Chua and Kang in 1976 [3.317], and finally extended by Chua himself to ReRAM devices in 2011 [3.318]. Briefly, memristive systems are two-terminal circuit elements characterized by two coupled equations: one for the *I–V* curve of the device and one for its internal state variable [3.319–3.321]. While in linear systems the first relationship expresses an Ohmic-type dependence, the second one is written as a time derivative in order to account for the previous history of the device.

However, in spite of the major technological advances and better understanding of the physics behind RS achieved during the last decade, simple and

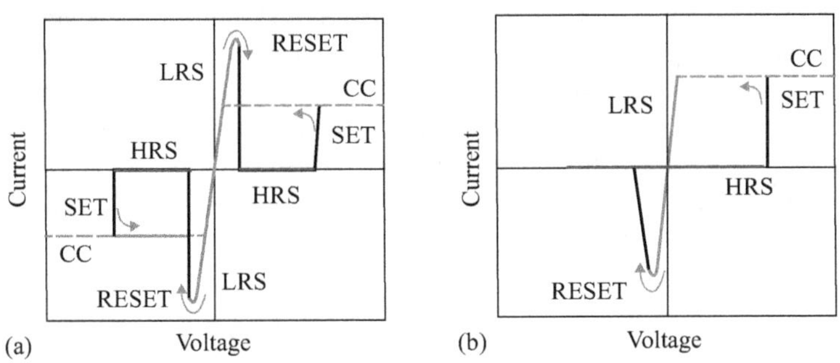

Figure 3.65 Schematic representation of (a) unipolar and (b) bipolar RS

flexible analytic models able to account for the wide variety of switching *I–V* curves exhibited by different dielectric films and electrode materials are hard to find in literature. While some of the available approaches are difficult to implement in circuit simulators because of the complexity of the physical processes involved or the mathematical constraints in the model equations and its derivatives [3.322–3.326], other models only focus the attention on the HRS and LRS *I–V*s separately, completely disregarding the gradual transition between both states that in many cases characterizes the RS phenomenon. Other approaches are exclusively aimed at describing the SET and RESET switching dynamics caused by the application of current or voltage pulses [3.327–3.329]. In addition, many SPICE-oriented models for RS have been recently reported [3.330–3.334], but their ability to accurately represent not only the electron transport characteristics in different materials but also their specific memory properties when subjected to arbitrary input signals has been seriously questioned [3.335]. Since the publication of Strukov's memristor model in 2008 [3.336], a number of simple approaches based on combinations of linear, nonlinear, and rectifying devices have been developed to describe the bistable conduction characteristics of electroformed MIM devices. Some of these models are reviewed in the next subsection.

Filamentary conduction models for resistive switching

In this section, different models proposed to deal with the bistable *I–V* characteristic exhibited by electroformed thin dielectric films in MIM structures are described. Given the large number of publications related to RS, it is worth pointing out that the model list reported below is by far not exhaustive. Models that are described by simple mathematical expressions or which admit a circuital representation will be mentioned here. Models that rely on the numerical solution of differential equations [3.323, 3.324, 3.337], iterative calculations [3.338], Monte Carlo simulations [3.339, 3.340], or which exclusively focus the attention on the HRS *I–V* curve adopting conventional electron transport mechanisms in dielectrics (Poole-Frenkel conduction, Schottky emission, space charge limited conduction, TAT are also deliberately omitted [3.341]).

The dopant-drift memristor model for TiO_2 layers proposed by Strukov *et al.* [3.336] represents a breakthrough in the field of RS devices. Although Strukov's model has been widely investigated, mainly from the theoretical viewpoint [3.342, 3.343], it is seldom applied in its original form to practical RS cases. In this regard, the application of the complementary series resistors model often requires the introduction of additional constraints in the state equation in the form of window functions [3.319, 3.344]. The two equations that describe this model are expressed as (see Figure 3.66a):

$$V = [(R_{LRS} - R_{HRS})x + R_{HRS}]I \tag{3.61}$$

$$\dot{x} = KIf(x, I) \tag{3.62}$$

where $0 \leq x \leq 1$ is the state variable ($x = w/D$), $\dot{x}$ its time derivative, K a constant related to the drift velocity of the oxygen deficiencies, R_{LRS} the LRS resistance, R_{HRS} the HRS resistance, and f the window function ($f = 1$ in Strukov's model).

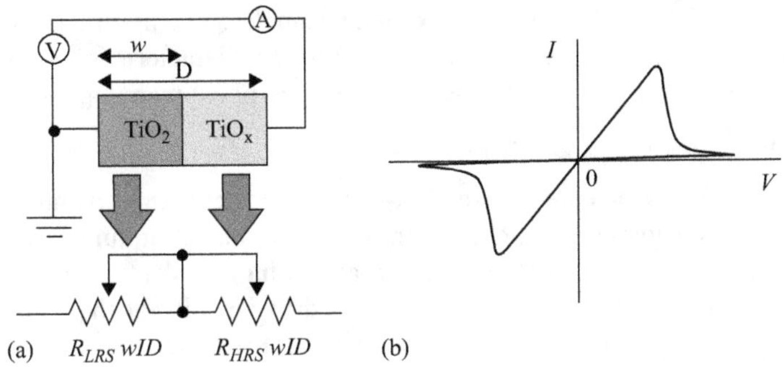

Figure 3.66 (a) Strukov's model for RS devices based on complementary resistors, (b) typical hysteretic I–V curve associated with the movements of vacancies within the device

In general, f is chosen so as to comply with $f(0,I) = f(1,I) = 0$ to ensure no drift of the dopant front beyond the boundaries of the device. One important aspect of Strukov's model is the absence of well-defined threshold voltages for the SET and RESET events. For a simple sinusoidal input signal the model yields a pinched hysteresis loop (Figure 3.66b). A recent study has revealed lacking of predictability of the models expressed by (3.61) and (3.62) regardless of the considered window function [3.335].

One of the earliest attempts to describe the HRS↔LRS transitions of the filamentary current by means of an equivalent electrical circuit comprising non-linear devices was proposed by Szot *et al.* [3.345], who attributed the switching of the electrical resistance in electroformed single-crystalline $SrTiO_3$ layers to a change in the transmission properties of individual dislocations. According to Szot's model the switching process is basically a consequence of the local modulation of the oxygen content related to the self-doping capability of TMO. In this case, RS is associated with an electrochemical closing and opening process of a single dislocation at the surface of the dielectric. An equivalent circuit formed by a network of resistors and diodes represents the electrical behavior of such nanowire. It is also assumed that a 3D orthogonal lattice of resistors connected to individually addressable elements, which can reversibly change between a diode and a resistor characterizes the inner network of dislocations. The idea of using an equivalent circuit approach was considered by Yang *et al.* [3.346]. In this case, the hysteretic *I–V* curves are modeled using a combination of memristors and rectifiers, which in turn can be used to construct a family of electronically reconfigurable circuit elements [3.299]. From the physical viewpoint, the device behavior is explained in terms of the coupled electron and ion dynamics driven by the applied electric field. Studying the bipolar switching *I–V* characteristics of micro- and nanoscale TiO_2 junction devices with Pt electrodes, the authors demonstrated that the HRS↔LRS transitions involve changes in the potential barrier heights at the Pt/TiO_2 interfaces due to the drift of positively charged oxygen vacancies. The modifications of these barriers alternately lead to Ohmic- or rectifying-type characteristics. In this model, the *I–V* curves are mathematically described by the phenomenological equation

$$I = w^n \beta \sin h(\alpha V) + \chi[\exp(\gamma V) - 1] \tag{3.63}$$

in which the first term represents a flux-controlled memristor: $\beta \sinh(\alpha V)$ is the approximation considered for LRS, which is related to electron tunneling through a thin insulating residual barrier. α, β, and n are fitting constants and w is the state variable of the memristor. While for $n = 1$ the drift velocity of the oxygen vacancies is directly proportional to the electric field, in the general case, n is used as a free parameter in the model which can be adjusted to modify the switching behavior. The second term in (3.63) represents the rectifying HRS which in this case adopts the form of a diode. χ and γ are fitting constants. A modification of Yang's model for WO_x-based MIM devices, which involves exponential voltage dependence in the state equation was proposed by Chang *et al.* [3.347]. Borghetti *et al.* [3.348] investigated the electrical transport characteristics of TiO_2-based MIM structures as a function of the temperature. A completely different behavior was reported after electroforming compared to the fresh device, which indicates the transition from area-distributed current flow to localized conduction. Again, two distinctive limiting behaviors were observed: Ohmic-like for the LRS and exponential for the HRS. This latest mode was attributed to tunneling through a potential barrier and the switching to voltage-induced changes in the oxygen vacancy concentration in the gap between the tip of the filamentary pathway and the adjacent metal contact. Similar arguments were used by Pickett *et al.* [3.349], but, in this case, the width of the tunneling barrier is identified as the dominant state variable and not the barrier height.

Hur *et al.* [3.350] proposed an equivalent circuit model for the bipolar RS *I–V* characteristics in Ta_2O_5/TaO_x stacks. The model is based on the modulation of the Schottky barrier height caused by the drift of oxygen vacancies. Bistable switching occurs as the conducting path is oxidized or reduced in the vicinity of the metal electrode–oxide layer interface. In the LRS, a resistor while for HRS, a diode, represents the doped Ta_2O_5 region close to the interface and a series resistance represent the resulting undoped Ta_2O_5. Many recently proposed models describe the HRS current in electroformed devices by means of hyperbolic sine dependence with the applied voltage. Occasionally, this consideration has also been extended to LRS [3.345]. This voltage dependence ultimately leads to pinched hysteretic loops, symmetric and Ohmic-type *I–V* curves for low applied biases, and exponential behavior of the conduction characteristics for large applied biases. In some cases, this choice has a physics foundation but in a vast majority this particular dependence is simply considered because it yields good fitting results [3.345, 3.351]. To our knowledge, Simmons and Verderber were the first to use hyperbolic sine dependence for the reversible memory phenomena in thin insulating films [3.352]. They proposed a model for the *I–V* curve of the form:

$$I(V) = K(V) \sin h[k(V)V] \tag{3.64}$$

where K and k are two functions of the applied voltage. Expression (14) was ascribed to DT in between defect sites and the transitions HRS↔LRS to the existence of a resonant discrete energy level within the insulator forbidden band. Guan *et al.* [3.334] have also implemented a SPICE compact model for RS which makes use of a sin $h(x)$-based expression. Inspired by the association of ReRAM operation with the conductive filament growth as a consequence of the movement of oxygen vacancies, the authors assumed the formation of a tunneling barrier of variable width. In their model, the change of the gap length g is associated with the probability for oxygen ions to overcome activation energy barriers following an Arrhenius-type law:

$$\dot{g} = v_0 \exp\left(-\frac{E_{a,m}}{kT}\right) \sin h\left(\frac{qa\gamma}{LkT}V\right) \qquad g \geq g_{\min} \tag{3.65}$$

where ν_0 is a velocity related to the attempt-to-escape frequency and $E_{a,m} = 1.2\text{eV}$ is the activation energy (migration barrier) for vacancy generation (oxygen migration) in a SET (RESET) process. L is the thickness of the switching material and a is the hopping distance. V is the voltage applied across the cell and g_{min} is the minimum gap size at which the tip of the filament is considered to be in contact with the electrode. γ is a local enhancement factor that takes into account the polarization of the material and the nonuniform potential distribution across the device. Equation (3.65) is derived from the Mott-Gurney ionic hopping current [3.335]. Assuming that the current exponentially depends on the tunneling distance and the field strength, the authors proposed that the RS I–V characteristic can be expressed as

$$I(g, V) = I_0 \exp\left(-\frac{g}{g_0}\right) \sin h\left(\frac{V}{V_0}\right) \tag{3.66}$$

where I_0, g_0, and V_0 are fitting parameters. Both the linear and exponential regions of the I–V curves can be captured by (3.66) by a proper selection of V_0. An interesting point of the model is the inclusion of the Joule heating effect in the formation of the gap and the cycle-to-cycle stochasticity in g.

Finally, it is worth mentioning that the QPC model for dielectric breakdown [3.354, 3.259] has also been used to model the RS effect [3.353]. As discussed in Section "Filamentary conduction models in SiO_2", according to this model, the current that flows through a filamentary pathway between metal electrodes (electron reservoirs) is governed by a tunneling barrier corresponding to the first quantized level associated with the confinement of the electron wavefunction. The height of this barrier determines the conduction mode: LRS for a wide constriction (low barrier) and HRS for a narrow constriction (high barrier). A similar argument holds in terms of the width of the barrier. If the barrier is represented by an inverted parabolic potential and the applied voltage drops symmetrically at both ends of the conductive bridge, the I–V characteristic can be obtained from the finite-bias Landauer approach as [3.259]

$$I(V) = \frac{4e}{h\alpha} \exp(-\alpha\varphi) \sin h\left(\frac{\alpha e}{2}V\right) \tag{3.67}$$

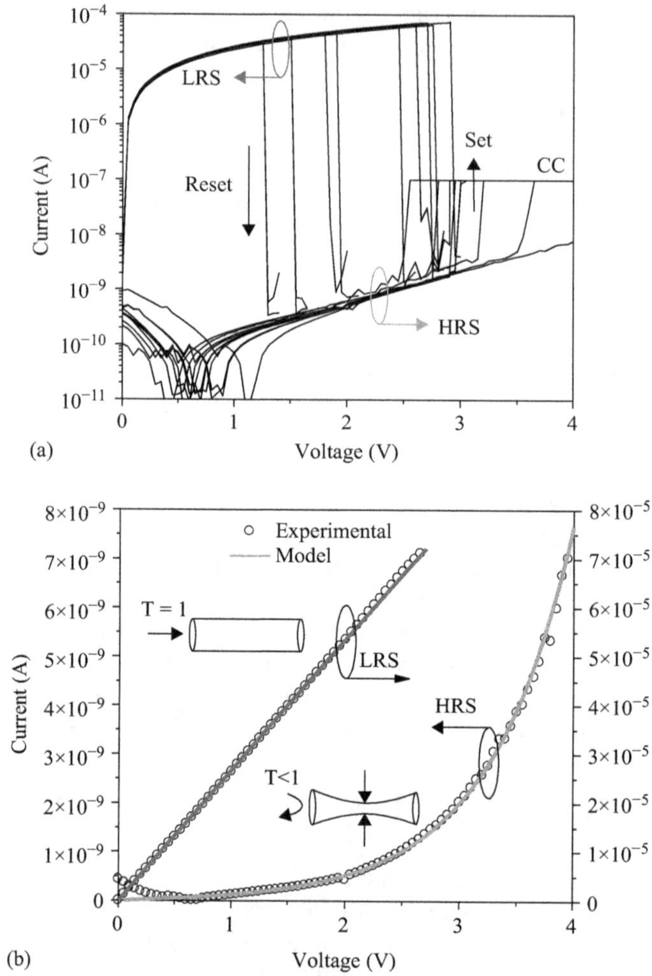

Figure 3.67 (a) Experimental HRS and LRS I–V *characteristics for HfO_2-based RS devices. (b) Experimental curves and model results using the QPC model for RS*

where e is the electron charge, h the Planck constant, α a constant proportional to the width of the barrier, and φ the potential barrier height measured from the equilibrium Fermi energy. Expression (17) strongly resembles Guan's model for the I–V curve (3.66), except that in this case the prefactor in the current and the slope of the curve are correlated. Remarkably, if the barrier width shrinks to zero in (3.67), $\alpha \to 0$ so that the current reads

$$I(V) = \frac{2e^2}{h} V \tag{3.68}$$

which corresponds to the electron transport characteristic of a monomode ballistic conductor (transmission probability $T = 1$). $G_0 = 2e^2/h = (12.9\ \text{k}\Omega)^{-1}$ is the quantum conductance unit. In this way, the cross-section area of the filament determines the RS current magnitude: the I–V characteristic follows an exponential law for a narrow constriction and a linear dependence as a function of the applied voltage for a wide constriction (see Figure 3.67). Although the QPC model does not make explicit reference to the ultimate cause behind the resistance change, the central idea is that it originates in a local atomic rearrangement driven by the external applied field. The idea that the RS filament electrically behaves as a nanowire with conductance values close to integer multiples of G_0 is supported by numerous experimental observations [3.258–3.266]. Nevertheless, it is worth pointing out that preferred atomic configurations for the filamentary path instead of true conductance quantization can also explain the observed phenomenology.

3.3 Alternative approaches for device characterization including magnetic fields

There is plethora of characterization techniques used to extract electrical parameters from semiconductor devices, as well as device physics and technology parameters. Just to mention a few, aside of the traditional I–V and C–V techniques, there is the charge pumping technique (CPT) first introduced in Reference 3.355, and later improved in Reference 3.356. The CPT is used to study the semiconductor–oxide interface and extract the interface trap density in MOSFET, which is then used for investigating the device degradation. There is "Deep-Level Transient Spectroscopy" (DLTS) introduced in 1974 in Reference 3.357, also used to study electrically active defects in semiconductors. DLTS uses also the application of a voltage or train of voltage pulses to study defects in the space charge region. The DLTS resolution is so good that it has been used to quantify the low defect concentrations that may affect the performance of spin-dependent transport FETs [3.358].

Regarding the reliability, the Bias Thermal Instability (BTI) is one of the most critical mechanisms degrading the MOSFET. The BTI concerns the electrical stability of Metal-Oxide device structures, and has to do with trapping and detrapping mechanisms. The nature of the energy traps involved in the trapping/detrapping is a function of the oxide and gate materials, therefore its experimental study requires of a wide variety of time-resolved characterization variations, which are well described in Reference 3.359.

The BTI happens in two different modalities, the NBTI for p-type and PBTI for n-type MOSFETs. The BTI phenomena is basically the activation of trap charges, under a bias or temperature stress, which alters the device electrostatics inducing a shift of the threshold voltage that degrades the current drive capability. Therefore, this is a characterization technique that requires the application of voltage and temperature as characterization variables.

Light is also used in combination with DLTS to determine deep energy levels in heavily doped semiconductors [3.360]. Besides thermal activation energy and capture cross-section, the photo-ionization energy of defects is also measured. The carrier density can also be determined from the reflectivity minimum wavelength λ_{min} when the material is exposed to optical plasma resonance experiments [3.361]. Moreover, optical infrared techniques, such as micro-Raman spectroscopy or X-ray micro-diffraction are used to characterize mechanical strain in sub-micrometer device structures [3.362, 3.363]. There are already commercial spectrometers with a spatial resolution around 200 nm [3.364]. However, with the current device dimensions below 20 nm and the large impact of nano-scaled mechanical strain on devices [3.365], there is an urgent need for developing device and material characterization for the nanoscale range.

A magnetic field B as an additional characterization variable, either applied along while recording electrical properties, or in conjunction with temperature or light, is another alternative to extract material or device properties. In that way, besides the approach used at low temperature and high magnetic fields to investigate fundamental quantum electronic properties of semiconductor materials and devices, such as the quantum Hall effect [3.366], the spin-dependent magneto-conductance [3.367], Landau levels splitting [3.368], Subnikov-de Haas oscillations [3.369], the fractional quantum Hall effect [3.370], the magnetic edge states [3.371], and many other relevant magnetic experiments on semiconductor materials and devices, which have resulted in two Nobel Prizes, the first awarded to Klaus von Klitzing in 1985 for the discovery of the integer quantum Hall effect, and the second awarded to R.B. Laughlin, H.L. Stormer, and D.C. Tsui in 1998 for the discovery of a new form of quantum fluid with fractionally charged excitations. There is the conventional or classic approach, at room temperature and low magnetic fields, that have been used to extract carrier mobility, the sign of the majority carriers, the carrier concentration, the Hall coefficient, and even the energy bandgap [3.372] and other electrical parameters.

Here we introduce experimental findings on the effect of the magnetic field on the channel and the gate oxide current, at room temperature, on 65 and 28 nm CMOS technologies, for a magnetic field ranging from −800 up to +800 mT, and applied at different orientations with respect to the channel plane.

First we introduce the classical Hall deflection effect on MOSFETs operated at room and 77 K temperatures as a vehicle to confirm the classical Lorentz's force.

3.3.1 Classical magneto transport in FET-based devices

The total force experienced by a charge moving under the influence of a cross electric E and magnetic fields B is given by

$$\vec{F} = q(\vec{E} + \vec{v} \times \vec{B}) \tag{3.69}$$

where v is the carrier velocity. From the classical approach, the previous equation that includes the Lorentz's force in the second term of the right side is converted into the Lorentz equation,

$$m^* \frac{d^2\vec{r}}{dt^2} + \frac{m^*}{\tau}\frac{d\vec{r}}{dt} = -q\left[\vec{E} + \vec{v} \times \vec{B}\right] \tag{3.70}$$

where m^* is the carrier effective mass, τ is the carrier relaxation time, and $\boldsymbol{r}$ is the position vector. The Lorentz equation, under isotropic, isothermal, and steady-state conditions, is applied to the charge transport of a MOSFET [3.373], which allows the calculation of both, the electron and hole current densities $\boldsymbol{J}_n$ and $\boldsymbol{J}_p$.

$$\vec{J}_n = -\sigma_n(\nabla\phi_n) - \sigma_n\frac{1}{1+\left(\mu_n^*\vec{B}\right)^2}\left\{\mu_n^*\vec{B}\times\nabla\phi_n\right\} - \sigma_n\frac{\mu_n^*}{1+\left(\mu_n^*\vec{B}\right)^2}\left\{\mu_n^*\vec{B}\times\left(\vec{B}\times\nabla\phi_n\right)\right\} \tag{3.71}$$

$$\vec{J}_p = -\sigma_p(\nabla\phi_p) - \sigma_p\frac{1}{1+\left(\mu_p^*\vec{B}\right)^2}\left\{\mu_p^*\vec{B}\times\nabla\phi_p\right\} - \sigma_p\frac{\mu_p^*}{1+\left(\mu_p^*\vec{B}\right)^2}\left\{\mu_p^*\vec{B}\times\left(\vec{B}\times\nabla\phi_p\right)\right\} \tag{3.72}$$

μ_n^* and μ_p^* are the electron and hole Hall mobilities, respectively. ϕ_n and ϕ_p are the quasi-Fermi potentials for electrons and holes, respectively, and σ_n and σ_p are the electrical conductivity for electrons and holes, respectively. The Hall mobility μ^* is related to the carrier mobility μ as $(r_H \cdot \mu)$, where r_H is the Hall coefficient. Along with the continuity and Poisson equations, a device under the presence of a magnetic field can be properly simulated with the drift-diffusion approximation, provided the device is sufficiently large [3.374].

Even when the semiconductor material is isotropic, the transport coefficients, under the presence of a magnetic field, show asymmetry [3.373], which depends on the way the electric driving force is oriented with respect to the magnetic field. In general this implies a nonsymmetric transport coefficient tensor L that can be written as

$$L = L_\perp(1 - P_B) + L_\parallel P_B \tag{3.73}$$

where P_B denotes the operation of projecting a vector along the direction of the magnetic field vector B.

In general, three transport coefficients can be defined: (1) isothermal Hall coefficients r_{Hn} and r_{Hp}, (2) isothermal Nernst coefficients N_n and N_p, and (3) Righi-Leduc coefficients RL_n and RL_p.

The r_{Hn} (r_{Hp}) coefficients characterize the transverse quasi Fermi level gradient caused by the magnetic field acting on the electron (hole) current. The N_n (N_p) coefficients characterize the transverse gradient of the quasi-Fermi potential caused by the deflection of an electron (hole) heat current down flowing a temperature gradient. And the RL_n (RL_p) coefficients represent the transverse temperature gradient caused by the deflection of an electron (hole) heat current down flowing a temperature gradient. The mathematical models for these three transport coefficients are as follows:

$$r_{Hn,p} = -\frac{\nabla\phi_{n,p}\cdot\left(\vec{B}\times\vec{J}_{n,p}\right)}{\left(\vec{B}\times\vec{J}_{n,p}\right)^2} \quad \text{with } \nabla T = 0 \tag{3.74}$$

$$N_{n,p} = -\frac{\nabla\phi_{n,p}\cdot\left(\vec{B}\times\nabla T_{n,p}\right)}{\left(\vec{B}\times\nabla T_{n,p}\right)^2} \quad \text{with } \vec{J}_{n,p} = 0 \tag{3.75}$$

$$RL_{n,p} = \frac{\nabla T\cdot\left(\vec{B}\times\vec{Q}_{n,p}\right)}{\left(\vec{B}\times\nabla T\right)\cdot\left(\vec{B}\times\vec{Q}_{n,p}\right)} \quad \text{with } \vec{J}_{n,p} = 0 \tag{3.76}$$

$\boldsymbol{Q}_{n,p}$ is the heat current density for electrons (holes).

In general, the electrical conductivity σ_n (σ_p) of the semiconductor device is modified by the presence of the magnetic field, which results in a magneto-conductivity tensor σ_B [3.375].

$$\sigma_{Bn,p} = \frac{\sigma_{n,p}}{1+(\omega_c\tau)^2}\begin{pmatrix} 1 & -\omega_c\tau & 0 \\ \omega_c\tau & 1 & 0 \\ 0 & 0 & 1+(\omega_c\tau)^2 \end{pmatrix} \tag{3.77}$$

A conclusion out of this isotropic model is that the conductivity, or the current, perpendicular to the magnetic field decreases by the factor $1/[1+(\omega_c\tau)^2]$. The off-diagonal elements in the magneto-conductivity tensor represent a current transverse to the electric field.

The classical approach represented by (3.77) is tested with an in-situ magnetic sensor and magnetic field generator as shown in Figure 3.68. The magnetic sensor is a two-drain MAGFET [3.376] fabricated in a three-metal layer 0.6 μm CMOS technology. It has an array of six interconnect lines on metal level 1 and 2. Three interconnect lines are on level 1. These three lines on level 1 and metal level 2 are 10 μm wide and are 10 μm separated each other. A current flowing through line L_2 on metal level 1 generates a magnetic field B that is sensed by the MAGFET configured as a transistor (drain 1 wired to drain 2), and biased at $V_d = 0.5$ V and $V_g = 3.0$ V. Previous to this experiment the MAGFET I_d current, configured as a transistor, was calibrated with an external magnetic field applied with a magnetometer. The differential drain current $\Delta I_d = (I_{d@B\neq 0} - I_{d@B=0})$ experienced by the MAGFET when the line current is swept from −60 to +60 mA is then correlated to the in situ generated magnetic field. The experimental results are shown in Figure 3.69.

The measured ΔI_d–B curve is fitted with a good correlation to a $1/[1+(\omega_c\tau)^2]$ function, which corroborates the classical magneto-conductivity tensor. The

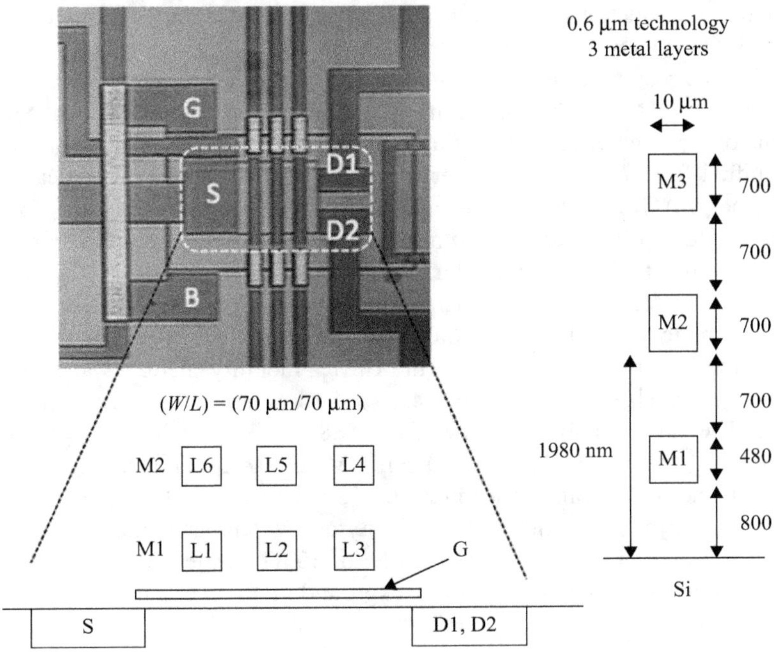

Figure 3.68 Test structure of a 0.6 μm technology (W/L) = (70 μm/70 μm) MAGFET with an interconnect line array as a magnetic field generator

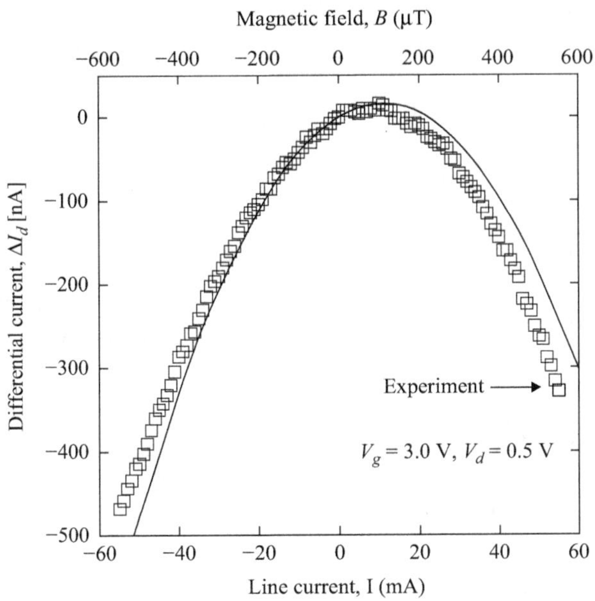

Figure 3.69 Measured (symbols) differential current ΔI_d versus line current (lower axis), and corresponding magnetic field B (upper axis). The continuous line corresponds to a $1/[1 + (\omega_c \tau)^2]$ fitting

interconnect line functions as a magnetic field generator in the +/−500 μT range, which is obviously a weak magnetic field that guarantees the applicability of (3.71) and (3.77). The magnetic field generated by the interconnect line is not fully perpendicular to the channel plane nor uniform on the total ($W*L$) area, which explains in part the difference between the fitting function and the experimental data, which is also shifted from $B=0$. We further proceed with an additional verification by using another MAGFET fabricated in an INAOE 10 μm CMOS technology [3.377]. This other $(W/L)=(100$ μm/125 μm) n-type MAGFET is characterized from room temperature down to 77 K showing an excellent linearity as shown in Figure 3.70.

The Lorentz force is proportional to the cross product of the carrier velocity and the magnetic field. Therefore the faster the particle moves, the higher the Lorentz force is. At low temperatures the carrier mobility increases and thus the velocity does, which results in an increase of the Lorentz force at cryogenic temperatures. The carrier mobility μ goes from 588 cm^2/Vs at room temperature up to 3021 cm^2/Vs at $T=77$ K for a 10 μm CMOS MAGFET technology with a $T_{ox}=60$ nm and bulk doping concentration of 2×10^{16} cm^{-3} [3.377]. This ends up in an increase of the magnetic field sensitivity at low temperatures. The magnetic resolution at room temperature is in the order of 50 μT, while at 77 K increases up to 1.5 μT. Filtering the magnetic noise background gives an extrapolated magnetic resolution of about 100 nano-Tesla (nT) at 77 K. This is already a good magnetic resolution to be used not only as a sensing device, but also as alternative device characterization technique.

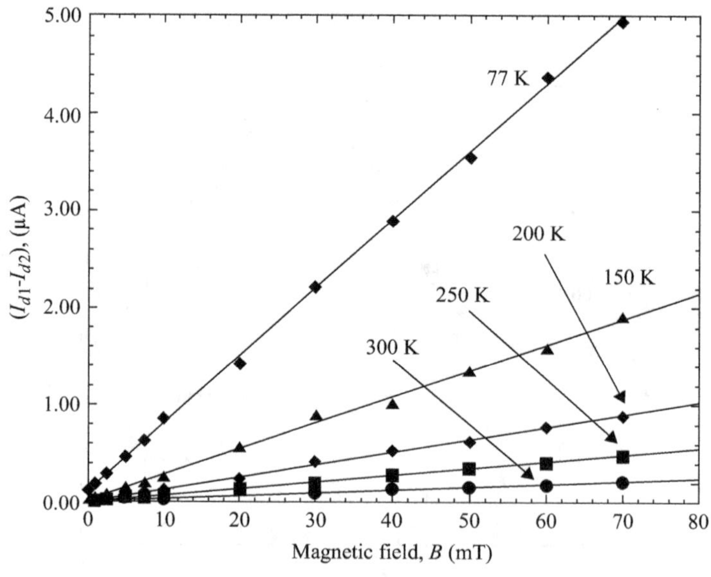

Figure 3.70 Measured (symbols) differential current ($I_{d1}-I_{d2}$) versus B with temperature T as a parameter, $V_g=5.0$ V, $V_{da}=V_{d2}=1.0$ V. The solid lines represent fittings

In the following sections some examples of an external magnetic field applied on different MOSFET devices, at room temperature, are introduced.

3.3.2 Magneto-tunneling and nonhomogeneous magneto conductance

One of the first experiments that can be conducted using a magnetic field, as an auxiliary characterization tool, is the study of current vector components. The cross vector product of carrier velocity and magnetic field is exploited as an auxiliary characterization technique, which allows research about the current vector components. This is the case of the gate oxide leakage current of MOSFETs. A schematic experimental setup for characterizing the gate oxide tunneling current I_g, with the magnetic field as a parameter, is shown in Figure 3.71.

The DUT, the MOSFET in this case, is placed in between the poles of a GMW 5403AC electromagnet [3.378], capable to generate AC and static magnetic fields up to 1.0 T, with a 1 mili-Tesla (mT) resolution, and up to 100 HZ switching frequency. The DUT can be fixed in any orientation with respect to the magnetic poles, so the *B* field enters perpendicular (*z*-axis) to the (*x*–*y*) channel plane, parallel to the (*x*–*y*) plane in the *y*-axis direction or the *x*-axis direction. Any other angles with respect to the (*x*–*y*) plane are also possible [3.379]. A p-type MOSFET of a high-k, metal-gate 28 nm technology is used as a DUT. The *B* field is applied in the *z*-axis direction, then the magneto-deflection effect on I_g, called $\Delta I_g = (I_{g@B\neq0} - I_{g@B=0})$, is recorded as a function of the gate voltage with the *B* field as a parameter. The experimental results are shown in Figure 3.72 [3.380].

The I_g current is assumed to flow in the *z*-axis direction, normal to the (*x*–*y*) channel plane and parallel to the *z*-axis applied *B* field. Ideally the charge velocity, of carriers flowing as a I_g current through the gate oxide, and its cross product with the *B* field should be zero, which implies $\Delta I_g = 0$. However, the experimental results of Figure 3.72 reveal the contrary effect. A positive $+B$ field, entering the channel

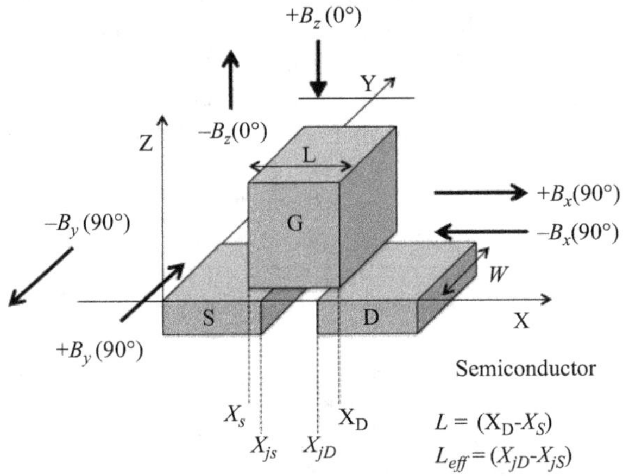

Figure 3.71 Schematic representation of the experimental setup

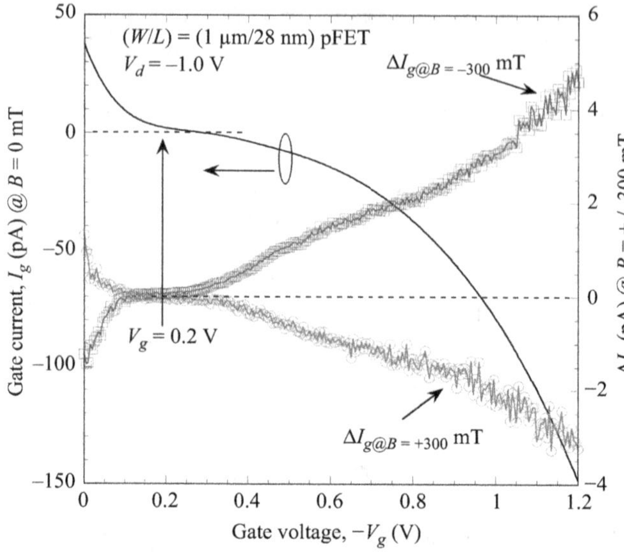

Figure 3.72 Measured I_g (left axis) and ΔI_g (right axis) for $V_d = 1.0$ V, $B = +/-300$ mT, for a p-type (1 μm/28 nm) MOSFET. Open circles corresponds to ΔI_g for $B = +300$ mT, while open squares corresponds to ΔI_d for $B = -300$ mT

plane, decreases I_g, whiles a negative $-B$ field increases I_g. This asymmetric behavior is shown for $-V_g > 0.2$ V, and switches sign when I_g changes sign at $-V_g < 0.2$ V, which confirms the Lorentz force effect on the I_g current. There is the question; is the action of the Lorentz force acting on the gate current density vectors, or is the B field acting on other electronic properties of the device structure? To be in a better position to propose a possible hypothesis, the effect of the B field on the channel current is shown in Figure 3.73 [3.381]. From these two last figures three distinctive bias points are observed: (1). 0.2 V $< -V_g <$ 0.6 V, (2). 0.6 V $< -V_g <$ 0.9 V, and (3). $-V_g >$ 0.9 V, where the slope $d\Delta I_g/dV_g$ has slight changes.

In the first region the $+B$ field increases the I_g flow from drain to gate, while for V_g voltages above region one the I_g current flows in the reverse direction from gate to channel. This confirms the magnetoconductance reversibility effect [3.382]. The first region corresponds to the subthreshold regime, the second region to the subthreshold-to-weak inversion regime. The third region belongs to the onset of the strong inversion regime. If the B field is generating a deflection force on the gate oxide carrier velocity, the slight change of the $d\Delta I_g/dV_g$ slope should be an indication of a gate oxide tunneling current density vector variation in the three different operating regimes.

The above observation suggests the increase of off-z-axis I_g components as the V_g voltage increases. On the other hand the asymmetry of the ΔI_gB curve for both polarities of B, suggests the off-z-axis I_g components are not uniform across the channel plane and semiconductor–oxide, and oxide–gate interfaces.

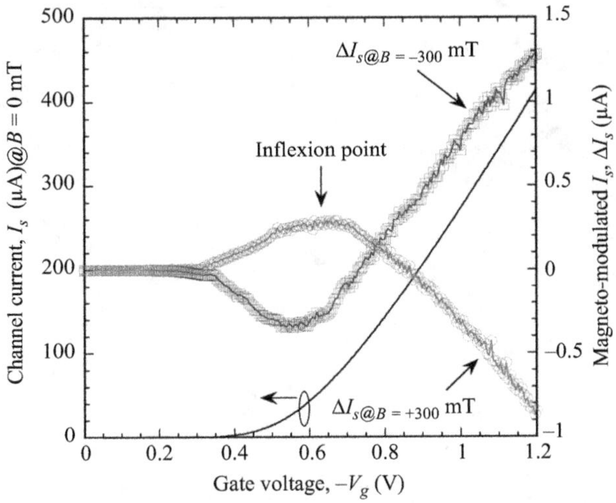

Figure 3.73 Measured channel current I_s (left axis) for $B = 0$ mT, and ΔI_s current (right axis) for $B = +/-300$ mT, for a p-type $(W/L) = (1\ \mu m/28\ nm)$ p-type MOSFET biased at $V_d = -1.0$ V. The open circles corresponds to ΔI_s for $B = +300$ mT, while the open squares corresponds to ΔI_s for $B = 300$ mT

The ΔI_s magneto-modulated current, shown in Figure 3.73, first increases for positive B fields in the 0.3 V $< -V_g <$ 0.7 V range, then rolls down to negative values for $-V_g >$ 0.8 V. When reversing the B field to negative values ΔI_s shows a complementary behavior. The change in the $d\Delta I_g/dV_g$ slope from region one to region two in Figure 3.72 coincides with the inflexion point of ΔI_s in Figure 3.73. This inflexion point is the diffusion-to-drift transition of the channel conductance. Then one assumes the B field has a differentiated effect on the diffusion and drift transport mechanisms [3.381].

The gate oxide tunneling current requires the existence of sufficient charges at the channel side, and a tunneling transmission coefficient different from zero. Therefore, the deflection of channel charges, evidenced by the ΔI_s modulation, changes the charge distribution in the channel, which is the charge source for gate oxide tunneling. This channel charge deflection effect, besides the Lorentz force acting on the gate oxide current density vector, is another cause for the ΔI_g effect.

A sweep of the magnetic field B from negative to positive values reveals information even more interesting. The ΔI_g versus B with V_g as a parameter and $V_d = 1.0$ V for the p-type MOSFET is shown in Figure 3.74.

An evident asymmetry with respect to B is observed for $-V_g = 1.2$ V. The ΔI_g–B asymmetry reduces as $-V_g$ goes from high to low values. All the curves cross each other through zero at $B_1 = -175$ mT, $B_2 = +0.75$ mT, and $B_3 = 200$ mT. At these three B points there is null magneto-conductivity effect, which can be interpreted as a good alignment of I_g with B along the z-axis. The positive values of ΔI_s in the 0.3 V $< -V_g <$ 0.75 V range for +300 mT and for the $-V_g >$ 0.75 V for

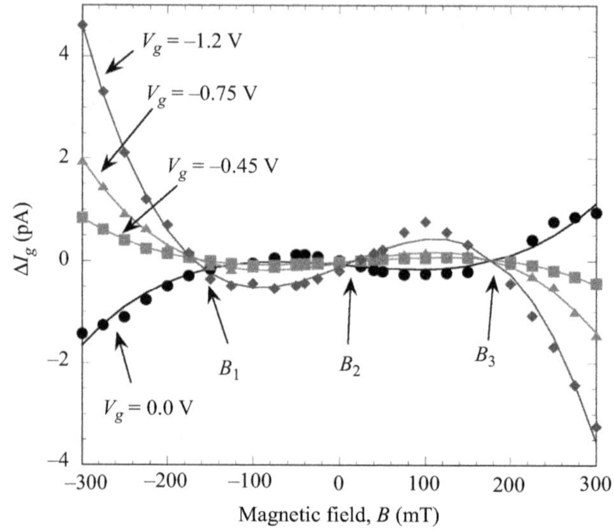

Figure 3.74 Measured ΔI_g current as a function of B for $V_d = 1.0$ V with V_g as a parameter

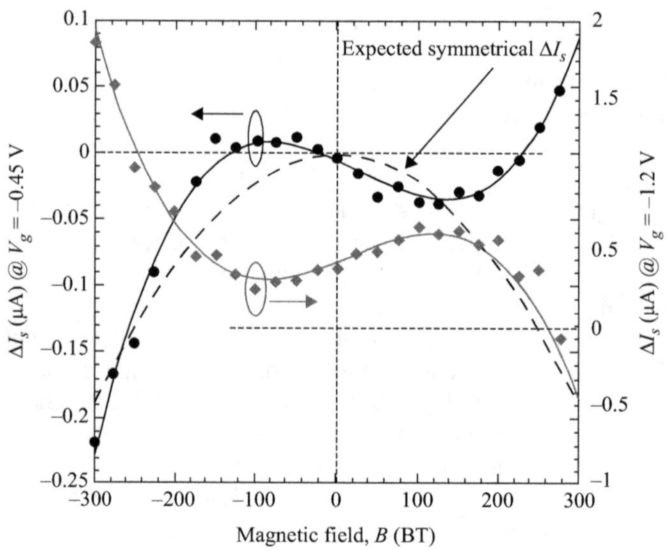

Figure 3.75 Measured ΔI_s versus the B field for $-V_g = 0.45$ V (left axis) and $-V_g = 1.2$ V (right axis)

−300 mT in Figure 3.73, denote a hole flow into different crystal orientations with different mobilities along the channel.

The ΔI_s–B characteristics for the diffusion ($-V_g = 0.45$ V) and drift ($-V_g = 1.2$ V) regions are shown in Figure 3.75.

According the magneto-conductivity tensor of (3.77) a parabolic-like behavior (dashed curve) of the ΔI_s–B curve is expected from the experimental data of Figure 3.75. However, there is an asymmetrical and monotonic behavior. This behavior is attributed to a multidimensional flow of holes both along the channel and through the gate oxide.

However, as the inversion channel is the source charge for gate oxide tunneling, the I_g current gets coupled to I_s. Thus, we get into a complex multidimensional Lorentz force as can be seen from Figure 3.76. Also as $(W/L) = (1\ \mu m/28\ nm) = 35$, this device classifies as a short Hall plate, and thus no Hall field or voltage is generated along the y-axis. Thus owing to the lack of a transverse Hall voltage, a transverse current component J_y is generated [3.383]. Considering that the B field is applied in the z-axis the space deflection of carriers in the inversion channel, of a thickness Z_{inv}, is restricted to the $(x–y)$ plane. This is not the case of carriers tunneling through the gate oxide. A carrier moving from source-to-drain on the $(x–y)$ plane can get enough energy to tunnel through the gate oxide to leave the $(x–y)$ plane and enter into a tri-dimensional (x,y,z) space. Once the carrier is moving through the oxide it will encounter interfaces 2, 3, and 4 if the oxide is a three-dielectric stack for instance. Interfaces 2 and 3 are in the bulk of the oxide, and interface 4 is at the border with the gate. This different dielectric layers might lead to some dispersion at the interfaces, which results in a nonlinear trajectory of the tunneling charge. Percolation is another mechanism that needs to be accounted for [3.384], which considers the trap-assisted current trough single and multiple trap paths, and the spatial distribution of these paths. Furthermore, a nonhomogeneous

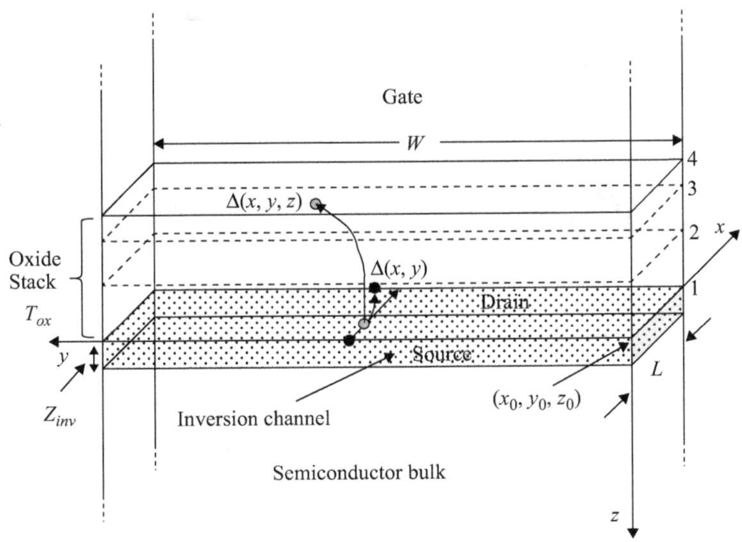

Figure 3.76 Geometric representation of the inversion channel and a charge moving from source-to-drain, on the (x–y) plane, and another charge tunneling through the oxide on the (y–z) plane

spatial distribution of filamentary leakage current paths has been probed to occur in HfO_2 oxides [3.385], which leads to nonhomogeneous spatial distribution of current leakage paths.

Because of the mechanical strain, used to enhance carrier mobility [3.386], the (1 μm/28 nm) MOSFET is not only a short Hall plate, but it has anisotropic conduction properties. Such an anisotropic transport properties are attributed to the nonhomogeneous distribution of the mechanical stress [3.379, 3.387]. Figures 3.77 and 3.78 show the simulated normalized mechanical stress across the width W and along the channel L of the MOSFET [3.379]. The stress is taken at the middle of the channel as a reference and normalized to its value at the middle of the channel.

The carrier mobility μ, the intrinsic concentration ni, the energy bandgap E_g, the density of energy states N_c (N_v), and consequently the electron concentration n are all functions of the stress s [3.388]. The E_g reduces linearly with the increase of s, while ni increases exponentially with the increase of s, and μ increases linearly with the increase of s.

$$E_g = E_{g0} - m_s s \tag{3.78}$$

Where $E_{g0} = 1.12$ eV, and m_σ ranges from 100 to 140 meV [10]. The intrinsic carrier concentration n_i is an exponential function of E_g.

$$n_i = \sqrt{N_c N_v}\, exp\left(-\frac{E_g}{kT}\right) \tag{3.79}$$

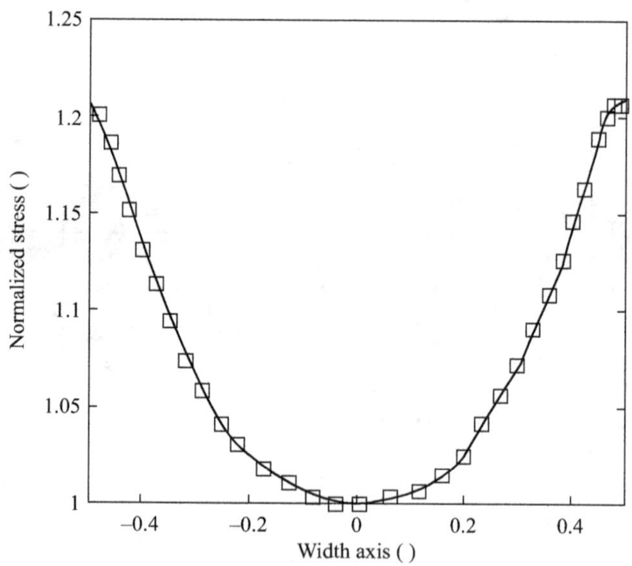

Figure 3.77 Calculated stress s, normalized to its value at the center of the channel, as a function of the width axis normalized to W

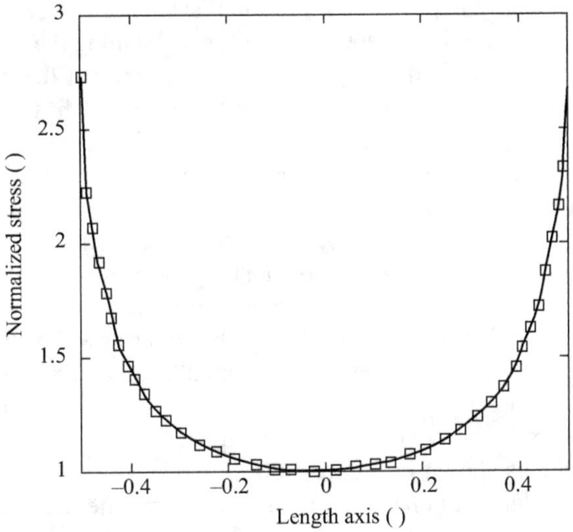

Figure 3.78 Calculated stress s, normalized to its value at the center of the channel, as a function of the length axis normalized to L

Where N_C and N_V are the effective density of states in the conduction and valence bands, respectively. k is the Boltzmann's constant, and T is the temperature. A reduction of E_g results in an exponential increase of n_i. On the other hand the carrier mobility μ follows the model.

$$\mu = \mu_0 + \mu_\mu s \tag{3.80}$$

where $\mu_0 = 600$ cm^2/Vs, and $\mu_\mu = 253$ cm^2/VsGPa.

As the stress s tensor is a function of both the x and y directions, with a minimum value in the middle of the channel, and with a larger value at the perimeter, one expects the electrical conductance σ_0 to have the same shape as that of s. The 28 nm CMOS technology uses also deep trenches to provide high-frequency isolation, and as pointed out in Reference 3.387 the stress distribution is also a function of the distance between the active region and the trench.

Other random or fluctuation fabrication effects, such as Random Dopant Fluctuation (RDF) [3.389], Line Edge Roughness (LER) [3.390], and Metal Gate Granularity (MGG) [3.391] with grain sizing from about 5 to 50 nm, together with the NonUniform Stress (NUS), contribute to a considerable change and space-dependence of the transport parameters in the active channel of the MOSFET [3.392].

Thus strained nanometric semiconductor devices cannot be considered as an isotropic Hall sample, which inhibits the use of the magneto-conductivity tensor of (3.77). Therefore a most generalized theoretical treatment is required, such as the one introduced in References 3.393 and 3.394, where semiconductors with anisotropic properties are considered.

Solving the Schrödinger-Poisson equation system in a self-consistent way, including the magnetic field is an option that allows obtaining the wavefunctions and energy with the magnetic field as a parameter. In the case of the gate oxide tunnel current, the GTS Framework commercial software [3.50] has been modified to account for the *B* field, temperature, and anisotropic semiconductor devices properties [3.395]. The GTS Framework simulation software contains the VSP, which is a simulation tool for 2D/3D quantum modeling of nano-scaled semiconductor device structures [3.396]. A screen window of the GTS Framework simulation is shown in Figure 3.79. The device structure is edited with the fabrication process parameters and geometry defined by the user. The doping profile, the gate oxide thickness geometry, the trap distribution, and even RDF and LER can be edited at will by the user. This way nonhomogeneities and anisotropy can be accounted for in addition to the *B* field.

The gate tunneling current density vectors, as well as the I_g spatial distribution, including device anisotropy, can then be studied at the semiconductor–oxide interface and oxide as shown in Figure 3.80.

Unlike a pure electrical characterization technique, the use of electromagnetic techniques, in conjunction with device simulation, gives additional information on

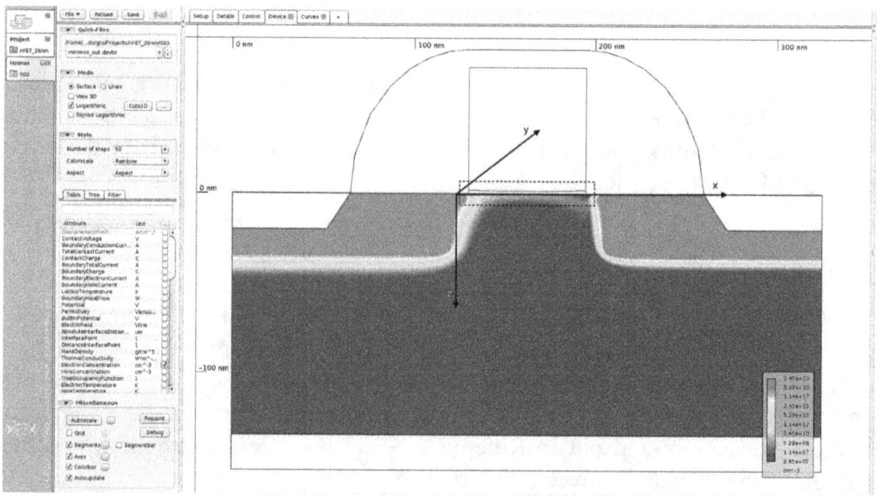

Figure 3.79 Screen view of the GTS Framework simulator showing a 28 nm MOSFET

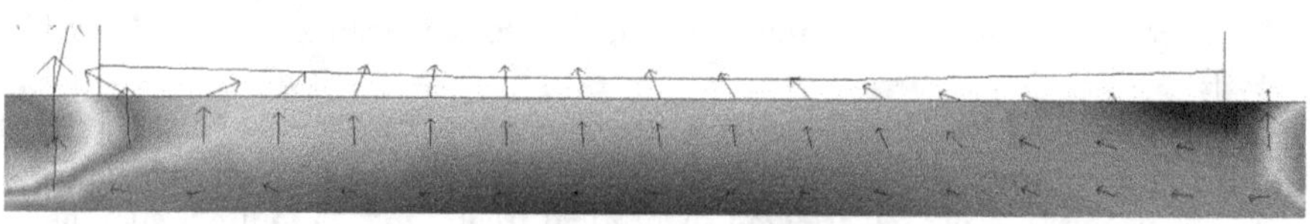

Figure 3.80 Simulated current density components at the semiconductor–oxide interface

the local space distribution of conductive properties. A first 2D modeling and simulation approach to account for the *B* field influence on MOSFETs with anisotropic properties is introduced in References 3.397 and 3.398. The energy and wavefunctions of the systems are calculated by solving, in this particular case, the time-independent effective mass Schrödinger equation,

$$\left[\frac{1}{2m^*}(\hat{p} - qA)^2 + V(r)\right]\psi(r) = E\psi(r) \tag{3.81}$$

where $\hat{p} = -i\hbar\nabla$ is the momentum operator, ψ is the wavefunction, *E* is the energy eigenvalue, *V(r)* is the potential energy, and *A* is the vector potential such that $B = \nabla \times A$. The electrostatic potential ϕ and its relation to the electron density distribution *n(r)* are calculated by means of the Poisson equation,

$$\nabla^2\phi = -\frac{q}{\epsilon}(N_b - n(r)) \tag{3.82}$$

where N_b is the bulk doping concentration. The electron density *n*(*r*) is then determined form the self-consistent solution of the Schrödinger-Poisson system, as

$$n(r) = \frac{m^*kT}{2\pi\hbar^2}\sum_i |\psi_i|^2 ln\left[1 + exp\left(\frac{E_F - E_i}{kT}\right)\right] \tag{3.83}$$

where E_F is the Fermi level, E_i is the energy eigenvalue, and ψ_i is the wavefunction of the corresponding *i* eigenvalue.

Once the E_i and ψ_i values are calculated, the direct gate oxide tunneling current can be calculated by using the Tsu-Esaki formulation [3.399]

$$J = \int_{E_{min}}^{E_{max}} TC(E_z)SF(E_z)\mathrm{d}E_z \tag{3.84}$$

where *TC* is the transmission coefficient, E_z represents the electron kinetic energy perpendicular to the gate oxide potential barrier, and represents the probability of a particle tunneling through the potential barrier that is calculated as

$$TC = \frac{J_t}{J_i} \tag{3.85}$$

where J_t is the transmitted current density and J_i is the incident current density. E_{min} is the lower energy and E_{max} is the higher value of the discretized energies in the potential well at the semiconductor side.

The supply function *SF*, which denotes the number of electrons in a certain energy level available for tunneling through the oxide potential barrier, is given by

$$SF = ln\left[\frac{1 + exp\left(E_{Fs} - \frac{E}{kT}\right)}{1 + exp\left(E_{Fg} - \frac{E}{kT}\right)}\right] \tag{3.86}$$

where E_{Fs} and E_{Fg} are the Fermi levels in the semiconductor and the gate segments, respectively. Both *TC* and *SF* becomes a function of the energy and the

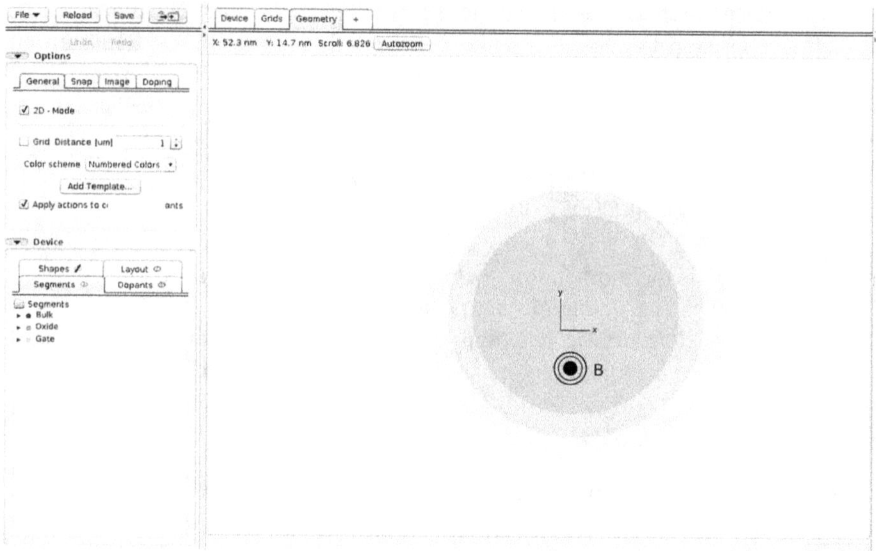

Figure 3.81 Simulated 30 nm Si circular nanowire. The inner circle is Si, the outer circle is a metal gate, and the circle in between is SiO_2

wavefunction, which in turns are B field dependent. A first attempt to probe the effect of the B field on the wavefunctions is tested with an n-type circular 30 nm Si nanowire with an equivalent oxide thickness $EOT = 2.0$ nm, and $N_D = 3 \times 10^{17}$ cm^{-3} as shown in Figure 3.81. The B field is perpendicular the page plane. Closed boundary conditions are first used, so the wavefunctions get confined in the silicon region and vanish at the Si–SiO_2 interface.

A B field of 0, 50, and 500 mT is applied on the nanowire and the probability density $|\psi|^2$ of wavefunctions of the second and sixth state are evaluated and shown in Figure 3.82.

The probability density $|\psi_6|^2$ of the sixth level wavefunction is shown in Figure 3.83.

A hybridization effect, or mixture of states, is observed as the strength of the B field increases. The B field perturbs the Hamiltonian and thus the states are no longer separated and begin to mix. The mixing results in new hybrid states, which in turn increase the probability to find charge carriers where they were not supposed to be. The hybridization between two degenerated electronic states produces repulsion in energy between each other, which lifts the degeneracy of the states [3.400, 3.401].

Furthermore, in Figure 3.84, the probability density of the second state $|\psi_2|^2$ and the density probability of the third state $|\psi_3|^2$ are shown.

For this case, both degenerated states begin to mix, and at sufficiently strong B fields exhibit a circular or ring shape [3.401].

In order to study the asymmetric I_g–B response of a MOSFET, under the influence of a B field applied parallel to the surface and perpendicular to the I_g and

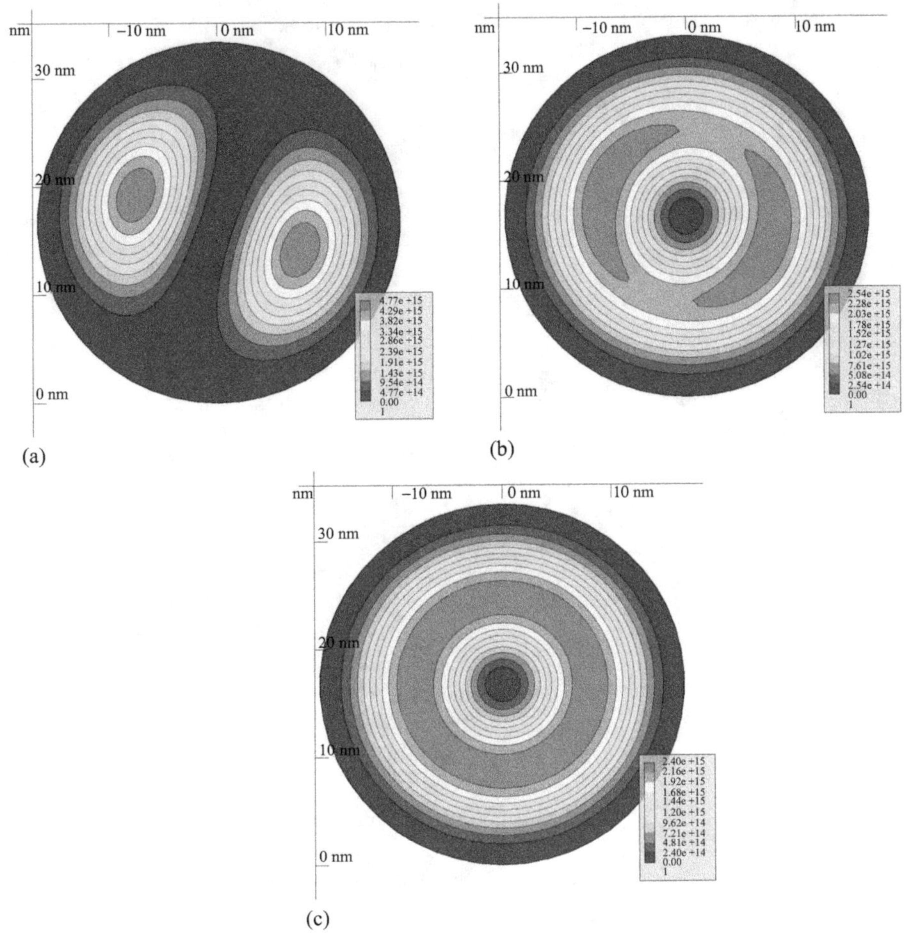

Figure 3.82 Probability density $|\psi_2|^2$ of the second level wavefunction of a cross-section of the circular nanowire under a B field of (a) 0 mT, (b) 50 mT, and (c) 500 mT

I_d current lines, a 28 nm n-type Si strained MOSFET is used as a benchmark. It has an HfO_2 $EOT = 2.0$ nm, a $(W/L) = (1\ \mu m/28\ nm)$, a metal gate, and a bulk doping concentration of 2×10^{18} cm^{-3}. Figure 3.85 shows the simulated magneto-modulated gate current $\Delta I_g = (I_{g@B\neq 0} - I_{g@B=0})$ under the action of a static magnetic field $B = +/-400$ mT. The simulated results are compared to measurements.

A good agreement between experiments and simulations is observed up to $V_g =$ 1.0 V. For $V_g > 1.0$ V ΔI_g rolls down, which cannot be reproduced by the simulator.

As mentioned earlier a strained MOSFET may have nonuniform conductive channel properties that can be recreated by means of a nonuniform gate oxide

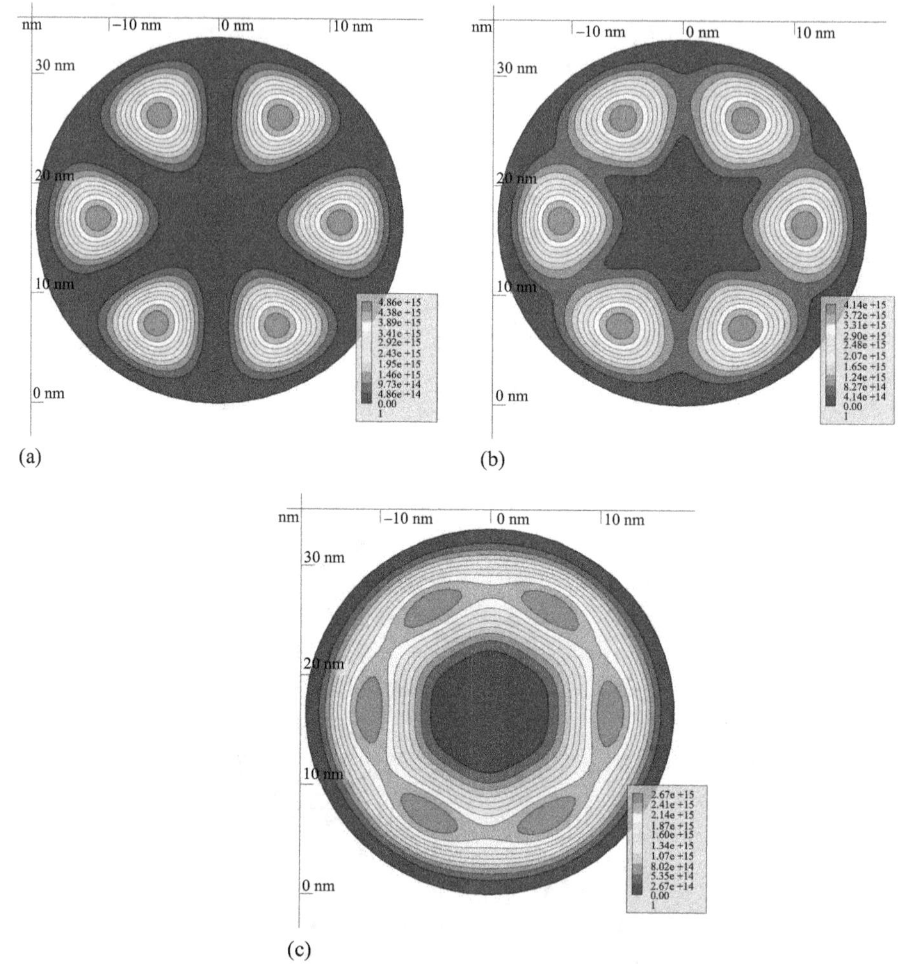

Figure 3.83 Probability density $|\psi_6|^2$ of the sixth level wavefunction of a cross-section of the circular nanowire under a B field of (a) 0 mT, (b) 50 mT, and (c) 500 mT

thickness or by a nonuniform doping profile. Such a recreation is performed with gtsFramework and shown in Figure 3.86. Three different nonuniform oxide configurations, T_{ox1}, T_{ox2}, and T_{ox3}, are used. T_{ox1} has $T_{oxL} = 2.0$ nm and $T_{oxR} = 2.5$ nm, T_{ox2} has $T_{oxL} = 2.0$ nm and $T_{oxR} = 2.6$ nm, and T_{ox3} has $T_{oxL} = 2.0$ nm and $T_{oxR} = 2.7$ nm. The V_d voltage is set equal to V_s and 0.0 V to have a symmetric flow of gate current density lines.

As seen from Figure 3.86 the T_{ox3} configuration, which is the most unbalanced or asymmetric gate oxide, gives the higher ΔI_g percentage change. Such a ΔI_g–B asymmetry is shown as a function of T_{oxR}, for $T_{oxL} = 2.0$ nm, in Figure 3.87.

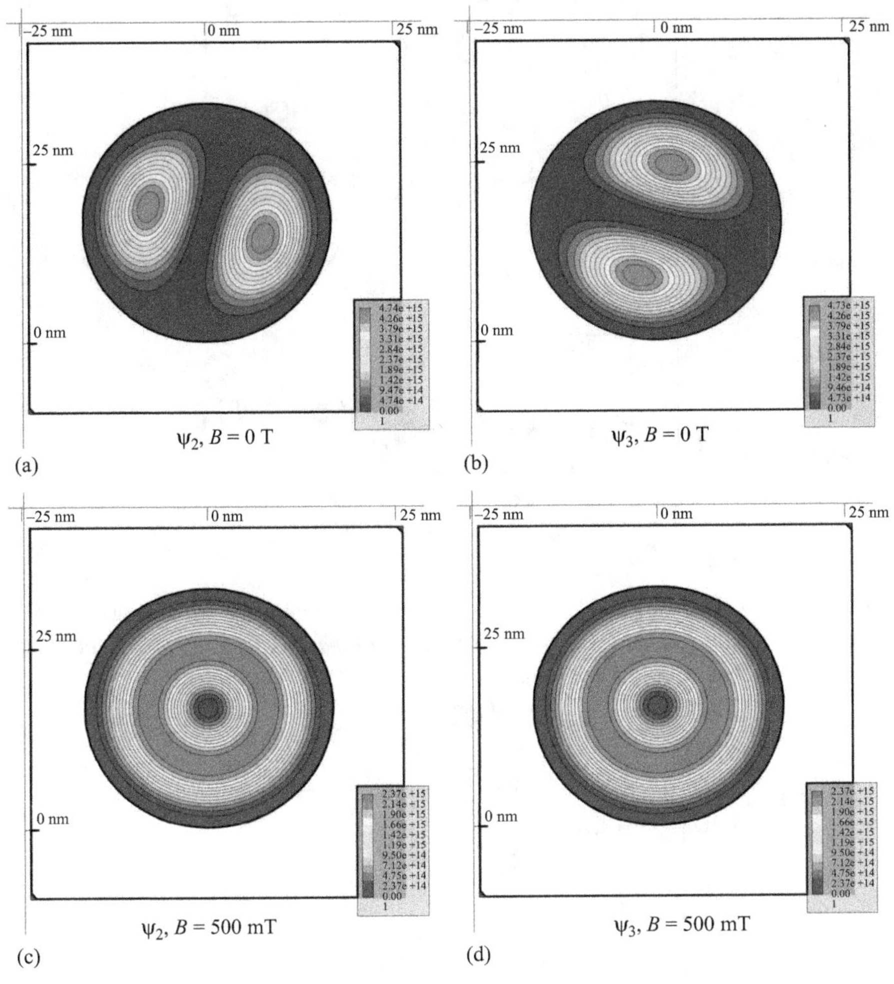

Figure 3.84 Hybridization of ψ_2 ((a) at B = 0 T, (c) at B = 500 mT) and ψ_3 ((b) at B = 0 T, (d) at B = 500 mT) states at B = 500 mT

As the gate oxide asymmetry increases the $\Delta I_g/I_g$ percentage change increases and saturates. This can be explained when looking at the transmission coefficient *TC* and supply function *SF* as a function of the *B* field in Figure 3.88. Owing to the nonuniform gate oxide thickness, the transmission coefficient *TC* increases for negative *B* fields and decreases for positive ones. A negative *B* field deflects carriers toward the side with a thinner gate oxide, while a positive one does to the right side with a thicker oxide. Therefore, a negative *B* field leads to a higher tunneling probability than a positive *B* field. Moreover, the supply function *SF* always decreases when a *B* field is applied, either positive or negative. A reduction of *SF* implies a decrease of the charge population with a probability to tunnel, which

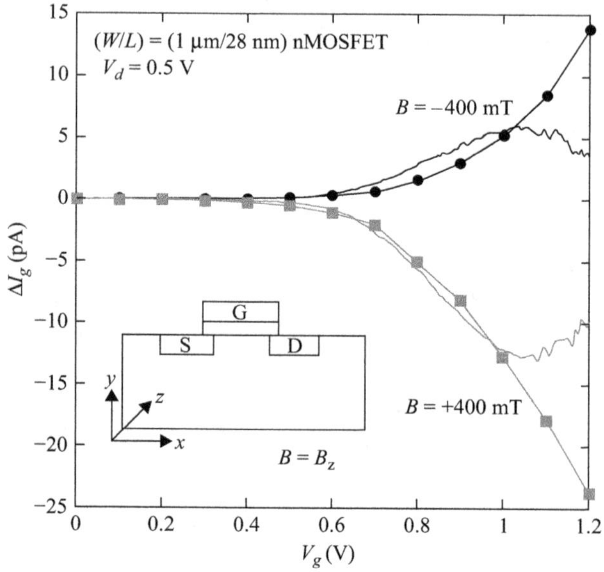

Figure 3.85 Simulated (symbols) ΔI_g magneto-modulated current and experiments (continuous line) for an n-type nMOSFET with an external magnetic $B = +/-400$ mT applied into the z-axis direction for this case

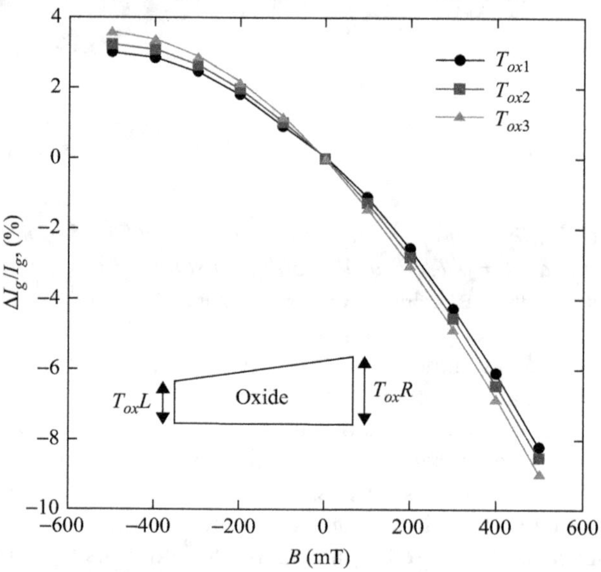

Figure 3.86 Simulated percentage of change of magneto-modulated gate current $\Delta I_g/I_g$ [%] for a 28 nm nMOSFET with $V_d = 0.0$ V and $V_g = 1.0$ V

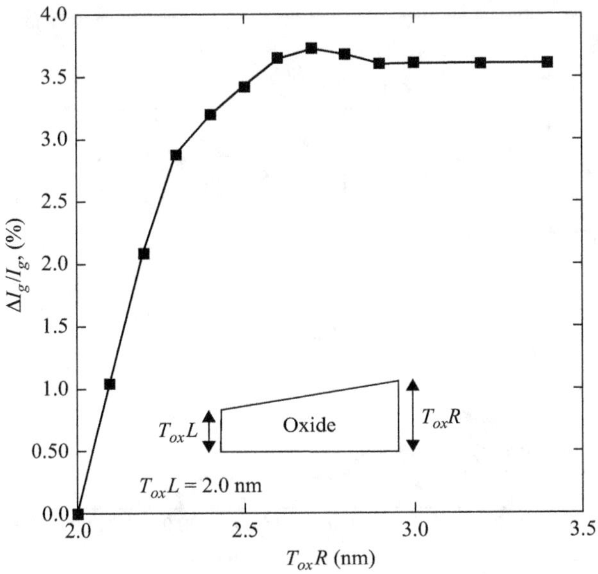

Figure 3.87 *Simulated $\Delta I_g/I_g$ percentage change as a function of T_{oxR} for the 28 nm nMOSFET biased at $V_d = 0.0$ V and $B = -500$ mT*

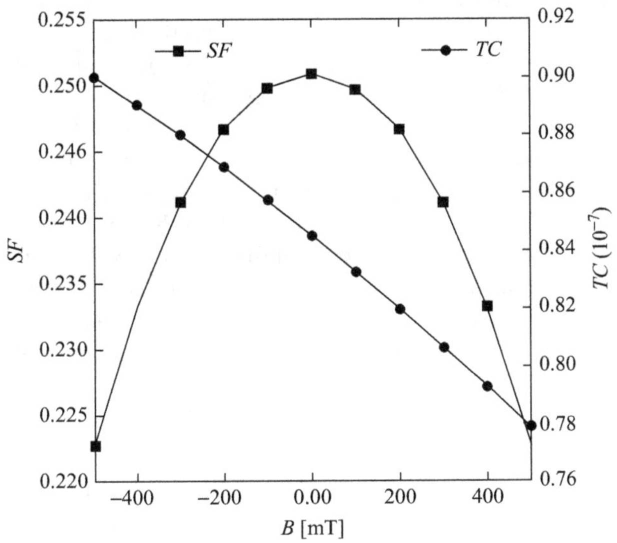

Figure 3.88 *Simulated supply function SF and transmission coefficient TC as a function of the B field, for oxide configuration T_{ox3} with $V_d = 0.0$ V, and $V_g = 1.0$ V*

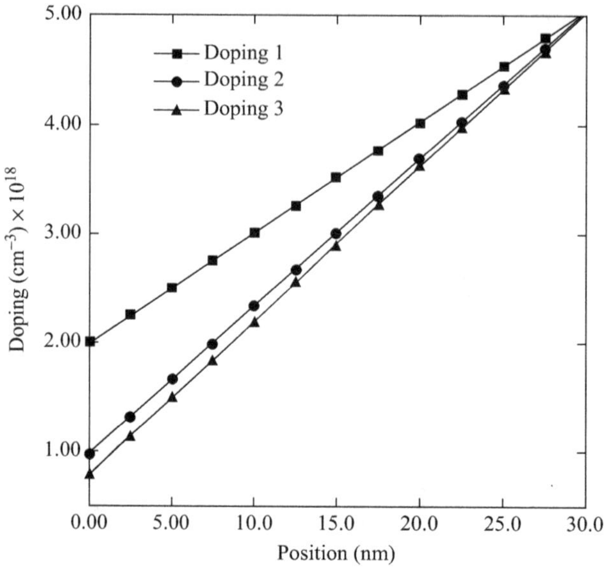

Figure 3.89 The three doping profiles used for the 28 nm nMOSFET simulation

results in a reduction of the gate oxide tunneling. Besides the gate oxide non-uniformity, the doping profile may be a source of nonhomogeneity. Three different doping profiles, shown in Figure 3.89, are proposed as way to test the doping nonuniformity. Doping profile 1 has a $D_{opL} = 2 \times 10^{18}$ cm^{-3} and $D_{opR} = 5 \times 10^{18}$ cm^{-3}, doping 2 has $D_{opL} = 1 \times 10^{18}$ cm^{-3} and $D_{opR} = 5 \times 10^{18}$ cm^{-3}, and doping 3 has $D_{opL} = 9 \times 10^{17}$ cm^{-3} and $D_{opR} = 5 \times 10^{18}$ cm^{-3}. Figure 3.90 shows the percentage change of $\Delta I_g/I_g$ for the B field swept from negative to positive values with the doping profile as a parameter. For a positive B field the current lines are deflected to the right side in the direction of the drain where the doping is larger. A larger doping implies an increase of the threshold voltage, which results in a smaller inversion layer charge population. Therefore, the supply function decreases leading to a reduction of the tunneling current. Doping profile 3, which is the less uniform doping shows the most pronounced asymmetrical ΔI_g.

3.3.3 Conductance properties under constant and time-dependent magnetic fields

The off-power consumption of integrated circuits is a critical issue of today's most advanced electronics. The gate oxide leakage current is one of the contributors to the off-power, as well as the subthreshold off-current I_{off}, and the reverse-biased drain-bulk junction leakage current I_{db}. In this section we examine the influence of the B field on both I_{off} and I_{db}. At the end of the section the effect of a switching B field on a gated-resistor, and the influence of the B field on the charge pumping characterization are introduced.

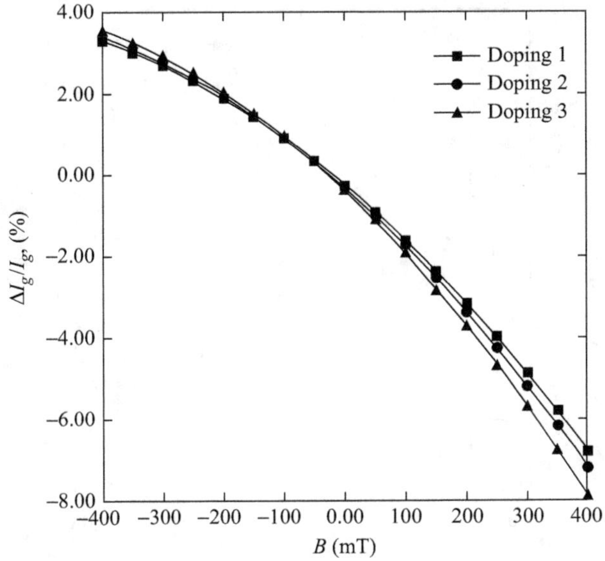

Figure 3.90 Simulated percentage change of $\Delta I_g/I_g$ as a function of the B field for a nMOSFET with $V_d = 0.0$ V and $V_g = 1.0$ V

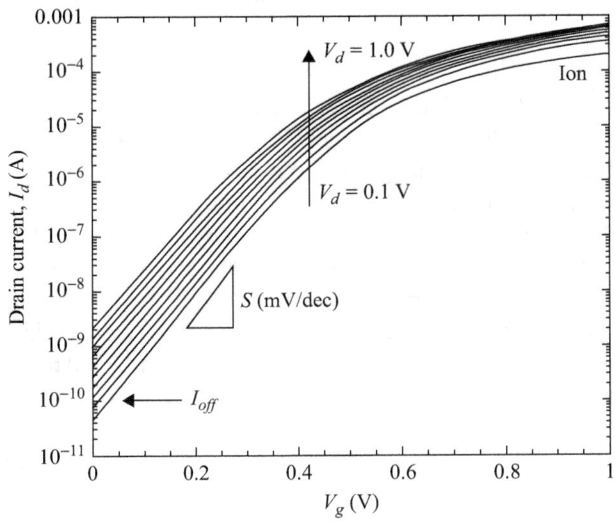

Figure 3.91 The I_d–V_g characteristics of a (W/L) = (1 μm/28 nm) NFET with V_d as a parameter, B = 0 mT

The regular I_d–V_g characteristics of a $(W/L) = (1$ μm/28 nm) are shown in Figure 3.91. In this case no B field is applied, but the V_d voltage is sweep as a parameter from 0.1 to 1.0 V in steps of 0.1 V. The off current I_{off} and subthreshold slope S as a function of the B field are plotted in Figure 3.92. Something quite

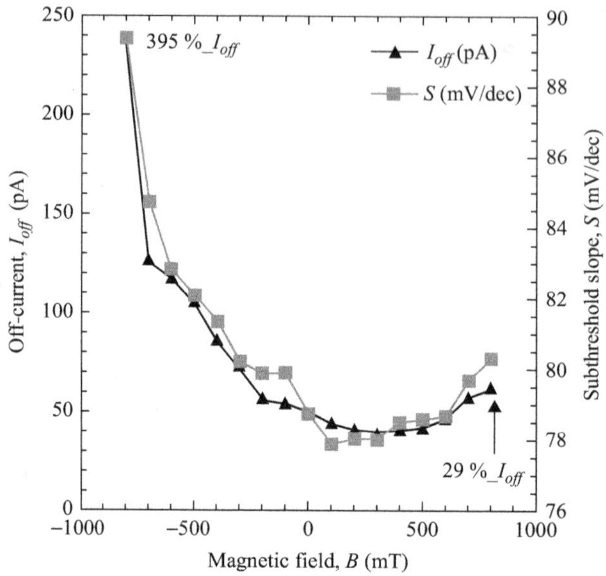

Figure 3.92 Measured off current I_{off} and subthreshold slope S as a function of the B field for a (W/L) = (1 μm/28 nm) NFET biased at $V_d = 0.1$ V
© 2015 IEEE. Reprinted with permission from Reference 3.397

unexpected is observed from Figure 3.92. There is no parabolic reduction of the I_{off} current as that shown by the control vehicle in Figure 3.68. On the contrary there is a 395 per cent increase of I_{off} for $B = -800$ mT, and of about 29 per cent increase for $B = +800$ mT. The subthreshold slope S follows the same trend as that of I_{off}.

In the subthreshold regime the electrons move by diffusion, so the driving force is the electron gradient ($q\mathrm{d}n/\mathrm{d}x$) along the channel. Under the bias conditions used for this experiment and the geometry of the FET device, the carrier mobility μ in the channel is only about 200 $\mathrm{cm^2/Vs}$ as extracted from simulations. This gives a marginal deflection of about 1.0 nm for a B field sweeping from −800 to +800 mT, or in other words, a sweep of about 1.83 degrees when varying B from −800 to +800 mT. The application of a B field perpendicular to the surface of the FET device, sweeps the electrons to the left for negative B fields, and sweeps them to the right for positive B fields. This is done thanks to the Lorentz's force that makes the function of a scanner mapping the conductance properties of the FET device along the width axis. The increase of the I_{off} current and subthreshold slope, when deflecting electrons toward the edges in the width direction, corroborates the presence of a larger mechanical stress in the edges than in the center of the channel. The I_{off}–B asymmetry is also an indication of the asymmetrical mechanical stress on the width axis.

Next the reverse-biased drain-bulk junction leakage current I_{db} is measured at room temperature with no B field applied. The measured results and its corresponding fitting to TAT [3.402] and direct BBT [3.403] are shown in Figure 3.93.

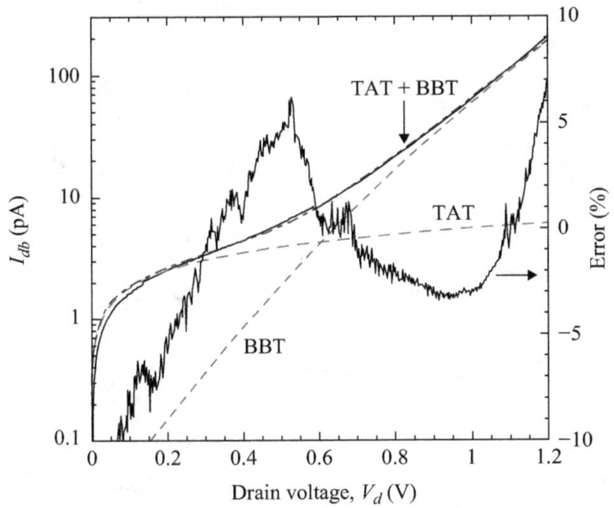

Figure 3.93 Measured I_{db} current of a (W/L) = (1 μm/28 nm) NFET (black continuous line). Red dashed line corresponds to the calculated I_{dBBT} components, while green dashed line corresponds to I_{dTAT} component. The error between the measured and calculated current is shown at the right axis

Two regions are identified, the first one for $V_d < 0.6$ V, where TAT is the dominant mechanism, and the second region for $V_d > 0.6$ V where BBT dominates. The GTS Framework program was used to fit the experiments to the model that accounts for both TAT and BBT tunneling mechanisms. The match between the models and the experimental data is within a 10 per cent error, which from the qualitative point of view is good enough. Once the electrical model has been validated, the magnetic field is applied on the FET device, while the I_{db}–V_d characteristics are recorded. The B field is applied perpendicular to the (x–y) channel plane. The experimental results of the I_{db}–V_d characteristics for $B = -800$, 0, and +800 mT are shown in Figure 3.94. Around $V_d = 0.8$ V there is a crossing point where the magneto-conductance coefficient changes sign. For $V_d < 0.8$ V the negative B field increases I_{db} while for $V_d > 0.8$ I_{db} decreases. The opposite happens for positive B fields.

This experiment indicates the TAT mechanism decreases when B goes from negative to positive values, while the BBT mechanism increases when sweeping B from negative to positive values. In other words, a positive B field sends electrons to the right where the BBT is more pronounced, while negative B fields send electrons to the left side where TAT is dominant. In the previous experiment a larger stress s at the left side than at the right side of the FET device has been assumed. Therefore, one can assume that a larger stress increases the TAT process, while it decreases the BBT mechanism. Adding the corresponding effect of the mechanical stress s on the

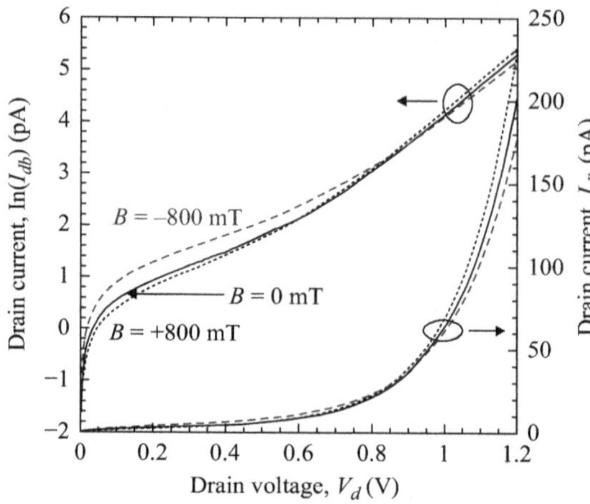

Figure 3.94 Measured I_{db}–V_d characteristics for $B = -800$, 0, and $+800$ mT

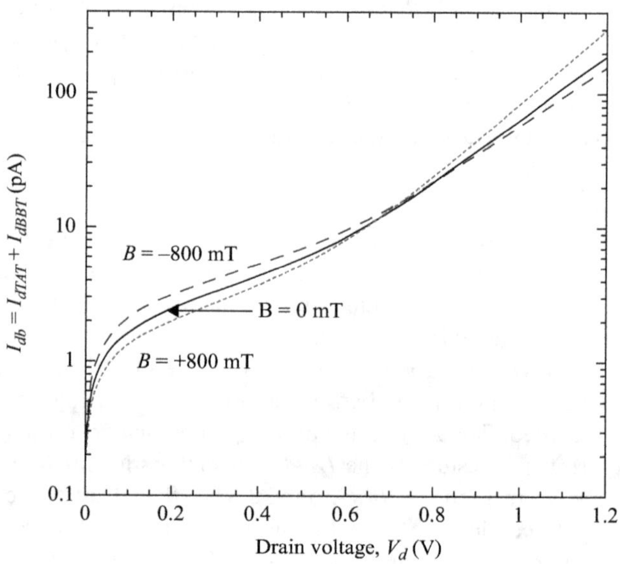

Figure 3.95 Calculated I_{db}–V_d curve for three different B values, –800, 0, and +800 mT. The dashed line corresponds to $B = -800$ mT, the dotted line to $B = +800$ mT, and the continuous line to $B = 0$ mT

energy bandgap E_g, carrier mobility μ, carrier density n, and on the effective phonon energy $\hbar\omega_0$ [3.404], which is related to the TAT process,

$$\hbar\omega_0[eV] = 0.0645\ eV \mp 2.32x10^{-7}s[MPa] \tag{3.87}$$

the effect of the B field on the I_{db}–V_d can be calculated as shown in Figure 3.95.

The I_{db}–V_d curves were calculated for three different $\hbar\omega_0$ energies, 64.186, 64.245, and 64.310 meV, which resembles that of the experimental data shown in Figure 3.94. This suggests that for each value of the applied magnetic field B corresponds a value of $\hbar\omega_0$. It is readily observed that for negative B fields, where current flow deflects toward the left side the I_{dTAT} current increases, while for positive B fields, i.e., current flow deflected toward the right side the I_{dTAT} current decreases. This is an indication of the correlation between asymmetrical magneto-conductance and the mechanical stress.

In order to have a better appreciation of the influence of the B field on the I_d–V_d characteristics at different V_d voltages, a figure of merit called magneto-conductance (MC) coefficient $\Delta I_d = (I_{d@B\neq 0} - I_{d@B=0})/I_{d@B=0}$ is defined. The ΔI_d–B characteristics for $V_d = 0.1$, 0.1, 0.4, 0.8, 0.9 V, and 1.2 V are shown in Figure 3.96. At $V_d =$ 0.1 V in the TAT region, the ΔI_d–B has a negative slope, i.e., the MC has a negative coefficient. At $V_d = 1.2$ V in the BBT region, the ΔI_d–B has a positive slope. i.e., the MC has a positive coefficient. In between TAT and BBT, the MC coefficient ΔI_d evolves from a negative slope to a positive slope passing by a sort of quasi-symmetrical shape. In between TAT and BBT both mechanisms contribute as can be seen by the change of the ΔI_d–B slope around $B = 0$ mT. The modification of the tunneling mechanism caused by the application of an external B field is evident.

So far we have shown the effect of a static B field o the electrical behavior of MOSFET devices, now a $(W/L) = (2\ \mu\text{m}/65\ \text{nm})$ nMOSFET fabricated with a

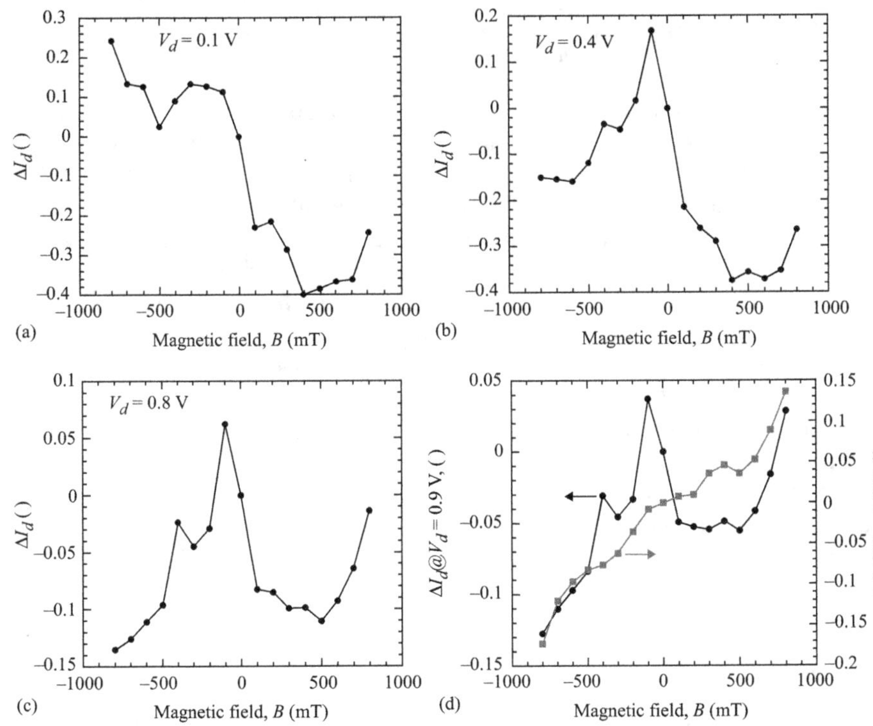

Figure 3.96 Measured ΔI_d coefficient for $V_d = 0.1$, 0.4, 0.8, 0.9, and 1.2 V [3.379]

65 nm technology, and with a $T_{ox}=2$ nm nitride oxide [3.405], is tested under dynamic conditions. For this a square-wave magnetic B field is applied. A steady-state B field induces a two-dimensional space modulation of the electron distribution on the channel, while a transient magnetic field induces an electromotive force along the channel axis. The space modulation of the channel electron concentration is proportional to the magnitude of the B field, while the induced electromotive force voltage V_{em} is proportional to the time-varying rate (dB/dt). Both, the space- and time-modulation of the electron channel concentration induce a surface potential modulation, which causes an image force oxide barrier modulation that results in gate oxide injection and channel current interference. The steady-state results can be validated using the Minimos-NT simulation tool version that accounts for the B field [3.373, 3.374], while the transient effects are validated with a Minimos-NT-Spice macro-model. The experimental results of square-wave B field applied on the MOSFET are shown in Figure 3.97.

The measured gate current I_{gate} shows a positive peak $+\Delta I_g$ for negative transients of B_z and a negative peak $-\Delta I_g$ for positive transients (see Figure 3.97a). A positive transient implies a B_z field going from negative to positive value. The frequency of the pulsed B_z is of 5 Hz with a (dB_z/dt) rate equal to 10 mT/ms. The magnitude of the magnetically modulated gate current peak $+\Delta I_g$ is a function of the magnitude of B_z and the rate (dB_z/dt) as shown in Figure 3.98.

The fitting gives the following model.

$$\Delta I_g = \Delta I_{g0} + \left[C + D\left(\frac{\partial B}{\partial t}\right)\right] \cdot B \tag{3.88}$$

where ΔI_{g0} ranges from 0.18 to 0.27 nA, $C=-0.00156$ nA/mT, and $D=0.00565$ nA-s/T^2 [3.405]. The space- and time-modulation of the channel electron concentration results in a linear modulation of the oxide potential barrier Φ_B at a rate of 1.1×10^{-5} eV/mT. The peak value Δn_s of this electron modulation is a linear increasing function of B as shown in the right axis of Figure 3.98. The larger the magnitude of B the larger the amount of surface electrons that are deflected either to the right or to the left along the width axis. When B switches in time (dB/dt) this Δn_s charge is induced as a gate current $\Delta I_g=(\Delta Q n_s/dt)$. The induced ΔI_g current increases ($+\Delta I_g$) or decreases ($-\Delta I_g$) with respect to its value at $B=0$ because the induced electromotive voltage $V_{em}=-(dB/dt)\cdot(W\cdot L)$ also makes the Δn_s charge to be larger at the drain or source side, which in turns results in an induced $+\Delta I_g$ or a reduced $-\Delta I_g$. Because of the induced V_{em} voltage the effective drain-to-source voltage becomes $V_{deff}=(V_d +/- V_{em})$, which also causes a modulation of the channel current ΔI_d as seen in Figure 3.97b. For this case the induced drain current variation ΔI_d is not only dependent on (dB/dt) but on the magnitude of B as well. In the linear regime the effective drain current I_{deff} is roughly approximated by the following equation:

$$I_{deff} = \frac{1}{\left[1 + \left(\frac{q}{m^*}\cdot \tau \cdot B\right)^2\right]}\left(\frac{W}{L}\right)\mu C_{ox}\left(V_g - V_T\right)V_d \tag{3.89}$$

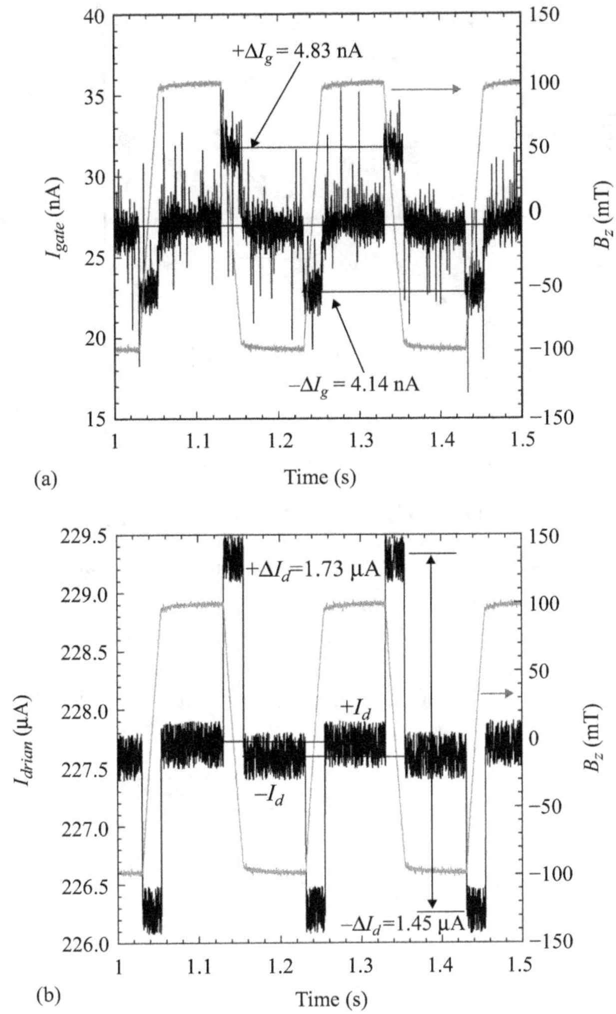

Figure 3.97 (a) Measured I_{gate} current versus time and applied B field on the right axis, (b) measured I_{drain} current versus time and applied B field on the right axis. The MOSFET is biased at $V_d = 1.0$ V and $V_g = 1.0$ V. The B field is pulsed from -100 to $+100$ mT

where τ is the scattering time and V_T is the threshold voltage. Notice from Figure 3.97b that for the steady-state value of $+B_z = 100$ mT I_d increases to $+I_d$, while for $-B_z = 100$ mT I_d reduces to $-I_d$. These variations are in the range of hundreds of nanoamperes and follow the model of (3.89). The I_{deff}–B curvature depends on the scattering time τ, which is a complex function of both V_g and V_d. For the particular case of $V_g = 1.0$ V and $V_d = 0.1$ V τ is equal to 1.029 ps, which is in the range (0.8–1.2 ps)

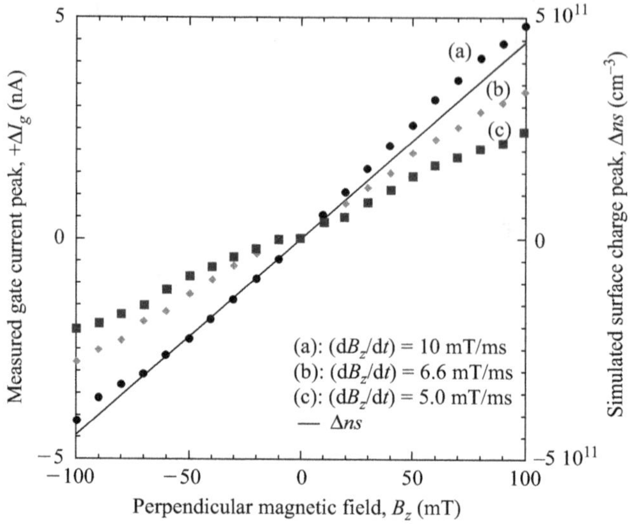

Figure 3.98 Measured $+\Delta I_g$ current versus B_z for three different (dB_z/dt) rates. Closed symbols correspond to experimental results, while continuous lines are Minimos-NT simulations of the modulated surface charge peak Δns

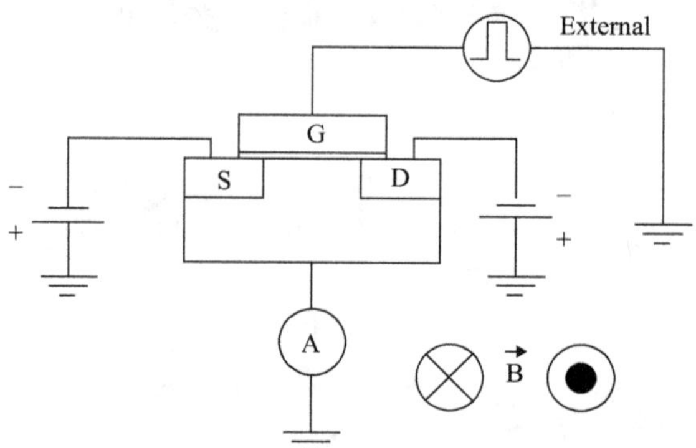

Figure 3.99 Schematic representation of the charge-pumping technique

reported for a two-dimensional electron gas on (111) silicon [3.406]. The use of time-dependent magnetic fields, in conjunction with electrical characterization, serves the purpose to investigate space- and time-dependent conductivity modulation of semiconductor devices, and thus gives some insight of the channel charge correlation with the channel and gate currents when a magnetic field interferes with the device.

Finally, we introduce a modified charge-pumping technique [3.407] that serves the purpose to determine the density of states at the semiconductor–oxide interface of MOSFETs. Figure 3.99 shows the schematic representation of the charge-pumping characterization settings.

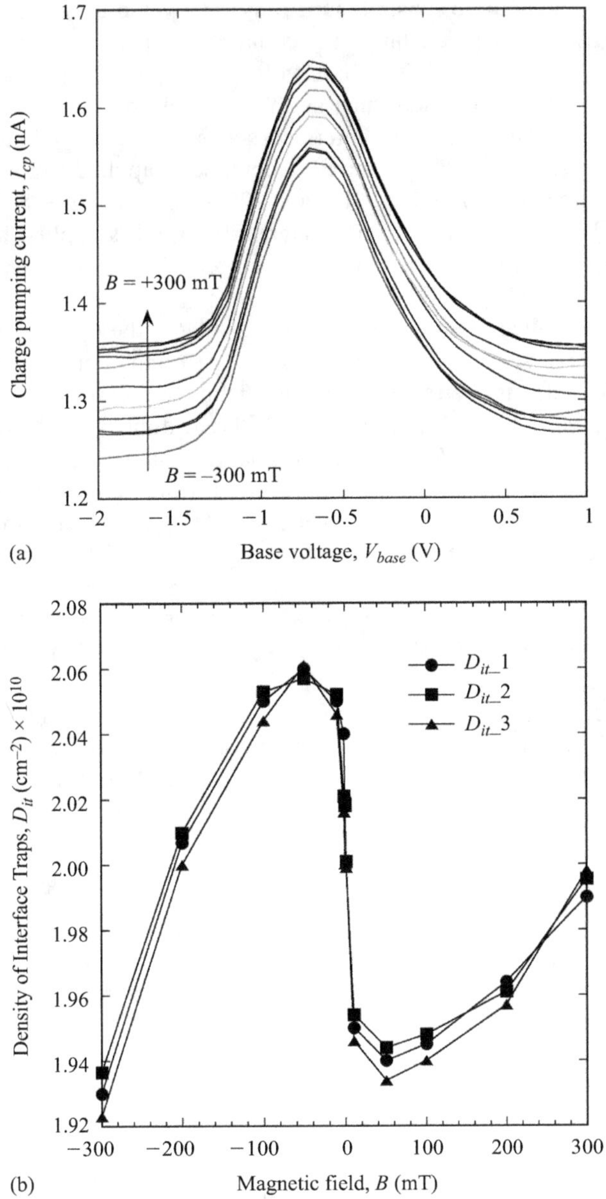

Figure 3.100 *(a) Measured charge-pumping current I_{cp} versus base voltage V_{base} with the B field as a parameter, and (b) extracted density of interface traps D_{it} versus B field for three different areas. D_{it}_1 = (7 × 7), D_{it}_2 = (10 × 10), and D_{it}_3 = (20 × 20) μm^2*

The source and drain to substrate junctions are reverse-biased, while the gate is voltage pulsed between inversion and accumulation operation regimes. The substrate current I_{cp} is measured. The base voltage V_{base} of the gate square voltage sweeps from low to high values such as the channel goes from accumulation to inversion regimes. The density of interface states D_{it} is then extracted from the measured I_{cp} versus V_{base}. The electrical technique is applied to three nMOSFET with different areas (7 × 7), (10 × 10), and (20 × 20) μm^2. The SiO_2 gate oxide $T_{ox} = 60$ nm. However, in this case a magnetic B field is applied parallel to the surface and perpendicular to the gate oxide current. The experimental results are shown in Figure 3.100 [3.408].

The B field applied in the z-axis direction displaces the I_{cp}–V_{base} curves up and down in current as seen from Figure 3.100a. The nonmonotonic response of I_{cp} versus the B field is better appreciated from the extracted D_{it} shown in Figure 3.100b. More than 20 devices with different dimensions have been tested, and all of them show the *"N-shape"* D_{it}–B characteristics.

The magnetic field B into the z-axis exerts a Lorentz force on the I_g current lines moving across the gate oxide, but it also affects the potential distribution at the semiconductor-oxide interface. It is the surface potential, or the energy at the semiconductor–oxide interface, which at the end defines the electrostatics at the interface and thus the I_{cp}–V_{base} characteristics. This is a preliminary approach that is backed by the extracted threshold voltage V_T as a function of the B field as shown in Figure 3.101.

The extracted V_T voltage does not follow an "N-shape." It shows a maximum shift of 1.5 mV at around $B = -100$ mT that coincides with the maximum of D_{it}, and then rolls down for negative B fields as D_{it} does. At positive B fields the V_T

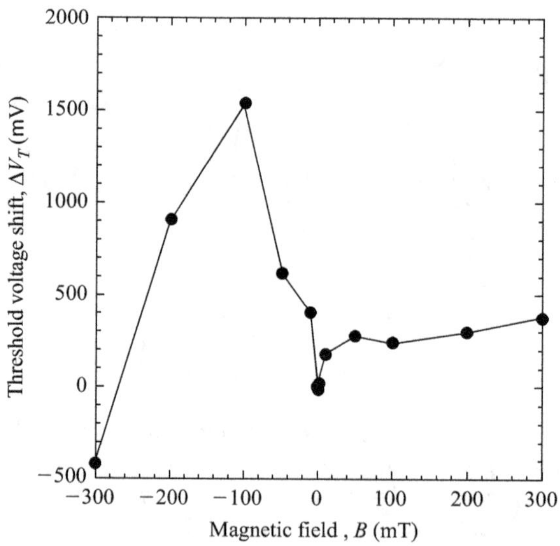

Figure 3.101 Extracted threshold voltage V_T versus B field for a (10 μm × 10 μm) nMOSFET biased at $V_d = 0.1$ V

increases but at a slower pace when compared to D_{it}. The magneto-modulation of the threshold voltage has to do with a change in the surface potential, which backs up the hypothesis that the D_{it}–B "N-shape" is related to the surface potential magneto-modulation rather to a potential effect of the B field on the interface states distribution. However, we have not ruled out any other potential explication as this work is in its preliminary stage.

3.4 Conclusions

As a reference and starting point we have done a shallow analysis of the basic electrical characterization techniques and models for conventional FET technologies. The electrical characteristics of a conventional SiO_2 0.5 μm devices fabricated with local oxidation for device insulation technique, have been compared to an nitride/oxynitride stacked SiO_2 65 nm fabricated with shallow trench device insulation techniques, with a metal-gate high-k 28 nm strained device technology, and with a 24 nm FinFET device technology. The supply voltage has a progressive reduction from 3.3, 1.2, 1.0, and 0.8 V. The superior characteristics of high-k, metal-gate, strained FET technologies are demonstrated by the current drive capability shown in Figure 3.102.

Different basic compact device models were also reviewed, and emphasis was put on the Si2 and the CMC, which accomplished three industrial models: (1) BSIM, (2) PSP, and (3) HiSIM. For small-signal performance and low overdrive voltages, the BSIM4 is accurate enough, which makes it a good option for LNAs, mixers, and other low-power circuits. For correct linearity estimation the PSP

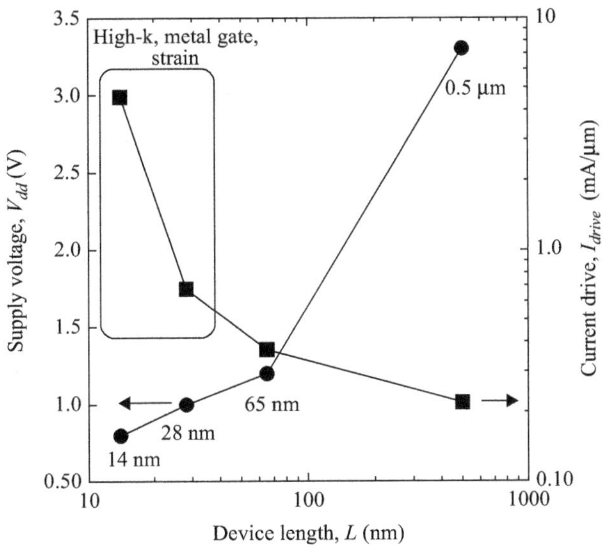

Figure 3.102 Supply voltage V_{dd} and current drive capability I_{drive} for 0.5 μm, 65 nm, 28 nm, and 14 nm FinFET technologies

model seems to be a reliable choice. For large signal analysis, HiSIM is a good model candidate.

The use of combined DC-RF measurements were introduced as an alternative to extract parasitic components with using a single one-channel-length FET device. The techniques showed to be useful for extracting capacitive and resistive parasitic components from DC up to 50 GHz.

The effect of the drain-bulk reverse-biased BTBT parasitic current was examined in terms of its impact on the FET RF performance. It was shown that this parasitic current degrades the output conductance of RF FET devices, which prompts from adding this parasitic current to the RF FET models.

Nonconventional bias conditions have been also tested to extract additional information from FET devices. This is the case of a floating-gate n-type FET from a metal-gate high-k 28 nm strained technology (Figure 3.103).

A memory effect is measured when the forward drain voltage is swept from low to high and back from high to low values. Three regions, *I*, *II*, and *III* are observed when sweeping the V_d voltage from 0 to -1.2 V. In general, any charge accumulation near the transistor channel leads to threshold voltage shifts. Thus when the density of trapped charge changes during a measurement, a hysteresis effect is observed. If enough charge is trapped in a short period of time a negative differential resistance effect (NDR) is observed as the one in region II. The floating gate voltage is measured by setting a zero gate current $I_g = 0$ A as a voltmeter [3.59]. As a reference example, as similar with a flash memory, an abrupt current roll down occurs when a floating quantum point charges/discharges with a control

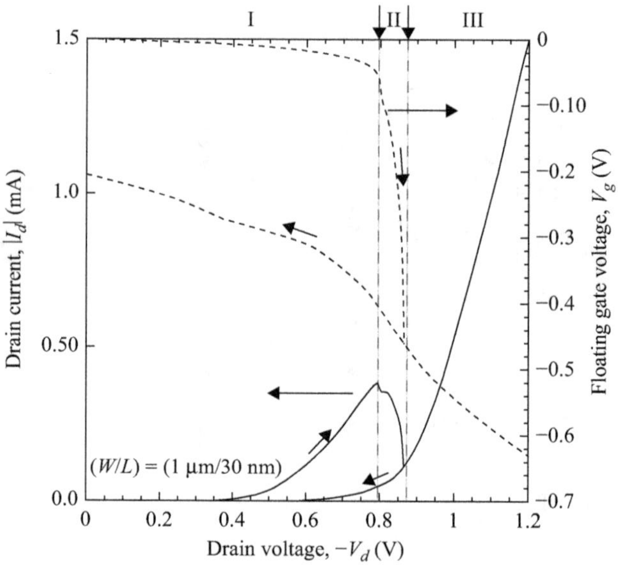

Figure 3.103 Measured I_d current versus forward bias $-V_d$ voltage for $V_s = V_b = 0.0$ V, with a current gate source $I_g = 0$ A

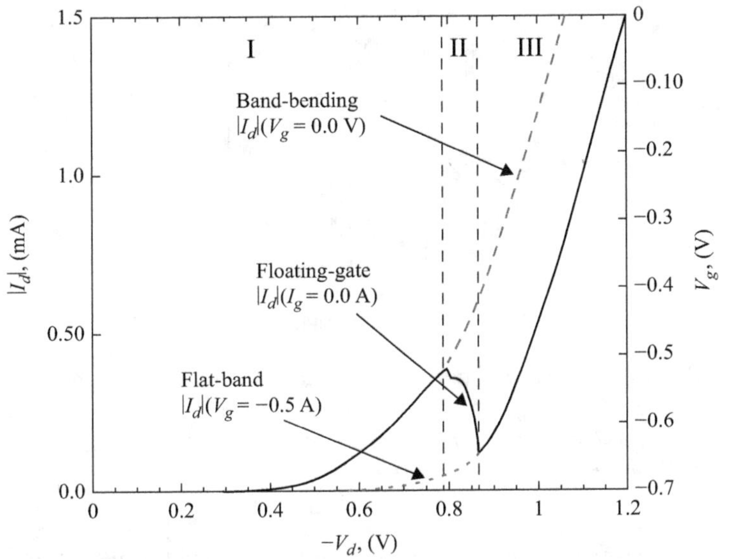

Figure 3.104 *Continuous I_d curve corresponds to measured floating gate condition ($I_g = 0.0$ A), dashed line corresponds to band-bending condition ($V_g = 0.0$ V), and dotted line to flat-band condition ($V_g = -0.5$ V)*

gate [3.409]. The remarkable difference is the number of trapped electrons needed to block the conduction channel. The NDR effect observed in a high-impedance or floating-gate MOSFET is explained as treating the drain, source, and bulk contacts of an n-MOSFET as the emitter, collector, and base of a Bipolar Junction Transistor (BJT), respectively.

The dashed curve in Figure 3.104 is the measured I_d with all other MOSFET terminals (source, gate, and bulk) grounded. In this case, current start to flow at the channel surface (region I) and in the bulk at higher V_d voltages (region III) since the potential barrier is lower at the substrate–oxide interface due to undesired trapped charges and the built-in potential between gate and bulk. The dotted curve is measured under flat-band condition ($V_g = -0.5$ V) so that only a volumetric, under the semiconductor–oxide interface, current flows (region III). A BJT at the surface (region I) works in parallel with another bulk BJT (region III). When the gate is floating (continuous line), the FET transistor switches from the surface BJT to the bulk BJT triggering the NDR effect (region II). The V_g voltage is a measure of these facts. In region I the energy bands remain bent ($V_g = 0.0$ V), while in region II, V_g drops abruptly, indicating a surface-band straightening process, which continues until the flat-band condition is reached.

The alternative way of measuring or characterizing a FET device, in this case using a floating or high-impedance gate condition, can be explained by assuming

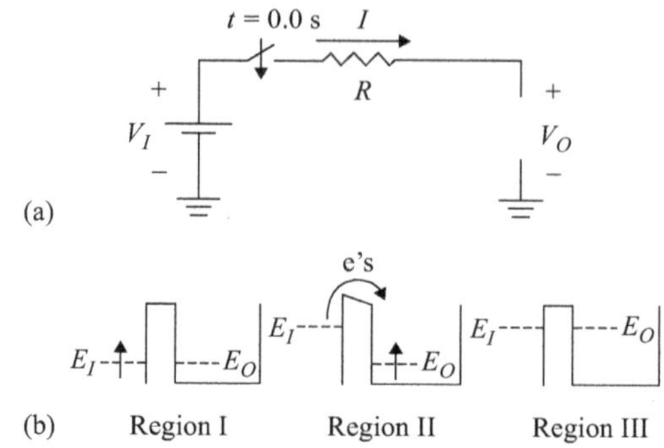

Figure 3.105 (a) Schematic diagram of a floating port V_o. (b) Band diagrams of the floating port with quasi-Fermi level E_o separated by a potential barrier from the input port with quasi-fermi level E_I for each of the three operation regions

that the system has reached equilibrium before carrying out the measurements. This means that all contacts (source, drain, gate, and bulk) are grounded, and after some time the gate must be also at zero. The relaxing process is accelerated when the gate is grounded and disconnected to release it from any charge excess. Figure 3.105 shows the schematic representation of a floating gate FET.

When the switch is closed at $t = 0$ s, charges start to flow across the resistance and start to build up on the other side, reducing the electric field inside the circuit down to its minimum value according to Gauss law. This happens almost instantaneously and very quickly $V_I = V_o$. In contrast, the charging process is much slower if a potential barrier separates the input reservoir from the floating port, both with quasi-Fermi levels E_I and E_o, respectively (see Figure 3.105b). Initially, when E_I starts raising as a result of a negative potential applied to that contact, very few electrons can be thermally activated to get over the barrier. As a consequence, E_I remains unchanged and a potential drop between input and floating port is established. This situation corresponds to region I in Figure 3.104. As E_I is further raised, electrons gain enough energy to surpass the barrier at a larger rate and the charging process of the floating port is now evident, which corresponds to region II. This happens until $E_I = E_o$, a moment when the floating port is considered to be full of charges (region III). This nonconventional bias condition serves to study the charging and discharging process at the semiconductor-oxide system.

The BTI mechanism, which is one of the major concerns regarding the long-term performance of FETs, has been also reviewed. BTI predicts the ability of FET devices to withstand any change or shift in threshold voltage, mobility, current drive, and transconductance capability. The density of states D_{it} has been correlated to threshold voltage, where the D_{it} is in turn correlated to the semiconductor–oxide interface. Therefore, with the incorporation of high-k dielectrics, the physical and electronic quality of the high-k/Si interface reduces, which makes advanced FET prone to an increase in degradation.

From a physics-based approach the degradation by BTI initially starts by the continuous trapping of inverted charge, which locally modifies the energy gap of the gate oxide, and thus creating interface traps. From an atomistic point of view the continuous breaking of initially passivated S–H chemical bonds at the oxide–semiconductor interface does this. It has been suggested, due to the high band offset (~5 eV), that La_2O_3, in comparison to HfO_2, could help to greatly minimize PBTI. For n-type FETs, because of La_2O_3 large band offset for electrons in the conduction band of Si as compared to HfO_2, the gate-leakage current density reduces significantly. The reduction of PBTI and gate leakage current is obtained even with $EOT = 1$ nm.

Not only the band offset of the gate oxide matters for NBTI degradation, but the metal gate work function as well. The energy barrier seen by electrons coming from the metal electrode and being injected into the high-k layer by a negative bias (gate injection condition) depends on the metal work function. The replacing of a gate electrode from a low to high work function material (aluminum to ruthenium, for instance) reduces the gate injection by up to two orders of magnitude. But also leads to thinner EOT, when compared to the aluminum gate case, which is indirect evidence that any IL (with low-k) developed at the metal/high-k interface could be minimized. In this sense, the use of inert metal electrodes for high-k gate oxide FETs has a great potential to further scale down the *EOT* of these devices. The chemical stability of metal gates with the underlying high-k oxide is also of the outmost importance for the reproducible behavior of flat-band voltage and stable threshold voltage for FET devices. There is a wide variety of metal gates that can be used for the stacked metal/high-k systems, among them are Al, TaN, Ti, W, and Ru. The thermodynamic stability of the meta electrodes when subjected to a PMA is monitored by the observation of a displacement of the original *C–V* characteristics after the PMA. From all these metals, Tungsten (W) is the more stable in terms of producing similar *C–V* characteristics after the device has followed a pre-metallization and post-deposition annealing processes. The W/La_2O_3–IL/pSi system structure in an n-type FET has shown a higher endurance during BTI degradation.

It is also of a high importance to identify the dominant current components, either electrons or holes, in the total gate leakage current (I_g). The charge type identification is a powerful tool to correlate precision conduction mechanisms during BTI stressing of the device and therefore, enabling a better estimation of lifetime predictions for advanced FET devices.

The possible replacement of HfO_2 by La_2O_3 for nodes below 14 nm (logic technology) will depend on the ability of La-based and La-silicate IL (sometimes combined with HfO_2) high-k materials to aggressively scale down *EOT* below the 0.6 nm regime. However, additional processing of p-type FET devices and the extraction of their performance and reliability, when metal gates with high work functions are used, is required in order to consider La_2O_3 introduction into any CMOS process.

The hot carrier (HC) mechanism has proven to be very persistent as technology progresses and scales down from the micrometer to the nanometer scale. Power supply voltage will decrease for devices scaling down to the sub-14-nm regime, which apparently will reduce HC, but it will also lead to an increase of ballistic and quasi-ballistic transport, and carrier-to-carrier scattering, which in turns may exacerbate HC effects.

From a conventional approach for non-nanometric devices, the carriers gain kinetic energy when moving through a high electric field close to the drain region. When their mean energy is larger than the thermal energy at the lattice temperature, their kinetic energy is assumed to be distributed with a thermal-like Maxwellian distribution at an effective temperature higher than that of the lattice, then the name "hot" carrier. Such a distribution is on a steady state with the local electric field, and thus its effective temperature depends on the electric field.

For nano-scaled devices, quasi-ballistic transport, at least near the drain region, becomes more substantial, and then a new model approach named "Energy-Driven Model" emerges in contraposition to the "Lucky Electron Model" used for non-nanometric devices. Thus, the assumption of a Maxwellian-like energy distribution in steady state, with the local electric field, progressively breaks down as the size of the high electric field region scales below 100 nm, and as the technology power supply also scales down. Quasi-ballistic transport in the high field region results in a rather shallow carrier EDF up to the total energy available in the high field region and then a steeper tail at higher energy occurs. *This energy "knee" leads to a hot carrier damage rate that is dependent on the total available energy, not the peak electric field.*

The carrier-to-carrier scattering is an event through which carriers moving along the FET channel gain energy. Monte Carlo or other simulation techniques have been employed by many authors to predict that at drain voltages below about 3 V, electrons heat by carrier-to-carrier scattering should dominate the high energy tail of the EEDF used to calculate impact ionization. There is an experimental evidence, for n-type FETs, that the HC damage rate follows a quadratic dependence on the drain current over much of the gate voltage range, which implies that the knee of the electron-to-electron scattering tail does drive the damage rate.

The effect of temperature on the mean free path, a critical factor for HC, decreases with scaling, and becomes small below about $L = 50$ nm. Thus for nano-scaled devices the HC rate temperature dependence will be driven by other parameters. Basically, as the ballistic transport becomes dominant, the dominance of phonon scattering vanishes. This explains, up to some extent, the evolution from a negative to a positive temperature dependence of the impact ionization with scaling.

A simple heat diffusion description breaks down for self-heating effects at scales smaller than ~20–30 nm. This is because the active device volume decreases, which make it more thermally isolated from the substrate that induces an enhanced and localized self-heating effect. In nano-scaled devices phonon transport is ballistic, where optical phonons are generated by HC and have low group velocity, which tend to accumulate locally. Since the local phonon EDF is not thermal, the concept of "lattice temperature" no longer applies, contrary to larger devices where heat flow is dominated by the diffusion of acoustic phonons. Self-heating (SH) then exacerbates HC degradation. For a device in static operation condition SH and HC there may pose a serious reliability issue for analog applications.

Moreover, HC DC stress is regularly done in static on condition under elevated voltages. The result of this is an excessive local SH, leading to an unknown and unpredictable acceleration of HC, and potential activation of the BTI degradation mechanism. These considerations may necessitate methodology changes for HC estimation and qualification. Degradation and reliability data taken from DC stress conditions need to be corrected for SH, which may not be a trivial task to do accurately. Therefore, DC characterization methods should be modified to account for AC HC stress using on-chip pulse generators or oscillators.

The final outcome of degradation: filamentary conduction has been analyzed as that corresponding to a breakdown or forming event of a thin oxide layer. The dielectric breakdown of the gate oxide may cause partial or total loss of the transistor function as well as an increase in standby power consumption. The oxide breakdown has been related to the formation of a percolation pathway through the dielectric film, which is then correlated to the formation of a chain of traps, defects, vacancies, metal atoms, etc.

The current runaway and the energy dissipation dynamics occurring during the percolation damage is a result of a random distribution of defects inside the oxide layer, where the rough estimation of the defect size, involved in the formation of the breakdown path, is in the range of 0.5–1.5 nm. As shown in Figure 3.55b, in some cases, the breakdown spots are visible through an optical microscope.

A wide variety of percolation models have been developed, among them some based on the memristor device concept that deals with the reversible dielectric breakdown. Other models have been served as the basis for single or multifilamentary conduction, such as ReRAMs, electronic synapses, and neuromorphic circuits.

The dielectric breakdown has been conventionally defined as the gradual increase of the dielectric conductance, with a change that can be abrupt depending on the oxide thickness, device area, and stress conditions. However, the gradual loss of insulating capacity is no longer valid for ultra thin oxides (<5 nm) since the distinction between insulating and conducting is not well defined in this gate oxide thickness range.

The oxide breakdown has been subdivided into three SBD and another three HBD modes. SBD is classified into analog SBD (A-SBD), characterized by random noise, and digital SBD (D-SBD) characterized by a random telegraph noise. The third SBD type is the one called MB, which is characterized by a lower current than that corresponding to a typical SBD. On the other edge HBD is correlated with and without thermal effects, i.e., with and without lateral propagation of damage. The third type of HBD is referred as nonlinear HBD (NL-HBD), which is an *I–V* curve similar to that of the high-HBD, but with a reduction of its slope for voltages exceeding some threshold. The SBD, which can progress with time is called PBD. The PBD is a major reliability issue for hyper-thin oxides (<2 nm). For these hyper-thin oxides the tunneling current makes difficult to identify HBD. As for the SBD it becomes harder to identify for larger areas and thinner oxides.

A crucial problem that arises from SBD and HBD is whether a transistor remain functional or not after a gate oxide breakdown event. The answer is not simple, even if the device is able to survive after SBD, there may still be some impact of subsequent stress on the already broken down device. A digital CMOS circuit, for instance, could tolerate many HBDs without affecting its logical functionality. This means that depending on the particular application, reliability criteria may be, in some extent, relaxed.

Within the approach of RS, the first oxide breakdown event upon the application of electrical stress is called electroforming. The electroforming, from the statistical point of view, is fully consistent with the percolation theory of dielectric breakdown. After this, the application of increasing and decreasing voltage sweeps leads to a pinched hysteretic behavior of the *I–V* characteristics, which has been often interpreted in terms of the memristor concept. The memristor concept used in combination with rectifiers, has been used to construct a family of electronically reconfigurable circuit elements.

Finally, we would like to close out the conclusions by mentioning that a relatively weak magnetic field *B* ($\leq$1 T) can be applied when doing electrical characterization as an auxiliary tool. This is the case when measuring the gate oxide tunneling current (I_g). The gate oxide current density is supposed to flow mostly perpendicular to the channel plane. If there were vector components off the normal to the channel plane, a pure electrical measurement would not be able to detect them. According to the magneto transport theory a *B* field applied parallel to a current flow would result in a null cross product vector, thus resulting in a null effect of the *B* field on the I_g current. However, this is not the case as shown in Figure 3.106. An evident effect of the *B* field on the I_g current is observed. The

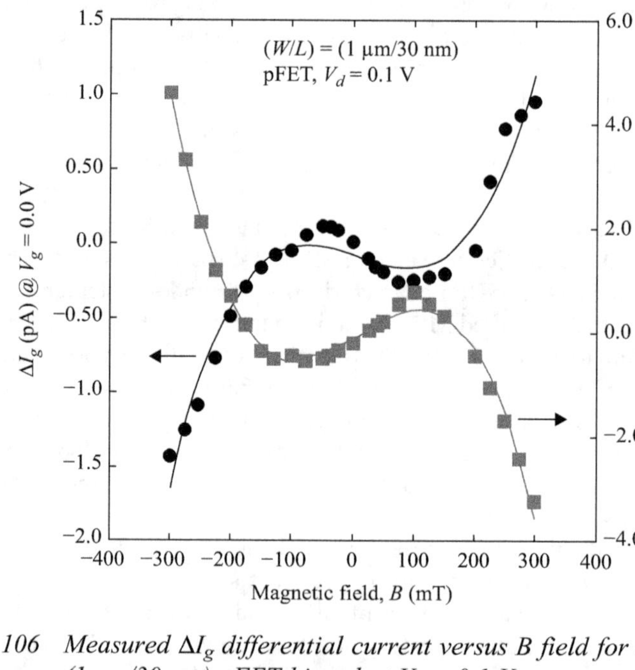

Figure 3.106 Measured ΔI_g differential current versus B field for a (W/L) = (1 μm/30 nm) pFET biased at V_d = 0.1 V

measured differential gate current $\Delta I_g = (I_{g@B\neq 0} - I_{g@B=0})$ versus the B field is an asymmetric function.

The effect shown in Figure 3.106 can be interpreted as if were gate current density vectors off the normal component to the channel plane. The B field, applied perpendicular to the channel plane, produces a vector cross product that measures this vector components as ΔI_g. The B field also affects the energy distribution at the semiconductor–oxide interface, which may be also another factor influencing on ΔI_g. Overall, the B field, applied at any direction, serves as an auxiliary tool to investigate carrier transport effects that a simple electrical current characterization technique does not.

Acknowledgment

Miranda acknowledges Prof. J. Suñé and Mr J. Blasco from the Universitat Autònoma de Barcelona, Spain, for their support in the realization of this manuscript.

References

[3.1] Sheu, B. J., Scharfetter, D. L., Ko, P.-K., and Jeng, M.-C., "BSIM: Berkeley short-channel IGFET model for MOS transistors". IEEE Journal of Solid-State Circuits 1987;SC-22(4):558–566.

[3.2] Cheng, Y. and Hu, C., "MOSFET modeling & BSIM3 User's guide". Kluwer Academic Publishers, 2002, ISBN: 0-306-47050-0.

[3.3] Cheng, Y., Chen, K., Imai, K., and Hu, C., "A unified MOSFET channel charge model for device modeling in circuit simulation". IEEE Transactions on Computer-Aided Design of Integrated Circuits and Systems 1998;17 (8):641–644.

[3.4] Fang, F. F. and Fowler, A. B., "Hot electron effects and saturation velocities in silicon inversion layers". Journal of Applied Physics 1970;41 (4):1825–1831.

[3.5] Huang, J. H., Liu, Z. H., Jeng, M. C., Ko, P. K., and Hu, C., "A physical model for MOSFET output resistance". In: Proceedings of the IEDM, 1992, pp. 569–572.

[3.6] Chung, J. E., Ko, P.-K., Hu, C., "A model for hot-electron-induced MOSFET linear-current degradation based on mobility reduction due to interface-state generation". IEEE Transactions on Electron Devices 1991; 38(6):1362–1370.

[3.7] Hurkx, G. A. M., Klaassen, D. B. M., and Knuvers, M. P. G., "A new recombination model for device simulation including tunneling". IEEE Transactions on Electron Devices 1992;39(2):331–338.

[3.8] Nagel, L. W and Pederson, D. O., SPICE (Simulation Program with Integrated Circuit Emphasis), Memorandum (ERL-M382), University of California, Berkeley, 1973.

[3.9] Leblebici, D. and Leblebici, Y., "Fundamentals of high-frequency CMOS analog integrated circuits". Cambridge University Press, 2009, ISBN: 978-511-53976-3.

[3.10] http://www.keysight.com/en/pc-1297149/ic-cap-device-modeling-software-measurement-control-and-parameter-extraction?cc=US&lc=eng.

[3.11] http://www.silvaco.com/tech_lib_TCAD/simulationstandard/2004/apr/a2/a2.html.

[3.12] http://www-device.eecs.berkeley.edu/bsim/.

[3.13] Enz, C. C., Krummenacher, F., and Vittoz, E. A., "An analytical MOS transistor model valid in all regions of operation and dedicated to low-voltage and low-current applications". Special Issue of the Analog Integrated Circuits and Signal Processing Journal on Low-Voltage and Low-Power Design 1995;8:83–114.

[3.14] Mavredakis, N. and Bucher, M., "Inversion-coefficient based design of RF CMOS low-noise amplifiers". In: 13th IEEE International Conference on Electronics, Circuits and Systems ICECS, 2006, pp. 74–77.

[3.15] Tan, L., Hakim, M. M. A., Connor, S., *et al.*, "Characterisation of CMOS compatible vertical MOSFETs with new architectures through EKV parameter extraction and RF measurement". In: 10th International Conference on Ultimate Integration of Silicon 2009, ULIS, 2009, pp. 165–168.

[3.16] Fadhuile, F., Taris, T., Deval, Y., Enz, C., and, Belot, D., "Design methodology for low power RF LNA based on the figure of merit and the inversion coefficient". In: 21st IEEE Conference on Electronics, Circuits and Systems (ICECS), 2014, pp. 478–481.

[3.17] http://ekv.epfl.ch/model.

[3.18] Antonopoulos, A., Bucher, M., Papathanasiou, K., *et al.*, "CMOS small-signal and thermal noise modeling at high frequencies". IEEE Transactions on Electron Devices 2013;60(11):3726–3733.

[3.19] Bucher, M., Lallement, C., and Enz, C., "An efficient parameter extraction methodology for the EKV MOST model". In: Proceedings of the 1996 IEEE International Conference on Microelectronics Test Structures, 1996, Vol. 9, pp. 145–150.

[3.20] Gildenblat, G., Li, X., Wu, W., *et al.*, "PSP: an advanced surface-potential-based MOSFET model for circuit simulation". IEEE Transactions on Electron Devices 2006;53(9):1979–1993.

[3.21] Thakker, R. A., Patil, M. B., and Anil, K. G., "Parameter extraction for PSP MOSFET model using hierarchical particle swarm optimization". Engineering Applications of Artificial Intelligence 2009;22:317–328.

[3.22] "The PSP model in RF CMOS design". White paper published in http://www.fujitsu.com/downloads/MICRO/fma/formpdf/rfpdk_0309.pdf.

[3.23] Scholten, A. J., Smit, G. D. J., De Vries, B. A., *et al.*, "The new CMC standard compact CMOS model PSP: advantages for RF applications". In: IEEE Radio Frequency Integrated Circuits Symposium, 2008, pp. 247–250.

[3.24] http://www.nxp.com/models/simkit/mos-models/model-11.html.

[3.25] https://www.synopsys.com/tools/Verification/AMSVerification/CircuitSimulation/HSPICE/Pages/default.aspx.

[3.26] Miura-Mattausch, M., Sadachika, N., Navarro, D., *et al.*, "HiSIM2: advanced MOSFET model valid for RF circuit simulation". IEEE Transactions on Electron Devices 2006;53(9):1994–2007.

[3.27] Si2 and CMC: http://www.si2.org/.

[3.28] http://psp.ewi.tudelft.nl/.

[3.29] http://home.hiroshima-u.ac.jp/usdl/HiSIM2/.

[3.30] Arora, N., "MOSFET modeling for VLSI simulation". In: International Series on Advances in Solid State Electronics and Technology, World Scientific Publishing Co., 2007, Chapter 9. ISBN: 13 978-981-256-862-5.

[3.31] Grabinski, W. and Gneiting, T. (Eds.), "Power/HV MOS devices compact modeling". Springer, 2010, Chapter 2. ISBN: 978-90-481-3045-0.

[3.32] Zhang, T., "Low power CMOS RF-transceiver circuits for K-band wireless localization systems". PhD dissertation, Berlin University, Germany, 2014.

[3.33] Torres-R., E., Torres-T., R., Valdovinos-F., G., and Gutiérrez-D., E. A., "A method to determine the gate bias-dependent and gate bias-independent component of MOSFET series resistance form S-Parameters". IEEE Transactions on Electron Devices 2006;53(3):571–573.

[3.34] Torres-R., E., "Characterization and modeling of bulk losses for CMOS high-frequency applications". PhD dissertation, INAOE, Mexico, 2003.

[3.35] Biesemans, S., Hendriks, M., Kubicek, S., and De Meyer, K., "Practical accuracy analysis of some existing effective channel length and series resistance extraction methods for MOSFETs". IEEE Transactions on Electron Devices 1998;45(6):1310–1316.

[3.36] Lim, K. Y. and Zhou, X., "A physically based semi-empirical series resistance model for deep-submicron MOSFET space I–V modeling". IEEE Transactions on Electron Devices 2000;47(6):1300–1302.

[3.37] Biesemans, S., Kubicek, S., and De Meyer, K., "Test structure to investigate the series resistance components of source/drain structure". IEEE Electron Device Letters 1997;18(10):477–479.

[3.38] Hwang, C. Y., Kuo, T. C., and Woo, J. C. S., "Extraction of gate dependent source/drain resistance and effective channel length in MOS devices at 77 K". IEEE Transactions on Electron Devices 1995;42(10):1863–1865.

[3.39] Liou, J. J. and Schwierz, F., "RF MOSFET: recent advances, current status and future trends". Solid-State Electronics 2003;47:1881–1895.

[3.40] Torres-T., R., Murphy-A., R., and Decoutere, S., "MOSFET bias dependent series resistance extraction from RF measurements". Electronics Letters 2003;39:1476–1478.

[3.41] Aur, S., Yang, S.-H., and Tran, T., "MOSFET asymmetry and gate-drain source overlap effects on hot carrier reliability". In: Proceedings IEEE 43rd Annual International Reliability Physics Symposium, 2005, pp. 714–715.

[3.42] Lee, W.-C. and Hu, C., "Modeling CMOS tunneling currents through ultrathin gate oxide due to conduction- and valence-band electron and hole tunneling". IEEE Transactions on Electron Devices 2001;48(7):1366–1373.

[3.43] Solomon, P. M., Frank, D. J., Jopling, J., *et al.*, "Tunnel current measurements on P/N junction diodes and implications for future device design". In: IEDM Technical Digest, 2003, pp 233–236.

[3.44] Sasse, G. T., Kuper, F. G., and Schmitz, J., "MOSFET degradation under RF stress". IEEE Transactions on Electron Devices 2008;55(11): 3167–3174.

[3.45] Vemuru, S. R., "Layout comparison of MOSFETs with large W/L ratios". Electronics Letters 1992;28(25):2327–2329.

[3.46] Torres-R., E., Torres-T., R., and Gutiérrez-D., E. A., "Analysis of the impact of the drain-junction tunneling effect on a microwave MOSFET from S-parameter measurements". Solid-State Electronics 2009; 53:145–149.

[3.47] Yamaguchi, K, Teshima, T., and Mizuta, H., "Numerical analysis of an anomalous current density assisted by locally generated deep traps in pn junctions". IEEE Transactions on Electron Devices 1999;46(6): 1159–1165.

[3.48] Xiang-W., J. and Shu-S., L., "Quantum confinement effects and source-to-drain tunneling in ultra-scaled double-gate silicon n-MOSFETs". Chinese Physics B2012;21(2):027304-1–027304-8.

[3.49] SOda, S. and Ferry, D., "Silicon nanoelectronics". Taylor & Francis Group, 2006, Chapter 3.

[3.50] www.globaltcad.com.

[3.51] Wang, J. and Lundstrom, M., "Does source-to-drain tunneling limit the ultimate scaling of MOSFETs?". In: Proceedings of the IEDM, 2002, pp. 707–710.

[3.52] Kawaura, H., Sakamoto, T., and Baba, T., "Observation of source-to-drain direct tunneling in 8 nm gate electrically variable shallow junction metal-oxide- semiconductor field-effect transistors". Applied Physics Letters 2000;76(25):3810–3812.

[3.53] Muci, J., Lugo-Muñoz, D. C., Latorre-Rey, A. D., *et al.*, "A new-integration-based procedure to separately extract series resistance and mobility degradation in MOSFETs". Semiconductor Science and Technology 2009;24(10):1–6.

[3.54] Torres-T., R., Murphy-A., R., and Torres-J., A., "An improved substrate loss model to determine MOSFET drain, source and substrate elements". Microwave and Optical Technology Letters 2004;43:126–130.

[3.55] Hauser, J. R., "Small signal properties of field effect devices". IEEE Transactions on Electron Devices 1965;12:605–618.

[3.56] Taur, Y. and Ning, T. K., "Fundamentals of modern VLSI devices". Cambridge University Press, 1998, pp. 94–95.

[3.57] Vega-G., Victor, H., Gutiérrez-D., Edmundo, A., and Guarin, F., "A negative differential resistance effect implemented with a single Mosfet from 375 K down to 80 K". In: 43rd ESSDERC, Bucharest, Romania, 2013.

[3.58] Gutiérrez-D., E. A., Deen, M. J., and Claeys, C. L., "Low temperature electronics, physics, devices, circuits and applications". Academic Press, New York, 2001, p. 112.

[3.59] Vega González, V. H., "Analysis of quantum electromagnetic effects in low dimensional MOSFETs". PhD dissertation, INAOE, Mexico, 2014.

[3.60] Sun, J. P., Haddad, G. I., Mazumder, P., and Schulman, J. N., "Resonant tunneling diodes: models and properties". Proceedings of the IEEE 1998;86:p. 64.

[3.61] Kimizuka, N., Yamaguchi, K., Iniai, K., *et al.*, "NBTI enhancement by nitrogen incorporation into ultrathin gate oxide for 0.10-um gate CMOS generation". In: IEEE VLSI Technology Symposium, 2000, pp. 92–93.

[3.62] Schroder, D. K., "Semiconductor material and device characterization". IEEE Press and Wiley Interscience 2006, 3rd ed., p. 711.

[3.63] Mahapatra, S. (Ed.), "Fundamentals of bias temperature instability in MOS transistors". Springer Series in Advanced Microelectronics 139, 2016.

[3.64] Schroder, D. K., "Negative bias temperature instability: what do we understand?" Microelectronics Reliability 2007;47:841–852.

[3.65] Campbell, J. P., Lenahan, P. M., Cochrane, C. J., Krishnan, A. T., and Krishnan, S., "Atomic-scale defects involved in the negative-bias temperature instability". In: IEEE Transactions on Device and Materials Reliability, Vol. 7, No. 4, December 2007, pp. 540–557.

[3.66] Stathis, J. H. and Zafar, S., "The negative bias temperature instability in MOS devices: a review". Microelectronics Reliability 2006;46:270–286.

[3.67] Huard, V., Denais, M., and Parthasarathy, C., "NBTI degradation: from physical mechanisms to modeling". Microelectronics Reliability 2006;46:1–23.

[3.68] Alam, M. A. and Mahapatra, S. A., "Comprehensive model of PMOS NBTI degradation". Microelectronics Reliability 2005;45:71–81.

[3.69] Mistry, K., Allen, C., Auth, C., *et al.*, "A 45 nm logic technology with high-k metal gate transistors, strained silicon, 9 Cu interconnect layers, 193 nm dry patterning, and 100% Pb-free packaging". In: Proceedings IEEE International Electron Devices Meeting, 2007, pp. 247–250.

[3.70] Greene, B., Narasimha, S., Cai, J., *et al.*, "High performance 14 nm SOI FinFET CMOS technology with 0.0174 μm^2 embedded DRAM and 15 levels of Cu metallization". In: Proceedings of the IEEE International Electron Devices Meeting, 2014, pp. 381–383.

[3.71] Natarajan, S., Agostinelli, M., Akbar, S., *et al.*, "A 14 nm logic technology featuring 2nd-generation FinFET, air-gapped interconnects, self-aligned double patterning and a 0.0588 μm^2 SRAM cell size". In: Proceedings of the IEEE International Electron Devices Meeting, 2014, pp. 371–373.

[3.72] Bedoya, C., "MOCVD of lanthanum oxides from $La(tmhd)_3$ and $La(tmod)_3$ precursors: a thermal and kinetic investigation". Chemical Vapor Deposition 2006;12: 46–53.

[3.73] Hattori, T., Yoshida, T., Shiraishi, T., *et al.*, "Composition, chemical structure, and electronic band structure of rare earth oxide/Si(100) interfacial transition layer". Microelectronics Engineering 2004;72:283–287.

[3.74] Wu, Y. H., Yang, M. Y., Chin, A., Chen, W. J., and Kwei, C. M., "Electrical characteristics of high quality La_2O_3 gate dielectric with equivalent oxide thickness of 5 Å". Electron Device Letters 2000;21(7):341–343.

[3.75] Iwai, H., Ohmi, S., Akama, S., *et al.*, "Advanced gate dielectric materials for sub-100 nm CMOS". In: IEDM Technical Digest, 2002, pp. 625–628.

[3.76] Sanghun J., Walker, F. J., Billman, C. A., McKee, R .A., and Hyunsang H., "Epitaxial $SrTiO_3$ on silicon with EOT of 5.4 Å for MOS gate dielectric applications". In: IEDM Technical Digest, 2002, pp. 955–957.

[3.77] Yamada, H., Shimizu, T., Kurokawa, A., Ishii, K., and Suzuki, E., "MOCVD of high-dielectric-constant lanthanum oxide thin films". Journal of the Electrochemical Society 2003;150(8):G429–G425.

[3.78] Lee, J. C., Kang, C., Rhee, S., *et al.*, "Hafnium-based high-k dielectrics". In: Symposium on VLSI Technology Digest of Technical Papers, 2005, pp. 122–125.

[3.79] Garros X., Reimbold, G., Duret, D., *et al.*, "Interface states in HfO_2 stacks with metal gate: nature, passivation, generation". In: Proceedings of the IEEE 43rd Annual International Reliability Physics Symposium, 2005, pp. 55–60.

[3.80] Chang, C., Hu, C., and Brodersen, R. W., "Quantum yield of electron impact ionization in silicon". Journal of Applied Physics 1985;57 (2):302–309.

[3.81] Molina, J., Tachi, K., Kakushima, K., *et al.*, "Effects of N_2-based annealing on the reliability characteristics of tungsten/La_2O_3/silicon capacitors". Journal of The Electrochemical Society 2007;154–155:G110–G116.

[3.82] Chen G., Chuah K. Y., Li M.-F., *et al.*, "Dynamic NBTI of PMOS transistors and its impact on device lifetime". Proceedings of the International Reliability Physics Symposium, 2003, pp. 196–202.

[3.83] Jeppson, K. O. and Svensson, C. M., "Negative bias stress of MOS devices at high electric fields and degradation of MNOS devices". Journal of Applied Physics 1977;48:2004–2014.

[3.84] Ogawa, S. and Shiono, N., "Generalized diffusion–reaction model for the low-field charge build up instability at the Si–SiO_2 interface". Physics Review 1995;51:4218–4230.

[3.85] Park, H., Rahman, M. S., Chang, M., *et al.*, "Effect of high pressure deuterium annealing on electrical and reliability characteristics of MOSFETs with high-k gate dielectric". In: Proceedings of the IRPS, 2005, pp. 646–647.

[3.86] Molina, J., De La Hidalga, F. J., Rosales, P., *et al.*, "Reliability characteristics of W–La_2O_3 structures compared with those of HfO_2-based gate oxides". In: Proceedings of the IEEE (EDSSC'08), 2008, pp. 1–4.

[3.87] Molina, J., Torres, A., Calleja, W., *et al.*, "Degradation and breakdown of W–La_2O_3 stack after annealing in N_2". Japanese Journal of Applied Physics 2008;47:7076–7080.

[3.88] Mavrou, G., Galata, S., Tsipas, P., *et al.*, "Electrical properties of La_2O_3 and HfO_2/La_2O_3 gate dielectrics for germanium metal-oxide-semiconductor devices". Journal of Applied Physics 2008;103: 014506.

[3.89] Ando, T., "Ultimate scaling of high-k gate dielectrics: higher-κ or interfacial layer scavenging?". Materials 2012;5:478–500.

[3.90] Robertson, J. and Wallace, R. M., "High-K materials and metal gates for CMOS applications". Materials Science and Engineering R 2015; 88(2–15):1–41.

[3.91] International Technology Roadmap for Semiconductors. http://www.itrs.net/ITRS 1999–2014 Mtgs,Presentions&Links/2013ITRS/2013Chapters/2013ExecutiveSummary.pdf, 2013.

[3.92] International Technology Roadmap for Semiconductors. http://www.itrs.net/Links/2013ITRS/2013Chapters/2013ERD.pdf, 2013.

[3.93] Hu, C. M., Tam, S. C., Hsu, F. C., Ko, P. K., Chan, T. Y., and Terril, K. W., "Hot-electron induced MOSFET degradation – model, monitor, and improvement". IEEE Transactions on Electron Devices 1985;32 (2):375–385.

[3.94] Rauch, S. E., III and La Rosa, G., "The energy driven paradigm of NMOSFET hot carrier effects". In: Proceedings of the IRPS'05, 2005, pp. 708–709.

[3.95] Shockley, W., "Problems related to p-n junctions in silicon". Solid-State Electronics 1961;2:35–67.

[3.96] Townsend, J., "The theory of ionization of gases by collision". Constable, London, 1910.

[3.97] Ko, P., Muller, R., and Hu, C., "A unified model for hot-electron currents in MOSFETs". In: International Electon Devices Meeting '81 Technical Digest, 1981, pp. 600–603.

[3.98] Taur, Y. and Ning, T., "Fundamentals of modern VLSI devices". Cambridge University Press, Cambridge, 1998, pp. 156 ff.

[3.99] Jakumeit, J. and Ravaioli, U., "Influence of electron-electron scattering on the hot electron distribution in ultra-short Si-MOSFETs". Physica B 2002;314(1–4):363–366.

[3.100] Chang, M. Y., Dyke, D. W., Leung, C. C., and Childs, P. A., "Modeling of gate currents in MOSFETs operating at low drain voltages". In: Proceedings of European Solid-State Device Conference '96, 1996, pp. 263–265.

[3.101] Bude, J. and Mastrapasqua, M., "Impact ionization and distribution functions in sub-micron nMOSFET technologies". IEEE Electron Device Letters 1995;16(10):439–441.

[3.102] Venturi, F., Sangiorgi, E., and Ricco, B., "The impact of voltage scaling on electron heating and device performance of submicrometer MOSFET's". IEEE Transactions on Electron Devices 1991;38(8): 1895–1904.

[3.103] Childs, P. A. and Dyke, D. W., "Hot carrier quasi-ballistic transport in semiconductor devices". Solid-State Electronics 2004;48:765–772.

[3.104] Scrobahaci, P. and Ting-Wei, T., "Modeling of the hot electron sub-population and its application to impact ionization in submicron silicon devices-part II: numerical solutions". IEEE Transactions on Electron Devices 1994;41:1197.

[3.105] Ghetti, A., Selmi, L., and Bez, R., "Low-voltage hot electrons and soft-programming lifetime prediction in nonvolatile memory cells". IEEE Transactions on Electron Devices 1999;46:696–702.

[3.106] Sano, N., Tomizawa, M., and Yoshii, A., "Temperature dependence of hot carrier effects in short-channel Si-MOSFETs". IEEE Transactions on Electron Devices 1995;42 (12):2211 -2216.

[3.107] Mietzner, T., Jakumeit, J., and Ravaioli, U., "Local iterative Monte Carlo analysis of electron-electron interaction in short-channel Si-MOSFETs". IEEE Transactions on Electron Devices 2001;48(12):2323–2330.

[3.108] Keldysh, L., "Kinetic theory of impact ionization in semiconductors". Soviet Physics Journal of ETP 1960;10(3):509–518.

[3.109] Kamakura, Y., Mizuno, H., Yamaji, M., Morifuji, M., Tanaguchi, K., and Hama-guchi, C., "Impact ionization model for full band Monte Carlo simulation". Journal of Applied Physics 1994;75(7):3500–3506.

[3.110] Zanchetta, S., Todon, A., Abramo, A., Selmi, L., and Sangiorgi, E., "Analytical study of the impact of HALOs on short channel and hot carrier effects in MOSFETs". Solid-State Electronics 2002;46(3):429–434.

[3.111] Pang, Y. S. and Brews, J. R., "Analytical subthreshold surface potential model for pocket n-MOSFETs". IEEE Transactions on Electron Devices 2002;49(12):2209–2216.

[3.112] Taur, Y. and Ning, T., op. cit., pp. 150, 151.

[3.113] Childs, P. and Leung, C., "New mechanism of hot carrier generation in very short channel MOSFETs". Electronics Letters 1995;31(2):139–141.

[3.114] Childs, P. A. and Leung, C. C., "A one-dimensional solution of the Boltzmann transport equation including electron-electron interactions". Journal of Applied Physics 1996;79(1):222–227.

[3.115] Chang, M. Y., Dyke, D. W., Leung, C. C., and Childs, P. A., "High-energy electron-electron interactions in silicon and their effect on hot carrier energy distributions". Journal of Applied Physics 1997;82(6):2974–2979.

[3.116] Ventura, D., Gnudi, A., and Baccarani, G., "Short-range electron-electron scattering in silicon with a non-parabolic band structure". Numerical Functional Analysis and Optimization 1995;16:565–581.

[3.117] Fischetti, M. and Laux, S., "Monte Carlo study of sub-band-gap impact ionization in small silicon field-effect transistors". In: Proceedings of the International Electron Devices Meeting '95, 1995, pp. 305–308.

[3.118] Guérin, C., Huard, V., and Bravaix, A., "The energy-driven hot-carrier degradation modes of nMOSFETs". IEEE Transactions on Device and Materials Reliability 2007;7:225.

[3.119] Rauch, S.E., Guarin, F.J., and La Rosa, G., "Impact of E-E scattering to the hot carrier degradation of deep sub-micron NMOSFETs". IEEE Electron Device Letters 1998;19 (12):463–465.

[3.120] Rauch, S. and Guarin, F., "The energy driven hot carrier model". In: Grasser, T. (Ed.), "Hot Carrier Degradation in Semiconductor Devices". Springer, 2015, pp. 51–54.

[3.121] Crowell, C. and Sze, S., Applied Physics Letters 1966;9:242.

[3.122] Arfaoui W., Federspiel X., Mora P., *et al.*, "Energy-driven hot-carrier model in advanced nodes". 2014 IEEE International Reliability Physics Symposium, 2014, XT.12.1–XT.12.5.

[3.123] Pop, E. and Goodson, K., "Thermal phenomena in nanoscale transistors". In: Proceedings of the Intersociety Conference on Thermal Thermo-mechanical Phenomena in Electronic Systems, 2004, pp.1–7.

[3.124] Rauch, S., *et al.*, "High V_{GS} PFET DC hot carrier mechanism and its relation to AC degradation". IEEE Transactions on Device and Materials Reliability 2010;10:40–46.

[3.125] La Rosa, G., Ng, W., Rauch, S., Wong, R., and Sudijono, J., "Impact of NBTI induced statistical variation to SRAM cell stability". In: Proceedings of the International Reliability Physics Symposium '06, 2006, pp. 274–282.

[3.126] Strong, A., Wu, E., Vollertsen, R., *et al.*, "Reliability wearout mechanisms in advanced CMOS technologies". IEEE Press, Wiley & Sons, 2009.

[3.127] Miranda, E. and Suñé, J., "Electron transport through broken down ultra-thin SiO_2 layers in MOS devices". Microelectronics Reliability 2004;44:1–23.

[3.128] Ha, S. and Ramanathan, S., "Adaptive oxide electronics: a review". Journal of Applied Physics 2011;110:071101.

[3.129] Lanza, M. A., "Review on resistive switching in high-k dielectrics: a nanoscale point of view using conductive atomic force microscope". Materials 2014;7:2155–2182.

[3.130] Clark, R., "Emerging applications for high-K materials in VLSI technology". Materials 2014;7:2913–2944.

[3.131] Degraeve, R., Kaczer, B., De Keersgieter, A., and Groeseneken, G., "Relation between breakdown mode and location in short-channel nMOSFETs and its impact on reliability specifications". IEEE Transactions on Device and Materials Reliability 2001;1:163–169.

[3.132] O'Dwyer, J., "The theory of dielectric breakdown in solids". Oxford University Press, London, 1964.

[3.133] Nafría, M., Suñé, J., and Aymerich, X., "Breakdown of thin gate silicon dioxide films – a review". Microelectronics Reliability 1996;36:871–905.

[3.134] Dumin, D. (Ed.), "Oxide reliability: a summary of silicon oxide wearout, breakdown, and reliability". World Scientific, 2002.

[3.135] Wu, E., Stathis, J., and Han, L., "Ultra-thin oxide reliability for ULSI applications. Semiconductor Science Technology 2000;15:425–435.

[3.136] Suehle, J., "Ultrathin gate oxide reliability: physical models, statistics, and characterization". IEEE Transactions on Electron Devices 2002;49:958–971.

[3.137] Ribes, G., Mitard, J., Denais, M., *et al.*, "Review on high-K dielectrics reliability issues". IEEE Transactions on Device and Materials Reliability 2005;5:5–19.

[3.138] Rodríguez, R., Nafría, M., Miranda, E., Suñé, J., and Aymerich, X., "Analysis of the degradation and breakdown of thin SiO_2 films under static and dynamic tests using a two-step stress procedure". IEEE Transactions on Electron Devices 2000;47:2138–2145.

[3.139] Scarpa, A., Paccagnella, A., Montera, F., *et al.*, "Ionizing radiation induced leakage current on ultra-thin gate oxides". IEEE Transactions on Nuclear Science 1997;44:1818–1825.

[3.140] McPherson, J. and Khamankar, R., "Molecular model for intrinsic time-dependent dielectric breakdown in SiO_2 dielectrics and the reliability implications for hyper-thin gate oxide". Semiconductor Science and Technology 2000;15:462–470.

[3.141] Chen, I., Holland, S., Young, K., Chang, C., and Hu, C., "Substrate hole current and oxide breakdown". Applied Physics Letters 1986;49 (49):669–671.

[3.142] DiMaria, D. and Stasiak, J., "Trap creation in silicon dioxide produced by hot electrons". Journal of Applied Physics 1989;65:2342–2356.

[3.143] Nissan-Cohen, Y., Shappir, J., and Frohman-Bentchkowsky, D., "Trap generation and occupation dynamics in SiO_2 under charge injection stress". Journal of Applied Physics 1986;60:2024–2030.

[3.144] Fleetwood, D., Winokur, P., Reber, R., *et al.*, "Effects of oxide traps, interface traps, and "border traps" on metal-oxide-semiconductor devices". Journal of Applied Physics 1993;73:5058–5074.

[3.145] Degraeve, R., Groeseneken, G., Bellens, R., *et al.*, "New insights in the relation between electron trap generation and the statistical properties of oxide breakdown". IEEE Transactions on Electron Devices 1998;45:904–911.

[3.146] Stathis, J., "Percolation models for gate oxide breakdown". Journal of Applied Physics 1999;86:5757–5766.

[3.147] Lopez-V., J., Jimenez-T., J., Cartujo, P., Bausells, J., and Carceller, J., "Analysis of the effects of constant-current Fowler-Nordheim-tunneling injection with charge trapping inside the potential barrier". Journal of Applied Physics 1991;70:3712–3720.

[3.148] Miranda, E., Redin, G., and Faigón, A., "Modeling of the I-V characteristics of high-field stressed MOS structures using a Fowler-Nordheim-type tunneling expression". Microelectronics Reliability 2002; 42:935–940.

[3.149] Olivo, P., Nguyen, T., and Ricco, B., "High-field-induced degradation in ultra-thin SiO_2 films". IEEE Transactions on Electron Devices 1988;35:2259–2264.

[3.150] Chou, A., Lai, K., Kumar, K., Chowdhury, P., and Lee, J., "Modeling of stress-induced leakage current in ultra-thin oxides with the trap-assisted tunneling mechanism". Applied Physics Letters 1997;70(70):3407–3409.

[3.151] Vincent, E., Bruyère, S., Papadas, C., and Mortini, P., "Dielectric reliability: in deep-submicron technologies: from thin to ultrathin oxides". Microelectronics Reliability 1997;37:1499–1506.

[3.152] Depas, M., Vermeire, B., Mertens, P., Meuris, M., and Heyns, M., "Wear-out of ultra-thin gate oxides during high-field electron tunneling". Semiconductor Science Technology 1995;10:753–758.

[3.153] Buchanan, D., Stathis, J., Cartier, E., and DiMaria, D., "on the relationship between stress induced leakage currents and catastrophic breakdown in ultra-thin SiO_2 based dielectrics". Microelectronics Engineering 1997;36:329–333.

[3.154] Rodríguez, R., Miranda, E., Pau, R., Suñé, J., Nafría, M., and Aymerich, X., "Monitoring the degradation that causes the breakdown of ultrathin (<5 nm) SiO_2 gate oxides. IEEE Electron Device Letters 2000;21: 251–253.

[3.155] Blasco, J., Ghenzi, N., Suñé, J., Levy, P., and Miranda, E., "Equivalent circuit modeling of the bistable conduction characteristics in electroformed thin dielectric films". Microelectronics Reliability 2015;55:1–14.

[3.156] Depas, M., Nigam, T., and Heyns, M., "Definition of dielectric breakdown for ultra thin (<2 nm) gate oxides". Solid-State Electronics 1997;41:725–728.

[3.157] Degraeve, R., Kaczer, B., and Groeseneken, G., "Reliability: a possible showstopper for oxide thickness scaling.". Semiconductor Science Technology 2000;15:436–444.

[3.158] Klein, N., "The mechanism of self-healing electrical breakdown in MOS structures". IEEE Transactions on Electron Devices 1966;13:788–805.

[3.159] Klein, N., "Advances in electronics and electron physics". In: Marton, L. (Ed.), Academic, New York, 1969, Vol. 28.

[3.160] Klein, N., "Switching and breakdown in films". Thin Solid Films 1971;7:149–155.

[3.161] Shatzkes, M., Av-Ron, M., and Anderson, R., "On the nature of conduction and switching in SiO_2". Journal of Applied Physics 1974;45:2065–2077.

[3.162] Hickmott, T., "Low-frequency negative resistance in thin anodic oxide films". Journal of Applied Physics 1962;33:2669.

[3.163] Dearnaley, G., Morgan, D., and Stoneham, A., "A model for filament growth and switching in amorphous oxide films". Journal of Non-Crystalline Solids 1970;4:593–612.

[3.164] Solomon, P., "Breakdown in silicon oxide – a review". Journal of Vacuum Science & Technology 1977;14:1122–1130.

[3.165] Wolters, D., "Instabilities in silicon devices". In: Barbottin, G. and Vapaille, A. (Eds.), North-Holland, Amsterdam, 1986.

[3.166] Suñé, J., Farrés, E., Placencia, I., *et al.*, "Nondestructive multiple breakdown events in very thin SiO_2 films". Applied Physics Letters 1989;55:128–130.

[3.167] Nafría, M., Suñé, J., and Aymerich, X., "Exploratory observations of post-breakdown conduction in polycrystalline-silicon and metal-gated thin-oxide metal-oxide-semiconductor capacitors". Journal of Applied Physics 1993;73:205–215.

[3.168] Fukuda, H., Yasuda, M., and Iwabuchi, T., "Oxide wearout phenomena of ultrathin SiO_2 film during high-field stress". Electronics Letters 1992;28:1516–1518.

[3.169] Okada, K., Kawasaki, S., and Hirofuji, Y., "New experimental findings on stress induced leakage current of ultra thin silicon dioxides". In: Proceedings of the Extended Abstract of SSDM'94, 1994, pp. 565–567.

[3.170] Lee, S., Cho, B., Kim, J., and Choi, S., "Quasibreakdown of ultrathin gate oxide under high field stress". In: Proceedings of the IEEE IEDM'94, 1994, pp. 605–608.

[3.171] Fu, K., "Partial breakdown of the tunnel oxide in floating gate devices". Solid-State Electronics 1997;41:774–777.

[3.172] Depas, M., Nigam, T., and Heyns, M., "Soft breakdown of ultra-thin gate oxide layers". IEEE Transactions on Electron Devices 1996;43:1499–1504.

[3.173] Lombardo, S., La Magna, A., Gerardi, C., Alessandri, M., and Crupi, F., "Soft breakdown of gate oxides in metal–SiO_2–Si capacitors under stress with hot electrons". Applied Physics Letters 1999;75:1161–1163.

[3.174] Sakura, T., Utsunomiya, H., Kamakura, Y., and Taniguchi, K., "A detailed study of soft- and pre-soft-breakdowns in small geometry MOS structures". In: Proceedings of the IEEE IEDM'98, 1998, pp. 183–186.

[3.175] Cellere, G., Larcher, L., Valentini, M., and Paccagnella, A., "Micro breakdown in small-area ultrathin gate oxides". IEEE Transactions on Electron Devices 2002;49:1367–1374.

[3.176] Lombardo, S., La Magna, A., Spinella, C., Gerardi, C., and Crupi, F., "Degradation and hard breakdown transient of thin gate oxides in metal–SiO_2–Si capacitors: dependence on oxide thickness". Journal of Applied Physics 1999;86:6382–6391.

[3.177] Lombardo, S., La Magna, A., Crupi, I., Gerardi, C., and Crupi, F., "Reduction of thermal damage in ultrathin gate oxides after intrinsic dielectric breakdown". Applied Physics Letters 2001;79:1522–1524.

[3.178] Monsieur, F., Vincent, E., Pananakakis, G., and Ghibaudo, G., "Wear-out, breakdown occurrence and failure detection in 18–25 Å ultrathin oxides". Microelectronics Reliability 2001;41:1035–1039.

[3.179] Hosoi, T., Lo Ré, P., Kamakura, Y., and Taniguchi, K., "A new model of time evolution of gate leakage current after soft breakdown in ultra-thin gate oxides". In: Proceedings of the IEEE IEDM'02, 2002, pp. 155–158.

[3.180] Linder, B., Lombardo, S., Stathis, J., Vayshenker, A., and Frank, D., "Voltage dependence of hard breakdown growth and the reliability implication in thin dielectrics". IEEE Electron Device Letters 2002;23:661–663.

[3.181] Miranda, E., Cester, A., and Paccagnella, A., "Logistic model for leakage current in electrical stressed ultra-thin gate oxides". Electronics Letters 2003;39:741–742.

[3.182] Monsieur, F., Vincent, E., Roy, D., *et al.*, "A thorough investigation of progressive breakdown in ultra-thin oxides. Physical understanding and application for industrial reliability assessment". In: Proceedings of the IEEE IRPS'02, 2002, pp. 50–60.

[3.183] Satake, H. and Toriumi, A., "Dielectric breakdown mechanism of thin-SiO_2 studied by the post-breakdown resistance statistics". IEEE Transactions on Electron Devices 2000;47:741–744.

[3.184] Miranda, E., Suñé, J., Rodríguez, R., Nafría, M., and Aymerich, X., "A function-fit model for the soft breakdown failure mode". IEEE Electron Device Letters 1999;20:265–267.

[3.185] Suñé, J. and Miranda, E., "Post soft breakdown conduction in SiO_2 gate oxides". In: Proceedings of the IEEE IEDM'00, 2000, pp. 1–4.

[3.186] Alam, M., Weir, B., and Silverman, P., "A study of soft and hard breakdown-part I: analysis of statistical percolation conductance". IEEE Transactions on Electron Devices 2002;49:232–238.

[3.187] Crupi, F., Degraeve, R., Groeseneken, G., Nigam, T., and Maes, H., "on the properties of the gate and substrate current after soft breakdown in ultrathin oxide layers". IEEE Transactions on Electron Devices 1998;45:2329–2334.

[3.188] Miranda, E., Suñé, J., Rodríguez, R., Nafría, M., and Aymerich, X., "Detection and fitting of the soft breakdown failure mode in MOS structures". Solid-State Electronics 1999;43:1801–1805.

[3.189] Pompl, T., Wurzer, H., Kerber, M., Wilkins, R., and Eisele, I., "Influence of soft breakdown on NMOSFET device characteristics". In: Proceedings of the IEEE IRPS'99, 1999, pp. 82–87.

[3.190] Synder, E. and Suehle, J., "Detecting breakdown in ultrathin dielectrics using a fast voltage ramp". In: IEEE International Reliability Workshop Final Report, 1999, pp. 118–123.

[3.191] Weir, B., Silverman, P., Monroe, D., *et al.*, "Ultra-thin gate dielectrics: they break down, but do they fail?". In: Proceedings of the IEEE IEDM'97, 1997, pp. 73–76.

[3.192] Alers, G., Weir, B., Frei, M., and Monroe, D., "J-ramp on sub-3 nm dielectrics: noise as a breakdown criterion". In: Proceedings of the IEEE IRPS'99, 1999, pp. 410–413.

[3.193] Roussel, P., Degraeve, R., Van den Bosch, G., Kaczer, B., and Groeseneken, G., "Accurate and robust noise-based trigger algorithm for soft breakdown detection in ultrathin gate dielectrics". IEEE Transactions on Device and Materials Reliability 2001;1:120–127.

[3.194] Schmitz, J., Tuinhout, H., Kretschmann, H., and Woerlee, P., "Comparison of soft-breakdown triggers for large-area capacitors under constant voltage stress". IEEE Transactions on Device and Materials Reliability 2001;1:150–157.

[3.195] Bruyère, S., Vincent, E., and Ghibaudo, G., "Quasi-breakdown in ultra-thin SiO_2 films: occurrence characterization and reliability assessment methodology". In: Proceedings of the IEEE IRPS'00, 2000, pp. 48–54.

[3.196] Suñé, J., Wu, E., Jiménez, D., Vollertsen, R., and Miranda, E., "Understanding soft and hard breakdown statistics, prevalence ratios, and energy dissipation during breakdown runaway". In: Proceedings of the IEEE IEDM'01, 2001, pp. 117–120.

[3.197] Alam, M., Weir, B., and Silverman, P., "A study of soft and hard breakdown-part II: principles of area, thickness, and voltage scaling". IEEE Transactions on Electron Devices 2002;49:239–246.

[3.198] Brière, O., Halimaoui, A., and Ghibaudo, G., "Breakdown characteristics of ultra thin gate oxides following field and temperature stresses". Solid-State Electronics 1997;41:981–985.

[3.199] Leroux, C., Blachier, D., Brière, O., and Reimbold, G., "Light emission microscopy for thin oxide reliability analysis". Microelectronics Engineering 1997;36:297–300.

[3.200] Wu, E., Nowak, E., Aitken, J., Abadeer, W., Han, L., and Lo, S., "Structural dependence of dielectric breakdown in ultra-thin gate oxides and its relationship to soft breakdown modes and device failure". In Proceedings of the IEEE IEDM'98, 1998, pp. 187–190.

[3.201] Lombardo, S., "Intrinsic dielectric breakdown of ultra-thin gate oxides". Microelectronics Engineering 2001;59:33–42.

[3.202] Weir, B., Alam, M., Silverman, P., *et al.*, "Ultra-thin gate oxide reliability projections". Solid-State Electronics 2002;46:321–328.

[3.203] Porti, M., Blüm, M., Nafría, M., and Aymerich, X., "Imaging breakdown spots in SiO_2 films and MOS devices with a conductive atomic force microscope". IEEE Transactions on Device and Materials Reliability 2002;2:94–101.

[3.204] Linder, B., Stathis, J., Wachnik, R., *et al.*, "Gate oxide breakdown under current limited constant voltage stress". In: Symposium on VLSI Technology Digest of Technical Papers, 2000, p. 214.

[3.205] Radhakrishnan, M., Pey, K., Tung, C., and Lin, W., "Physical analysis of hard and soft breakdown failures in ultrathin gate oxides". Microelectronics Reliability 2002;42:565–571.

[3.206] Alam, M., Weir, B., Bude, J., Silverman, P., and Monroe, D., "Explanation of soft and hard breakdown and its consequences for area scaling". In: Proceedings of the IEEE IEDM'99, 1999, p. 449.

[3.207] Suñé, J., Wu, E., Jiménez, D., and Lai, W., "Statistics of soft and hard breakdown in thin SiO_2 gate oxides". Microlectronics Reliability 2003;43:1185–1192.

[3.208] Okada, K., "Extended time dependent dielectric breakdown model based on anomalous gate area dependence of lifetime in ultra thin silicon dioxides". Japanese Journal of Applied Physics 1997;36:1443–1447.

[3.209] Lombardo, S., Stathis, J., and Linder, B., "Breakdown transients in ultrathin gate oxides: transition in the degradation rate". Physics Review Letters 2003;90:167601–167604.

[3.210] Huang, C. and Hwu, J., "Enhancement in soft breakdown occurrence of ultra-thin gate oxides caused by photon effect in rapid thermal post-oxidation annealing". Solid-State Electronics 2000;44:1405–1410.

[3.211] Tomita, T., Utsunomiya, H., Sakura, T., Kamakura, Y., and Taniguchi, K., "A new soft breakdown model for thin thermal SiO_2 films under constant current stress". IEEE Transactions on Electron Devices 1999;46:159–164.

[3.212] Suñé, J., Mura, G., and Miranda, E., "Are soft breakdown and hard breakdown of ultrathin gate oxides actually different failure mechanisms?". IEEE Electron Device Letters 2000;21(2):167–169.

[3.213] Crupi F., Iannaccone G., Crupi I., Degraeve R., Groeseneken G., and Maes H., "Characterization of soft breakdown in thin oxide NMOSFETs based on the analysis of the substrate current". IEEE Transactions on Electron Devices 2001;48:1109–1113.

[3.214] Mizubayashi, W., Yoshida, Y., Miyazaki, S., and Hirose, M., "Statistical analysis of soft breakdown in ultrathin gate oxides". In: Symposium on VLSI Technology Digest of Technical Papers, 2001, pp. 95–96.

[3.215] Kaczer, B., Degraeve, R., De Keersgieter, A., Van de Mieroop, K., Simons, V., and Groeseneken, G., "Consistent model for short-channel nMOSFET after hard gate oxide breakdown". IEEE Transactions on Electron Devices 2002;49:507–513.

[3.216] Kaczer, B., Degraeve, R., De Keersgieter, A., Groeseneken, G., and Crupi, F., "Understanding nMOSFET characteristics after soft breakdown and their dependence on the breakdown location". In: Proceedings of the ESSDERC'02, 2002, pp. 139–142.

[3.217] Crupi, F., Kaczer, B., Degraeve, R., De Keersgieter, A., and Groeseneken, G., "A comparative study of the oxide breakdown in short-channel nMOSFETs and pMOSFETs stressed in inversion and in accumulation regimes". IEEE Transactions on Device and Materials Reliability 2003;3:8–13.

[3.218] Pompl, T., Wurzer, H., Kerber, M., and Eisele, I., "Investigation of ultrathin gate oxide reliability behavior by separate characterization of soft breakdown and hard breakdown". In: Proceedings of the IEEE IRPS'00, 2000, pp. 40–47.

[3.219] Chen, T., Tse, M., Sun, C., and Fung, S., "Post-breakdown conduction instability of ultrathin SiO_2 films observed in ramped-current and ramped-voltage measurements". Japanese Journal of Applied Physics 2002;41:3047–3051.

[3.220] Okada, K. and Taniguchi, K., "Electrical stress-induced variable range hopping conduction in ultrathin silicon dioxides". Applied Physics Letters 1997;70:351–353.

[3.221] Houssa, M., Nigam, T., Mertens, P., and Heyns, M., "Model for the current-voltage characteristics of ultrathin gate oxides after soft breakdown". Journal of Applied Physics 1998;84:4351–4355.

[3.222] Halimaoui, A., Brière, O., and Ghibaudo, G., "Quasi-breakdown in ultrathin gate dielectrics". Microelectronics Engineering 1997;36: 157–160.

[3.223] Goguenheim, D., Bravaix, A., Vuillaume, D., *et al.*, "Experimental study of the quasi-breakdown failure mechanism in 4.5 nm-thick SiO_2 oxides". Microelectronics Reliability 1999;39:165–169.

[3.224] Bruyère, S., "Etude des mecanismes de degradation et de defaillance des oxydes ultra-minces – Application a la fiabilite des technologies CMOS sub-0.25 μm". PhD. Thesis, Institut National Polytechnique de Grenoble, 2000.

[3.225] Cester, A., Paccagnella, A., Suñé, J., and Miranda, E., "Post-radiation-induced soft breakdown conduction properties as a function of temperature". Applied Physics Letters 2001;79:1336–1338.

[3.226] Ceschia, M., Paccagnella, A., Sandrin, S., *et al.*, "Low field leakage current and soft breakdown in ultra-thin gate oxides after heavy ions, electrons or X-ray irradiation". IEEE Transactions on Nuclear Science 2000;47:566–573.

[3.227] Blasco, J., Castán, H., García, H., *et al.*, "Single-parameter model for the post-breakdown conduction characteristics of HoTiOx-based MIM capacitors". Microelectronics Reliability 2014;54:1707–1711.

[3.228] Miranda, E., Suñé, J., Rodríguez, R., *et al.* "Soft breakdown conduction in ultrathin (3–5 nm) gate dielectrics". IEEE Transactions on Electron Devices 2000;47:82–88.

[3.229] Samanta, P., Man, T. Y., Chan, A. C. K., and Chan, M., "Experimental evidence of two conduction mechanisms for direct tunneling stress-induced leakage current through ultrathin silicon dioxide gate dielectrics". Semiconductor Science and Technology, 2006, 21(10), 1393–1401.

[3.230] Umeda, K. and Taniguchi, K., "Hot-electron-induced quasi-breakdown of thin gate oxides". Journal of Applied Physics 1997;82:297–302.

[3.231] Bearda, T., Woerlee, P., Wallinga, H., and Heyns, M., "Charge transport after hard breakdown in gate oxides". Japanese Journal of Applied Physics 2002;41:2431–2436.

[3.232] Avellán, A., "Charakterisierung von MOS-Transistoren vor und nach Gateoxiddurchbruch". PhD. Thesis, Technische Universitât Hamburg-Harburg, 2003.

[3.233] Mott, N. and Davis, E., "Electronic processes in non-crystalline materials". Clarendon, Oxford, 1979.

[3.234] Cester, A., Bandiera, L., Ceschia, M., Ghidini, G., and Paccagnella, A., "Noise characteristics of radiation-induced soft breakdown current in ultrathin gate oxides". IEEE Transactions on Nuclear Science 2001;48:2093–2100.

[3.235] Avellán, A., Miranda, E., Sell, B., and Krautschneider, W., "Experimental study and modeling of the temperature dependence of soft breakdown conduction in ultrathin gate oxides". In: Proceedings of the IEEE IRPS'03, 2003, pp. 580–581.

[3.236] Ghibaudo, G., Riess, P., Bruyere, S., *et al.*, "Emerging oxide degradation mechanisms: stress induced leakage current (SILC) and quasibreakdown (QB)". Microelectronic Engineering, 1999, 49, 41–50.

[3.237] Houssa, M., Nigam, T., Mertens, P., and Heyns, M., "Soft breakdown in ultrathin gate oxides: correlation with the percolation theory of nonlinear conductors". Applied Physics Letters 1998;73:514–516.

[3.238] Houssa, M., Vandewalle, N., Nigam, T., Ausloos, M., Mertens, P., and Heyns, M., "Analysis of the gate voltage fluctuations in ultra-thin gate oxides after soft breakdown". In: Proceedings of the IEEE IEDM'98, 1998, pp. 909–912.

[3.239] Vandewalle, N., Ausloos, M., Houssa, M., Mertens, P., and Heyns, M., "Non-Gaussian behavior and anticorrelations in ultrathin gate oxides after soft breakdown". Applied Physics Letters 2002;74:1579–1581.

[3.240] Yoshida, T., Miyazaki, S., and Hirose, M., "Analytical modeling of quasi-breakdown of ultrathin gate oxides under constant current stressing". In: Proceedings of the Extended Abstract of SSDM'96, 1996, pp. 539–541.

[3.241] Omura, Y. and Komiya, K., "Transport characteristics of post-hard breakdown thin silicon oxide films and consideration of physical models". Journal of Applied Physics 2002;91:4298–4306.

[3.242] Loh, W., Cho B., and Li, M., "Evolution of quasi-breakdown in thin gate oxides". Journal of Applied Physics 2002;91:5302–5306.

[3.243] Riccó, B., Ya Azbel, M., and Brodsky, M., "A novel mechanism for tunneling and breakdown in thin SiO_2 films". Physics Review Letters 1983;51:1795–1798.

[3.244] Cheung, K., Colonell, J., Chang, C., *et al.*, "Energy funnels-A new oxide breakdown model". In: Symposium on VLSI Technology Digest of Technical Papers, 1997, pp. 145–146.

[3.245] Uno, S., Sakura, T., Kamakura, Y., and Taniguchi, K., "I–V characteristics of ultra thin oxide films after soft breakdown". In: Proceedings of the SISC'99, 1999, unpublished.

[3.246] Ting, D., "An embedded quantum wire model of dielectric breakdown". Applied Physics Letters 1999;74:585–587.

[3.247] Nigam, T., Crupi, F., Degraeve, R., Heyns, M., Groeseneken, G., and Maes, H., "An inelastic quantum tunneling model for current conduction after soft-breakdown". In: Proceedings of the SISC'98, 1998, unpublished.

[3.248] Nigam, T., "Growth kinetics, electrical characterization and reliability study of sub-5 nm gate dielectrics". PhD Thesis, Katholieke Universiteit Leuven, 1999.

[3.249] Grabert, H. and Devoret, M, "Coulomb blockade phenomena in nanostructures". Plenum Press, New York, 1992.

[3.250] Suñé, J., Miranda, E., Nafría, M., and Aymerich, X., "Point contact conduction at the oxide breakdown of MOS devices". In: Proceedings of the IEEE IEDM'98, 1998, pp. 191–194.

[3.251] Miranda, E. and Suñé, J., "Mesoscopic approach to the soft breakdown failure mode in ultrathin SiO_2 films". Applied Physics Letters 2001;78:225–227.

[3.252] Miranda, E. and Suñé, J., "Modeling the conduction characteristics of broken down gate oxides in MOS structures". Microelectronics Reliability 2000;40:1599–1603.

[3.253] Datta, S., "Electronic transport in mesoscopic systems". Cambridge University Press, 1997.

[3.254] Cester, A., Bandiera, L., Suñé, J., Boschiero, L., Ghidini, G., and Paccagnella, A., "A novel approach to quantum point contact for post soft breakdown conduction". In: Proceedings of the IEEE IEDM'01, 2001, pp. 13.6.1–13.6.4.

[3.255] Suehle, J., Wang, B., Miyahara, T., Vogel, E., Bernstein, J., and Conley, J., "Heavy-ion-induced soft breakdown of thin gate oxides". IEEE Transactions on Nuclear Science 2001;48:1913–1916.

[3.256] Crupi, F., Ciofi, C., Iannaccone, G., Neri, B., and Lombardo, S., "Current noise at the oxide hard-breakdown". Microelectronics Engineering 2001;59:43–47.

[3.257] Schroeder, D. and Avellán, A., "Physical explanation of the barrier height temperature dependence in metal-oxide-semiconductor leakage current models". Applied Physics Letters 2003;82:4510–4512.

[3.258] Miranda, E., Walczyk, C., Wenger, C., and Schroeder, T., "Model for the resistive switching effect in HfO_2 MIM structures based on the transmission properties of narrow constrictions". IEEE Electron Device Letters 2010;31:609–611.

[3.259] Walczyk, C., Walczyk, D., Schroeder, T., *et al.*, "Impact of temperature on the resistive switching behavior of embedded HfO_2-Based RRAM devices". IEEE Transactions on Electron Devices 2011;58:3124–3131.

[3.260] Degraeve, T., Roussel, Ph., Goux, L., *et al.*, "Generic learning of TDDB applied to RRAM for improved understanding of conduction and switching mechanism through multiple filaments". In: IEEE International Electron Device Meeting, 2010, pp. 28.4.1–28.4.4.

[3.261] Tsuruoka, T., Hasegawa, T., Terabe, K., and Aono, M., "Conductance quantization and synaptic behavior in a Ta_2O_5-based atomic switch". Nanotechnology 2012;23:435705.

[3.262] Chen, C., Gao, S., Zeng, F., *et al.*, "Conductance quantization in oxygen-anion-migration-based resistive switching memory devices". Applied Physics Letters 2013;103:043510.

[3.263] Long, S., Lian, S., Cagli, C., *et al.*, "Quantum-size effects in hafnium-oxide resistive switching". Applied Physics Letters 2013;102:183505.

[3.264] Zhu, X., Su, W., Liu, Y., *et al.*, "Observation of conductance quantization in oxide-based resistive switching memory". Advanced Materials 2012;24:3941.

[3.265] Mehonic, A., Rajitoarea, A., Cueff, S., *et al.*, "Quantum conductance in silicon oxide resistive memory devices". Scientific Reports 2013;3:2708.

[3.266] Miranda, E., Kano, S., Dou, C., Kakushima, K., Suñé, J., and Iwai, H., "Nonlinear conductance quantization effects in CeOx/SiO_2-based resistive switching devices". Applied Physics Letters 2012;101:012910.

[3.267] Lee, S., Cho, B., Kim, J., and Choi, S., "Quasi-breakdown of ultrathin gate oxide under high field stress". In: Proceedings of the IEEE IEDM'94, 1994, pp. 605–608.

[3.268] Henson, W., Yang, N., and Wortman, J., "Observation of oxide breakdown and its effects on the characteristics of ultra-thin-oxide nMOSFET's". IEEE Electron Device Letters 1999;20:605–607.

[3.269] Pompl, T., Wurzer, H., Kerber, M., and Eisele, I., "Influence of gate oxide breakdown on MOSFET device operation". Microelectronics Reliability 2000;40:37–47.

[3.270] Zhang, J., Yuan, J., Ma, Y., Chen, Y., and Oates, A., "Experimental evaluation of device degradation subject to oxide soft breakdown". Solid-State Electronics 2001;45:1521–1524.

[3.271] Cester, A., Cimino, S., Paccagnella, A., Ghidini, G., and Guegan, G., "Collapse of MOSFET drain current after soft breakdown and its dependence on the transistor aspect ratio W/L". In: Proceedings of the IRPS'03, 2003, pp. 189–195.

[3.272] Weir, B., Alam, M., and Silverman, P., "Soft breakdown at all positions along the n-MOSFET". Microelectronics Engineering 2001;59:17–23.

[3.273] Tsai, M., Lin, H., Lee, D., and Huang, T., "Post-soft-breakdown characteristics of deep submicron nMOSFETs with ultrathin gate oxide". IEEE Electron Device Letters 2001;22:348–350.

[3.274] Kaczer, B., Degraeve, R., Rasras, M., Van de Mieroop, K., Roussel, P., and Groeseneken, G., "Impact of MOSFET gate oxide breakdown on digital circuit operation and reliability". IEEE Transactions on Electron Devices 2002;49:500–506.

[3.275] Stathis, J., Rodríguez, R., and Linder, B., "Circuit implications of gate oxide breakdown". Microelectronics Reliability 2003;43:1193–1197.

[3.276] Yeoh, T., Kamat, N., Nair, R., and Hu, S., "Gate oxide breakdown model in MOS transistors". Proceedings of the IEEE-IRPS, 1995, pp. 149–155.

[3.277] Crupi, F., Kauerauf, R., Degraeve, R., Pantisano, L., and Groeseneken, G., "A novel methodology for sensing the breakdown location and its application to the reliability study of ultrathin Hf-silicate gate dielectrics". IEEE Transactions on Electron Devices 2005;52:1759–1765.

[3.278] Fernández, R., Martín-Martínez, J., Rodríguez, R., Nafría, M., and Aymerich, S., "Gate oxide wear-out and breakdown effects on the performance of analog and digital circuits". IEEE Transactions on Electron Devices 2008;55:997–1004.

[3.279] Nicollian, P., Cakici, T., Krishnan, A., Reddy, V., and Seshadri, A., "Device characteristics and equivalent circuits for NMOS gate-to-drain soft and hard breakdown in polysilicon/SiON gate stacks". IEEE Transactions on Electron Devices 2011;58:1170–1175.

[3.280] O'Connor, R., Hughes, G., Degraeve, R., and Kaczer, B., "Progressive breakdown in ultrathin SiON dielectrics and its effect on transistor performance". Microelectronics Reliability 2005;45:869–874.

[3.281] Miranda, E., Kawanago, T., Takamasa, T., Kakushima, K., and Iwai, H., "Analysis and simulation of the postbreakdown I–V characteristics of n-MOS transistors in the linear response regime". IEEE Electron Device Letters 2013;34:798–800.

[3.282] Alam, M., Varghese, D., and Kaczer, B., "Theory of breakdown position determination by voltage- and current-ratio methods". IEEE Transactions on Electron Devices 2008;55:3150–3158.

[3.283] Miranda, E., "Analytic model for the post-breakdown conductance of sub-5-nm SiO_2 gate oxides". IEEE Electron Device Letters 2005;26:673–675.

[3.284] Ortiz-Conde, A., García Sánchez, F., and Muci, J., "Exact analytical solutions of the forward non-ideal diode equation with series and shunt parasitic resistances". Solid State Electronics 2000;44:1861–1864.

[3.285] Waser, R. and Aono, M., "Nanoionics-based resistive switching memories". Nature Materials 2007;6:833–840.

[3.286] Waser, R., Dittmann, R., Staikov, G., and Szot, K., "Redox-based resistive switching memories – Nanoionic mechanisms, prospects, and challenges". Advanced Materials 2009;21:2632–2663.

[3.287] Chung, A., Deen, J., Lee, J. S., and Meyyappan, M., "Nanoscale memory devices". Nanotechnology 2010;21:412001.

[3.288] Kim, K. M., Jeong, D. S., and Hwang, C. S., "Nanofilamentary resistive switching in binary oxide system; a review on the present status and outlook". Nanotechnology 2011;22:254002.

[3.289] Yang, Y. C., Pan, F., Liu, Q., Liu, M., and Zeng, F., "Fully room-temperature-fabricated nonvolatile resistive memory for ultrafast and high-density memory application". Nano Letters 2009;9:1636–1643.

[3.290] Yang, J. J., Zhang, M. X., Pickett, M. D., *et al.*, "Engineering nonlinearity into memristors for passive crossbar applications". Applied Physics Letters 2012;100:113501.

[3.291] Huang, Y., Huang, R., Pan, Y., *et al.*, "A new dynamic selector based on the bipolar RRAM for the crossbar array application". IEEE Transactions on Electron Devices 2012;59:227722–227780.

[3.292] Kim, S., Choi, S. H., Lee, J., and Lu, W. D., "Tuning resistive switching characteristics of tantalum oxide memristors through Si doping". ACS Nano 2014;8:10262–10269.

[3.293] Dittmann, R., Muenstermann, R., Krug, I., *et al.*, "Scaling potential of local redox processes in memristive $SrTiO_3$ thin-film devices". Proceedings of the IEEE 2012;100:1979–1990.

[3.294] Peng, H. Y., Pu, L., Wu, J. C., *et al.*, "Effects of electrode material and configuration on the characteristics of planar resistive switching devices". Applied Physics Letters Mater 2013;1:052106.

[3.295] Zhirnov, V. V., Meade, R., Cavin, R. K., and Sandhu, G., "Scaling limits of resistive memories". Nanotechnology 2011;22:254027.

[3.296] Zhang, L., Hsu, Y. Y., Chen, F. T., *et al.*, "Experimental investigation of the reliability issue of RRAM based on high resistance state conduction". Nanotechnology 2011;22:254016.

[3.297] Chiu, F. C., Li, P. W., and Chang, W. Y., "Reliability characteristics and conduction mechanisms in resistive switching memory devices using ZnO thin films". Nanoscale Research Letters 2012;7:178.

[3.298] Long, S., Cagli, C., Ielmini, D., Liu, M., and Suñé, J., "Analysis and modeling of resistive switching statistics". Journal of Applied Physics 2012;111:074508.

[3.299] Yang, J. J., Borghetti, J., Murphy, D., Stewart, D. R., and Williams, R. S., "A family of electronically reconfigurable nanodevices". Advanced Materials 2009;21:3754–3758.

[3.300] Shkabko, A., Aguirre, M. H., Marozau, I., Lippert, T., and Weidenkaff A., "Measurements of current-voltage-induced heating in the Al/ $SrTiO_3-xNy$/Al memristor during electroformation and resistance switching". Applied Physics Letters 2009;95:152109.

[3.301] Chang, Y. F., Chen, P. Y., Chen, Y. T., *et al.*, "Study of polarity effect in SiOx-based resistive switching memory". Applied Physics Letters 2012;101:052111.

[3.302] Yang, R., Li, X. M., Yu, W. D., *et al.*, "Stable bipolar resistance switching behaviour induced by a soft breakdown process at the Al/ $La0.7Ca0.3MnO_3$ interface". Journal of Physics D: Applied Physics 2009;42:175408.

[3.303] Guo, H. X., Gao, L. G., Xia, Y. D., *et al.*, "The growth of metallic nanofilaments in resistive switching memory devices based on solid electrolytes". Applied Physics Letters 2009;94:153504.

[3.304] Yu, S. and Wong, H. S. P., "Compact modeling of conducting-bridge random-access memory (CBRAM)". IEEE Transactions on Electron Devices 2011;58:1352–1360.

[3.305] Wu, X., Li, K., Raghavan, N., *et al.*, "Uncorrelated multiple conductive filament nucleation and rupture in ultra-thin high-K dielectric based resistive random access memory". Applied Physics Letters 2011;99:093502.

[3.306] Liu, X., Biju, K. P., Lee, J., *et al.*, "Parallel memristive filaments model applicable to bipolar and filamentary resistive switching". Applied Physics Letters 2011;99:113518.

[3.307] Tanaka, H., Kinoshita, K., Yoshihara, M., and Kishida, S., "Correlation between filament distribution and resistive switching properties in resistive random access memory consisting of binary transition-metal oxides". AIP Advances 2012;2:022141.

[3.308] Kinoshita, K., Tanaka, H., Yoshihara, M., and Kishida, S., "Insight into distribution and switching of resistive random-access memory filaments based on analysis of variations in memory characteristics". Journal of Applied Physics 2012;112:044503.

[3.309] Hur, J. H., Kin, K. M., Chang, M., *et al.*, "Modeling for multilevel switching in oxide-based bipolar resistive memory". Nanotechnology 2012;23:225702.

[3.310] Liu, M., Wang, A. W., He, X., Liu, Q., and Guan, W., "Multilevel resistive switching with ionic and metallic filaments". Applied Physics Letters 2009;94:233106.

[3.311] Wang, F. Z., Helian, N., Wu, S., Lim, M. G., Guo, Y., and Parker, M. A., "Delayed switching in memristors and memristive systems". IEEE Electron Device Letters 2010;31:755–757.

[3.312] Jiménez, D., Miranda, E., Tsurumaki-Fukuchi, A., Yamada, H., Suñé, J., and Sawa, A., "Multilevel recording in Bi-deficient Pt/BFO/SRO heterostructures based on ferroelectric resistive switching targeting high-density information storage in nonvolatile memories". Applied Physics Letters 2013;103:263502.

[3.313] Lee, S. B., Chang, S. H., Yoo, H. K., and Kang, B. S., "Stabilizing the forming process in unipolar resistance switching using an improved compliance current limiter". Journal of Physics D: Applied Physics 2010;43:485103.

[3.314] Chen, A., "Switching control of resistive switching devices". Applied Physics Letters 2010;97:263505.

[3.315] Long, S., Lian, X., Cagli, C., *et al.*, "A model for the set statistics of RRAM inspired in the percolation model of oxide breakdown". IEEE Electron Device Letters 2013;34:999–1001.

[3.316] Chua, L., "Memristor-the missing circuit element". IEEE Transactions on Circuit Theory 1971;18:507–519.

[3.317] Chua, L. and Kang, S., "Memristive devices and systems". Proceedings of the IEEE 1976;64:209–223.

[3.318] Chua, L., "Resistance switching memories are memristors". Applied Physics A 2011;102:765–783.

[3.319] Joglekar, Y. N. and Wolf, S. J., "The elusive memristor: properties of basic electrical circuits". European Journal of Physics 2009;30:661–676.

[3.320] Shin, S., Kim, K., and Kang, S. M., "Compact models for memristors based on charge–flux constitutive relationships". IEEE Transactions on Comput-Aided Design of Integrated Circuits and System 2010;29:590–598.

[3.321] Eshraghian, K., Kavehei, O., Cho, K. R., *et al.*, "Memristive device fundamentals and modeling: applications to circuits and systems simulation". Proceedings of the IEEE 2012;100:1991–2007.

[3.322] Sánchez, M. J., Rozenberg, M. J., and Inoue, I. H., "A mechanism for unipolar resistance switching in oxide nonvolatile memory devices". Applied Physics Letters 2007;91:252101.

[3.323] Ielmini, D., "Modeling the universal set/reset characteristics of bipolar RRAM by field- and temperature-driven filament growth". IEEE Transactions on Electron Devices 2011;58:4309–4317.

[3.324] Larentis, S., Nardi, F., Balatti, S., Gilmer, D. C., and Ielmini, D., "Resistive switching by voltage-driven ion migration in bipolar RRAM—part II: modeling". IEEE Transactions on Electron Devices 2012;59:2468–2475.

[3.325] Nardi, F., Larentis, S., Balatti, S., Gilmer, D. C., and Ielmini, D., "Resistive switching by voltage-driven ion migration in bipolar RRAM—part I: experimental study". IEEE Transactions on Electron Devices 2012;59:2461–2467.

[3.326] Ramu, A. T. and Strukov, D. B., "Thermal modeling of resistive switching devices". IEEE Transactions on Electron Devices 2013;60: 1938–1943.

[3.327] Lu, Y., Gao, B., Fu, Y., *et al.*, "A simplified model for resistive switching of oxide-based resistive random access memory devices". IEEE Electron Device Letters 2012;33:306–308.

[3.328] Ielmini, D., Nardi, F., and Balatti, S., "Evidence for voltage-driven set/reset processes in bipolar switching RRAM". IEEE Transactions on Electron Devices 2012;59:2049–2056.

[3.329] Hashem, N. and Das, S., "Switching-time analysis of binary-oxide memristors via a nonlinear model". Applied Physics Letters 2012;100: 262106.

[3.330] Benderli, S. and Wey, T. A., "on SPICE macromodelling of TiO_2 memristors". Electronics Letters 2009;45:377–379.

[3.331] Biolek, Z., Biolek, D., and Biolkova, V., "SPICE model of memristor with nonlinear dopant drift". Radio Engineering 2009;18:210–214.

[3.332] Rak, A. and Cserey, G., "Macromodeling of the memristor in SPICE". IEEE Transactions on Comput-Aided Design of Integrated Circuits and System 2010;29:632–636.

[3.333] Abdalla, H. and Pickett, M. D, "SPICE modeling of memristors". In: International Symposium on Circuits and Systems (ISCAS), 2011, pp. 1832–1835.

[3.334] Guan, X., Yu, S., and Wong, H. S. P., "A SPICE compact model of metal oxide resistive switching memory with variations". IEEE Electron Device Letters 2012;33:1405–1407.

[3.335] Linn, E., Siemon, A., Waser, R., and Menzel, S., "Applicability of well-established memristive models for simulations of resistive switching devices". IEEE Transactions on Circuits and System I 2014;61: 2402–2410.

[3.336] Strukov, D. B., Sniden, G. S., Stewart, D. R., and Williams, R. S., "The missing memristor found". Nature Letters 2008;453:80–83.

[3.337] Bersuker, G., Gilmer, D. C., Veksler, D., *et al.*, "Metal oxide RRAM switching mechanism based on conductive filament microscopic properties". In: Proceedings of the IEEE IEDM, 2010, Vol. 19, pp. 6-1-4.

[3.338] Gao, B., Kang, J. F., Chen, Y. S., *et al.*, "Oxide-based RRAM: unified microscopic principle for both unipolar and bipolar switching". In: Proceedings of the IEEE IEDM'11, 2011, Vol. 17.4, pp. 1–4.

[3.339] Makarov, A., Sverdlov, V., and Selberherr, S., "Stochastic model of the resistive switching mechanism in bipolar resistive random access memory: Monte Carlo simulations". Journal of Vacuum Science & Technology B 2011;29(03):1–5.

[3.340] Sun, P., Liu, S., Li, L., and Liu, M., "Simulation study of conductive filament growth dynamics in oxide-electrolyte-based ReRAM". Journal of Semiconductors 2014;35:104007.

[3.341] Chiu, F., "A review on conduction mechanisms in dielectric films". Advances in Materials Science and Engineering 2011(578168): 18 p.

[3.342] Ho, Y., Huang, G. M., and Li, P., "Dynamical properties and design analysis for nonvolatile memristor memories". IEEE Transactions on Circuits and Systems I: Regular Papers 2011;58:724–36.

[3.343] Kvatinsky, S., Friedman, E. G., Kolodny, A., and Weiser, U. C., "TEAM: ThrEshold Adaptive Memristor Model". IEEE Transactions on Circuits and Systems I: Regular Papers 2013;60:211–221.

[3.344] Shang, Y., Fei, W., and Yo, H., "Analysis and modeling of internal state variables for dynamic effects of nonvolatile memory devices". IEEE Transactions on Circuits and Systems I: Regular Papers 2012;59: 1906–1918.

[3.345] Szot, K., Speier, W., Bihlmayer, G., and Waser, R., "Switching the electrical resistance of individual dislocations in single-crystalline $SrTiO_3$". Nature Materials 2006;5:312–320.

[3.346] Yang, J. J., Pickett, M. D., Li, X., Ohlberg, D. A. A., Stewart, D. R., and Williams, R. S., "Memristive switching mechanism for metal/oxide/metal nanodevices". Nature Nanotechogy 2008;3:429–433.

[3.347] Chang, T., Jo, S. H., Kim, K. H., Sheridan, P., Gaba, S., and Lu, W., "Synaptic behaviors and modeling of a metal oxide memristive device". Applied Physics A 2011;102:857–863.

[3.348] Borghetti, J., Strukov, D. B., Pickett, M. D., and Yang, J. J., "Electrical transport and thermometry of electroformed titanium dioxide memristive switches". Journal of Applied Physics 2009;106:124504.

[3.349] Pickett, M. D., Strukov, D. B., Borghetti, J. L., *et al.*, "Switching dynamics in titanium dioxide memristive devices". Journal of Applied Physics 2009;106:074508.

[3.350] Hur, J. H., Lee, M. J., Lee, C. B., Kim, Y. B., and Kim, C. J., "Modeling for bipolar resistive memory switching in transition-metal oxides". Physical Review B: Condensed Matter 2010;82:155321.

[3.351] Rossel, G., Miejer, G. I., Brémaud, D., and Widmer, D., "Electrical current distribution across a metal–insulator–metal structure during bistable switching". Journal of Applied Physics 2001;90:2892–2898.

[3.352] Simmons, J. and Verderber, R., "New conduction and reversible memory phenomena in thin insulating films". Proceedings of the Royal Society A 1967;301: 77–102.

[3.353] Kund, M., Beitel, G., Pinnow, C.-U., *et al.*, "Conductive bridging RAM (CBRAM): An emerging non-volatile memory technology scalable to sub 20 nm". In: IEDM Technical Digest 2005, pp. 754–757.

[3.354] Miranda, E., Jimenez, D., and Suñé, J., "The quantum point-contact memristor". IEEE Electron Device Letters 2012;33:1474–1476.

[3.355] Brugler, J. S. and Jespers, P. G. A., "Charge pumping in MOS devices". IEEE Transactions on Electron Devices 1969;16(3):297–302.

[3.356] Groeseneken, G., Maes, H., Beltran, N., and De Keersmaecker, R. F., "Reliable approach to charge-pumping measurements in MOS transistors". IEEE Transactions on Electron Devices 1984;31(1):42–53.

[3.357] Lang, D. V., "Deep-level transient spectroscopy: a new method to characterize traps in semiconductors". Journal of Applied Physics 1974;45 (3023):3023–3032.

[3.358] Johnson, B. C., McCallum, J. C., Willems van Beveren, L. H., and Gauja, E., "Deep level transient spectroscopy study for the development of ion-implanted silicon field-effect transistors for spin-dependent transport". Thin Solid Films 2010;518:2524–2527.

[3.359] Grasser, T., "Bias temperature instability for devices and circuits". Springer, 2014. ISBN 978-1-4614-7909-3.

[3.360] Mooney, P. M., "Photo-deep level transient spectroscopy: a technique to study deep levels in heavily compensated semiconductors". Journal of Applied Physics 1983;54:208–213.

[3.361] Abe, T. and Nishi, Y., "Non-destructive measurement of surface concentrations and junction depths of diffused semiconductor layers". Japanese Journal of Applied Physics 1968:397–403.

[3.362] De Wolf, I., Senez, V., Balboni, R., *et al.*, "Techniques for mechanical strain analysis in sub-micrometer structures: TEM/CBED, micro-Raman spectroscopy, X-ray micro-diffraction and modeling". Microelectronics Engineering 2003;70:425–435.

[3.363] Sorianello, V., Colace, L., Assanto, G., and Nardone, M., "Micro-Raman characterization of Germanium films evaporated on various substrates". Microelectronic Engineering 2011;88:492–495.

[3.364] http://www.lotis-tii.com/files/NF_eng.pdf.

[3.365] Sverdlov, V., "Strain-induced effects in advanced MOSFETs". Springer-Verlag/Wien, 2011.

[3.366] von Klitzing, K., Dorda, G., and Pepper, M., "New method for high-accuracy determination of the fine-structure constant based on quantized hall resistance". Physics Review Letters 1980;45:494.

[3.367] Frustaglia, D., "Spin-dependent transport in mesoscopic quantum systems". PhD dissertation, Technical University Dresden, Germany, 2001.

[3.368] Landau, L. D. and Lifshitz, E. M., "Quantum mechanics: nonrelativistic theory". Pergamon Press, 1977.

[3.369] Lifshitz, E. M. and Kosevich, A. M., "Theory of the Shubnikov-de Haas effect". Journal of Physics and Chemistry of Solids 1958;4(1–2):1–10.

[3.370] Stormer, H. L., Tsui, D. C., and Gossard, A. C., "The fractional quantum Hall effect". Review of Modern Physics 1999;71:S298–S305.

[3.371] Halperin, B. I., "Quantized Hall conductance, current-carrying edge states, and the existence of extended states in a two-dimensional disordered potential". Physical Review B 1982;25(4):2185–2190.

[3.372] Popovic, R. S., "Hall effect devices". 2nd ed., Institute of Physics, 2004, Chapter 6.

[3.373] Rodríguez-T., R., "Three-dimensional simulations of split-drain MAGFET". PhD dissertation, Technical University Vienna, Austria, 2003.

[3.374] Klima, R., "Three-dimensional device simulation with Minimos-NT". PhD dissertation, Technical University Vienna, Austria, 2002.

[3.375] Yu, P. and Cardona, M., "Fundamentals of semiconductors, Physics and Materials Properties". 4th ed., Springer, 2010, pp. 233–234.

[3.376] Popovic, R. S. and Baltes, H. P., "A CMOS magnetic field sensor". IEEE Journal of Solid-State Circuits 1983;SC-18:426–428.

[3.377] Gutiérrez-D., E. A., Murphy-A., R. S., Torres-J., A., *et al.*, "A sub-mT cryogenic silicon magnetic sensor". In: IEEE ESSDERC'98 Proceedings, Bordeaux, France, 1998, pp. 188–191.

[3.378] www.gmw.com/electromagnets/dipole/5403/5403_AC-Syst.html.

[3.379] Póndigo De Los, A. E., Gutiérrez-D., E. A., Molina-R, J., Guarin, F., "Non-homogeneous space mechanical strain induces asymmetrical magneto-tunneling conductance in MOSFETs". In: Proceedings of the ESSDERC'2014 Conference, Venice, Italy, 2014, pp. 90–93.

[3.380] Gutiérrez-D., E. A., Póndigo de los-A., E., Vega-G., V. H., and Guarin, F., "Identification of vector gate current components in strained 28 nm high-k pFETs". In: IEEE Semiconductor Interface Specialists Conference SISC'11, Arlington, VA, USA, 2011, Vol. 42, p. P11.

[3.381] Gutiérrez-D., E. A., Póndigo-de los-A., E., Vega-G., V. H., and Guarin, F., "Observation of asymmetric magnetoconductance in strained 28-nm Si MOSFETs". IEEE Electron Device Letters 2012;33(2):254–256.

[3.382] Yu, P. Y. and Cardona, M., "Fundamentals of semiconductors". Springer-Verlag, NewYork, 1996, Chapter 5, pp. 222–223.

[3.383] Roumenin, C. S., Handbook of sensors and actuators, "Solid State Magnetic Sensors". Elsevier, 1994, Vol. 2, pp. 30–33.

[3.384] Suñé, J., Wu, E. Y., and Tous, S., "Physics-based percolation model of oxide breakdown". ECS Transactions 2007;8(1):177–183.

[3.385] Miranda, E., Jiménez, D., Suñé, J., *et al.*, "Nonhomogeneous spatial distribution of filamentary leakage current paths in circular area Pt/HfO_2/Pt capacitors". Journal of Vacuum Science & Technology B 2013; 31:01A107.

[3.386] Chang, W.-T., Chih-Chung, W., Lin, J.-A., and Yeh, W.-K., "External stresses on tensile and compressive contact etching stop layer SOI MOSFETs". IEEE Transactions on Electron Devices 2010;57(8):1889 –1894.

[3.387] Xu, N., Wang, L. T.-N., Neureuther, A. R., and King Liu, T.-J., "Physically based modeling of stress-induced variation in nanoscale transistor performance". IEEE Transactions on Device and Materials Reliability 2011;11(3):378 –386.

[3.388] Gopinath, V., Aronowitz, S., Palankovski, V., and Selberherr, S., "Effects of stress-induced bandgap narrowing on reverse-bias junction behavior". In: Proceedings of the European Solid State Device Research Conference, 2002, pp. 631–634.

[3.389] Asenov, A., "Random dopant induced threshold voltage lowering and fluctuations in sub-0.1 μm MOSFETs: A 3-D "atomistic" simulation study". IEEE Transactions on Electron Devices 1998;45(12):2505 –2513.

[3.390] Kaya, S., Brown, A. R., Asenov, A., Magot, D., and Linton, T., "Analysis of statistical fluctuations due to line edge roughness in sub-0.1 μm MOSFETs". Simulation of Semiconductor Processes and Devices 2001:78 –81.

[3.391] Patsalasa, P. and Logothetidis, S., "Optical, electronic, and transport properties of nanocrystalline titanium nitride thin films". Journal of Applied Physics 2001;90(9):4725 –4734.

[3.392] Wang, X., Brown, A. R., Idris, N., Markov, S., Roy, G., and Asenov, A., "Statistical threshold-voltage variability in scaled decananometer bulk HKMG MOSFETs: a full-scale 3-D simulation scaling study". IEEE Transactions on Electron Devices 2011;58(8):2293 –2301.

[3.393] Balázs, J. and Podör, B., "on the theory of galvanomagnetic effects in semiconductors with anisotropic scattering". Acta Physica Acaddemiae Scientiarum Hungaricae, Tomus 1970;29(1):35–44.

[3.394] Balázs, J. and Podör, B., "Transverse magnetoresistance in plastically deformed n-type Germanium". Physica Status Solidi 1970;37(119): 119–126.

[3.395] Rodríguez, R. and Gabriela, A, "A 2D simulation methodology for thermo-magnetic effects on tunneling mechanisms of nano-scaled MOS devices". PhD dissertation, INAOE, Mexico, 2015.

[3.396] Karner, M., Gehring, A., Holzer, S., *et al.*, "A multi-purpose Schrödinger-Poisson Solver for TCAD applications". Journal of Computational Electronics 2007;6:179–182.

[3.397] Rodríguez-Ruiz, G. A., Gutiérrez-D., E. A., Sarmiento-R., L. A., *et al.*, "Thermo-magnetic effects in nano-scaled MOSFET: an experimental, modeling, and simulation approach". Journal of the Electron Devices Society 2015;3(2):78–84.

[3.398] Rodríguez-Ruiz, G. A., Gutiérrez-D., E. A., Sarmiento-R., L. A., *et al.*, "Experimental and simulation results of magnetic modulation of gate oxide tunneling current in nanoscaled MOS transistors". IEEE Electron Device Letters 2015;36(4):387–389.

[3.399] Rtsu, R. and Lesaki, L., "Tunneling in a finite superlattice". Applied Physics Letters 1973;22(11):562–564.

[3.400] Shen, S.-C., Zhu, J.-B., Mu, Y.-M., and Liu, P.-L. "Wavefunction mixture and composition for hybridized Zeeman states of P in Si." Physical Review B 1994;49(8):5300–5305.

[3.401] Lei, W., Wibbelhoff, O., Notthoff, C., *et al.*, "Magnetic field induced modification of the wavefunctions in InAs quantum dots" Physica E2008;40(6):1870–1872.

[3.402] Hurkx, G. A. M., Klaassen, D. B. M., Knuvers, M. P. G., and O'Hara, F. G., "A new recombination model describing heavy-doping effect and low-temperature behavior". In: IEDM Proceedings, 1989, pp. 307–310.

[3.403] Hurkx, G. A. M., de Graaff, H. C., Kloosterman, W. J., and Knuvers, M. P. G., "A new analytical diode model including tunneling and avalanche breakdown". IEEE Transactions on Electron Devices 1992;39(9):2090–2098.

[3.404] De Wolf, I., "Micro-Raman spectroscopy to study local mechanical stress in silicon integrated circuits". Semiconductor Science Technology 1996;11:139 –154.

[3.405] Gutierrez-D., Edmundo, A., Molina-R., J., Garcia-R., P. J., Martinez-C., J., and Guarin, F., "Magnetic field induced gate leakage current in 65 nm nMOS transistors". In: Published in the Proceedings of the IEEE ESSDERC'09 Conference, Athens, Greece, 2009, pp. 185–188.

[3.406] Gold, A., "Scattering times of the two-dimensional electron gas on silicon (111) with a density dependent effective mass". Journal of Physics: Condensed Matter 2007;19:1–12.

[3.407] Groeseneken, G., Maes, H. E., Beltrán, N., and De Keersmaecker, R. F., "A reliable approach to charge-pumping measurements in MOS transistors". IEEE Transactions on Electron Devices 1984;ED-31(1):42–53.

[3.408] Gutiérrez-D. and Edmundo A., Private communication.

[3.409] Futatsugi, T., Nakajima, A., and Nakao, H., "Silicon single-electron memory using ultra-small floating gate". FUJITSU Scientific and Technical Journal 1998;34:142–152.

Chapter 4
The semiconductor device technology and its societal impact

Edmundo A. Gutiérrez-D.[1], *Fernando Guarín*[2] *and Pedro J. García-R.*[3]

4.1 Introduction

This chapter brings together the social implications and ubiquitous impact of semiconductor devices in today's life throughout every corner of the world. It is our intention to show tangible examples on how high-technology electron devices have improved the quality and conditions of human life, while helping to preserve a sustainable and greener environment. We present here a compilation of several different applications, where high-technology or state-of-the-art electron devices make it possible to save vast amount of energy in several areas; from more energy-efficient lighting, control of motors, scaling down to the plethora of electronic systems that are prevalent in modern society and have changed the manner in which we interact with others at the local and global levels. We intend to show how new materials or fabrication steps, incorporated in the manufacture of high-technology electron devices, have been combined to produce solutions that reach far beyond the microprocessor's mainstream. Applications beyond the highly integrated processors, which are the main driving force of high-technology electron devices, have been shown to be highly effective in several fields, such as energy conversion, biosignal monitoring, human health, medicine, and drinking water monitoring and decontamination, to list just a few examples.

4.2 Energy-efficient electron devices and the sustainable and green environment

Let us start our journey by considering the environmental implications brought about by the significant advances in our ability to switch power. In a relatively not too

[1]Department of Electronics, National Institute for Astrophysics, Optics, and Electronics, Puebla, Mexico
[2]GlobalFoundries, Inc., East Fishkill, NY, USA
[3]Instituto de Ingeniería, Universidad Veracruzana, Boca del Río, Veracruz, Mexico

distant past early 1900s, we were in the era of mechanical switches and relays that could only achieve speeds in the range of seconds. A major improvement was achieved by the introduction of the bipolar transistor in around 1950, but it was bulky for power applications and had relatively low current gain so it was replaced by power metal–oxide–semiconductor field-effect transistors (MOSFETs) around 1970. While today's semiconductor switches (MOS transistors) can achieve rates that are nine orders of magnitude faster (billions of switching events per second), these speeds, however, can be achieved only for very small current loads. In the realm of higher power applications, we can achieve an increase of six orders of magnitude (millions of switching events per second) by using the insulated-gate bipolar transistor or IGBT, a power semiconductor device pioneered in around 1980 by J. Baliga [4.1] that is primarily used as an electronic switch in just about all of today's modern appliances and many more advanced systems, from lighting, electric washers, air conditioners (anything with an electric motor) to electric cars, airplanes, bullet trains, and medical applications. All of these modern advances we associate with a high standard of living and convenience demand large amounts of energy and drive large increases to the generated amount of CO_2 emissions. The IGBT's pervasive utilization has played a significant role in reducing our carbon footprint by greatly diminishing the "on" power dissipation while reducing switching power losses and cost.

One of its major applications was to replace the mechanical distributor for the spark-based ignition systems used to fire automotive combustion engines with the far more energy-efficient electronic ignition system in the late 1980s. According to Baliga's estimates assuming a 10 per cent improvement in fuel efficiency, a cumulative worldwide savings of 1.48 trillion gallons of gasoline has been obtained [4.2].

Thanks to the IGBTs small foot print and large power handling capability compact fluorescent lights (CFL) are now possible, there are 20 billion CFLs used today around the globe and they can be credited with an annual worldwide energy savings of 1000 GW in 2015 [4.3].

It has been calculated that "the improved efficiency gained by using the IGBT has resulted in saving over one trillion gallons of gasoline and reducing electrical energy consumption by more than 50,000 terrawatt-hours (equivalent to not having to build 600 one-gigawatt coal-fired power plants). This has saved consumers $15 trillion while reducing carbon dioxide emission by more than 75 trillion pounds" (Table 4.1) [4.1].

The social impact of the IGBT constitutes a remarkable feat for the tangible contribution of semiconductor devices to our society [4.2].

Another significant contribution of semiconductor devices to society that have enabled a new source of energy-efficient lighting are LED lamps fabricated using Gallium Nitride GaN [4.4]. They emit a bright white light that lasts longer than incandescent light bulbs and are more energy-efficient. They have made continuous improvements, achieving higher efficiencies with higher luminous flux per unit electrical input power. The most recent record is about 300 lm/W, equivalent to 16 regular light bulbs or close to 70 fluorescent lamps. Considering that an estimated 25 per cent of the global electricity consumption is used for lighting purposes, white LEDs contribute to significant energy savings. The LED is expected to have

Table 4.1 Energy savings enabled by the use of IGBT's and CFL's

IGBT-enabled application	**Cumulative gasoline or energy savings**		**Consumer cost savings**		**Utility cost savings**	
	United States	**World**	**United States**	**World**	**United States**	**World**
Electronic ignition system	318 B gallons	1477 B gallons	\$0.654 T	\$9.125 T	–	–
Adjustable speed motor drive	26,170 TWh	56,910 TWh	\$2.537 T	\$11.38 T	\$0.398 T	\$1.12 T
Compact fluorescent lamp	1550 TWh	16,120 TWh	\$0.156 T	\$3.224 T	\$0.26 T	\$2.98 T
Total			\$3.35 T	\$23.73 T	\$0.665 T	\$4.10 T

a large societal impact for many people around the world that lack access to electricity grids. Thanks to the GaN LEDs' low power requirements, they can be powered by solar power in remote corners of the world [4.4].

The design and mass production of microprocessors, which constitute the computing core inside of several computer-based platforms, such as personal computers (PCs), laptops, smart phones, and car and plane electronic control systems, have been the driving force behind the miniaturization of the metal-oxide-semiconductor field-effect-transistor, known as a MOSFET or simply FET. The smaller the FET the higher its speed and integration density, which results in the processing of huge amounts of electrical signals at a very high speed. An example of this is the Broadwell microprocessor by Intel [4.5]. The Broadwell is fabricated using a 14-nm manufacturing process integrating 1.3 billion transistors in an area of about 82 mm^2. This represents an integration density of 0.15 million of transistors per square centimeter. This microprocessor would allow driving screens larger than 13 inches with resolutions of up to 3200 × 1800 pixels, with a battery life that can extend for up to 11 hours, thus improving image quality and human mobility. As a size-scale comparison, in the diameter of a human hair, which ranges from 20 to 180 μm, 1428 to 12850 14-nm transistors can be accommodated. In terms of relative size, a 14-nm scale corresponds to the size range of antibodies [4.6, 4.7] and viruses, and it is even smaller than bacteria like the *Escherichia coli* (1 μm) or the cancer cell (10 μm). The Intel Broadwell microprocessor is fabricated using 14-nm FinFET transistor technology [4.12]. This FinFET is comparable in size to antibodies, smaller compared to viruses, bacteria, and cancer cells. Therefore, if materials or device structures, with different material composition, can be manipulated at that scale, then an interaction of materials or device structures with biology elements is possible. As referred in Chapter 1, there are already more than 20 different semiconductor, dielectric, and metal materials, as well as more than 14 different device-structures, incorporated into the manufacturing process, which combined or used as separate elements, have a potential application with sensor or actuator potential functionalities.

Some already implemented examples are those based on Micro-Electro-Mechanical systems (MEMs), finding applications in biomedicine [4.13], or the "lab-on-chip" concept [4.14], where microfluidics combines with the synthesis of smart particles as a system for drug delivery into the human body. An example of the alternative use of gate oxide dielectric layers for water decontamination has been introduced in Reference 4.15, where TiO_2 nanoparticles that have a photo-catalytic UV reaction with *E. coli* in water are prepared as thin films using an atomic layer deposition (ALD) system that is regularly used to deposit high-k FET gate dielectrics. A slightly more extensive review on the use of nanotechnologies for environmental remediation can be found in Reference 4.16, where drinking water cleaning, removal of air pollutants, and clean-up of industrially contaminated sites were addressed. Another example of the use of the ALD technique is in the manufacturing of nanoporous alumina membranes, where ALD is used to modify the pore size in the range of 2 nm, which then can be potentially used as an implantable sensor or water purification–filtering membrane [4.17].

Besides the high-density integration, switching speed, and reduced short-channel effects, provided by the FinFET technology, there is the energy efficiency concern. AMD [4.18], for instance, has announced the Graphics Core Next, which is a graphic processing unit implemented with a 14/16-nm FinFET that will basically double the energy efficiency when compared to previous versions based on planar or bulk FET technology. In Reference 4.19, an analysis based on various combinational and sequential circuits forecasts an average energy reduction of 10 and 1000 times in the super-threshold regime (strong inversion), and an average 16 and 3000 times in the near-threshold regime (weak inversion). These are results compared to 14 and 45 nm bulk CMOS technologies, respectively.

High-density integration and high-speed information processing are two key elements achievable with FinFET technology. However, there is still a third feature required to get a high-performance and energy-efficient system; the storage and retrieval of information. When FinFETs are used as access transistors in spin-transfer torque magnetic random access memory (STT-MRAM), the FinFET-accessed STT-MRAM has an average improvement of 116, 75, 27, 14, 22, and 43 per cent, in area, read/write latency, read/write energy consumption, and leakage power, when compared to its CMOS counterpart [4.20].

For digital circuit, downscaling of FET structures brings benefits in terms of a high speed of information processing, information density, and information storage. On the analog circuit arena, as the transistor gain reduces with the channel length, the downscaling tends to degrade analog circuit performance. However, for transceiver systems used in optical interconnects, the FinFET-based amplifier in a 30 Gb/second system, an energy saving of more than 0.1 pJ/bit (16%) can be achieved when compared to fully-depleted SOI technology [4.21]. Other examples of analog FinFET can be found in Reference 4.22, where by re-engineering of the source-drain extension (SDE), the intrinsic gain and cutoff frequency are improved nearly twice compared to devices with abrupt SDE regions. This improvement provides good opportunities for ultra-low-voltage analog applications. On the other hand, experiments conducted at cryogenic temperatures on 90-nm triple-gate FinFETs [4.23] shows a maximum

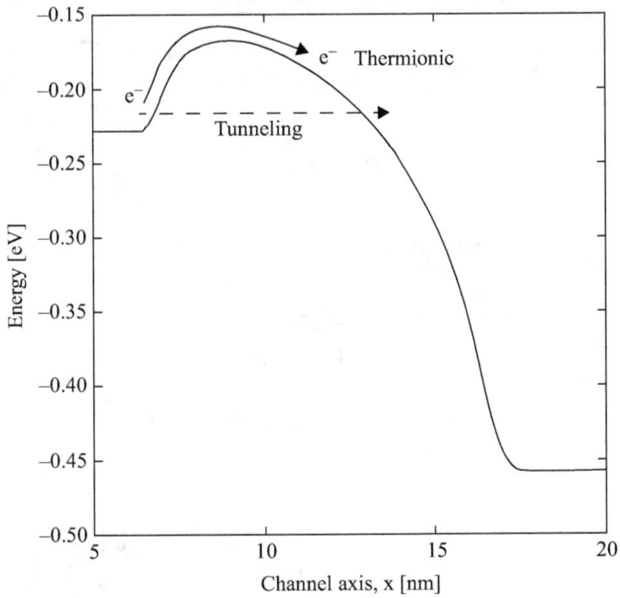

Figure 4.1 Schematic representation of an energy barrier along the channel of a MOSFET device, with the thermionic injection of electrons passing over the energy barrier, and a probable tunneling injection through the potential barrier

voltage gain of 36.2 dB at 300 K with a roll down to 35.3 dB at 350 K, and also a roll down to 33 dB at 100 K. The maximum transconductance per width ($g_{m,max}/W$) increases from about 1.8 [μS/μm] at 300 K up to 4.5 [μS/μm] at 100 K.

The variety of FET device structures, such as bulk CMOS, SOI-FET, and FinFET are based on the conventional thermionic injection of electrons over the source-to-drain surface energy barrier as shown in Figure 4.1.

When the channel length or the potential barrier width approaches 10 nm or less, then some electrons are able to tunnel through, but still the majority of carriers are injected into the channel by the thermionic process. A thermionic process then still dominates the injection of electrons into the channel from the source, in the subthreshold regime, where the subthreshold slope S is [4.24]

$$S = \frac{\partial V_g}{\partial \psi_s}\frac{\partial \psi_s}{\partial(log_{10}I_d)} \cong \left(1 + \frac{C_d}{C_{ox}}\right)\ln 10\frac{kT}{q} \tag{4.1}$$

with

$$\frac{kT}{q}\ln 10 \cong 60\,\mathrm{mV/decade} \tag{4.2}$$

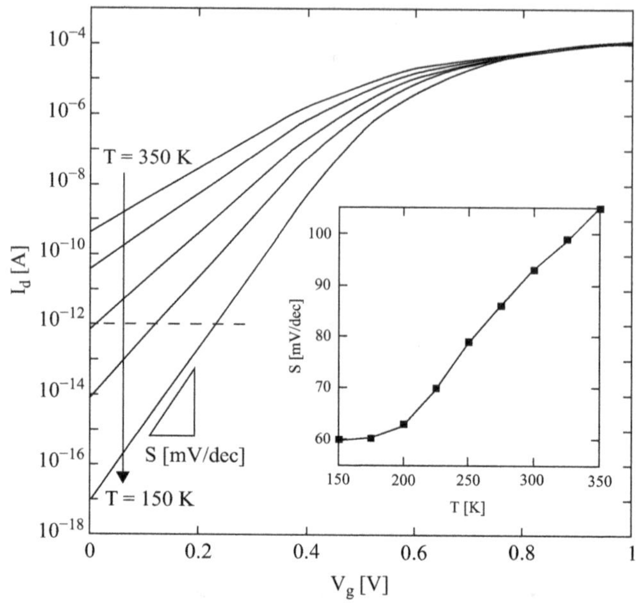

Figure 4.2 Measured I_d-V_g characteristics of a (W/L) = (1 μm/30 nm) n-MOSFET biased at $V_d = 0.1$ V at 5 different operating temperatures T. For the sake of clarity, the subthreshold regime below $V_g < 0.3$ V has been artificially extended to Id currents lower than 1 pA. The inset shows the temperature dependence of the subthreshold slope S

at T = 300 K. C_d and C_{ox} are the depletion and gate oxide capacitances, respectively. ψ_s is the surface potential. The thermionic injection set a minimum of 60 mV/decade as shown by (4.2). Figure 4.2 shows the subthreshold regime of a (W/L) = (1 μm/30 nm) n-type high-k metal gate strained n-MOSFET at different temperatures.

As expected the increase in temperature T enhances the diffusive process, and the subthreshold slope degrades. The smallest the subthreshold slope, the most energy efficient the transistor is. Figure 4.3 shows the I_d-V_g characteristics of a 14-nm two-fins FinFET and a high-k metal gate 28-nm bulk FET.

The FinFET exceeds the bulk FET by 27 per cent in the subthreshold slope, and has a larger drive current capability than its bulk counterpart. The FinFET is more energy efficient than the bulk FET. However, there is still a chance to improve the energy efficiency by doing another change to the FET device structure. The additional gain on energy efficiency is achieved by the use of band-to-band-tunneling at the source/channel junction as shown in Figure 4.4.

The tunnel FET operation [4.24] is based on band-to-band tunneling at the n^+-intrinsic junction as shown at the lower left side in Figure 4.4. For low values of V_g (dashed line), the minimum of the conduction band edge E_C does not overlap with the maximum of the valence band E_V, preventing electrons to move into the channel. However, as the V_g voltage increases, the energy bands bend until E_C and

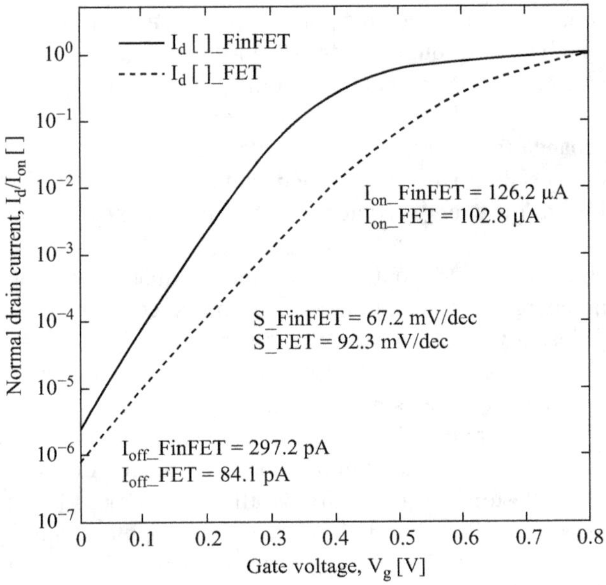

Figure 4.3 Measured normalized drain current I_d/I_{on} versus gate voltage for $V_d = 0.1$ V at $T = 300$ K. The continuous line corresponds to a FinFET, while the dashed line is for a bulk FET

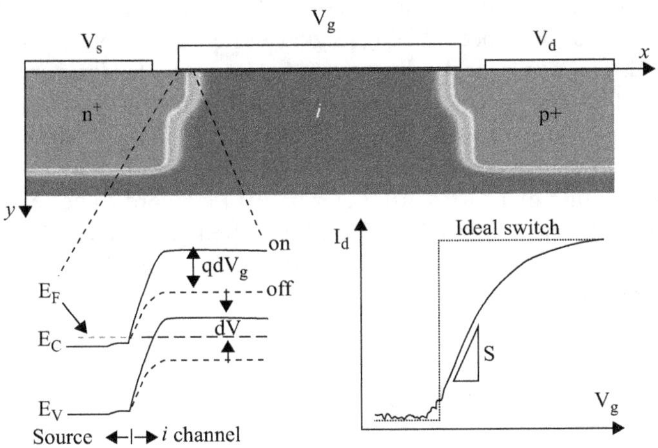

Figure 4.4 The tunnel-FET concept. Above the gated n^+-i-p^+ structure, below at the left the energy band bending at the source-channel junction

E_V overlaps (continuous line). At this point, electrons are injected from the source into the intrinsic channel through band-to-band-tunneling.

An excellent review on the potential applications of tunnel FETs toward less than 0.5 V beyond CMOS logic is found in Reference 4.25, where several options for high-performance tunnel FETs are projected for supply voltages under 0.5 V, with on currents exceeding 100 μA/μm when using narrow bandgap heterojunctions.

In general terms the energy efficiency needs to be looked at from the device up to the system level.

The evolution of the electronics has an inescapable dependence on the triad space, time, and energy. Space and time has been two components of the triad that have prevailed during this early evolution that accounted for the major transition from vacuum tubes to solid-state devices, the concept of integrated circuit (IC) [4.26], and one of the major developments at the system level, the solid-state flight-computer used on the Apollo 11 mission. The Apollo mission is one of the first emblematic examples of the ubiquitous societal impact of the electronics. It represented a big milestone for humanity by allowing humans for the first time to land on the moon in 1969. The Apollo Guidance Computer, which provided guidance, navigation, and control of the flights to the moon, was the first flight computer that used ICs [4.27] based on the resistor-transistor-logic (RTL). Of note, 2800 ICs from the Fairchild Company were flat-packed with a 16-bit CPU capacity running at 2.048 MHz with a power consumption of 55 W.

It took only 19 years from the concept of the first transistor in 1947 to the first commercial IC in 1966. The middle of the 1960s is the launching stage for the integrated electronics as demonstrated by the first mainframe computers using ICs [4.28]. The early adventure of the integrated electronics extends until the 1970s, 1980s, and 1990s with the subsequent development of several computers and ICs based on bipolar transistors, such as high-speed analog–digital systems, and optical fiber network systems [4.29]. The bipolar technology extended beyond 20 ps/gate and up to cutoff frequencies of 100 GHz, which was later combined with CMOS to form a BiCMOS technology intended to reduce the power consumption aiming at telecommunication applications. In 2008 a speed record of 618 GHz for peak f_{max} was achieved at an operating temperature of 12 K [4.30]. Later in 2014 a record ac performance of 0.8 THz for a SiGe HBT operating at $T = 4.3$ K is reported in Reference 4.31. A $T = 300$ K projected roadmap based on TCAD simulations targets an f_{max} of 910 GHz for a 32-nm emitter width [4.32]. The incorporation of SiGe heterostructures in the bipolar technology since the 1990s has been beneficial for increasing the speed close to the TeraHertz (THz) regime, where automotive radars, and sub-mm-wave and near-THz imaging systems look as a potential application. However, the SiGe HBT by itself is a high-power consumption technology.

An option for high-speed devices and low-power consumption circuits and systems is the FET technology, born in 1960 when Dawong Kahng filled the patent on the electric field–controlled semiconductor device [4.33]. In 1964 Robert H. Norman filled his patent for a 20-bit shift register composed of 120 p-type transistors [4.34]. In 1963 F. Wanlass filled the patent on complementary FETs, i.e., the combined use of p- and n-type FETs (CMOS) [4.35]. In 1968 the RCA company introduced the first commercial CMOS ICs, the so-called 4000 family [4.36]. The 4000 family ICs implemented a variety of logic functions, such as AND, OR, and NAND, NOR. At this very beginning the CMOS ICs operated from a 3 to 15 V supply range. It took only 8 years for the FET technology to go from the transistor concept to the first IC. In 1971 Intel introduced the first integrated microprocessor [4.38], the 4004 a 4-bit 740 kHz microprocessor with 2238 transistors.

Then in 1979 Intel introduced the 8088, a 16-bit microprocessor built with a 3-μm nMOS technology, that later was followed by the 80C88 and other variants with speeds of up to 8 MHz [4.40]. A mature commercial product, the PC by IBM, is launched in 1981 [4.39]. The IBM PC is based on the Intel 8088 microprocessor. The launching of the IBM PC is a milestone that marks the beginning of a very fast evolution of the CMOS technology and its multiple applications at the circuit and system levels. It took only 21 years since the issue of the FET patent concept until a mature and ubiquitous application, such as the PC, was released to the market. The evolution of microprocessors has been accompanied by a parallel development of memories as shown in the excellent review by C.-T. Sah [4.41]. The review covers the 1928–1986 period of time, where the three-dimensional development of the one-transistor dynamic random access memory cell is covered. From then on, there has been an exponential explosion in terms of downsizing, number of transistors, and transistors per square millimeter as shown in Figure 4.5.

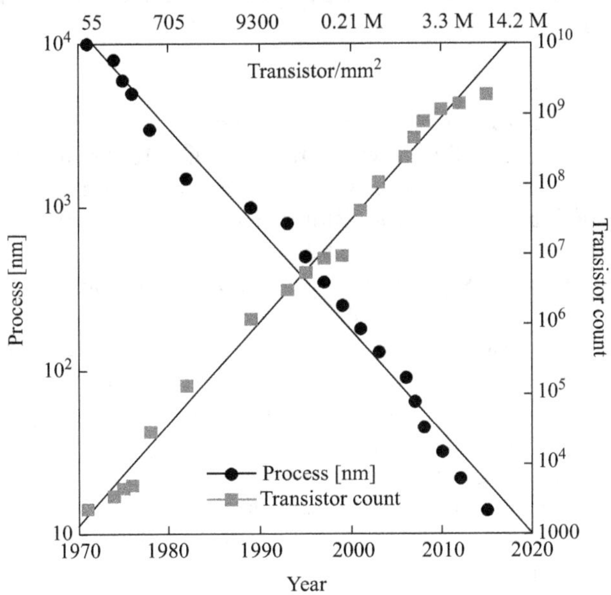

Figure 4.5 Time evolution of the fabrication process, measured in nanometers, the transistor count per microprocessor (right axis), and the transistor density per square millimeter (upper axis). Starting the year 2000 the upper axis scales to millions. Data collected from multiple sources

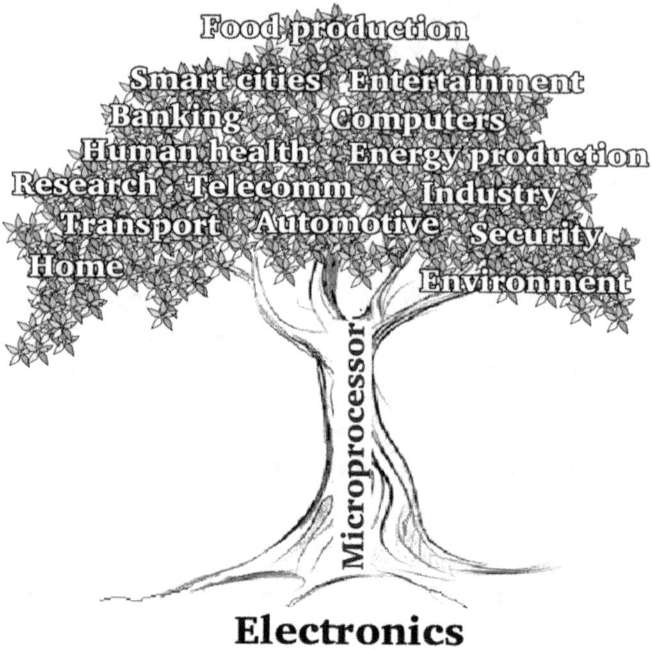

Figure 4.6 A very fruitful technology tree planted on a very productive soil—the Silicon

From 1970 to 2015 the minimum fabrication feature, for a microprocessor technology, has exponentially decreased from 10 μm (10,000 nm) to 14 nm. The number of transistors per microprocessor chip goes from about 2300 transistors up to 1.9×10^9, which represents a transistor density increase from 55 transistors per square millimeter up to 14.2 million of transistors per square millimeter. This is an amazing rate increase of 890×10^3 transistor/mm^2 per decade.

The explosion of the computer-platform–based systems over the last 20 years emerged as pivotal impacting the society in a large extent. Thus electronics together with its associated controlling software can be found everywhere; at home in different domestic appliances, such as smart TV's, fridges, washing machines, etc., at the industry in automation processes; at the amazing self-driving car [4.42], a vehicle that can autonomously drive on roads, at any moment in the mobile devices, such as tablets or smart phones; at hospitals as different monitoring and surgery instruments and biomedicine; at telecommunications and as many other aspects of life as shown in Figure 4.6.

The microprocessor, as the main driving force of miniaturization during the past two decades, has been the trunk of the very fruitful Electronics tree (Figure 4.6), which has extended its benefits far beyond the microprocessor itself. However, in spite of that huge societal benefit, it also comes with some detrimental

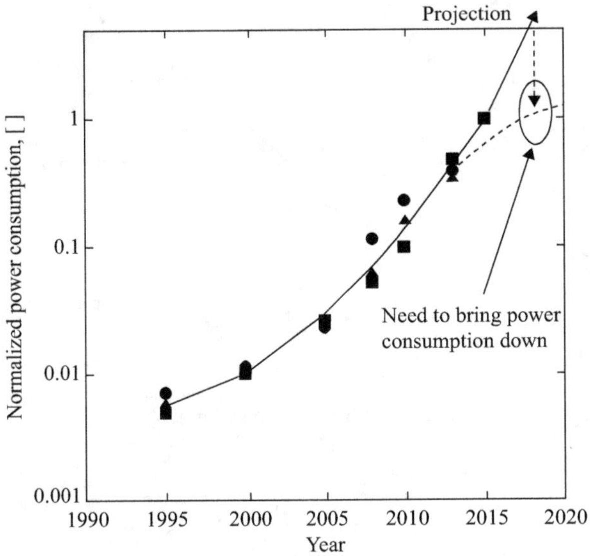

Figure 4.7 Normalized power density (W/mm^2) of several different commercially available microprocessors at 2015. The arrow indicates a projected trend beyond 2015, and the dotted line indicates a need to keep the power dissipation limited

impact on the energy consumption and the environment. Figure 4.7 shows the evolution of power consumption of some of the most relevant microprocessors during the past decades since 1995 until 2015 [4.43, 4.44]. There has been an exponential increase in the demand of energy consumption as the microprocessors advance toward more complex architectures and processing speeds. There is a myriad of different types of microprocessor for different applications, such as servers, laptops, PCs, and mobile devices, with different performance and power consumption, but in general, there is a trend toward an increase of power consumption. In some specific cases, such as mobile applications, the speed is sacrificed in favor of the mobility or long-term battery operation as shown in Figure 4.8.

The most performance-demanding microprocessors are those that require faster transistors and thus advanced technologies with an inherent larger leakage power. On the other hand, the almost or always-on devices mobile devices, such as smart phones or tablets, are designed with the purpose of improving its mobility or long-term battery operability.

The energy efficiency of a range of several different computers, from PCs to mainframes, has been doubling every 1.57 years from 1946 to 2009 [4.43], and every 2.7 from 2009 on [4.44]. As quoted by J. Koomay and S. Naffziger in Reference 4.44 the slight decline in energy efficiency is attributed to leakage

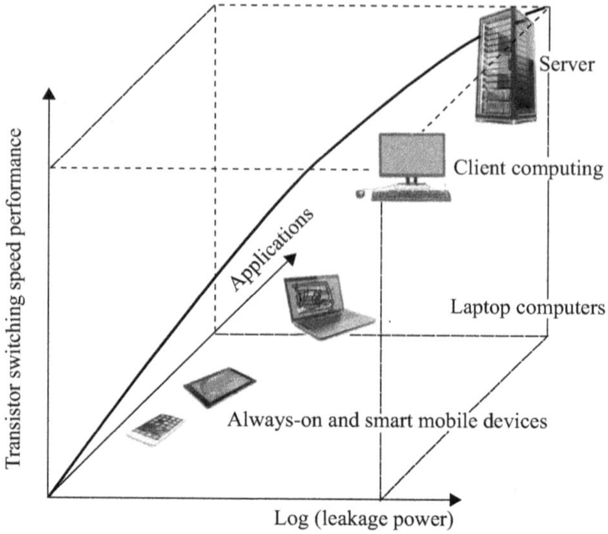

Figure 4.8 The speed–energy–application triad

current of miniaturized transistors. The data shown in Figure 4.9 belong to computers operating at peak-output efficiency. However, because of power management today most computers run at computational peak only a fraction of their operation, which reduces power consumption. Additional strategies that help keeping energy efficiency accounts for multicore dynamic management [4.45], dynamic sleep-mode for mobile devices and desktop computers [4.46], and architectural innovation such as putting graphical processing units with and accelerated processing units together in the same silicon die, allowing sharing power and thermal management infrastructure that also helps energy efficiency as proposed by the AMD company [4.47]. As quoted by a white paper by the AMD company [4.47] "if all computers sold in U. S. A. were Energy Star certified, it would save \$1 billion and reduce greenhouse gas emissions by 15 billion pound annually, the equivalent emissions of 1.4 million vehicles" [4.48], which denotes the relevant importance of energy efficiency techniques, strategies, and architectures at the device, circuit, and system level.

The moving or transport of the electron, which produces an electric current and a potential, is the substantial mechanism used for the implementation of any digital or analog function. This has been accomplished so far with bipolar- or FET-based transistors. The electric current and its associated scattering and heat dissipation effects are the mechanisms that contribute to energy waste. If one could avoid the energy dissipative mechanisms of a moving electron, then one would be able to create a high energy-efficient electronic system. This may be accomplished with the use of spintronics. Spintronics has to do with the control and manipulation of

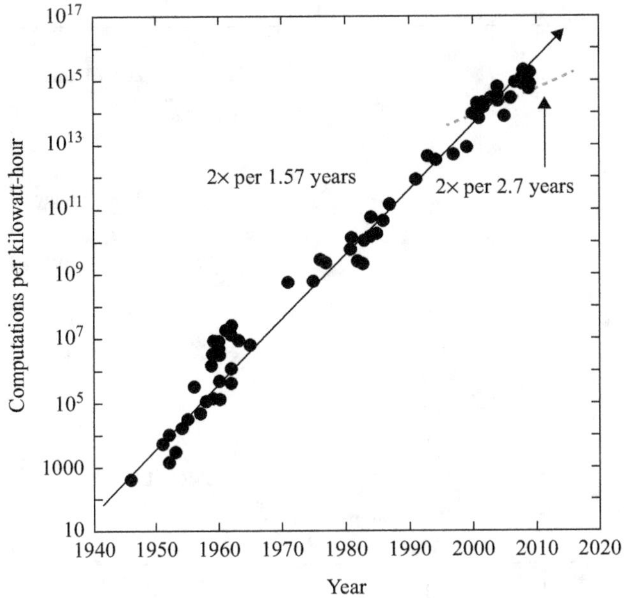

Figure 4.9 Evolution of computing efficiency at peak performance per kilowatt-hour

the electron spin, or in other words with the detection and manipulation of the spin state without the need of applying an electric current. A paramount difference with charge-based devices is that spin-based devices are nonvolatile, are very fast, and have a good endurance [4.49]. Figure 4.10 shows the analysis of energy versus delay of inverters with fanout-4. The 15 devices shown in Figure 4.10 are grouped into three different classes depending on whether they are controlled by charge or spin, or a by an exciton. Table 4.2 shows the device classification.

The conduction mechanism differs from class to class. The most common and widely used is the conduction that takes place over a potential barrier by diffusion or drift. The tunneling via band-to-band is the mechanism through which carriers are injected into the channel. The graphene pn junction device works in function of reflection and transmission controlled by a gate. The BisFET is a graphene-based device that functions by tunneling in between a graphene bilayer. Holes are injected into one graphene monolayer and electrons into the other monolayer. Then they bind into excitons that relax into a Bose–Einstein condensate.

A fanout-4 inverter is used as a benchmark to test the energy-delay characteristics of the 15 different device topologies shown in Table 4.2. The results are shown in Figure 4.10. The graphene nanoribbon technology is the most efficient in

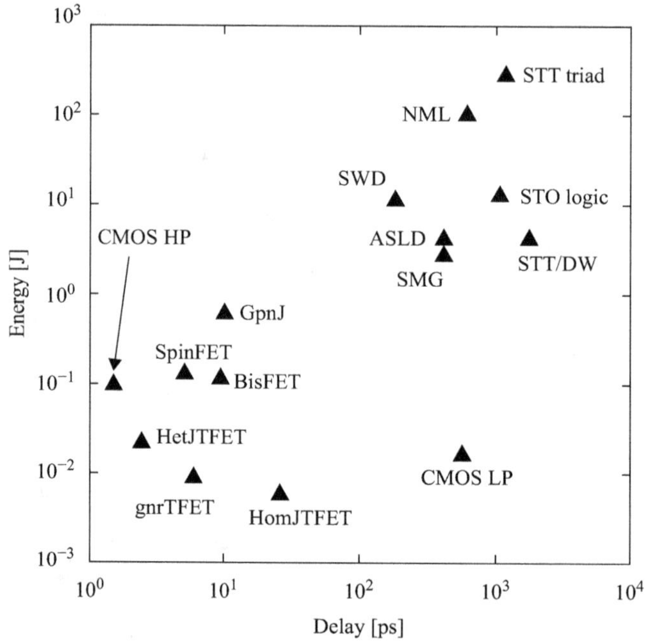

Figure 4.10 Calculated energy-delay plot for several different charge- and spin-based inverters with fanout-4. Spin-based devices use current-controlled switching with $V_{dd} = 0.01$ V. CMOS HP, high power; CMOS LP, low power, STT, spin-transfer torque, BisFET, bilayer pseudospin FET, NML, nanomagnetic logic, SWD, spin wave device, STO, spin torque oscillator, ASLD, all spin logic device, SMG, spintronic majority gate, STT/DW for spin torque domain wall, SpinFET, spin-based FET, GpnJ, graphene pn-junction, HetJTFET, heterojunction tunnel FET, gnrTFET, graphene nanoribbon TFET, and HomJTFET, homojunction TFET

terms of energy efficiency and speed, followed by the heterojunction tunnel FET devices.

Among the spin-based devices a nonvolatile magnetic flip-flop, which is operational between 4×10^{10} A/m^2 and 10^{12} A/m^2, has been modeled and simulated with a switching speed in the range of tens of nanoseconds to picoseconds [4.50]. Such a CMOS-compatible topology reduces integration space and is extendable to more complex circuits, such as shift registers. Despite spintronics still faces many challenges, the results on flip-flops [4.50] are very promising for future applications combined with CMOS.

Table 4.2 Classification of electronic, spintronic, and orbitronic devices

© IEEE. Reused with permission from Reference 4.49

Device	Class	Mechanism
CMOS HP	Electronic	Barrier
CMOS LP	Electronic	Barrier
HomJTFET	Electronic	Tunneling
HetJTFET	Electronic	Tunneling
gnrTFET	Electronic	Tunneling
GpnJ	Electronic	Refraction
SpinFET	Spintronic	Spin drift
STT/DW	Spintronic	Domain wall
SMG	Spintronic	Domain wall
STT triad	Spintronic	Nanomagnet
STO logic	Spintronic	Nanomagnet
ASLD	Spintronic	Spin diffusion
SWD	Spintronic	Spin wave
NML	Spintronic	Nanomagnet
BisFET	Orbitronic	Exciton

4.3 Applications to safe and green environment

We leave now the discussion on energy consumption and efficiency at the device and circuit level, and move to a global level discussion by looking at the expansion of urbanization, and its impact on energy consumption and efficiency. This will be followed by a discussion on how technology, especially electronics, can help to improve energy efficiency, food production, and alleviate greenhouse gas emissions.

According to the Department of Economic and Social Affairs of the United Nations [4.51], 54 per cent of the world's population resided in urban areas in 2014, and by 2050, 66 per cent of the world's population will be expected to live in urban areas. On the other hand, the rural population is now close to 3.4 billion and is expected to decline to 3.2 billion by 2050. With these figures, 2.5 billion people will be added to world's urban population by 2050.

Just to mention a few urban areas, the six world's largest cities account for 149 million of inhabitants as shown in Figure 4.11. Tokyo is the largest one with 38 million inhabitants, followed by Delhi with 25, Shanghai with 23, and Mexico City, Mumbai, and Sao Paulo with 21 million each. These large cities require of a very complex supply system for water, transportation, energy, food, and multiple services, such as schools, hospitals, and public offices. Mexico City, for instance, operates the second busiest publicly owned urban transport system in North America after New York City [4.52]. According to the study in Reference 4.52, Mexico City transport system moves about twice the number of passengers than New York's city; more than 3.5 million cars circulate a day that translates into

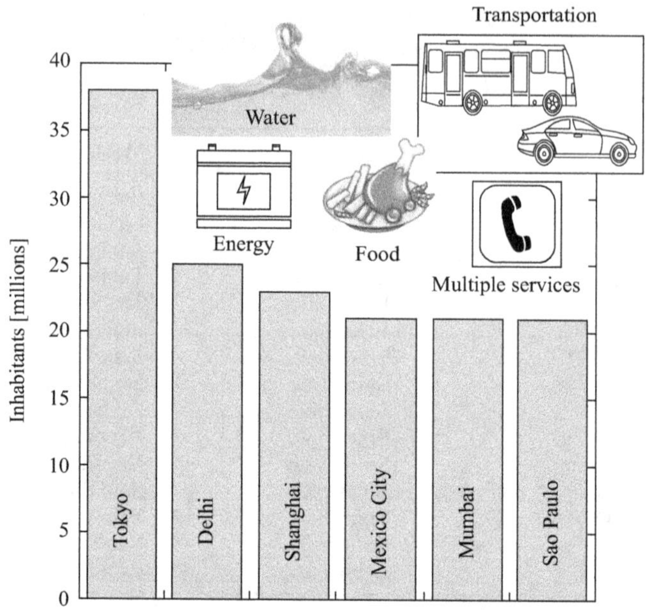

Figure 4.11 Number of inhabitants in the six world's largest cities

30 million vehicular trips a day. Mexico City's urban area is about 741,000 hectares (1 hectare = 10,000 m^2), where the number of industries amount to 35000.

The number of inhabitants and industries around these large cities need to be supplied with water, energy (in its different forms), food, and very critical services, such as waste water sewage, emergency, and communications. This requires an effective energy administration service that propels the other city's functions. This energy is basically consumed by four sectors: (1) industry, (2) services, (3) transportation, and (4) household and appliances. When looking at different information sources, there is discrepancy in the percentage energy consumption by sector, but in general, the industry is the one with the largest consumption, followed by transportation, residential accounting for household and appliances, and commercial and services.

The energy consumption by sector is shown in Figure 4.12. The numbers reflect an average percentage value for North America, South America, Europe, and Asia. However, they might vary from region to region depending on the climate, economy, and industry development conditions. In regions with extreme weather conditions the use of cooling/heating systems accounts almost for the 40 per cent of the energy delivered to residential and commercial buildings. In the United States, for instance, the residential sector splits its energy consumption in 41.5 per cent for space heating, 34.6 per cent for appliances, electronics, and lightning, 17.7 per cent for water heating, and 6.2 per cent for air conditioning [4.53]. In Mexico, in regions with mild climate conditions, for instance, the residential sector consumes 40 per cent for lightning, 29 per cent for the freezer, 13 per cent for the television,

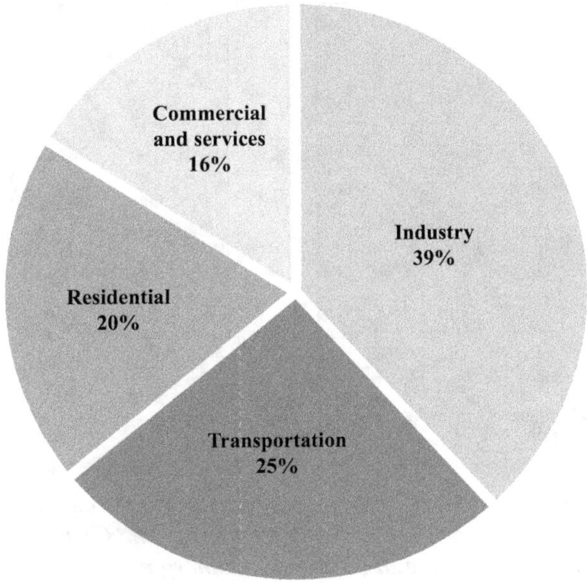

Figure 4.12 Average energy consumption by sector. Graph assembled for multiple open information sources

7 per cent for other appliances such as computers, 6 per cent for the iron, and a 5 per cent for the washing machine [4.54]. Then the question is how energy-efficient are the components, instruments, systems, and strategies used to provide energy, housing, transportation, food, water, and general services to the population? The answer to this question could be intimately linked to the planning for economic prosperity of mankind.

> *The pace at which population grows, food production changes, energy is produced, city and industry infrastructure evolves, education and information progresses, environment and natural resources are preserved, and the science and technology is paired to them, will determine the success or failure of mankind.*

For the above to be a successful story, the first thing that we need to do is to monitor, sense, or measure, as much as possible, the variables involved in the human-to-environment activity; otherwise, we will not be able to change or feedback what we cannot measure. An excellent socioeconomic analysis of energy use and prosperity is done in Reference 4.55, where the figure of merit gross domestic product (GDP) per exajoule (10^{18} Joules = EJ) of energy consumed is introduced. As a reference, the total world energy consumption in 2010 was of 500 EJ. The energy productivity index (EPI), measured in billions of euros (€) of GDP per exajoule of energy consumed, put Hong Kong at number 1 with an EPI of 456, an economy almost 100 per cent based on services. Germany, which has a strong industry infrastructure, is at number 17 with an

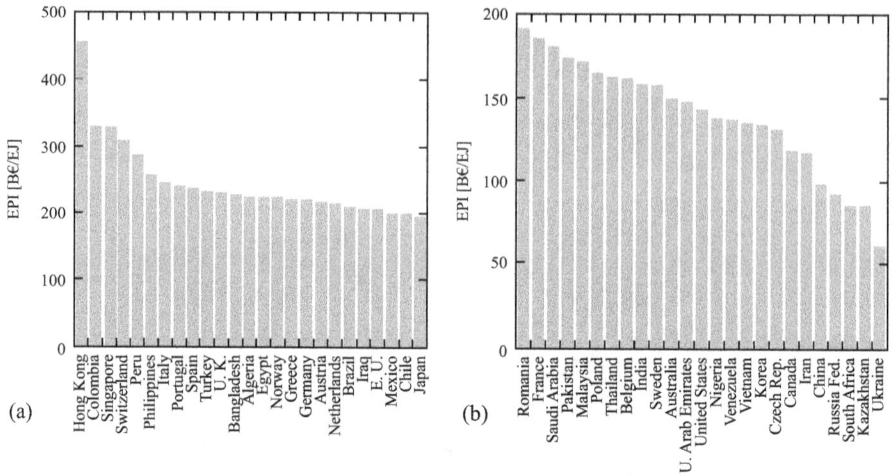

Figure 4.13 Top 50 countries in the energy productivity index (EPI) from Reference 4.55. The EPI index measures the GDP in billions of euros (B€) per exajoule of consumption energy

EPI of 220 that shows countries can balance a high industry development with energy productivity. The top 50 countries of the EPI [4.55] are shown in Figure 4.13. The GDP these countries produce relative to the amount of energy they consume is an example of the vast opportunities for improvement yet exist. We should, anyway, recognize the extreme socio, economical, and political complexity required to influence on the variables that determine the EPI and the rate at which it changes with time. However, we examine, from the technological point of view, options for moving toward an energy-efficient society. These options comprise the use of "Internet of Things (IoT)," "smart-cities," and technology-assisted food production.

The IoT, which is a term originally coined by K. Ashton in 1999 [4.56], is envisioned as the collection of different interconnected (via Internet) objects (things) that would save society time and money. If time and money, in general, a society spends in producing goods, accessing services of any kind, communicating among individuals or groups, generating information, etc., is optimized, then energy is saved as well. Here is where the electronics, signal processing (computing), and the communications, come into play together with transducers (devices with the ability to monitor signals and actuate) to create energy-, time-, and money-efficient conditions. Thus, as a natural extension of the IoT concept, the idea was immediately extended to the human being urban environment: the city. Then the "smart city" term came in play as well [4.57]. The "smart city" term was adopted since 2005 with the objective of applying complex information systems to integrate, in an efficient way, the operation of urban infrastructure and services. Figure 4.14 shows the schematic representation of a "smart city," where the elements of a city, such as infrastructure, services, environment, and people, are interconnected through either a wireless or a wired communication network.

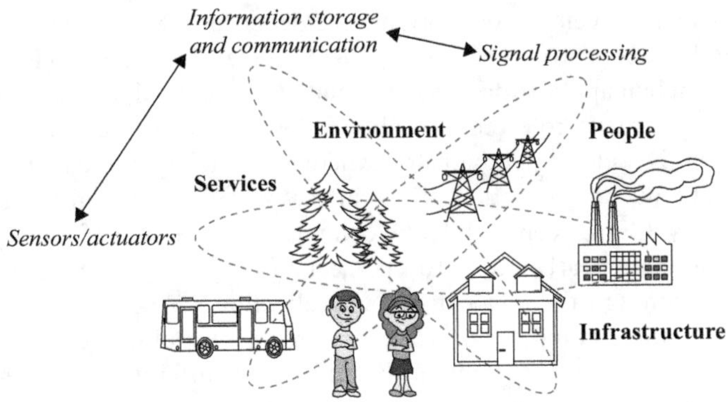

Figure 4.14 The interconnected city or "smart city" scheme

The "smart" term should be used very carefully. The smartness of the system comes from the ability to dynamically collect several city variables, like temperature, humidity, pollution, drinking water inflow and consumption/waste, electricity distribution, car traffic, medical service availability, emergency situations, and many others, and be able in a short timeframe to respond and act for the city operation correctness. For that to happen, a city should have a sensor network distributed all over the city, with every and all of these sensors continuously (dynamically activated to reduce energy consumption) collecting and sending information to a data storage center, which then is processed by a signal processing unit. Once the signal is processed, it is sent back for the actuator to respond if needed.

The "smart city" scheme shown in Figure 4.14 is very complex in the sense that captures a huge amount of information from a network of sensors that should wirelessly connect with the information storage and communication center. Then it should be processed, a decision taken based on a city operation model, and then an order sent back to the actuator. The actuator can be a traffic light, a water pipeline valve, a public lightning switch, an emergency action, etc. All of these require energy to operate, and thus a careful analysis should be done in a way that no energy consumption overhead, instead of energy consumption reduction, is created. If possible, most of the sensors should be battery-operated and be supplied with solar cell or any other kind of renewable energy. An initial approach toward this goal has been initiated at the city of Amsterdam, the so-called Almere Smart Society [4.58]. The deployment of information and communications technology (ICT) at the city of Amsterdam is focused on promoting a more efficient urban management, innovation and economic growth, strong social cohesion, and sustainable development. The smart connections can also generate substantial cost savings in running the city. For example, the local urban management processes will be supported by an intelligent digital infrastructure for the exchange of information, services, and applications between all municipal departments in areas, such as public safety, traffic and mobility, waste management, and the coordination of relief efforts in the event of disruptions, incidents, or disasters in the city. This project is led by a consortium of Cisco, IBM, Liander, Living PlanIT, and Philips.

For the efficient application of such an initiatives technology, not only plays an enabling and synergetic role but also allows doing more for less. An example of this is found in the automotive industry, where the efficiency of cars has improved by 40 per cent since the inception of ICs in cars in 1978. The lightning efficiency has increased by 339 per cent, while the computing processing efficiency has gone up to an exorbitant nearly 3.0×10^6 per cent [4.59]. Another example of ICT deployment is through the concept of "virtualization." According to References 4.59 to 4.61, a single computer or server that replaces multiple operating systems and applications, which were previously running in different computers, reduces the energy consumption drastically. It is reported that more than 1.2 million servers that have been virtualized until 2006 have saved 8.4 billion kWh of electricity a year. This energy saving amounts for more than the heating, ventilation, and cooling electricity consumed in New England in a year [4.62].

The use of technology in different industry sectors, including agriculture, has resulted in an increase of energy efficiency as shown in Figure 4.15. The semiconductor industry, through the IC technology evolution, has dramatically

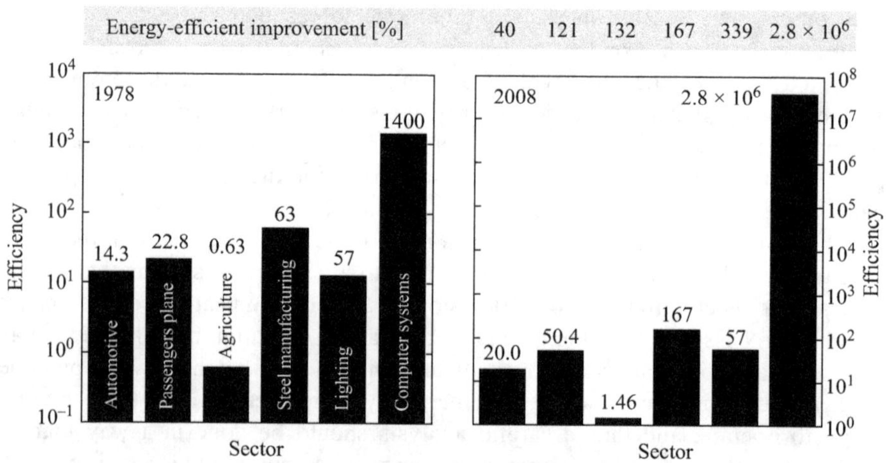

Figure 4.15 Energy-efficiency improvement for different industry sectors. The units for automotive, are; miles per gallon of gas, the units for Passengers plane, are: revenue passenger miles per gallon, the units for Agriculture, are: units of output per unit energy use, the units for Steel manufacturing, are: pounds of steel per MBtu, the units for Lighting, are: Incandecent light bulb-13 lumens per watt for 1978, and Compact fluorescent bulb-57 lumens per watt for 2008, the units for Computer systems, are: 1400 instructions per second per watt for 1978, and 4e7 instructions per second per watt. The numbers above the bars indicate the magnitude for the different sectors, and the bar above the two histograms indicate the percentage in energy efficiency from the period of time from 1978 to 2008, for each of the 6 industry sectors. Data taken from Reference 4.59

impacted the computer industry sector with an astonishing nearly 3.0×10^6 energy efficiency improvement. Henceforth, semiconductor-device–based technology should be considered as the foundations for the development of solid-state sensors and actuators [4.63], as well as energy conversion systems, such as solar cells [4.64], thermoconverters [4.65], and energy-harvesting systems [4.66, 4.67]. Energy harvesting is a technology that captures energy from the ambient environment, such as heat, mechanical vibration, and sunlight, for instance. The electrical power obtained from such an energy conversion mechanism is minimal, but enough to keep maintenance-free battery-operated wireless-connected remote sensors.

Coming back to the "smart city" management, an option to administrate the sensor network and combine that with the "smart city" model is by using graph theory [4.68]. The graph theory, which is widely used in mathematics and computer sciences, studies graphs as mathematical structures to model pairwise relations between multiple objects. Graph theory finds applications in multiple fields, such as social and information systems, electronic and computer systems, and in biological and physical systems for instance. In electronic circuit theory, where a complex circuit network may have multiple solutions, graph theory is used to find out the optimal solution based on the conditions imposed by the user. In this way the energy consumption of a circuit can be minimized at the expense of signal speed processing or the other way around. Another situation may happen when, for instance, signal resolution is preferred over signal range, and a reconfiguration of the circuit is required for. From this perspective, a city can be viewed as a circuit with multiple nodes, paths, flows, and types of signals. The water pipeline in a city with pipelines of different cross sections and lengths, and lay on a terrain with altitude gradients, can be viewed as an electric circuit (Figure 4.16), where the water pressure is the potential, the water flow is the current, and the water valves are switches or transistors. This way water consumption and leakage can be characterized as it is done with electric circuits.

Water flow is an example where graph theory can be applied to study and optimize water consumption and distribution, but it can be applied also for energy

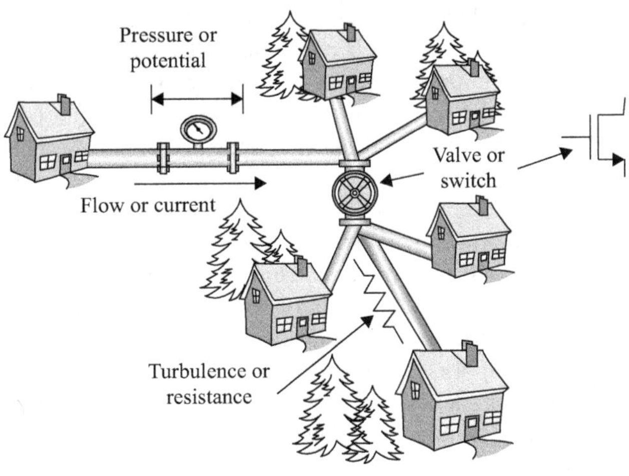

Figure 4.16 Analogy between a city water pipeline and an electric circuit

flow as shown in Reference 4.68, where at a scale of few hundred buildings, a conceptual model is developed as an unified tool to compare energy efficiency scenarios regarding a broad range of technological aspects of energy demand and supply. The preliminary results indicate the model needs to be extended at a larger scale and on real cases, in order to assess the robustness and efficiency of the suggested algorithm.

The water and energy flow consumption and distribution is not the only multiple-solution problem a city faces these days, but also the rapid increase in the number of private vehicles and public transportation (of any kind), which represents a major challenge to the city, in terms of traffic management, to reduce traffic congestion, accidents, and air pollution. In Reference 4.69, a communication-oriented perspective on traffic management for smart cities has been proposed. Twenty-four major European and international projects, with different technological approaches, aiming at improving the Traffic Management Systems (TMS) are reviewed in terms of architecture, safety, efficiency, sustainability, energy awareness, reliability, security, and innovative services. These proposed TMS include different technological approaches that span from the use of wireless sensor networks (WSN), data sensing and gathering, data fusion, processing, and aggregation, data exploitation (DE), service delivery (SD), which is the product delivered to the end-user (drivers, authorities, private companies, etc.). The DE uses the acquired information from the data processing phase to compute optimal routes for the vehicles, short-term traffic forecasts, etc. The graph theory may also find its application in the core model or the algorithm that needs to take an optimal decision for the SD. The efficiency in economy, energy consumption, and CO_2 pollution reduction, relies on an optimal decision. According to Reference 4.69, road traffic congestions have cost 200€ billion in Europe (2% of GDP) and about $101 USD billion in the United States. With the aggregated delay of 4.8 billion hours and 1.9 billion gallons of fuel wasted worldwide. There are still open questions on how to arrive at the optimal solution in terms of energy-, economy-, time, and pollution-less–efficiency. These challenges represent plenty of opportunities at the hardware (new materials, sensors/actuators, computing frames, etc.) and the software (models, algorithms, architecture, logics, etc.) levels for the research and innovation. This is testified by the recent efforts on improving energy efficiency of WSNs using graph theory [4.70, 4.71], where innovative designs of efficient power management and power failure diagnosis are developed. Energy-efficiency techniques are used to improve the reliability of WSNs links based on three-phases secure data aggregation. This three-phase approach optimizes the transmission distance and thus improves energy efficiency, which alleviates the limited availability of energy within network nodes. These models optimize the position of the sensors in terms of energy consumption and power delivery.

A practical example of the use of graph theory has been applied in Mexico, where a new methodology for the optimal design of water network sectorization has been proposed [4.72]. This is focused on improving the management and security of multiple-source water supply systems. In particular, the network sectorization problem concerns the definition of isolated district meter areas, each supplied by its own water source (or water sources), which is completely disconnected from the rest of the water system through boundary valves. This approach is based on minimizing the dissipated power in the water network and has been tested on two existing water distribution networks (WDN), one in Italy, and the second one in San Luis Rio Colorado, Mexico. This is a proposal that leaves the traditional passive approach and moves to a proactive and smart approach based on the possibility of inserting remotely controlled gate valves and monitoring devices into the WDN. According to Reference 4.72 the low cost and availability of new remotely controlled monitoring and management devices can accelerate the upgrading of WDNs and its aligning to other network systems, such as the one for the electricity, the gas, and the transport, which already envisions a general smart city systems for multiple services (water, energy, transport, security, etc.). The water network partitioning for both district metering and sectorization can assist in modernizing water supply system for achieving water balance distribution, applying pressure control techniques, and protecting users and the WDN from malicious attacks. The effectiveness of the graph theory–based approach applied to the Mexican city of San Luis Rio Colorado, located at the border with United States and with a population of nearly 180,000 inhabitants, was confirmed by showing a minimization of the dissipated power, while maintaining the level of service for the users. This methodology can be applied to a large water system, and it offers water utilities a design tool that is based on performance indices that outperforms the in-use conventional empirical trial-and-error approaches.

So far we have taken an overview of different aspects of energy-efficiency at the material, device, circuit, and system level, and its potential societal implications at the urban environment, such as energy, water, and transport services under the concept of "IoT" and "Smart City." Now it is time to board the energy-efficient sustainable green food chain (GFC) topic and its relationship to high-technology semiconductor technology. We will start from the triple bottom line PPP context introduced in Reference 4.73. The PPP context refers to production of food based on the intersection of three elements: People, Planet, and Profit. This implies a positive impact on society and natural environment while keeping long-term economic benefits for the food producer.

The food production can be subdivided into four categories: (1) agricultural food, (2) animal meat and derivatives, (3) processed food, and (4) drinks. In the production of all of them, there are common factors: the use/waste of water, energy, transportation, storage and conservation, distribution, and their associated pollution (Figure 4.17). The whole GFC is by its own nature a very complex and multidisciplinary research area that deserves special attention. Although the food transport accounts for over 80 per cent of goods moved in England (for instance, Reference 4.73) and that 25 per cent of Europe's road transport CO_2 emissions are generated by large good vehicle, here we will only touch a few examples where technology can help to make the GFC a sustainable and environment- and economy-friendly activity for the mankind. We will basically focus on the food production, and we will not touch the transport, storage and conservation, and distribution as shown in Figure 4.17.

Livestock in particular ruminants like the cows account for up to one third of methane emission worldwide [4.74]. Methane as a greenhouse gas has a pollution

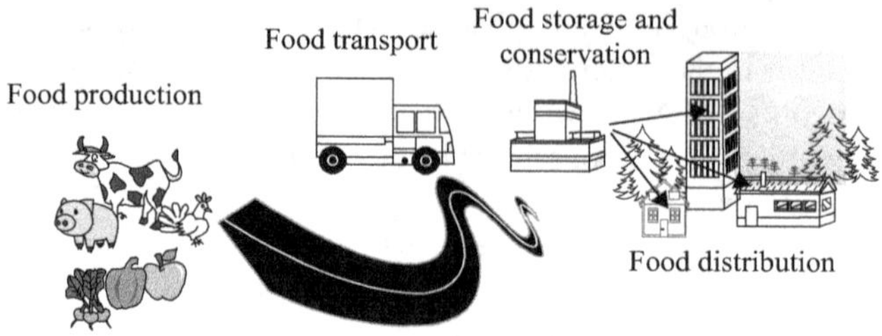

Figure 4.17 Schematic representation of the food chain

potential 25 times that of CO_2 [4.75]. Thus, methane contamination contributes as a large CO_2 equivalent contamination from agriculture. Unfortunately, methane (CH_4) is not the only gas emission from animal husbandry, ammonia (NH_3) and nitrous oxide (N_2O) are other gasses emitted by animals. NH_3 emission cause eutrophication and acidification and thus plays an important role in the decline of biodiversity and dying of forests [4.76]. The methane emission from ruminants is a chemical process that has to do with the food digestion chemistry in the rumen of the animal. According to Reference 4.77, there exists multiple factors that influence methane emissions from cattle and include the following: level of feed intake, type of carbohydrate in the diet, feed processing, addition of lipids or ionophores to the diet, and alterations in the ruminal microflora. Manipulation of these factors can reduce methane emissions from cattle. Cows exhale gasses as burps while they are grazing and roaming around the pasturelands, which is not easy to monitor and control. However, researchers from the Sustainable Agriculture Flagship in Australia [4.78] are installing tiny wireless devices that will monitor chemical, temperature, and pressure conditions inside the rumen of the cows. The evaluation of the wireless remote sensing system is described in References [4.79–4.81].

The system is schematically described in Figure 4.18. It consists of an electronic circuitry with sensing capability, signal conditioning, a preprocessing analog-to-digital unit, and a transmitter/receiver stage. Everything is inside of a capsule (bolus) in the rumen of the animal. The system is battery-operated and connects with an external computer via a wireless channel. The physiological variables such as pH or temperature are continuously monitored. The information is then processed at a computer, which based on a livestock feed model, adjusts the cattle diet, so the pH and temperature goes into a range where methane emission reduces. This requires the farmer to check the wireless systems and the cattle diet in a continuous basis until an optimal livestock feed and pH–temperature relationship is found.

The Veracruz state in Mexico, which is located in the tropical area, has a considerable cattle population, which is exposed to climate changes that affect the milk and meat production and quality. The measured impact of climate changes and inappropriate cattle feed results in a 10 to 25 per cent reduction in milk production for instance [4.81]. Therefore, besides control of methane emission, a pH and temperature control is required to keep cattle healthy and in an adequate milk production level. Off-the-shelf electronic components were used for the design and fabrication of the

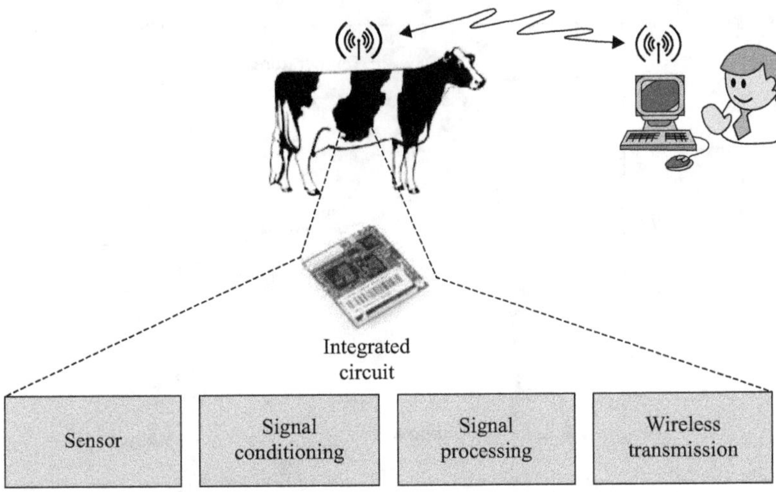

Figure 4.18 A wireless electronic system for pH, temperature, and pressure remote monitoring of cattle

wireless system. For temperature monitoring a precision centigrade solid-state LM35DZ Zener diode [4.82] was used as temperature sensor, with an accuracy of $\pm 1/4\,^{\circ}$C, from room temperature up to 150°C. For the pH variable an ion-sensitive field-effect-transistor (ISFET) [4.83] fabricated by the Microelectronics Institute of Barcelona, Spain [4.84] was used. The ISFET is a modified version of an FET device. There is no gate electrode and the oxide is exposed to the chemical agent, which changes the potential at the oxide–semiconductor interface and the channel current at the same time. The pH level is then associated to the channel current variation. The signal conditioning circuit was designed and developed at the National Institute for Astrophysics, Optics and Electronics (INAOE) at Puebla, Mexico [4.85]. It is a temperature-compensated preamplifier that includes a self-calibration system for the pH sensor. This signal-conditioning block is connected to an Arduino Fio microprocessor [4.86] that converts the analog signal from the signal-conditioning block into a digital signal. Then the digital signal is transmitted with an Xbee wireless module [4.87] and received at the outside by the computer system. An IEEE 802.15.4 communication protocol is used for the transmission/reception of data. The wireless system covers a 30 m^2 area, and is powered with a 72-hour continuous use of Li-ion rechargeable battery. The measured values for the pH and temperature over a time span of 6 hours, from 8:00 to 14:00 hours, are shown in Figure 4.19.

The wireless system went on at 8:00 hours, the cow was fed at 9:30 hours, and the measurements were stopped at 14:00 hours. According to Reference 4.88 a prototype with these characteristics should be capable to show a pH range in between 5.4 and 6.8 with a $\pm$0.1 pH resolution, if is to be used for monitoring acidity or alkalinity of ruminal contents. For optimum fermentation of the diet and fiber digestion, ruminal pH should be between 6.0 and 6.4. Thus a prototype with

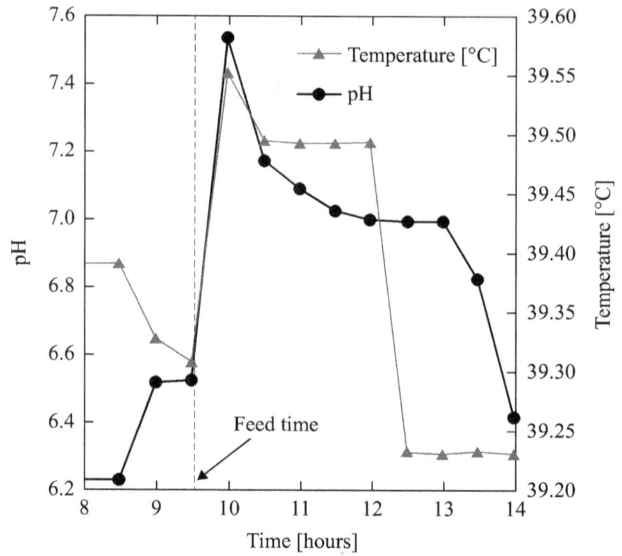

Figure 4.19 Measured pH and temperature versus time

these characteristics is within the expected range, which shows that off-the-shelf electronics, sensors, and computing capability can be used to help farmers to increase milk production at this very preliminary stage.

Besides cattle grazing, there are other activities done in the rural field, as is the case of agricultural crop. Agricultural activities use about 85 per cent of available freshwater resources worldwide [4.89]. Therefore, there is an immediate need to look at water management and its implications in a sustainable water-agricultural production system. The optimization of water for use in agricultural crops, through the use of electronics, sensors, computing, and wireless communication systems, is a very valuable and efficient approach as demonstrated in Reference 4.90. An automated irrigation system using a WSN and GPRS (general packet radio service) module has been put in place in San Jose del Cabo, Baja California Sur (Mexico). As shown in Figure 4.20, the system has a distributed WSN of soil-moisture and temperature sensors placed in the root zone of the plants. A wireless information unit (WIU) handles information from the wireless sensor units (WSU), triggers actuators, and transmits data to a web application. An algorithm, which resides in a microprocessor installed in the WIU, with threshold values of temperature and soil moisture controls the water administration to the crop.

In order to save energy and have autonomy the system is powered by photovoltaic panels. It has a duplex communication link based on a cellular–Internet interface that gives access to data inspection and allows irrigation scheduling to be programmed in a remote way through a web page. As shown in a 1-week period of time in Figure 4.21, the system benefits of the high solar radiation existing in the Baja California peninsula.

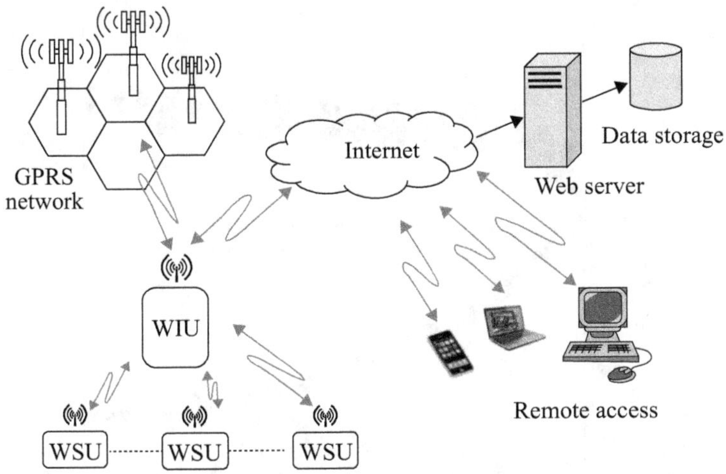

Figure 4.20 Schematic configuration of the automated irrigation system with a WNS

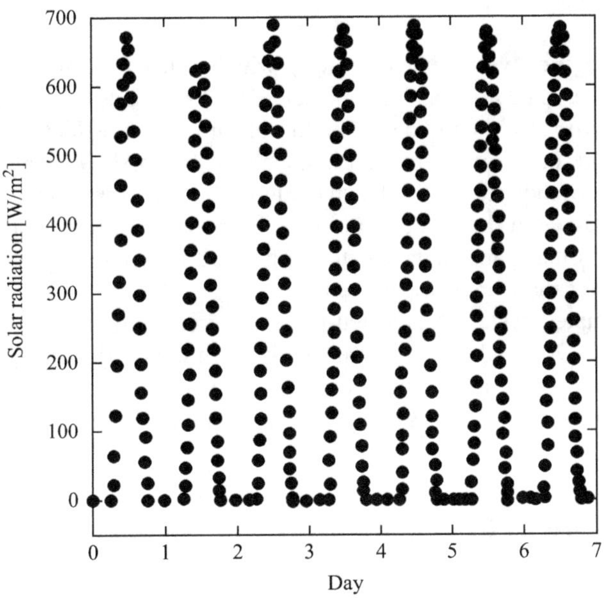

Figure 4.21 Solar radiation along the experiment of the charge–discharge cycle of the WSU during a period of time of a week

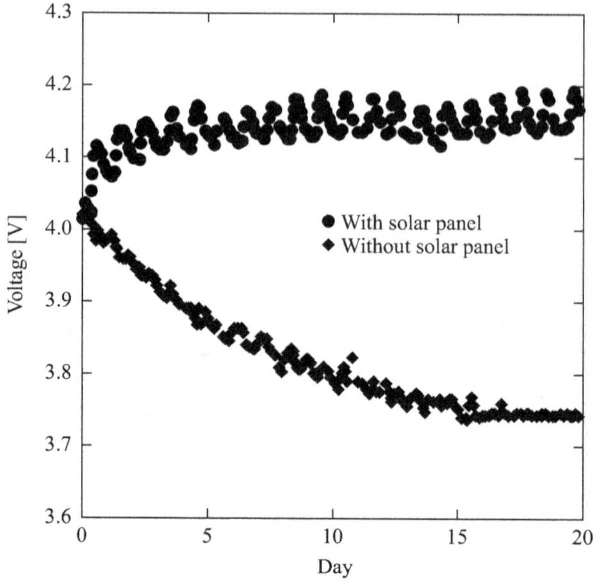

Figure 4.22 Battery charge–discharge cycle of the WSU

The solar energy is converted into electricity, which is then used by the WSU and WIU through a voltage regulator enabled for a period of 20 seconds by the microcontroller. This is enough time for the radio modem to wake up and transmit the data. This way, the total average power consumption is kept at 0.455 mAh. This allows a charge–discharge cycle of the batteries as shown in Figure 4.21. The data in Figure 4.22 are for a period of 20 days in the winter with the solar panel connected and disconnected. Thus the photovoltaic panel and the batteries provide sufficient energy to maintain the WSU running for the whole crop season.

The automated system was tested in a sage crop field for 136 days and water saving of up to 90 per cent compared with traditional irrigation practices if the agricultural zone were achieved. The WIU and WSU cost of 1,900 USD plus its energy autonomy makes this system useful in water limited and geographically isolated areas.

4.4 Applications to human health and medicine

4.4.1 Introduction

The pace of life in society has always gone hand-in-hand with technology. As a glance, for people born in the 1960s or 1970s, it was common to turn on the TV a half an hour in advance respect to the scheduled start of the show. One could see a tiny light spot right in the center of the phosphor-coated screen that was opening until the image appeared completely [4.91]. There was plenty of time in a day. Nowadays, a product like TV vacuum tube, although it may have interesting attributes, would not succeed in the market because it would not cover user expectations. Just turning on our electronic equipment and noting a timeout greater than 1 minute, which it may take to enable all parts of its system, it is sufficient reason to panic. Then, society wishes products that allows them to be interconnected at anytime, anywhere, and with the highest transmission quality. Semiconductors are present in most of the modern electronic devices that we use. They control the computers we use to conduct business, the phones and mobile devices we use to communicate, the cars and planes that get us from place to place, the machines that diagnose and treat illnesses, the military systems that protect us, and the electronic gadgets we use to listen to music, watch movies, and play games, just to name a few.

Technology is moving so quickly, and in so many directions, that it becomes challenging to even pay attention—we are victims of "next new thing" fatigue. Yet technology advancement continues to drive economic growth and, in some cases, unleash disruptive change. Economically disruptive technologies—like the semiconductor microchip, the Internet, or steam power in the industrial revolution—transform the way we live and work, enable new business models, and provide an opening for new players to upset the established order [4.92]. Business leaders and policy makers need to identify potentially disruptive technologies, and carefully consider their potential, before these technologies begin to exert their disruptive powers in the economy and society.

The power of new technologies is everywhere. They change how businesses make money and how we live and work, sometimes with amazing speed.

Electronic gadgets have supplemented recreational activities that need power and real energy just to do it, but today we can now play games and adventures by just sitting and playing any video game console [4.93–4.95]. Some gadgets are used for security purposes like the spying camera, surveillance camera, detectors, biometrics, and other gadgets that are used for security purposes [4.96]. With this innovation and advancements the security is safer and better. It is also used in some works like computers that is widely used through the aid of the Internet and all the things, whereabouts and happenings are just nearer and easy one click to make us more updated every day. There are also several professions that have been evolved because of this digital era that gadgets are widely used. A microscopic specimen can be enlarged up to 100,000 times through the scanning and transmission of electron microscope, which is a very powerful tool. It allows scientists to study even minute pathogens such as the AIDS virus. In physics, some laser devices are developed and used in medical purposes like cataract treatment [4.97]. Ultrasonic devices are also an effective tool in determining the sex and position of unborn fetus, which illustrates the concept of the interrelatedness [4.98].

On the other hand, social media was practically unknown a decade ago, yet almost one and a half billion people now has Facebook accounts; in fact, entirely new ways of socializing and interacting with friends, family, and colleagues have become the norm. Considering mobile devices market, by 2017 nearly more than 10 units will have downloaded 77 billion apps. By 2020 there will be

50 billion networked devices, according to estimates from Cisco, since less than 1 per cent of objects are today connected to the Internet [4.99]. Around the world, hundreds of millions of people have been lifted out of poverty as developing nations have adopted the technologies that drove growth in advanced economies in earlier times. Today, technologies such as the mobile Internet are helping to accelerate economic development, allowing millions of people in remote areas of developing regions to leapfrog into the 21st-century global economy. We notice the effects of new technologies as they rapidly change our work routines, the way we spend our leisure time, or the products and services we use (often, increasingly, and free). We experience the benefits of new technologies in profound ways when they save or extend our lives or those of our loved ones [4.100].

Semiconductors has pervaded our lives for the past 50 years, with massive penetration into health, mobility, security, communications, education, entertainment, and virtually every aspect of human lives [4.101–4.104].

The shift from the microelectronics era, where semiconductor devices were measured in microns (1 millionth of a meter) to the new era of nanoelectronics where they shrink to dimensions measured in nanometers (1 billionth of a meter) will make the semiconductor sector even more pervasive than it is today. It will allow building much more intelligence and far greater interactivity into many more electronic items around us.

Technology will play a part in virtually every aspect of our lives, from personal health and traffic control to public security. In the past decades, as the main stream, mainly Moore's law, with two focused development arenas, namely, IC miniaturization down to nanodimension, powers these progresses and SoC (Systems on Chip)-based system integration. While microelectronics community continues to invent new solutions around the world to keep Moore's law alive and even to go "beyond Moore" with disruptive technologies, there are ever-increasing awareness, R&D effort and business drivers to push the development and application of "More than Moore" (MtM) that are based upon or derived from silicon technologies but do not simply scale with Moore's law (with typical examples as RF, power, sensor and actuator, heterogeneous integration, etc.). This emerging trend is partially trigged by the increasing social needs for high-level system integration including nondigital functionalities, the necessity to speed up the innovative product creation and to broaden the product portfolio of existing wafer fabs, and the limiting cost and time factors of advanced SoC development.

Semiconductors not only enable traditional technologies to become smarter and more effective, they also have led directly to the creation of entirely new industries, such as mobile telecommunications, which is reinventing entertainment, media, commerce, and many other sectors of the global. The pace at which mobile phones have been adopted around the world is unmatched in the history of technology, according to the World Bank, and its growth is in large part due to semiconductors. It is estimated that in the United States alone form 1960 to 2007, more than 35 per cent of the communication-equipment industry's growth was attributable to semiconductors [4.105]. And as semiconductors grow, more powerful and energy efficient, so too do the mobile phones and networks they support, at an increasingly lower cost. This makes mobile telephony available to literally billions of people, many of whom live below the poverty line. In many parts of the world, a mobile phone is more stable than a home address. The revolutionary impact of the smartphone across both emerging and advanced economies is clear. And it's only the tip of the iceberg: mobile devices—including not only smartphones, but tablets and new wearable devices like Google Glass and Pebble's smart watch—lie at the epicenter of the hyperconnected age and represent a new frontier in innovation, social, and economic development.

Semiconductors have become critical components in virtually every aspect of automotive operations, and the transition from mechanical to electronic systems has created one of the fastest-growing market segments for the industry at 8 per cent annual growth. This transition has made automobiles safer and more dependable than ever before. The penetration of semiconductors into all facets of the car—from the engine and transmission systems to the airbags, cruise control, and braking componentry—has been dramatic. Ford Motor Co. estimated that in 2010, 30 to 45 per cent of a car's value came from the electronics embedded in the vehicle [4.106, 4.107]. Semiconductors are also making automotive transit time sustainable. Achieving emissions reductions in conventional drive trains requires constant monitoring and correction of engine performance. Powertrain microcontrollers and power modules help make engines more efficient by reducing fuel-injection losses and optimizing gear ratios and shifting, which reduce energy waste from hydraulics and friction. Some carmakers now offer advanced driver assistance systems, which integrate adaptive cruise control, lane-drift warning system, blind-spot detection, low-speed collision avoidance, and rear-facing cameras to help drivers avoid potentially dangerous conditions. Hybrid and all-electric vehicles have even higher semiconductor content: the Volt (a plug-in hybrid car) uses 10 million lines of software code and 100 electronic controllers, and each Volt on the road has its own IP address.

The video game market is among the biggest of all entertainment branches. With 14.8 billion USD in sales, its size has more than doubled between 2006 and 2012 [4.100]. These sales figures comprise revenues from console and computer games. Gamers in both markets need a complement to play the game, either a computer with the sufficient hardware power or a console. Whereas in the console market the console itself is sufficient, a PC gamer needs to consider minimum system requirements of the game. The game will not run smoothly if the computer is not power enough. Moreover, technological progress is different for consoles and computers. Whereas consoles make discrete quality jumps from one generation to the following, hardware power of computer rises more incrementally over time. The attractiveness of a computer gamer is largely driven by how realistic the game's graphics, physics, and artificial intelligence is. However, the more realistic a game becomes, the more demanding a game becomes computationally. Only consumers with a sufficiently powerful PC will be able to play the game.

We often think of semiconductor simply as commercial products. Yet scientists and researchers have been for years searching for technological solutions to help improve medical diagnosis and the treatment of patients [4.108, 4.109]. Indeed, the rapidly growing processing power of semiconductors is both improving the patient's experience today and changing the future of health care. Thanks to the increasing power and sophistication of RFID devices and other tracking tools [4.110, 4.111], semiconductors are already giving nurses and doctors far more powerful abilities to monitor patients, find medicines within the hospital, and track expensive diagnostic equipment. As hospital system merge their data into electronic record-keeping, they can easily check to ensure drugs do not cause adverse interactions; manage patient charges and reimbursements; and make sure that specialists treating a patient are aware of what other doctors and nurses are seeing during their interactions. These initiatives will have significant positive effects on patient outcomes, but they are only the most visible current developments. As rapid improvements in semiconductor electronic merge with nanotechnology and wireless sensor networks, a new generation of micromachines becomes feasible. These include new forms of pressure sensors for blood pressure; micropumps for infusion drug delivery; ultrasound sensors for medical imaging, and a whole range of microactuators, silicon microphones, and microelectrodes to improve the performance of hearing aids [4.112–4.118]. At the same time, powerful large diagnostic equipment that used to reside only in a hospital or a doctor's office will increasingly be available on mobile phones. Such devices put more control in the hands of patients and allow them to quantify their healthy behaviors more accurately. The ability to use wireless technology to monitor patients remotely is a tremendous breakthrough for the health care industry. Telemedicine initiatives not only improve care for the chronically ill but also can make health care available to millions of patients in remote regions where access to doctors and nurses is scarce. The power of semiconductors also puts society on the verge of medical revolution in which designer medicines will be created for individual patients based on their unique genetic profile. This is becoming feasible because powerful sequencing machines run by semiconductors can decode an individual's genetic profile in a single day.

In 1965, Gordon Moore published his legendary paper "Cramming more components onto integrated circuits" [4.119]. In this paper, he described that the number of components in ICs had doubled every year since the invention of the IC in 1958 until 1965, and that this trend would continue for at least 10 more years. Now, almost five decades later, industry is still following Moore's law, even though Moore himself has predicted on many occasions its end. Indeed, one needs to point out that since the development of the 130 nm technology node, it has become more and more difficult to scale down the traditional MOSFET structure in order to keep up with the pace of Moore's law. Especially in the last decade, many innovations such as high-k/metal gate and strain engineering have been introduced to ensure the required performance improvement with every new technology node. However, these material innovations are not enough to control ever-increasing off-state leakage problem in these advanced devices. New devices architectures such as multigate devices have emerged as a means to solve this issue by increasing the gate control through geometry. Although proposed already in the beginning of the 1980s, multigate devices (Delta FET, FinFET, trigate, etc.) and more generally fully depleted transistors have long been considered the exotic devices that somehow were no real contenders to replace the planar bulk MOSFET. With Intel's announcement to introduce trigate architecture on bulk substrate at the 22 nm technology node in 2011, a new era of MOSFET scaling arrived [4.120]. It is expected that more companies will follow and introduce multigate devices at future technology nodes.

4.4.2 Invasive and noninvasive biosensors

A device or procedure that requires penetration into the body is referred to as invasive. A different set of constraints must be considered when designing sensors for use on human compared to those in the automotive, telecommunications, or environmental fields. Foreign materials in the human body are often rejected and attacked by the immune system. This can lead to dangerous complications for the device and the individual. The complications that go along with designing invasive sensors can make the process very difficult. Some biosensors justify a requirement for being invasive. Two challenges are addressed in general—packaging and electrical configuration methods for biocompatibility and filtering methods for signal noise reduction. Packaging for biocompatibility must encapsulate the device in a biocompatible material while maintaining functionality. Sending a current through an in vivo device must be done in a manner to avoid unintentional discharge or unnecessary heating effects.

Sensors are implanted to provide information that was previously inaccessible [4.121, 4.122]. This information along with the knowledge of the relationship between the sensor and the measured object provides information regarding the condition of the measured object. Sensors are implanted practically everywhere to measure the condition of objects. One of the first and still the largest areas using biosensors is the medical field [4.123–4.128]. The demand to save lives inspires the development of sensors that can measure bio signals from vital signs both in surgery and during transport to high-powered sensors that can practically create a pinpoint image of any location in the human body. The use of these devices is not limited to the medical field however. Biosensors are used in many laboratories to monitor much more than just vital signs.

Biosensors are not commonly created for permanent use in a single subject, but rather for many subjects for a variety of tests [4.129, 4.130]. This need to repeatedly apply and remove the sensor that makes noninvasive sensors desirable. These noninvasive sensors must find ways of effectively collect data from biosignals detectable outside or on the exterior of the body. Noninvasive biosensors collect signals either form a distance or, more commonly, are fixed directly on a human subject for data collection.

The key factors driving both research and market of biomedical electronics are aging populations, rising health care costs, the need for access to medical diagnoses and treatment in emerging and remote regions and in homes, and the fast development of biotechnologies [4.131]. The applications of biomedical electronics in research, design, and development of biomimetic devices/systems, instruments, and appliances that treat intractable neurological disorders, restore health and extend life, and enable biotechnology development is an exciting area of future growth for the electronics industry. It is an area that link engineering, biology, and medicine. A few examples of biosignal sensing for human body application are shown in Figure 4.23.

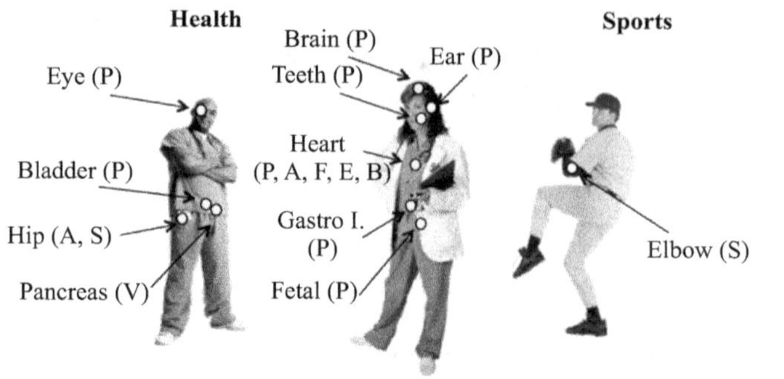

Figure 4.23 A few examples of sensors for human biosignal monitoring

The major future trends in biomedical electronics are portability, miniaturization, connectivity, security, and reliability [4.132, 4.133]. Portability requires accurate biosignal sensors/actuators, efficient system power management, ultra-low power electronics, and energy harvesters. Miniaturization requires advanced integration like CMOS ICs or heterogeneous integration of CMOS, MEMS, and/or flexible technologies. Connectivity requires low-power RF wireless communication technologies. Data security requires more hardware and software tools to support medical data security in RF transmission and storage. Reliability requires enforcement of regulations and standards.

According to document 10993 of the International Organization of Standardization (ISO) [4.134], medical devices can be divided in to four categories: non-contact devices, surface-contacting devices, external communicating devices, and implant devices.

The advent of portable computing devices and miniature sensing devices presents new opportunities for personal health care. Formerly, most medical sensing devices were used in a hospital setting under the care of trained medical and technical personnel. Physiology monitoring is expanding explosively in the personal health and fitness space, with the rapid advance of innovative personal fitness monitoring technologies [4.135]. These devices collect health data related for many purposes, by patients with chronic medical conditions (such as blood sugar sensors for diabetics), people seeking to change behavior (losing weight or quitting smoking), or athletes wishing to monitor their condition and performance. The resulting data may be used directly by the person, or shared with others: physician for treatment, insurance company, etc. A key element for this is remote monitoring. The convergence of pervasive wireless networks, cloud technology, miniaturization, and noninvasive biosensors is rapidly making the concept of monitoring patients as they go about their daily lives a reality [4.136, 4.137]. Wearable health-monitoring devices represent an exciting opportunity in health care. With 1.91 billion smartphones in the world (2015), and an explosion of other devices getting connected to the web, people are unconsciously generating more data about themselves than ever before. How that data get used is still an open question. Supplied with tools to track their own conditions, patients could be empowered to take responsibility for their own health. And rather than getting small glimpses during infrequent checkups, physicians could be provided with a steady stream of data about how their patients are doing. Today, just one in 10 American adults own a fitness tracker, but these devices should become more widespread over the next decade alongside an explosion of sensors that can monitor everything from steps to breathing to heart rate and apps that can sense the onset of chronic illnesses or stress [4.138].

Telemedicine and remote care are being practiced in diagnostics. However, their communication with the biology is mainly in digital, whereas biology is analog. On the other hand, the potential in semiconductors is much more. With a bio-inspired vision, it is possible to develop systems on a microchip than can mimic biological process [4.139, 4.140]. Bio-inspired technology takes advantage of the semiconductor device's capabilities more than a switch. It encourages implementing intelligence at a point of measurement. With this approach, only the information needed is transmitted. Going one step further, by mimicking the physiology of biological systems, we may have diagnosis and provide therapy at the same time at the point of care. Over the past few decades, the semiconductor industry has enabled significant breakthroughs. From the early massive computers dealing with a few hundreds of bits, to the miniaturized portable though powerful units dealing with voluminous data in our hands, we have seen an impressive impact on our lives, both individually and socially [4.141, 4.142]. Expensive bulky instruments of yesterday, demanding people to stay at the hospitals for monitoring, are being replaced by small, friendly wearable devices remotely controlled at the convenience of both clinicians and patients [4.111–4.144]. Advances in microfabrication and bio/medical engineering techniques are now enabling a large variety of miniaturized implantable systems for sensing, health monitoring, and deficiency treatments. This progress is driving physicians and patients to express an increasing desire to miniaturized implantable devices as they are offering less invasive implementation procedures, greater comfort for the patient, improved performance, and often provide innovative measurements and treatments. Telemedicine is not a new term anymore. In the era of Big Data, we see numerous applications for mobile devices like smartphones, tablets, and smart watches, collecting and transferring huge amount of data from body sensors to digital processors and databases. Besides the fact that biology needs its common language when interfaced with instrumentation, this approach is far from optimum.

The semiconductor industry has perceived that medical monitoring is physiological [4.145]. There is less convincing evidence in metabolic monitoring while biology is fundamentally chemical. From cells to tissues and organs, biochemistry is playing a significant role in controlling body functions through changes in concentrations of ions, enzymes, and hormones. For example, we may measure the glucose level and inject the appropriate amount of insulin for a diabetes patient automatically and continuously, or we may monitor the hormones secretion accordingly stimulate a nerve to control appetite. An ion-sensitive field-effect transistor (ISFET) acting like a chemical sensor allows us to create such a connection on a semiconductor microchip. An ISFET is basically a MOSFET but controlled by the ionic concentration of the analyte to which it is exposed [4.146, 4.147]. In a digital mode, MOSFET acts like a switch controlled by the gate, either ON and passing current or OFF and blocking it. In an analog fashion, the channel current varies by the change in the terminal's voltages. The principle in the flow of electrons in the transistor channel is similar to the flow of ions in an analyte. In ISFETs, ionic concentration modulates the channel current. By further modifications of the ISFET insulating membrane, it may become sensitive to different ions, or by immobilizing biomarkers, it may help detecting specific biomolecules. Consequently, the circuit currents and voltages can trace the changes on the top of the chip in the analyte. Other interesting diagnostic, patient monitoring, and therapy applications come from Texas Instrument™ [4.148]. Medical equipment such as digital stethoscopes, patient monitoring, ECG, EEG, and pulse oximetry have all become more portable through improvement in battery management technologies, and proliferation of wireless communications technologies like Bluetooth® and Zigbee®.

4.4.3 Gastric applications

In our body, the pancreas is the organ that regulates the secretion of insulin by its beta cells. Insulin is a hormone that is in charge of carbohydrate and fat metabolism. The lack of insulin results in high sugar levels in the blood (as in diabetes). Diabetes causes frequent urination and an increase in thirst and appetite. In the long term, it impairs heart, kidney, and eye function. The prevalence of diabetes for all age groups worldwide was estimated to be 2.8 per cent in 2000 and 4.4 per cent in 2030. The total number of people with diabetes is projected to rise from 171 million in 2000 to 366 million in 2030 [4.149]. The prevalence of diabetes is higher in men than women, but there are more women with diabetes than men. The urban population in developing countries is projected to double between 2000 and 2030. The most important demographic change to diabetes prevalence across the world appears to be the increase in the proportion of people >65 years of age. A treatment for diabetes is the injection of insulin, and it is crucial to take the right amount. While the deficiency of insulin leads to hyperglycemia (high glucose in the blood), and an overdose may severely drop the sugar level (hypoglycemia) and create problems such as seizures, unconsciousness, and even brain damage and death [4.150]. Thus, it is absolutely vital to keep one's insulin level in a salutary mid-range. An alternative might be organ transplant, but it comes with operation complications and risks [4.151]. An artificial pancreas is a bio-inspired closed-loop system that mimics the actual function of normal pancreas beta cells, and it runs novel physiological control algorithms on a microchip to keep the insulin level in a healthy appropriate range [4.152, 4.153].

Along with the microchip, which runs the processing and operation live and continuous, a glucose sensor and an insulin pump are implanted onto the abdomen. ISFET also enables measuring the glucose and insulin levels. The system communicates wirelessly with a wearable monitoring system. Taking into account body activity, as well as vital signs like temperature and heart and respiration rates, the chip not only monitors the glucose level but also calculates the optimum amount of insulin required and drives the pump accordingly to dispense the necessary dose. An artificial pancreas is an example of personalized diagnostics and therapy at the point of care. The vagus nerve is in charge of transmitting satiety signals to the brain. Controlling its function can help controlling appetite, which in perspective would help managing obesity [4.154, 4.155]. Obesity is the result of and imbalance between the food taken in and the energy consumed, when the remaining unused calories are stored as fat. The causes of underlying obesity are diverse such as genetic inheritance, sedentary life-style, unhealthy diet, and medical treatment side effects. Nonetheless, it is a common vital risk factor as it increases the chance of getting cancer, heart disease, high blood pressure, diabetes, and musculoskeletal disorders [4.156, 4.157]. Obesity is a growing problem in both developing and developed countries, particularly Mexico and the United States. Many dietary regimes have been designed requiring a strong commitment with no guarantee of long-term weight loss. The most effective alternative is bariatric surgery, which is not recommended for everyone due to its entailing risks. In the past, physicians observed that people who had undergone vagotomy, a procedure involving cutting of the vagus nerve, would lose weight. However, vagotomy as a treatment for obesity has been discarded due to its secondary effects. Neuronal engineering has helped significantly in the development of prosthetics like bio-inspired cochlear and retina implants that replace the dysfunctional elements. Neuronal signals are generated by changes in the flow of ions at the membranes of the cells. As a result, an electric potential called action potential is generated and travels through the nerve. For recording neuronal activities and stimulation, conventional methods take an electrical approach using cuff electrode or microelectrode arrays. Then apply a burst of high-frequency pulses to block the action potentials on a nerve. In the long term, it may cause pain and peripheral neuropathy. In recording, it causes artifacts-interfering measurements. As an innovative development, i2move, a project at the Centre of Bio-inspired Technology at the Imperial College London, is taking an electrochemical approach using ISFETs along with a combination of cuff electrodes and microelectrodes arrays [4.158, 4.159]. Electrochemical recording may help eliminate the artifacts by measuring the ionic activities in the generation of action potentials. In addition, by a modification of ISFET sensing membrane, hormonal secretions can be monitored for responses to appetite control. On an implemented microchip, all of the measurements are processed for an algorithm stimulation of the nerve to control appetite. While all the processing is done locally on the intelligent microchip through wireless communication, the whole procedure may be monitored by external equipment. Nevertheless, this strategy can be utilized for other applications including in epileptic studies, drug monitoring, and prosthetics.

4.4.4 DNA applications

Recently, there have been much research interest in developing fully electronic DNA (deoxyribonucleic acid) microarrays based on postprocessed silicon semiconductor technologies as a tool in the biological and medical communities

[4.160–4.162]. DNA is a long molecule, located at the nucleus of every cell in the body, and contains the genetic code inherited from our parents. Segments of DNA that code for a particular function or attribute are called genes. A specific variation at this code may result in a malfunction of an organ or system. Learning the code and the genes can help evaluate body organs' strength as well as the chance of illnesses. DNA is a twisted double-stranded chain of about 3 billion nucleotide bases long. The genetic information is encrypted in the sequence of these nucleotides on the strands. The genes only occupy a fraction of the DNA. Recently, there has been much research interest in developing fully electronic DNA microarrays based on post-processed silicon semiconductor technologies [4.163, 4.164]. This has the advantage of requiring no fluorescent tags (i.e., label-free) and no optical methods of detection.

Current research has shown that sensors based on the field-effect can be used to achieve this purpose quite efficiently [4.165, 4.166]. The term BioFET refers to a field-effect transistor (FET), the gate material and contact of which have been replaced by an electrolytic solution and a reference electrode. The surface of the exposed gate insulator is functionalized, and then probe DNA oligonucleotides are immobilized on this surface. On successful hybridization, since the DNA carries a negative charge [4.167], a negative shift in the threshold voltage of the FET is detected.

Electronic versions of DNA biosensors are very useful in the design of low power and portable devices. Such devices can be used for early detection of various bacterial and viral pathogens that can affect food and water supplies [4.168]. Examples of pathogens that can affect water supplies are *Escherichia coli* and *Campylobacter jejuni*. Food-borne pathogens include foot and mouth diseases, classical swine fever, and salmonella. Using electronic DNA sensors, early detection of pathogens can be deployed in rural areas where access to sophisticated labs and expensive optical microarray readers are infeasible due to the high cost of optical microarray technologies. Experiments have shown that BioFET threshold shifts in the range of several millivolts are possible for dense enough DNA probe density [4.165]. However, the amount of threshold shift can depend in a significant manner on various parameters or experimental conditions other than the DNA density. Examples include electrolyte concentration, DNA strand length, and pH of the solution and surface adsorption affinity, which can all lead to different shifts in the threshold voltage. To predict the amount of voltage shift in response to DNA hybridization, it is necessary to include all these effects when modeling the BioFET. The integration capabilities of ICs technology may have significant potential and opportunities in the development of the biosensing and bioimaging systems for emerging applications in health and environmental sciences.

4.4.5 Smart skin

Today's industry and personal medical care both strongly demand accurate, reliable, robust, low power, and low cost methods to sense changes in the environmental and the condition of the body [4.169]. This is where the concept of smart skin comes in. Smart skin can monitor changes in environmental parameters, such as temperature, strain, and the presence of ambient gas, and communicate these parameters' changes wirelessly or wired [4.170]. The smart skin concept can also be extended to that of wearable electronic devices for continuous monitoring and reporting of critical biosignals [4.171]. There are a lot of challenges for the state of the art of smart skin, such as expensive fabrication methods, a lack of flexibility and mobility, and the large area fabrication method.

The skins scavenge energy using ambient electromagnetic, solar, thermal, mechanical, or radio frequency identification (RFID)/radar-based interrogation techniques. In short, these smart skins could prove to be the ultimate sensing tool that could potentially allow for the mass implementation of a perpetual wireless network even in extremely rugged environments. Carbon nanomaterials, such as single-walled carbon nanotubes CNTs, multiwalled CNTs, and graphene, have been considered and investigated as candidates for the quick and precise detection of various chemicals [4.172–4.174]. Because of their large surface area, carbon nanomaterials have the ability to physically and chemically absorb the chemicals on their surface; these carbon nanomaterials then alter their properties, which is the foundation for the chemical sensor applications. The absorption of the chemical compound causes changes in the material and electrical properties such as dc resistance, real and imaginary parts of the impedance, and an effective dielectric constant. These electrical changes can determine the presence of various chemicals by measuring electrical quantities such as changes in the current, resonant frequency, and amplitude of backscattered power. Their lightweight, low cost, outstanding electrical conductivity and ease of fictionalization targeted for a broad range of chemicals make these carbon materials ideal candidates for the development of a wide spectrum of portable and wearable sensors.

4.4.6 Cardiac and renal applications

Accelerometers have been the first sensor that is an actual component to be integrated in a cardiac implant [4.175–4.178]. Other types of sensors were just features of the implant electronics that performed electrical measurements such as impedance sensing for respiration rate or muscle spontaneous electrical activity sensing. After years of development mainly for the automotive industry, MEMS accelerometers have achieved a high level of reliability, miniaturization, and reduced power consumption. Hence, they have been introduced in cardiac systems for patient's activity monitoring and can easily detect if a patient is resting on his bed or climbing the stairs.

This upgrade has been a major improvement for implants as they have been able to automatically adapt the stimulation pace to the patient's activity. A new level of improvement has been reached with the introduction on the market of SonRtip™ lead [4.179]. The letter device includes an accelerometer at the distal end of the lead which measures the "sound" of the heart through signals PEA (pulseless electrical activity) occurring repeatedly during the cardiac cycle. PEA leads to a loss of cardiac output and the blood supply to the brain is interrupted, as a result a person loses consciousness and stops breathing spontaneously. This measurement allows the implant to optimize the resynchronization therapy in real time, the patient being at rest or exercising. The localization of this sensor through its miniaturization avoids repetitive tests at a physician's office or in a hospital. In the near future, a merged and upgraded version of this device is expected to be seen through the integration of new generation MEMS accelerometer, providing enhanced performances, multi-axial measurements, and further reduced dimensions for minimal invasiveness. According to the manual data report of 2013USRDS (Unite States Renal Data System), it indicates that the reported rates of incident end-stage renal disease (ESRD) across the globe noticed an important trend. In 2011, the top three countries with the highest rates of reported incident ESRD are Mexico, the United States, and Taiwan. Hemodialysis therapy continues to be the most common method to treating ESRD worldwide. Venous needle dislodgement has been reported to be a potentially serious complication during hemodialysis therapy. A commercial blood leakage detector, HEMOdialert products, specific for hemodialysis therapy is currently available which requires less than 1 ml of blood, and the blood leaking condition can be detected in 1 to 2 seconds. The sensing method is based on the changes of the voltage signal in the sensor. This device includes a detector having two spaced apart electrodes; each electrode is connected to a signal generator source via a lead.

The device also includes a signal processing unit that detects a change of state across the electrodes produced by the introduction of a fluid and an alarm actuated by the change of state. The electrodes are encased in a flexible nonconductive material and can be reused after cleaning. The price of such product is not affordable for general patients, and thus can be popularized. Although these products have an alarm, the loudness of the alert is limited by distance. A recent innovation, a photo-interrupter, is used as a sensor for detecting the blood leakage, combined with Bluetooth 4.0 for the function of wireless transmission. Furthermore, the blood leakage detector can be integrated on a bracelet, which is a simple and inexpensive way to monitor the leakage of blood during hemodialysis treatment [4.180]. If the detector senses a leakage of blood, an alert signal will be transmitted to the health care station via the wireless transmission.

4.4.7 Neuronal applications

Deep brain stimulation (DBS) has demonstrated significant therapeutic benefits in (1) treating symptoms relating to neurological motor disorders including Parkinson's disease, essential tremor, and dystonia, and (2) providing relief from chronic pain [4.181–4.183]. DBS has been also used for treatment-refractory neuropsychiatric indications including Tourette's syndrome, obsessive-compulsive disorder, and for treatment-resistant depression. More recently, DBS has been proposed as a potential treatment of severe drug and alcohol addiction. Considerable research involving laboratory animals has been conducted in recent years to study the underlying principles of DBS and the mechanisms through which its therapeutic effects are mediated. These are not yet fully understood but research suggests that DBS directly changes brain activity in a controlled manner. DBS is an invasive neuronal circuitry-based neurosurgical intervention [4.184]. Understanding of brain circuity in terms of neural networks for various neurophysiological interactions and neurologic disorders is essential to define specific relay nodes which may be targeted within this neural network for electrical perturbation by DBS. The current common targets or nodes for DBS are the subthalamic nucleus for Parkinson's disease, the globus pallidus pars internus for dystonia and Parkinson's disease, and the ventralis intermedius nucleus of the thalamus for essential tremor. However, a bottleneck in establishing the therapeutic mechanisms and benefits of DBS has been the lack of portable DBS system that enable long-term brain stimulation in freely moving laboratory animals. A typical DBS system consists of three major components including an implantable pulse generator (IPG), electrodes, and a programmer.

The IPG is the main part of a DBS system. It is a signal generator, which is implanted in the subclavicular or chest region of patients. It delivers electrical pulses to the electrode through and extension lead. The extension lead is an insulated wire that connects the IPG and the electrodes. The electrodes are inserted in the targeted region of the brain to deliver balanced biphasic pulse into the brain. The programmer is used for IPG settings. It communicates with the IPG to set amplitude, frequency, duration, and polarity of the generated signals. In the existing DBS practices, complications including migration or misplacement of the leads, lead fractures, and skin erosion may happen due to the long extensions wires. Battery malfunction and electrode displacement can also cause complications. Moreover, the battery needs to be surgically replaced on a regular basis as it has a limited lifespan. To reduce the difficulties caused by the battery and the long wires, an antenna can be employed near the electrodes. Wireless transmission and reception of control and power signals can be accomplished with such an antenna. Most of the existing DBS devices employ complex circuitry and are, as a consequence, quite bulky. These devices are connected to the implanted electrode through long insulated wires that run from the device to the animal's head. The stimulator often needs to be disconnected from the electrode over the course of the study. Thus, in order to better reflect the longer duration stimulation applied to the brain in clinical trials, the laboratory animal would ideally receive this brain stimulation continuously without interruption for days and even weeks. When connected to the DBS device through long wires, behavioral tests and normal behaviors are substantially limited. The frequency of operation for a passive DBS device is also important. Huang *et al.* [4.185] reported on a miniaturized implantable planar PIFA at the medical implant communication service (MICS) band of 402 MHz, and the industrial, scientific, and medical (ISM) band of 433 MHz and 2.45 GHz for rectenna application. Gosalia *et al.* [4.186] investigated a data telemetry link for retinal prosthesis at microwave frequencies of 1.45 and 2.45 GHz. An intracranial pressure monitoring device implantable in the skull and operating at the 2.4 GHz ISM band was also demonstrated in Reference 4.187. We have selected the ISM band of 915 MHz for the DBS antenna because the frequency is higher than the MICS band of 402 MHz, thus offering small antenna size and high data rate. Moreover, the frequency of 915 MHz is lower than the ISM band of 2.4 GHz, therefore providing less dielectric loss inside biological tissues. Further progress in DBS depends on the wireless passive device design and high performance antenna design. A low power and lightweight portable micro-DBS device for laboratory animals was developed [4.188]. The animal can easily carry the device during the course of a clinical trial, and it can produce nonstop stimulation current pulses of desired characteristics for over 12 days on a single battery. The device consists of the following components: microcontroller, current source, stimulation electrode, power source, and printed circuit board.

The micro DBS device was tested in vitro at Mayo Clinic. The test was conducted by placing a stainless twisted wire part of the stimulation into a tank of physiologic saline solution. After the in vitro test, the device was used in a study to examine the role of DBS in reducing ethanol preference in alcohol preferring rats. The pair of wires connecting the stimulation board to the stimulating electrode was tunneled under the skin at the back of the neck and externalized for a short distance above the stimulating electrodes where it was connected. The microcontroller used in the device was 10-bit analog-to-digital converter channels. The advantage offered by this device includes ultra-low cost when mass-produced, ultra-low maintenance, and ease of use. Advances in biomedical engineering will continue to drive the development and refinement of implantable DBS devices. Designs of DBS leads with the capability of directional stimulation or DBS systems with feedback sensors that can detect changes in the electrical activity of the brain networks and concentration of relevant neurotransmitters and adjust the degree of stimulation as needed are promising future potentials. Compatibility of DBS leads with MRI (magnetic resonance imaging) is another area of great interest as MRI scans are often required for the assessments of other neurological conditions in patients with DBS systems. The development of rechargeable powering mechanisms capable of harnessing the body's mechanical energy is also of immense interest as current pulse generators (batteries) used for DBS have a limited lifespan.

4.4.8 Biometrology applications

When designing wearable systems to be used for physiological and biomechanical parameters monitoring, it is important to integrate sensors easy to use, comfortable to wear, and minimally obtrusive. Wearable systems include sensors for detecting physiological signs placed on body without discomfort and possibly with capability of real-time and continuous recording. The system should also be equipped with wireless communication to transmit signals, although sometimes it is opportune to extract locally relevant variables, which are transmitted with needed [4.189, 4.190]. Most sensors embedded into wearable systems need to be placed at specific body locations, e.g., motion sensors used to track the movements of body segments, often in direct contact with the skin, and physiological sensors such as pulse meters or oximeters. However, it is reasonable to embed sensors within pieces of clothing to make the wearable system as less obtrusive as possible. In general, such systems should also contain some elementary processing capabilities to perform signal preprocessing and reduce the amount of data to be transmitted. Mobile wearable device communications create new challenges compared to ordinary sensor networks and short-range communications. In mobile wearable communications, devices communicate with each other in a peer-to-peer fashion or client–server fashion and also communicate with aggregation points (e.g., smartphones, tablets, and gateway nodes).

Wearable devices are expected to integrate multiple radio technologies for various applications' needs with small power consumption and low transmission delays. These devices can hence collect, interpret, transmit, and exchange data among supporting components, other wearable devices, and the Internet. Such data are not limited to people's personal biomedical information but also to include human-centric social and contextual data. A key design consideration of future wearable devices is the ability to ubiquitously connect to smartphones or the Internet with very low energy consumption. Radio propagation and, accordingly, channel models are also different from those in other wireless technologies. A huge number of connected wearable devices require novel big data processing algorithms, efficient storage solutions, cloud-assisted infrastructures, and spectrum-efficient communications technologies. Since the 21st century, aging population is emerged as a preeminent worldwide phenomenon. The large proportion of elderly population leads to a series of medical problems, such as inadequate medical care resource and low quality of medical service. These incoming problems have strong impact on hospitals and nursing homes. It is getting more difficult to provide good health care programs for those patients. Moreover, insufficient amounts of doctors and nurses cannot monitor these large numbers of elderly people who have unstable health conditions. In order to solve these medical problems, a series of innovations in this field have been developed, i.e., wearable biomedical health care systems, which integrate the wearable computing technology and multiagent software architecture. These systems improve the traditional bedside patient monitoring system that is only used in the intensive care unit. In the past, many researchers were focused on the development of home automation and wearable computing fields, but few of them tried to integrate these two fields. The field of medicine is on the verge of transformation, where health care will be provided on a personal basis to prevent an illness rather than treat it post-trauma. The wearable wireless multisensory system can serve as a viable platform to continuously and non-invasively acquire physiological signals to track cardiorespiratory dynamics, and quantitatively assess apneic conditions for prediction of obstructive sleep apnea (OSA) episodes, which is a common sleep disorder that affects 24 per cent of adult men and 9 per cent of adult women and is symptomatic in a third of that population [4.191]. An OSA episode is marked by the obstruction of pharyngeal airways resulting in interruption of the airflow during sleep. A novel aspect of this development is due to the use of MEMS technology; the total footprint of the wireless unit is highly adjustable and remains lightweight, and hence highly wearable. The hardware platform contributes toward affordable, yet powerful, and early warning systems for sleep apnea treatment. The wireless (Bluetooth) platform along with sensors and microprocessor components are integrated into a customized garment to continuously monitor sleep apnea episodes.

It was embedded the multisensory platform as part of a sleepwear shirt to enhance wearability of the sensor suite. The fusion of information from VCG (Frank vectorcardiogram), heart sound, and respiration provide adequate information to track variations and detect transitions in cardiorespiratory dynamics during sleep. Similarly, Chen *et al.* [4.192] developed a system that adaptively tracks energy expenditure (EE), based on user's characteristics in order to assist them to manage their daily exercise. People are doing exercise to control their weight and keep their cardiovascular system healthy. Heart disease has been top ranking cause of death in the United States for several years []. The goal is to have balanced EE. Knowing how much EE of one's body make it easier to achieve a good body mass index (BMI) requirement. Previous researches point out that the heart rate is easily affected by the human emotion or is raised in emergency situation. There are still problems unsolved. First, some previous prototypes require multiple electrodes contacting with the skin when doing heart rate measurement. Second, the heart rate is measured based on electrocardiograph (ECG)-based technology. There is no ready to use ECG interface circuit in most of the available devices in the market. In this application, authors used the microphone (MEMS device) of the mobile device to get heart sound signal as well as a three-axis accelerometer to collect the motion status from user. A vest with a pocket in front of the chest was designed. It enables the users to place a mobile device. The heart rate is extracted from the recording heart sound using an algorithm extended to calculate the hearth rate while the user is doing moderate to intense exercise. Once abnormalities during exercise are detected, multiple automatic notification mechanisms are invoked to inform about it. Textiles, being a pervasive and comfortable interface, are an ideal substrate for integrating miniaturized electronics components, or, through seamless integration of electroactive fibers and yarns, they have the potentiality to become fully functional electronic systems. The idea of e-textile being a viable solution to implement truly wearable, smart platforms adds bidirectional interfaces with the human body. Nowadays, several research projects are under way to examine the appropriateness of e-textile solutions in different areas, such as neuro- and cerebrovascular rehabilitation, safety, and security. Intelligent biomedical clothes are based on conductive and piezoresistive fabric developed to work as textile sensors, i.e., working as transducer of vital signs to electrical signals to be sent to a unit of microprocessors. First, commercial products are now offered in the market. Most of them are direct replacement of the chest strap by a sensitized garment, such as the POLOTECH™ SHIRT from Ralph Lauren that uses silver fibers woven directly into the fabric read heart rate and breathing depth and balance, as well as other key metrics, which are streamed to a mobile device via a detachable, Bluetooth-enabled black box. Runners, cyclists, skaters, and athletes can be interested in testing cardiopulmonary response.

The ergospirometry systems allow the determination of a subject's metabolic response while exercising or working. One example of such a system is a wireless portable ergospirometry system Xycon™ mobile device by CareFUsio.

The system has a light battery operated portable ergospirometry that is mounted onto the subject's body via a comfortable vest. The breath-by-breath data are collected through a facemask or mouthpiece and is sent to a host computer wirelessly. Specific software allows online and postprocessing of data. In addition to vital signals measured from the athlete's body, there are plenty of devices in the market that help the user to control other different useful parameters during the sport practice, such as time, altitude, speed, distance, cadence, stride, etc. Different devices are used to asses and improve the running techniques and analyze individual performance. Recently, a wearable sensor that tracks strain on a pitcher's elbow is capturing attention. This season, 27 MLB teams and their minor league affiliates are trying out the device, called the mThrow, in a hope that it will help monitor pitchers' workloads, improve pitching mechanics, and prevent injuries [4.193]. Injuries to professional pitchers in the United States have become epidemic. The reconstructive procedures known as Tommy John surgeries, which repair the elbow's ulnar collateral ligament (UCL), have increased among major league players from 14 performed in 2002 to 31 in 2014, according to the blog Baseball Heat Maps. The mThrow consists of a compression sleeve with a small removable sensor worn in a pocket over the elbow [4.194].

The sensor's accelerometers and gyroscopes track arm movements, and the device wirelessly transmits the data to an app that calculates the stress caused by torque on the UCL. The app also tracks a number of other metrics, such as pitch count, arm speed, release point, elbow height at release, maximum shoulder rotation, and "arm slot"—an indication of whether the throw was a side-arm pitch or over the top. The app can gauge the amount of stress on a player's elbow joint by comparing the condition to similar cases in the company's database of pitcher workloads. The app can also compare stress levels created by, say, a fastball versus a curveball, and it can indicate when a pitcher's arm slot might be changing due to fatigue.

4.4.9 Ingestible sensors

The future of medicine is in devices you swallow, doctors prescribe pills for their patients that double as ingestible sensors, wirelessly reporting back on the body's vital signs [4.195, 4.196]. A simple version of this already exists: a biotech company called Proteus Digital Health has actually created a primitive version of just such a sensor to track what drugs patients have taken; it was approved by the US Food and Drug Administration (FDA) in 2012 [4.197]. This smart pill system consists of a pinhead-sized sensor embedded in a pill, and a battery-powered patch that monitors various health indicators, such as sleep, respiration, and heart rate. As the human body powers the sensor, no batteries are required.

Once it has been swallowed, the device transmits a signal to the patch, indicating the time of ingestion and type of pill swallowed. The patch periodically sends all of its data to a designed smartphone. Digital medicine, the use of wearable and implantable physiological sensors, mobile communications technology, and web-based patient communities in managing patient health represents a new and rapidly evolving paradigm in health care. Continuous measurements of physiological metrics and rapid sharing of data between patients and caregivers offers unprecedented opportunities for diagnosing disease, tailoring treatment to individual patient physiology and behavior, and responding to new information with little or no delay. Medication without adherence is one such problem. Recent studies have estimated that 30 to 50 per cent of drug prescription are never taken [4.198, 4.199], resulting in significant complications and deterioration of patient health.

The extent and impact of the problem are expected to grow as patients live longer lives, often requiring management of one or more chronic conditions with multiple medications. In Reference 4.200, a sensor is presented for detecting the ingestion of a pharmaceutical tablet or capsule. The microfabricated sensor is designed to be incorporated into every digital capsule during pharmaceutical manufacturing. Upon ingestion and contact with the fluid in the stomach, each sensor communicates a unique and private digital coder to identify the medication and dose. In combination with a wearable sensor patch and a mobile phone user interface, the sensor provides a system for real-time, continuous measurement of medication adherence. The system further allows direct correlation between drug ingestion, health-related behavior such as physical activity, and critical metrics of physiological response, such as heart rate, sleep quality, and blood pressure. The sensor consists of three functional components: (1) the active layer, (2) the IC, and (3) the insulating skirt disk. The IC is a 1 mm × 1 mm × 0.3 mm complementary metal-oxide-semiconductor (CMOS) chip. After production of the IC, the active layers are deposited directly on the silicon wafers using a sequence of microfabrication steps. The active layers are thin films of magnesium and cuprous chloride. The gold underlayer acts as current collector underneath the cuprous chloride. Upon contact with gastric fluid, these layers create a battery that activates and powers the device. The function of the skirt is to shape and amplify the electric field generated by the sensor. The skirt is made of standard pharmaceutical excipients (ethyl cellulose, hydroxypropyl cellulose, and triethyl citrate) and is a critical part of the sensor design. By enhancing the signal amplitude, the skirt allows the sensor to be only 300 μm thick. This makes it feasible to manufacture the device almost entirely on a silicon wafer without the need for extended electrical antennas and connections. Another important application is related to the obesity problem, which is a major contributor to several life-threatening chronic diseases such as hypertension, high cholesterol level, stroke, coronary heart disease, and diabetes. A variety of methods have been applied for treatment of obesity. Intragastric balloon (IGB), introduced in 1982 [4.201], is a less-invasive method that has been successfully applied for short-term treatment of obesity [4.202], [4.203]. In this method, a gas (or liquid)-filled balloon is inserted into patient's stomach to occupy the space and induce the sense of satiety and loss of appetite. Thus, by reducing the food intake, in combination with specific exercise programs, the patient can lose weight more effectively and naturally. After the balloon reaches the stomach, it will be filled with liquid or gas through the tube. Saline solution, air, and carbon dioxide are the most common materials for this application. Once the balloon is inflated to a desired volume, the tube is detached from the self-sealing valve on the balloon surface and removed through the patient's upper gastrointestinal tract. The volume of the balloon is determined by the amount of injected liquid or gas. The inflated balloon stays in the stomach for 1 to 6 months according to physician's prescription. Current IGB systems suffer from several major drawbacks. Requiring an endoscopic procedure or surgery to insert and exert the balloon from the stomach is the most important disadvantage of this method. These procedures are usually costly and are associated with undesired side effects and feeling of discomfort. Giddiness and nausea are among the most common side effects of these procedures. A less-invasive method for insertion, removal, and adjusting the volume of the balloon is highly desired. In Reference 4.204, a volume-adjustable pill for treatment of obesity is proposed. The control unit is in charge of interpreting received commands, adjusting operation of various capsule modules, and performing desired tasks through proper activation of actuation mechanisms. The wireless communication unit makes it possible for the in-body capsule and the external controller to communicate. It could be used to send the operational commands to the capsule and/or extract the data and status of the capsule. The actuation mechanism, acting as an interface between the electronic modules and mechanical sub-modules, is in charge of adjusting the capsule volume to a desired level based on received commands. In the developed prototype, a Zarlink ZL70101 transceiver in conjugation with a fabricated antenna was utilized as the wireless communication unit. Using this transceiver, a bidirectional communication between the capsule and the external communication unit at 433 MHz ISM band was established. MSP430 ultra-low power microcontroller from Texas instruments was used to implement the control unit of the system and control the operation of the capsule. The system was powered using Lithium Ion electrochemical batteries.

4.4.10 Health care platforms

Emergency medical services response time is an important aspect within the health care services. Authorities around the world are making tremendous efforts to reduce it to the minimum concerning both urban and rural environments. No matter how easy or complicated a situation can be, if the medical staff does not react on time, everything becomes doubtful and unsafe. When it comes to outdoor investments, medical services use helicopters, emergency cars, and other transportation machines that can help save time. In terms of indoor investments, in addition to modern infrastructure and a high degree of security, modern medical equipment represents a leading priority. Blessed with the huge technological boom from the last decade, the health care system from developed nations started to use innovative devices and wireless body sensors for surveillance, diagnosis, and treatment.

The health care systems such as in-home assistance, smart home care, and remote patient monitoring area greatly enriched by the adoption of the wireless sensor network technologies [4.205]. Two-thirds of consumers are expected to purchase a connected home device within the next 5 years, and the ownership of consumer wearable is expected to double year over year by 2016. However, there are also common diseases around the world, which can be prevented or better treated through continuous and ubiquitous health monitoring by the smartphone. Biomedical wireless technology, networks and sensing systems (BiowireleSS) are rapidly becoming an integral component of healthcare delivery.

A wide range of wireless technologies is being brought to bear on a long list of healthcare and pharmaceutical applications with the promise of utterly transforming healthcare, as we know it today. Where once a patient would have been confined to a hospital bed for extensive tests, to be monitored postsurgery, or for chronic conditions, in many cases, patients will soon be monitored at home, free to go about their normal activities, while a wireless device will transmit data to healthcare providers, who will be alerted if vital signs vary from normal ranges. Wireless medical applications include chronic disease management, post-surgical recovery, vital sign monitoring, proactive monitoring, eldercare, remote diagnosis, emergency communications, wellness and fitness, telemedicine, facilities monitoring, asset and staff locating, positioning, asset management, tracking pharmaceuticals, imaging, and video. Wireless technology for biomedical sensing systems potentially include wireless personal area network (WPAN)/WBAN/medical body area network (MBAN), wireless local area network (WLAN), Wi-Fi, WiMAX, ZigBee, Bluetooth, ANT, ultrawideband, e-textiles, radar, web conferencing, capsule endoscopy, implantable and ingestible sensors, and epidermal electronics, smart bandages, smartphone apps, RFID, real-time location system, indoor positioning systems, etc. [4.128]. WBAN integrates low-power sensor platforms incorporating micro- and nanosensors and CMOS ICs with body area network [4.206]. Due to the inherent low-power requirements, these networks are best suited for integration into wearable health monitoring units [4.207], which are usually incorporated with various types of implantable and transdermal sensors to measure various physiological parameters such as ECG, blood pressure, glucose, lactate, oxygen, pH, etc. WBANs are also employed in a hospital environmental for monitoring of critical care patients as well as in smart home care for real-time data acquisition and transmission over the secure internet to healthcare providers. RFID technology has become a very popular technology choice for a number of commercial, residential, and industrial as well as healthcare applications [4.208]. RFID tags are used in hospitals for inventory monitoring to provide a real-state status report to the staffs and the suppliers as well as for easy tracking of the positions of the patients and doctors within the hospital premises. WPANs using IEEE 802.15.4, which is the basis for Zigbee, ISA100.11a, WirelessHART, and MiWi specifications, have potential uses in healthcare [4.209–4.211]. These short-distance communication networks can be deployed inside a hospital room so that caregivers can monitor a patient remotely in real time from a nearby workstation. WPAN can also be used for interfacing multiple smart medical devices within the hospital premises and the data between the devices can be accomplished with minimum time overhead. Zigbee technology can be also combined with WBANs to constitute smaller networks for real-time monitoring of the physiological activities of the patients. The WLAN channels are also utilized to transfer patient data around the hospital and to provide communication between various smart medical devices.

4.4.11 Health care IoT applications

Intelligence hardware is bridging the last mile between the digital enterprise and the physical world. The IoT is a concept reflecting a connected set of anyone, anything, anytime, anyplace, any service, and any network. The IoT is a megatrend in next-generation technologies that can impact the whole business spectrum and can be thought of as the interconnection of uniquely identifiable smart objects and devices within today's Internet infrastructure with extended benefits [4.143]. Benefits typically include the advanced connectivity of devices, systems, and services that go beyond machine-to-machine scenarios. Therefore, introducing automation is conceivable in nearly every field. The IoT provides appropriate solutions for a wide range of applications such as smart cities, traffic congestion, waste management, structural health, security, emergency services, logistics, retails, industrial control, and healthcare. Medical care and healthcare represent the most attractive areas for the IoT. The IoT has the potential to give rise to many applications such as remote health monitoring, fitness program, chronic diseases, and elderly care. Compliance with treatment and medication at home and by health care providers is another important potential area. Therefore, various medical devices, sensors, and diagnostic and imaging devices can be viewed as smart devices or objects constituting a core part of the IoT. IoT-based health care services are expected to reduce costs, increase the quality of life, and enrich the user's experience. Ease of cost-effective interactions through seamless and secure connectivity across individual patients, clinics, and health care organizations is an important trend.

4.5 Conclusions

As has been shown with different examples through this chapter, the efficient use of energy is intimately linked to a sustainable and green environment, where the ubiquitous and continuous pervasiveness of semiconductor-based electronics, in the human environment, is a key piece in the energy efficiency and sustainable green environment equation.

Besides, electronics, in conjunction with sensors/actuators, computing, and communications, has pervaded our daily lives at any time anywhere: smart phone being one of the most and widespread examples seen today. In that sense, electronics can be compared to the pervasiveness of oil industry and its derivative products, such as combustibles, plastics, fertilizers, textile fibers, detergents, beauty products, and many other oil-based products that are also around us anywhere.

Energy or power management is behind every aspect of our lives, in the industry sector that produces cars, processed food, medicines, electronics, etc., in our homes, in the streets where car traffic is controlled, in our work office, and even for the energy production itself. Then I would like to quote Prof. Baliga's words from his book "The IGBT Devices Physics, Design and Applications of the Insulated Gate Bipolar Transistor" [4.2], where he talks about power management and delivery for motors, which accounts for two thirds of the world electricity consumption, lighting, which consumes one fifth of the world electricity, and transportation sector:

> I determined that society had derived a savings of over 50,000 terawatt hours in electricity consumption (equivalent to not building 600 coal-fired power plants) and over 1 trillion gallons of gasoline consumption. This had not only saved worldwide consumers more than $15 trillion but reduced carbon dioxide emissions by more than 75 trillion pounds over the period from 1990 to 2010.

The insulated-gate-bipolar-transistor (IGBT) is a key semiconductor device player for energy and delivery management for power electronics. The IGBT is an electronic switch that has replaced the traditional mechanical switches, and it can be found in cars, airplanes, motors, printers, home appliances, and much other electronic instrumentation.

The microprocessor technology has pushed the device miniaturization to the scale where about an average 10,000 times of a MOSFET channel length can be accommodated in the diameter of a human hair. In a time scale, for instance, the Intel Core i7 5960X microprocessor is capable of processing 283,310 millions of instructions per second at a clock frequency of 3.0 GHz. Within a second of time, a light–gas gun-launched projectile, like those used by the NASA [4.212], travels a distance of 8.5 kilometers. This amazing signal processing velocities requires the incorporation of tens of different materials, such as semiconductors, dielectrics, and metals into a manufacturing process. The incorporation of that vast variety of materials has propelled the development of other than microprocessor devices, such as MEMs with multiple applications to different sectors (automotive, biomedicine, aeronautics, oil industry, etc.), implantable biosensors, water decontaminants, nanomembrane filters, and a plethora of different kind of sensors, just to mention a few examples.

At a global level, the expansion of urbanization is posing a challenge in terms of energy efficiency and pollution as well. The rural population is declining, while the urban population is expected to be in the order of 66 per cent of the world's population by 2050. These large concentrations of population requires services of multiple nature, such as vehicle traffic management, water, food and energy supply, and many other such as schools, hospitals, and communication services. The industry nearby the urban centers consumes most of the energy, followed by transportation, residential, and commercial and services. The combined use of electronics, computing, communications, and sensors and actuators, in all these sectors, is a common factor. Therefore, the energy efficiency of electronic components, instruments and systems, is a key factor for green urban environments. The smart city is an initiative with the quest to find optimal solutions for the urban zones. It is based on the use of interconnected networks that share information of different nature, such as vehicle traffic, drinking water distribution, temperature, humidity, energy distribution and consumption, etc. It makes use of wireless interconnected sensors that sense variables, such as vehicle traffic, temperature, humidity, solar radiation, air pollution level, electricity distribution, drinking water inflow, and consumption/waste. These variables interplay in a complex way, which require the use graph theory for instance as a way to find the optimal solution for a multivalued problem. Signal processing and wireless communications are two key additional factors for this approach, which linked to devices, such as smart phones, sensors/actuators, for instance, form the IoT. If energy-efficient electronics and smart-city concepts are put in place, a considerable amount of energy in the urban zones can be saved, as has already happened in other sectors. For instance, the efficiency of cars has improved by 40 per cent since the inception of electronics, and the lighting efficiency increased by 339 per cent.

The graph theory is a valuable tool as has already been demonstrated in the administration of WSNs. The innovative and efficient design of power management and power failure diagnosis has been successfully applied in agriculture. In Mexico, for instance, the irrigation water has been reduced up to 90 per cent in a crop field when compared with traditional irrigation practices. The smart irrigation system has a temperature and humidity WSN, supplied with solar cells that continuously feed information to a signal processing unit, which based on a model sends back a signal to a water irrigation actuator. In this way, the system is almost fully energy autonomous as well.

Meat and milk production is another sector that can benefit of smart electronics. Livestock in particular ruminants, like cows, account for up to one third of methane emission worldwide. Methane is a greenhouse gas that has pollution potential 25 times that of CO_2. Ruminants emit methane, which is a function of feed intake, type of diet, feed processing, and alterations in the ruminal microflora. Both rumen temperature and pH are also key factors for the methane emission. If these two variables are kept into a safe window, then the quality and quantity of milk can be increased at the same time methane emission is reduced. Placing a temperature and pH WSN in the rumen of the cow allows the continuous monitoring of these variables. Then by correlating the feed intake and type of diet to these two variables, the cow feed process is adjusted to the point where methane emission is reduced. This way, WSNs and smart electronics contribute to food production and reduction of greenhouse gas pollutants.

The smart electronics in its form of portable and miniature wireless-connected sensing devices has opened new opportunities for medicine and healthcare. These devices collect health data related to many purposes, by patients with chronic medical conditions (such as blood sugar sensors for diabetics), people seeking to change behavior (losing weight or quitting smoking), or athletes wishing to monitor their condition and performance. The resulting data may be used directly by the person, or shared with others: physician for treatment, insurance company, etc. A key element for this is remote monitoring. A few examples of these applications are invasive and noninvasive biosensors, gastric monitoring, DNA measurement, smart skin applications, cardiac sensors, renal monitoring, neuronal systems, ingestible sensors, and healthcare platforms.

References

[4.1] Baliga, B. J., Adler, M. S., Love, R. P., Gray, P. V., Zommer, N. D., "The insulated gate transistor: a new three-terminal MOS-controlled bipolar power device," IEEE Transactions on Electron Devices, Vol. 31, 1984, pp. 821–828.

[4.2] http://people.engr.ncsu.edu/bjbaliga/IGBT_Applications_and_Social_ Impact.pdf.

[4.3] Baliga, B. J., "Social impact of power semiconductor devices (Invited Paper)", Proc. IEDM 2014, 2014, pp. 2.1.1–2.1.4.

[4.4] Nakamura, S., "Energy savings by LED lighting", In: CLEO 2015, OSA Technical Digest (online), Optical Society of America, paper JW5A.2.

[4.5] http://www.zdnet.com/article/intel-reveals-faster-processor-it-hopes-will-spearhead-pc-revival/, http://www.intel.com/content/www/us/en/processors/core/core-processor-family.html.

[4.6] Reth, M., "Matching cellular dimensions with molecular sizes", Nature Immunology, Vol. 14, Number 8, 2013, pp. 765–767.

[4.7] Antibody picture, http://phil.cdc.gov/phil/details.asp?pid=10000.

[4.8] Virus picture, http://www.public-domain-image.com/free-images/science/microscopy-images/influenza/3d-graphical-representation-of-flu-virus.jpg.

[4.9] Bacteria picture, http://phil.cdc.gov/phil/details.asp?pid=10048.

[4.10] Cancer cell picture, http://phil.cdc.gov/phil/details.asp.

[4.11] Human hair picture, http://www.sciencedaily.com/releases/2015/07/150729142412.htm.

[4.12] Natarajan, S., Agostinelli, M., Akbar, S., *et al.*, FinFET picture, "A 14nm logic technology featuring 2nd generation FinFET transistors, Air-gapped interconnects, self-aligned double patterning and a 0.0588 μm^2 SRAM cell size", Proc. IEDM 2014, pp. 71–73.

[4.13] Bhansali, S., Vasudev, A., "MEMS for Biomedical Applications", Woodhead Publishing, 1st edition, Cambridge, UK, 2012.

[4.14] Nguyen, N.-T., Shaegh, S. A. M., Kashaninejad, N., Phan, D.-T., "Design, fabrication and characterization of drug delivery systems based on lab-on-a-chip technology," Advanced Drug Delivery Reviews, Vol. 65, 2013, pp. 1403–1419.

[4.15] Molina, J., Sanchez-Salas, J. L., Zuniga, C., *et al.*, "Low-temperature processing of thin films based on rutile TiO_2 nanoparticles for UV photocatalysis and bacteria inactivation", Journal of Materials Science, Vol. 49, 2014, pp. 786–793.

[4.16] Rickerby, D., Morrison, M., "Report from the workshop on nanotechnologies for environmental remediation", JCR Ispra, nanoforum.org, 2007.

[4.17] Narayan, R. J., Shashishekar, P., Pellin, M. J., *et al.*, "Atomic layer deposition-based functionalization of materials for medical and environmental health applications", Philosophical Transactions of the Royal Society, Vol. 368, 2010, pp. 2033–2064.

[4.18] www.amd.com.

[4.19] Xie, Q., Lin, X., Wang, Y., Chen, S., Dousti, M. J., Pedram, M., "Performance comparisons between 7 nm FinFET and conventional bulk CMOS standard cell libraries", IEEE Transactions on Circuits and Systems, Vol. 62, Issue 8, 2015, pp. 761–765.

[4.20] Shafaei, A., Wang, Y., Pedram, P., "Low write-energy STT-MRAMs using FinFET-based access transistors", 32nd IEEE International Conference on Computer Design (ICCD), 2014, pp. 374–379.

[4.21] Li, J., Buckwalter, J., "Energy efficiency of optoelectronic interfaces in scaled FinFET and SOI CMOS technologies", IEEE Optical Interconnects Conference, 2015, pp. 18–19.

[4.22] Kranti, A., Armstrong, A., "Design and optimization of FinFETs for ultra-low-voltage analog applications", IEEE Transactions on Electron Devices, Vol. 54, No. 12, 2007, pp. 3308–3316.

[4.23] Pavanello, M. A., Martino, J. A., Simoen, E., Claeys, C., "Cryogenic operation of FinFETs aiming at analog applications", Cryogenics, Vol. 49, 2009, pp. 590–594.

[4.24] Ionescu, A., Riel, H., "Tunnel field-effect transistors as energy-efficient electronic switches", Nature, Vol. 479, 2011, pp. 329–337.

[4.25] Seabaugh, A. C., Zhang, Q., "Low-voltage tunnel transistors for beyond CMOS logic", Proceedings of the IEEE, Vol. 98, No. 12, 2010, pp. 2095–2110.

[4.26] http://www.computerhistory.org/semiconductor/timeline/.

[4.27] https://en.wikipedia.org/wiki/Apollo_Guidance_Computer.

[4.28] Potter, G. B., Mendelson, J., Sirkin, S., "Integrated scratch pads sire new generation of computers," Electronics Vol. 39, No. 7, 1966, pp. 118–126.

[4.29] Shin, H. J., Lu, P. F., Chin, K., Chuang, C. T., Wamock, J. D., Franch, R. L., "A 1.2 ns/1 ns 1 k × 16 ECL dual-port cache RAM," ISSCC Digest of Technical Papers, 1993, pp. 244–245.

[4.30] Yuan, J., Cressler, J. D., Krithivasan, R., *et al.*, "On the performance limits of cryogenically operated SiGe HBTs and its relation to scaling for terahertz speeds", IEEE Transactions on Electron Devices, Vol. 56, No. 5, 2009, pp. 1007–1018.

[4.31] Chakraborty, P. S., Cardoso, A. S., Wier, B. R., *et al.*, "A 0.8 THz f_{max} SiGe HBT operating at 4.3 K", IEEE Electron Device Letters, Vol. 35, No. 2, 2014, pp. 151–153.

[4.32] Cressler, J. D., A retrospective on the SiGe HBT: what we do know, what we don't know, what we would like to know better, IEEE 13th Topical Meeting on Silicon Monolithic Integrated Circuits in RF Systems (SIRF) 2013, pp. 81–83.

[4.33] Kahng, D., "Electric field controlled semiconductor device," U. S. Patent No. 3,102,230 (Filed May 31, 1960, issued August 27, 1963).

[4.34] Norman, R. H., Stephenson, H. E., "Shift register employing insulated gate field effect transistors," U. S. Patent 3454785 (Filed July 27, 1964. Issued July 8, 1969).

[4.35] Wanlass, F. M., "Low stand-by power complementary field effect circuitry." U. S. Patent 3,356,858 (Filed June 18, 1963. Issued December 5, 1967).

[4.36] https://en.wikipedia.org/wiki/4000_series.

[4.37] http://www.computerhistory.org/semiconductor/welcome.html

[4.38] http://www.intel4004.com/

[4.39] https://en.wikipedia.org/wiki/IBM_Personal_Computer

[4.40] http://www.intel.com/content/www/us/en/company-overview/intel-museum.html

[4.41] Sah, C.-T., "Evolution of the MOS transistor—from conception to VLSI", Proceedings of the IEEE, Vol. 76, No. 10, October 1988, pp. 1280–1326.

[4.42] http://spectrum.ieee.org/static/the-self-driving-car.

[4.43] Koomey, J. G., Berard, S., Sanchez, M., Wong, H., "Implications of historical trends in the electrical efficiency of computing", IEEE Annals of the History of Computing, Vol. 33, Issue 3, March 2010, pp. 46–54.

[4.44] Koomey, J., Naffziger, S., "More's law might be slowing down, but not energy efficiency", In: IEEE Spectrum, 2015. http://spectrum.ieee.org/computing/hardware/moores-law-might-be-slowing-down-but-not-energy-efficiency.

[4.45] Coskun, A. K., Ayala, J. L., Atienza, D., Rosing, T. S., "Modeling and dynamic management of 3D multicore systems with liquid cooling", 17th International Conference on Very Large Scale Integration (VLSI-SoC), 2009, pp. 35–40.

[4.46] Youssef, A., Anis, M., Elmasry, M., "Dynamic standby prediction for leakage tolerant microprocessor functional units", 39th Annual IEEE/ACM International Symposium on Microarchitecture, 2006, pp. 371–384.

[4.47] https://www.amd.com/Documents/energy-efficiency-whitepaper.pdf.

[4.48] http://www.energystar.gov/products/certified-products/detail/computers.

[4.49] Nikonov, D. E., Young, I. A., "Overview of beyond-CMOS devices and a uniform methodology for their benchmarking", Proceedings of the IEEE, Vol. 101, No. 12, December 2013, pp. 2498–2533.

[4.50] Windbacher, T., Mahmoudi, H., Sverdlov, V., Selberherr, S., "Rigorous simulation study of a novel non-volatile magnetic flip flop", Proc. of the SISPAD, 2013, pp. 368–371.

[4.51] http://esa.un.org/unpd/wup/Highlights/WUP2014-Highlights.pdf

[4.52] http://supernet.isenberg.umass.edu/visuals/FOMgt341-F11/mexico-city-project.pdf

[4.53] http://www.eia.gov/todayinenergy/detail.cfm?id=10271

[4.54] http://www.inegi.org.mx/

[4.55] http://www.ecofys.com/files/files/the-2015-energy-productivity-and-economic-prosperity-index.pdf

[4.56] http://www.rfidjournal.com/articles/view?4986.

[4.57] Harrison, C., Donnely, I. A., "A theory of smart cities", Proceedings of the 55th Annual Meeting of the ISSS, 2011, pp. 1–15.

[4.58] http://amsterdamsmartcity.com/projects/detail/id/30/slug/almere-smart-society.

[4.59] http://www.techceocouncil.org/clientuploads/reports/TCCsmartgreen.pdf

[4.60] http://www.gartner.com/it/page.jsp?id=499090

[4.61] "Electricity consumption by end use in New England households, 2001." New England Household Electricity Report, U.S. Energy Information Administration, October 11, 2005.

[4.62] http://www.eia.doe.gov/emeu/reps/enduse/er01_new-eng_tab1.html.

[4.63] IEEE International Conference on Solid-State Sensors and Actuators. Available at: http://www.journals.elsevier.com/sensors-and-actuators-a-physical/.

[4.64] http://phys.org/news/2015-02-high-efficiency-solar-cells-rooftop.html

[4.65] http://materialsforenergy.typepad.com/materials/2010/11/solid-state-thermoelectric-generation-capturing-waste-heat.html

[4.66] Kocolosky, M., Eger, C., McCarty, R., Hallinan, K., Kissock, K., "Industrial solid-state energy harvesting: mechanisms and examples", 2007 ACEEE Summer Study on Energy Efficiency in Industry, pp. 4-55–4-68.

[4.67] Tanaka, T., Suzuki, T., Kurihara, K., "Energy harvesting technology for maintenance-free sensors", Fujitsu Scientific & Technical Journal, Vol. 50, No. 1, January 2014, pp. 93–100.

[4.68] Pérez, D., Kämpf, J., Scartezzini, J.-L., "Simulation of urban energy flow: a graph theory inspired approach", CISBAT 2013, Lausanne, Switzerland, 2013, pp. 1071–1076.

[4.69] Djahel, S., Doolan, R., Muntean, G.-M., Murphy, J., "A communications-oriented perspective on traffic management systems for smart cities: challenges and innovative approaches", IEEE Communications Surveys & Tutorials, Vol. 17, No. 1, 2015, pp. 125–151.

[4.70] Kannadhasan, S., Karthikeyan, G., Sethupathi, V., "Graph theory based energy efficient clustering techniques in wireless sensor networks", Proceedings of the 2013 IEEE Conference on Information and Communications Technologies, ICT 2013, pp. 151–155.

[4.71] Ambekar, C., Kalyankar, V., Raina, V., Gund, M., "Energy efficient modeling for wireless sensor networks using random graph theory", International Journal on Trends in Engineering and Technology, Vol. 10, No. 2, 2014, pp. 10–18.

[4.72] Di Nardo, A., Di Natale, M., Santonastaso, G. F., Tzatchkov, V. G., Alcocer-Yamanaka, V. H., "Water network sectorization based on graph theory and energy performance indices", ASCE Journal of Water Resources Planning and Management, 2014, pp. 620–629.

[4.73] http://sfcplatform.eu/

[4.74] Storm, I. M. L. D., Hellwing, A. L. F., Nielsen, N. I., Madsen, J., "Methods for measuring and estimating methane emission from ruminants", Animals, Vol. 2, 2012, pp. 160–183.

[4.75] http://www.ipcc.ch/pdf/assessment-report/ar4/wg1/ar4_wg1_full_report.pdf

[4.76] Amon, B., Kryvoruchko, V., Amon, T., Zechmeister-Boltenstern, S., "Methane, nitrous oxide and ammonia emissions during storage and after application of dairy cattle slurry and influence of slurry treatment", Agriculture, Ecosystems and Environment, Vol. 112, 2006, pp. 153–162.

[4.77] Johnson, K. A. and Johnson, D. E., "Methane emissions from cattle", Journal of Animal Science, 1995, Vol. 73, No. 8, pp. 2483–2492.

[4.78] http://www.csiro.au/en/Research/AF

[4.79] Lin, X., "Evaluation of Kahne rumen sensors in fistulated sheep and cattle under contrasting feeding conditions", MSc Thesis at Massey University, Palmerston North, New Zealand, 2009.

[4.80] Kilic, U., "Use of wireless rumen sensors in ruminant nutrition research", Asian Journal of Animal Sciences, Vol. 6, No. 1, 2011, pp. 46–55.

[4.81] Díaz-Díaz, H., "Diseño e implementación de un sistema inalámbrico de medición in vivo de variables fisiológicas en rumen de Ganado (in Spanish)", Bachelor Thesis at Instituto Tecnológico de Veracruz, Mexico, 2015. "Design and implementation of an in vitro wireless systems for monitoring physiological variables in the cattle's rumen."

[4.82] http://www.ti.com/lit/ds/symlink/lm35.pdf

[4.83] Bergveld, P., "ISFET, theory and practice", IEEE Sensor Conference, Toronto, Canada, 2003, pp. 1–26.
[4.84] http://www.imb-cnm.csic.es/index.php/en/
[4.85] http://www.inaoep.mx
[4.86] https://www.arduino.cc/en/Main/ArduinoBoardFio
[4.87] https://www.arduino.cc/en/Main/ArduinoXbeeShield
[4.88] http://www.ecow.co.uk/wp-content/uploads/2013/09/farmBolus-results.pdf
[4.89] Jury, W. A., Vaux, H. J., "The emerging global water crisis: managing scarcity and conflict between water users," Advances in Agronomy, Vol. 95, 2007, pp. 1–76.
[4.90] Gutiérrez, J., Villa-Medina, J. F., Nieto-Garibay, A., Porta-Gándara, M. A., "Automated irrigation system using a wireless sensor network and GPRS module", IEEE Transactions on Instrumentation and Measurement, Vol. 63, No. 1, 2014, pp. 166–176.
[4.91] Wilcock, W. L., "Introductory review photo-electronic imaging 1958–1978", In: L. Marton and C. Marton, Eds., Advances in Imaging and Electronics Physics, vol. 52, 1st. ed. London: Proceedings of the Seventh Symposium held at The Blackett Laboratory, Imperial College, 1980, pp. 1–11.
[4.92] Manyika, J., Chui, M., Bughin, J., Dobbs, R., Bisson, P., Marrs, A., "Disruptive technologies: advances that will transform life, business, and the global economy", McKinsey Global Institute, Washington, DC, USA, Tech. Rep., 2013.
[4.93] Sharpe, B., Pary, V., Dubhthaigh, R., Barter, T., "Tackling obesities: future choices—future trends in technology and their impact on obesity", UK Government's Foresight Programme, London, UK, Tech. Rep. DIUS/PUB/2K/10/07/NP, 2007.
[4.94] Doi, M., Howell, J., Hirakawa, S., Personal and home electronics and our changing lifestyles", Proc. IEEE, Vol. 100, 2012, pp. 1646–1656.
[4.95] Kurkovsky, S., "Can mobile game development foster student interest in computer science?", International IEEE Consumer Electronics Society's Games Innovations Conference, London, 2009, pp. 92–100.
[4.96] De Silva, L., Morikawa, C., Petra, I., "State of the art of smart homes", Engineering Applications of Artificial Intelligence, Vol. 25, 2012, pp. 1313–1321.
[4.97] Stanescu, C. L, Pavaloiu, B., Tuduce, R., Cristea, P. D., "Software and hardware optimization for a laser surgical ophthalmic microscope", 20th International Conference on Systems, Signals and Image Processing (IWSSIP), Bucharest, 2013, pp. 99–102.
[4.98] Hossain, N., Tegnander, E., Dreier, J. M., Eik-Nes, S., Torp, H., Kiss, G., "Automatic measurement of biparietal diameter with a portable ultrasound device", IEEE International Ultrasonics Symposium (IUS), Chicago, 2014, pp. 459–462.
[4.99] Evans, D., "The Internet of Things How the Next Evolution of the Internet is Changing Everything", CISCO IBSG, San José, California, USA, Tech. Rep., 2011.

[4.100] Díaz, R., "The Mobile Economy 2015", GSMA, London, UK, Tech. Rep., 2015.

[4.101] Kahng, A., "Scaling: more than Moore's law", IEEE Design and Test of Computers, Vol. 27, 2010, pp. 86–87.

[4.102] Suresh, P., Daniel, J. V., Parthasarathy, V., Aswathy, R. H., "A state of the art review on the Internet of Things (IoT)", International Conference on Science, Engineering and Management Research (ICSEMR 2014), Chennai, 2014, pp. 1–8.

[4.103] Chen, A., Hutchby, J., "Emerging Nanoelectronic Devices", Wiley, 1st ed., UK., 2015.

[4.104] Sohbati, M. and Toumazou, C., "Personalized microchips for health care", IEEE Potentials, Vol. 34, 2015, pp. 33–37.

[4.105] Shelton, J., "Enabling the hyperconnected age: the role of semiconductors", GSMA, London, UK, Tech. Rep., May 2013.

[4.106] Morris, M., Kay, D., Newman, D., *et al.*, "Impact of ICT R&D on the Large-Scale Deployment of the Electric Vehicle", AEA Technology plc, London, U.K., Tech. Rep. ED57083-2, Nov. 2012.

[4.107] Charette, R., "This Car Runs on Code" [Online]. Available: http://spectrum.ieee.org/transportation/systems/this-car-runs-on-code.

[4.108] Divanis, G., "Toward smarter healthcare: anonymizing medical data to support research studies", IBM Journal of Research and Development, Vol. 58, January 2014, pp. 9:1–9:11.

[4.109] Heikenfeld, J., "Let Them See You Sweat", IEEE Spectrum, New York, USA, 2014, pp. 46–63.

[4.110] Kamoun, F., "RFID system management: state-of-the art and open research issues", IEEE Transactions on Network and Service Management, Vol. 6, September 2009, pp. 190–205.

[4.111] Amendola, S., Lodato, R., Manzari, S., Occhiuzzi, C., Marrocco, G., "RFID technology for IoT-based personal healthcare in smart spaces", IEEE Internet of Things, Vol. 1, 2014, pp. 144–152.

[4.112] Tu, T.-Y., Kao, Y.-H., Chao, P. C.-P., Lee, Y.-P., "Optimizing a new blood pressure sensor for maximum performance based on finite element model", IEEE Sensors, Valencia, 2014, pp. 1873–1876.

[4.113] Cheong, J., Ng, S. S. Y., Liu, X., *et al.*, "An inductively powered implantable blood flow sensor microsystem for vascular grafts", IEEE Transactions on Biomedical Engineering, Vol. 59, 2012, pp. 2466–2475.

[4.114] Chen, J., Cheng, X., Lin, P., Zhou, S., "Implantable ultrasound emitter array for medical applications", 19th IEEE International Conference on Micro Electro Mechanical Systems, Istanbul, January 2006, pp. 422–425.

[4.115] Rothermel, A., "Retinal implants for blind patients", In: B. Hoefflinger, Ed, Chips 2020—A Guide to the Future of Nanoelectronics, Sindelfingen: Springer-Verlag, 2012, pp. 367–382.

[4.116] You, X. and Pak, J., "Silk stabilized graphene FET enzymatic glucose biosensor", The 17th International Conference on Solid-State Sensors, Actuators and Microsystems, Barcelona, 2013, pp. 2443–2446.

[4.117] Kim, B., "Electrostatically driven micropump with peristaltically moving membrane", Micro & Nano Letters, Vol. 8, 2013, pp. 654–658.

[4.118] Hassarati, R. T., Tasche, C., Carter, P. M., Poole-Warren, L. A., Green, R. A., "Improving cochlear implant properties through conductive hydrogel coatings", IEEE Transactions on Neural Systems and Rehabilitation Engineering, Vol. 22, pp. 411–418, 2014.

[4.119] Moore, G., "Cramming more components onto integrated circuits", Electronics, Vol. 38, 1965, pp. 114–116.

[4.120] Ferain, I., Colinge, C., Colinge, J. P., "Multigate transistors as the future of classical metal-oxide-semiconductor field-effect transistors", Nature, Vol. 479, 2011, pp. 310–316.

[4.121] Ward, W., Duman, H., "A window to observe the foreign body reaction to glucose sensors", In: D. Cunningham and J. Stenken Eds., In Vivo Glucose Sensing, New Jersey: Wiley, 2010, pp. 87–112.

[4.122] Córcoles, E., Boutelle, M., "Biosensors and Invasive Monitoring in Clinical Applications", Sindelfingen: Springer-Verlag, 2013.

[4.123] Tuantranont, A., "Applications of Nanomaterials in Sensors and Diagnostic", Sindelfingen: Springer-Verlag, 2013.

[4.124] Serra, P., "Biosensors", Vukovar, Intech, 2010.

[4.125] Zaffino, R., Mir, M., Samitier, J., "Label-free detection of DNA hybridization and single point mutations in a nano-gap biosensor", Nanotechnology, Vol. 25, 2014, pp. 1–8.

[4.126] Urban, G., "Micro- and nanobiosensors—state of the art and trends", Measurement Science and Technology, Vol. 20, 2009, pp. 1–19.

[4.127] H. Cong, H., Xu, X., Yu, B., Yuan, H., Peng, Q., Tian, C., "Recent progress in preparation and application of microfluidic chip electrophoresis", Journal of Micromechanics and Microengineering, Vol. 25, 2015, pp. 1–11.

[4.128] Huei, D., Begg, R., Palaniswami, M., "Healthcare Sensor Network-Challenges Toward Practical Implementation", Boca Raton: CRC Press, 2012.

[4.129] Puers, R., Carta, R., Thoné, J., "Wireless power and data transmission strategies for next-generation capsule endoscopes", Journal of Micromechanics and Microengineering, Vol. 21, 2011, pp. 1–15.

[4.130] Receveur, R., Lindemans, F., Rooij, N., "Microsystem technologies for implantable applications", Journal of Micromechanics and Microengineering, Vol. 17, 2007, pp. R50–R80.

[4.131] Adams, J., Bakalar, R., Boroch, M., Mounib, E, L., Stuart, N., "Healthcare 2015 and Care Delivery—Delivery models refined, competencies defined" IBM, NY, USA, Tech. Rep., GBE03061-USEN-00, 2015.

[4.132] Salvatore, G., "High performance flexible electronics for biomedical devices" 36th Annual International Conference of the Engineering in Medicine and Biology Society (EMBC), Chicago, 2014, pp. 4176–4179.

[4.133] M. Kwiat, D. Stein, F. Patolsky, "Nanotechnology Meets Electrophysiology", Current Opinion in Biotechnology, Elsevier, Vol. 24, 2013, pp. 654–663.

[4.134] International Organization for Standardization. Available at: http://www.iso.org/iso/iso_catalogue/catalogue_tc/catalogue_tc_browse.htm?commid=54508

[4.135] Rodgers, M., Pai, V., Conroy, R., "Recent advances in wearable sensor for health monitoring", IEEE Sensors Journal, Vol. 15, 2015, pp. 3119–3126.

[4.136] Leijdekkers, P., Gay, V., Lawrence, E., "Smart homecare system for health tele-monitoring", First International Conference on the Digital Society, ICDS'07, 2007 pp. 3.

[4.137] Baig, M., Hosseini, H., Connolly, M., "Multiple physical signs detection and decision support system for hospitalized older adults", Physiological Measurement, Vol. 36, 2015, pp. 2069–2088.

[4.138] Olson, P., "Wearable tech is plugging into health insurance" [Online]. Available at: http://www.forbes.com/sites/parmyolson/2014/06/19/wearable-tech-health-insurance/

[4.139] Smith, C., Villanueva, A., Priya, S., "Aurelia aurita bio-inspired tilt sensor", Smart Materials and Structures, Vol. 21, 2012, pp. 1–5.

[4.140] Herzog, H., "μ-biomimetic flow-sensors—introducing light-guiding PDMS structures into MEMS", Bioinspiration & Biomimetics, Vol. 10, 2015, pp. 1–14.

[4.141] Triantafyllidis, A., Velardo, C., Shah, S. A., *et al.*, "Supporting heart failure patients through personalized mobile health monitoring", 4th International Conference on Wireless Mobile Communication and Healthcare, Athens, 2014, pp. 287–290.

[4.142] Snatoso, F., Redmond, S., "Indoor location-aware medical system for smart homecare and telehealth monitoring: state-of-the-art", IOP Physiological Measurement, Vol. 36, 2015, pp. 53–87.

[4.143] Riazul, S., Kwak, D., Kabir, M., Hossain, M., Kwak, K.-S., "The Internet of things for health care: a comprehensive survey", IEEE Access, Vol. 3, 2015, pp. 678–708.

[4.144] Dellaca, R. L., Gobbi, A., Pastena, M., Pedotti, A., Celli, B., "Home monitoring of within-breath respiratory mechanics by a simple and automatic forced oscillation technique device", Physiological Measurement, Vol. 31, 2010, pp. 11–24.

[4.145] Bonfiglio, A., De Rossi, D., Wearable Monitoring Systems, NY: Springer, 2011.

[4.146] Sohbati, M., Toumazou, C., "Dimension and shape effects on the ISFET performance", IEEE Sensor J., Vol. 15, 2015, pp. 1670–1679.

[4.147] Huang, X., Yu, H., Liu, X., Jiang, Y., Yan, M., Wu, D., "A dual-mode large-arrayed CMOS ISFET sensor for accurate and high-throughput pH sensing in biomedical diagnosis", IEEE Transactions on Biomedical Engineering, Vol. 62, 2015, pp. 2224–2233.

[4.148] Texas Instruments, [Online]. Available at: http://www.ti.com/lsds/ti/apps/medical/applications.page?DCMP=medical&HQS=Medical

[4.149] Wild, S., Roglic, G., Green, A., Sicree, R., King, H., "Global prevalence of diabetes—estimates for the year 2000 and projections for 2030", Diabetes Care, Vol. 27, 2004, pp. 1047–1053.

[4.150] Zhao, C., Fu, Y., "Statistical analysis based online sensor failure detection for continuous glucose monitoring in type I diabetes", Chemometrics and Intelligent Laboratory Systems, Vol. 144, 2015, pp. 128–137.

[4.151] N. Pinelli, Clrk, L. M., Carrington, A. C., Malinzak, L., Patel, A., "Pharmacist managed diabetes and cardiovascular risk reduction clinic in kidney transplant recipients: bridging the gap in care transition", Diabetes Research and Clinical Practice, Vol. 106, 2014, pp. e64–e67.

[4.152] Pagkalos, I., Herrero, P., Toumazou, C., Georgiou, P., "Bio-inspired glucose control in diabetes based on an analogue implantation of a β-cell model", IEEE Transactions on Biomedical Circuits and Systems, Vol. 8, 2014, pp. 186–195.

[4.153] Toumazou, C. and Georgiou, P., "Bio-inspired semiconductor for early detection and therapy", IEEE Asian Solid-State Circuits Conference, Jeju, Korea, November 2011, pp. 129–132.

[4.154] Dockray, G., "Enteroendocrine cell signaling via the vagus nerve", Current Opinion in Pharmacology, Vol. 13, 2013, pp. 954–958.

[4.155] Camilleri, M., Toouli, J., Herrera, M. F., *et al.*, "Intra-abdominal vagal blocking (VBLOC therapy): clinical results with a new implantable medical device", Surgery, Vol. 143, 2008, pp. 723–731.

[4.156] Tsai, Y., Wu, M., Hsu, Y., "Emerging health problems among women: inactivity, obesity, and metabolic syndrome", Gynecology and Minimally Invasive Therapy, Vol. 3, 2014, pp. 12–14.

[4.157] Mansur. R., Brietzke, E., McIntyre, R., "Is there a "metabolic-mood syndrome?—a review of the relationship between obesity and mood disorders", Neuroscience and Biobehavioral Reviews, Vol. 52, 2015, pp. 89–104.

[4.158] Sharkawy, M. E., Georgiou, P., Toumazou, C., "A silicon pancreatic islet for the treatment of diabetes", Proceedings of 2010 IEEE International Symposium on Circuits and Systems, Paris, 2010, pp. 3136–3139.

[4.159] C. Toumazou, C., Constandinou, T. G., Georgiou, P., Mc Leod, C. N., Nikolic, K., "Centre for Bio-Inspired Technology- Annual Report 2013", Imperial College London, London, UK, Tech. Rep., 2013.

[4.160] E. Anderson, *et al.*, "A label-free CMOS DNA microarray based on charge sensing", IEEE International Instrumentation and Measurement Technology Conference, Vancouver Island, 2008, pp. 1631–1636.

[4.161] Aldridge, S., "The incredible shrinking lab", IEEE Review, Vol. 51, 2005, pp. 34–38.

[4.162] Howlader, M., Deen, M. J., Suga, T., "Nanobonding: a key technology for emerging applications in health and environmental sciences", Japanese Journal of Applied Physics, Vol. 54, 2015, pp. 346–350.

[4.163] Stagni, C., Guiducci, C., Benini, L., *et al.*, "A fully electronics label-free DNA sensor chip", IEEE Sensors Journal, Vol. 7, 2007, pp. 577–585.

[4.164] Jafari, H. M., Abdelhalim, K., Soleymani, L., Sargent, E. H., Kelley, S. O., Genov, R., "Nanostructured CMOS wireless ultra-wideband label-free PCR-free DNA analysis SoC", IEEE Journal of Solid-State Circuits, Vol. 49, 2014, pp. 1223–1241.

[4.165] Shinwari, M., Deen, M. J., Selvaganapathy, P., "Analytic modelling of biotransistors", IET Circuits, Devices & Systems, Vol. 2, 2008, pp. 158–165.

[4.166] Wang, L., Estrela, P., Hug, E., *et al.*, "Fabrication of BioFET linear array for detection of proteins interactions", Microelectronic Engineering, Vol. 87, 2009, pp. 753–755.

[4.167] Shinwari, M., "Challenges in implementation of silicon-based DNA hybridization sensors", Saudi International Electronics, Communications and Photonics Conference, Fira, April 2013, pp. 1–4.

[4.168] A. Vacic, *et al.*, "Multiplexed SOI BioFETs", Biosensors and Bioelectronics, Vol. 28, 2011, pp. 239–242.

[4.169] Hu, J., Meng, H., Li, G., Ibekwe, S., "A review of stimuli-responsive polymers for smart textile applications", Smart Materials Structure, Vol. 21, 2012, pp. 1–23.

[4.170] Le, T., Lin, Z., Wong, C. P., Tentzeris, M. M., "Smart skin—could they be the ultimate sensing tool?", IEEE Nanotechnology M., Vol. 9, 2015, pp. 4–10.

[4.171] Rodgers, J., (2011), "Images & Video for smart skin: electronics that stick and stretch like a temporary tattoo", [Online]. Available at: http://news.illinois.edu/news/11/0808skin_electronics_JohnRogers.html.

[4.172] Ahmed, M., Gonenli, I. E., Nadvi, G. S., Kilaru, R., Butler, D. P., Celik-Butler, Z., "MEMS Sensors on flexible substrates towards a smart skin", IEEE Sensors, 2012, pp. 1–4.

[4.173] Nakamoto, H., Ootaka, H., Tada, M., Hirata, I., "Stretchable strain sensor based on a real change of carbon nanotube electrode", IEEE Sensors Journals, Vol. 15, 2015, pp. 2212–2218.

[4.174] Gan, L., Shang, S., Yuen, C. W. M., Jiang, S.-X., "Graphene nanoribbon coated flexible and conductive cotton fabric", Composites Science and Technology, Vol. 117, 2015, pp. 208–214.

[4.175] Castiglioni, P., Faini, A., Parati, G., Di Rienzo, M., "Wearable seismocardiography", Proc. 29th Annual International Conference of the IEEE EMBS, Lyon, 2007, pp. 3954–3957.

[4.176] Tjulkins, F., NGuye, A. T. T., Reassen, E., Hoivik, N., "MEMS-based implantable heart monitoring system with integrated pacing function", Electronic Components & Technology Conference, Orlando, May 2014, pp. 139–144.

[4.177] Ponmozhi, J., Frias, C., Marques, T., Frazao, O., "Smart sensors/actuators for biomedical applications: review", Measurement, Vol. 45, pp. 1675–1688, 2012.

[4.178] Lowrie, C., Desmulliez, M. P. Y., Hoff, L., Elle, O. J., Fosse, E., "MEMS three-axis accelerometer: design, fabrication and application of measuring heart wall motion", Symposium on Design, Test, Integration & Packaging of MEMS/MOEMS, Rome, 2009, pp. 229–234.

[4.179] http://www.livanova.sorin.com/products/cardiac-rhythm-management/heart-failure/sonr-technology.

[4.180] Chuang, H.-C., Shih, C.-Y., Chou, C.-H., Huang, J.-T., "The development of a blood leakage monitoring system for the applications in hemodialysis therapy", IEEE Sensors Journal, Vol. 15, 2015, pp. 1515–1522.

[4.181] Udupa, K., Chen, R., "The mechanisms of action of deep brain stimulation and ideas for the future development", Progress in Neurobiology, http://dx.doi.org/10.1016/j.pneurobio.2015.08.001, 2015.

[4.182] Boccard, S., Pereira, E., Aziz, T., "Deep brain stimulation for chronic pain", Journal of Clinical Neuroscience, Vol. 22, 2015, pp. 1537–1543.

[4.183] Oluigbo, C., Salma, A., Rezai, A., "Deep brain stimulation for neurological disorders", IEEE Reviews Biomedical Engineering, Vol. 5, 2012, pp. 88–99.

[4.184] Greenberg, B., Rauch, S., Haber, S., "Invasive circuitry-based neurotherapautics: stereotactic ablation and deep brain stimulation for OCD", Neuropsychopharmacology, Vol. 35, 2010, pp. 317–336.

[4.185] Huang. F.-J., Lee, C.-M., Chang, C.-L., Chen, L.-K., "Rectenna application of miniaturized implantable antenna design for triple-band biotelemetry communication", IEEE Transactions on Antennas and Propagation, Vol. 59, 2011, pp. 2646–2653.

[4.186] Gosalia, K., Lazzi, G., Humayun, M., "Investigation of a microwave data telemetry link for a retinal prosthesis", IEEE Transactions on Microwave Theory and Techniques, Vol. 52, 2004, pp. 1925–1933.

[4.187] Tofighi, M. R., Kawoos, U., Kralick, F. A., Rosen, A., "Wireless intracranial pressure monitoring through scalp at microwave frequencies", Electronics Letters, Vol. 42, 2006, pp. 148–150.

[4.188] Kouzani, A. Z., Abulseoud, O. A., Tye, S. J., Hosain, M. K., Berk, M., "A low power micro deep brain stimulation device for murine preclinical research", IEEE Journal of Translational Engineering in Health and Medicine, Vol. 1, doi: 10.1109/JTEHM.2013.2264093, 2013.

[4.189] Pantelopoulos, A., Bourbakis, N., "A survey on wearable sensor-based systems for health monitoring and prognosis", IEEE Transactions on Systems, Man, and Cybernetics—Part C: Applications and Reviews, Vol. 40, 2010, pp. 12.

[4.190] Bonato, P., "Wearable sensors and systems—from enabling technology to clinical applications", IEEE Engineering in Medicine and Biology Magazine, Vol. 29, 2010, pp. 26–36, 2010.

[4.191] Le, T. Q., Cheng, C., Sangasoogsong, A., Wondhamma, W., "Wireless wearable multisensory suite and real-time prediction of obstructive sleep apnea episodes", IEEE Journal of Translational Engineering in Health and Medicine, Vol. 1, doi: 10.1109/JTEHM.2013.2273354, 2013.

[4.192] Chen, H. H., Chen, Y. H., Chen, T. C., Chen, L. G., "Mobile energy expenditure tracking system based on heart rate and motion providing extra extension for personalized care", 33rd Annual International Conference of the IEEE EMBS, Boston, 5256–5259, 2011.

[4.193] Waltz, E., "A wearable turns baseball pitching into a science", IEEE Spectrum, pp. 16–17, 2015.

[4.194] Waltz, E., (2015). The mThrow wearable sleeve turns baseball pitching into a science [Online]. Available at: http://spectrum.ieee.org/biomedical/diagnostics/the-mthrow-wearable-sleeve-turns-baseball-pitching-into-a-science.

[4.195] Aydin, N., Arslan, T., Cumming, D., "Direct sequence CDMA based wireless interface for an integrated sensor microsystem", Proc. 4th Annual IEEE Conference on Information Technology Applications in Biomedicine, doi: 10.1109/ITAB.2003.1222556, 2003.

[4.196] Arshak, A., Waldron, D., Morris, D., Korostynska, O., Jafer, E., Lyons, G., "Review of the potential of a wireless MEMS and TFT microsystem for the measurement of pressure in the IG tract", Medical Engineering & Physics, Vol. 27, 2005, pp. 347–356.

[4.197] Kim, E., (2012). "Digital Pill" with chip inside gets FDA green light. [Online]. Available at: http://money.cnn.com/2012/08/03/technology/startups/ingestible-sensor-proteus/

[4.198] Osterberg, L. and Blaschke, T., "Adherence to medication", New England Journal of Medicine, Vol. 353, 2005, pp. 487–497.

[4.199] Nasseh, K., Frazee, S. G., Visaria, J., Viahiotis, A., Tian, Y., "Cost of medication nonadherence associated with diabetes, hypertension, and dyslipidemia", American Journal of Pharmacy Benefits, Vol. 4, 2010, pp. 41–47.

[4.200] Hafezi, H., Robertson, T. L., Moon, G. D., Au-Yeng, K.-Y., Zdeblick, M. J., Savage, M., "An ingestible sensor for measuring medication adherence", IEEE Transactions on Biomedical Engineering, Vol. 62, 2015, pp. 99–109.

[4.201] Nieben, O. and Harboe, H., "Intragastric balloon as an artificial bezoar for treatment of obesity", Lancet, Vol. 1, 1982, pp. 198–199.

[4.202] Dumonceau, J., "Evidence-based review of the bioenterics intragastric balloon for weight loss", Obesity Surgery, Vol. 12, 2008, pp. 1611–1617.

[4.203] Imaz, I., Martínez-Cervell, C., García-Alvarez, E. E., Sendra-Gutiérrez, J. M., González-Enríquez, J., "Safety and effectiveness of the intragastric balloon for obesity. A meta-analysis", Obesity Surgery, Vol. 18, 2008, pp. 841–846.

[4.204] Kencana, A. P., Rasouli, M., Huynh, V. A., *et al.*, "An ingestible wireless capsule for treatment of obesity", 32nd Annual International Conference of the IEEE EMBS, Buenos Aires, pp. 963–966, 2010.

[4.205] F. Lamonaca, Polimeni, G., Barbé, K., Grimaldi, D., "Health parameters monitoring by smartphone for quality of life improvement", Measurement, Vol. 73, 2015, pp. 82–94.

[4.206] Ivanov, S., Foley, C., Balasubramaniam, S., "Virtual groups for patient WBAN monitoring in medical environment", IEEE Transactions on Biomedical Engineering, Vol. 59, 2012, pp. 3239–3246.

[4.207] Kantoch, E., "Technical verification of integrating wearable sensors into BSN-based telemedical monitoring system", 12th International Conf. Machine Learning and Appl., Miami, pp. 432–435, 2013.

[4.208] Fosso, S., Anand, A., Carter, L., "A literature review of rfid-enabled healthcare applications and issues", International Journal of Information Management, Vol. 33, 2013, pp. 875–891.

[4.209] Lee, H. J., Lee, S. H., Ha, K. S., *et al.*, "Ubiquitous healthcare service using zigbee and mobile phone for elderly patients", International Journal of Medical Informatics, Vol. 78, 2009, pp. 193–198.

[4.210] Kim, A., Hekland, F., Petersen, S., Doyle, P., "When HART goes wireless: understanding and implementing the Wireless HART standard", IEEE International Conference on Emerging Technologies and Factory Automation, Hamburg, pp. 899–907, 2008.

[4.211] Chhajed, S., Sabir, M., Singh, K., "Wireless sensor network implementation using MiWi wireless protocol stack", IEEE International Advance Computing Conference, Gurgaon, pp. 239–244, 2014.

[4.212] http://www.nasa.gov/centers/wstf/laboratories/hypervelocity/.

Chapter 5
The silicon era and beyond

Edmundo A. Gutiérrez-D.[1] *and Fernando Guarín*[2]

5.1 Comments

Since silicon started to be widely used as the key element for semiconductor electron devices, the pervasive influence of electronic devices and circuits combined with computing and communications, and the ability to sense and actuate on the variables that define the human environment, has played a key role for societal development. From the very beginning of silicon when it was employed as the guidance computer system of the Apollo 11 mission in 1969, until the observance of the Higgs boson in 2012, which was awarded the 2013 Nobel Prize in physics for Francois Englert and Peter Higgs, for an amazing mechanism that contributes to the understanding of mass subatomic particles [5.1]. The use of electronics has influenced just about every aspect of the human environment (Figure 5.1).

The Large Hadron Collider at CERN [5.2, 5.3] (shown in Figure 5.1) with the ATLAS and CMS experiments contains tens of thousands of integrated circuits [5.6]. In that sense, electronics, through the use of integrated circuits and sensors of different nature and function, have revolutionized particle physics experiments. Electronics has been a cornerstone in the discovery of the Higgs boson. The detectors that sense the particles are giant cameras, about 40 m long by 20 m in diameter that take pictures of the collision mechanisms. These cameras contain millions of channels, often implemented as reverse-biased silicon pin diode arrays. The readout circuitry is also implemented by using integrated circuits. The power consumption management is key for the measurement quality, and radiation tolerance of the electronics, having to exceed space application requirements (limited to a few 100 krads or a few kilograys) by orders of magnitude. The electronics need to tolerate radiation levels well beyond 10 Mrad and still be operational. Silicon-integrated circuits and particle sensors have enabled the discovery of the Higgs boson.

The exploration of Mars was also heavily supported by electronics. For instance, the rover's science instruments rely on a powerful computing system [5.4, 5.7], which is a radiation-hardened version of the PowerPC microprocessor

[1]Department of Electronics, National Institute for Astrophysics, Optics, and Electronics, Puebla, Mexico
[2]GlobalFoundries, Inc., East Fishkill, NY, USA

Figure 5.1 Electronics has been a key element in several different scientific discoveries or applications, such as the Large Hadron Collider
2014 Image courtesy of CERN [5.2]

used in some models of Macintosh computers. All the electronics that control rover movement and instrument deployment is housed in the Warm Electronics Box, which protects and keep a "warm" temperature appropriate for the electronics to operate correctly [5.8]. The rover is an engineering marvel that keeps itself alive by self-monitoring its own "vital signs," such as temperature, energy consumption, mechanical balance, and other variables. The rover communicates with earth using X-band radio waves in both directions. The electronics, the computing, the communications, and the sensors and actuators are present in this marvel of science and engineering. The combination of all of these elements forms what is now known as mechatronics.

The drone technology [5.5] is entering our lives much in the same manner as the smart phones have done a few years ago. Nowadays we see different types of drones everywhere. From the military unmanned aerial vehicles to the miniature hummingbird-size flying object. The nano hummingbird [5.9] is a project sponsored by DARPA under the name of Nano Air Vehicle. This device mimics the flight of a hummingbird with a continuous hover endurance of 8 minutes without external power source. Drones are being used for city security surveillance as well in many cities around the world [5.10]. In Mexico City the Tlaloc II drone [5.12] has been used to explore a 2000-year-old tunnel under the Pyramid of the Sun [5.11]. This tiny drone, or robot, has moved through very narrow tunnels, under the pyramid, that are too small for a person to fit in. Tlaloc II is equipped with video cameras, scanner, and a mechanical arm to clear obstacles out of its way through the tunnels.

The drones are also an example of the mechatronics field, as well as the most recent development of commercial 3D printing machines [5.13]. The 3D printing machines will enable several industry sectors to become more competitive and efficient, as is the case of automotive where many components may consolidate into a single part with a faster product development cycle. In the health sector, prosthetic components for human beings are already being printed using 3D printers. Such is the case of a powered prosthetic leg [5.14], which is guided by electromyographic data. Thirteen mechanical sensors that measure inertia, load, position, angle, acceleration, velocity, and torque of the knee power these prosthetic devices and ankle joins. These devices are very effective for the rehabilitation process of above-knee amputated people.

The certification, required for an end-user product, of all these electronics-based applications constitute a process that begins with the semiconductor technology itself. Silicon technology is the core behind the integrated circuits, sensors, and actuators that together with computing and communications become essential pieces for the final assembly of a successful product. Therefore, Chapter 2 reviews device physics and modeling, while Chapter 3 introduces several different characterization techniques, reliability, degradation, and lifetime, which together with modeling, enables engineers the design and fabrication of integrated circuits from the transistor to the system level. The transistor model must reproduce the electrical characteristics and comply with several other features, such as reliability, degradation, lifetime, and failure prediction. The miniaturization of semiconductor devices, in particular FET devices, has brought complex nonlinear solid-state transport mechanisms that require a fundamental understanding and modeling, which in turns needs to be included in predictive simulation tools. A prediction of the electrical behavior is required for estimating reliability, lifetime, and a potential failure. The ultimate goal of reliability is to make sure that the life of a system, built with a semiconductor technology, will last beyond the target life. The failure rate should be also below a target failure rate during the operating lifetime of the system. The lifetime of a system, a personal computer or a laptop for instance, must be larger than the target failure rate. It is more frequent to find failures such as display malfunctioning, keyboard, power source, or computer port intermittent operation than a catastrophic failure of the processor. From the end-user perspective the target life of a computer is defined by other conditions, such as obsolescence due to software incompatibility or by the availability of upgraded, faster, and more energy-efficient computers, which is a result of Moore's law which projects the computing industry doubling the amount of transistors on a computer chip every 2 years. From a market or end-user's perspective the lifetime of a computer might be 2 years following the Moore's law trend. In general there are two different computer lifespans, the operational lifetime and the useful lifespan. The second one is defined by the obsolescence of hardware and software compatibility, and the first one can exceed the average 5 years. In any case, the semiconductor technology needs to be characterized and modeled in terms of reliability. However, during the development of a product, we cannot afford to spend years to study reliability. Therefore, the concept of aging test or accelerated life is used at the characterization laboratories to characterize and model device reliability and lifetime in a short timeframe without the need of waiting for years.

Let us examine, as an example, the reliability characterization of hot carriers (HCs). When a semiconductor device, an FET in this case, is operated under a bias condition determined by the source, gate, drain, and bulk voltages, as expected, a source current (Is) injects carriers at the source electrode and a drain (Id) current is collected at the drain side. However, depending on the bias condition, a gate (Ig) and a bulk (Ib) current appears as a result of electron–hole pairs generated by hot carriers, and tunneling through the gate oxide and the drain-bulk reversed junction (Figure 5.2).

The absolute value of the gate oxide current Ig and bulk current Ib is shown in Figure 5.3b, while the Id-Vg characteristics with the stress time as a parameter are shown in Figure 5.3a. A voltage stress is applied during a period of time. The applied stress voltage exceeds that of the maximum supply voltage allowed by

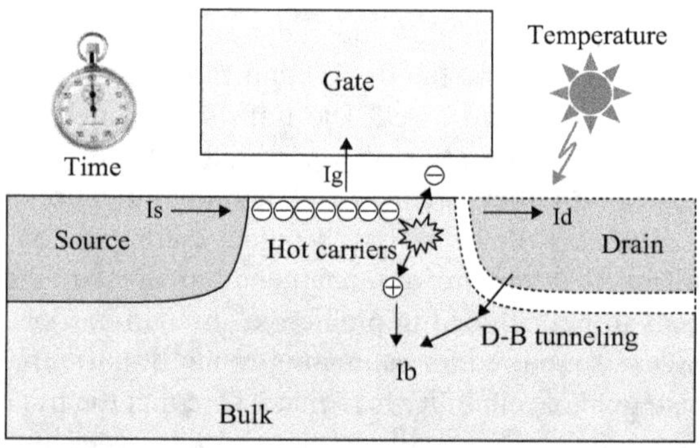

Figure 5.2 An FET device with the two major parasitic charge flows through the gate oxide (Ig) and the bulk (Ib)

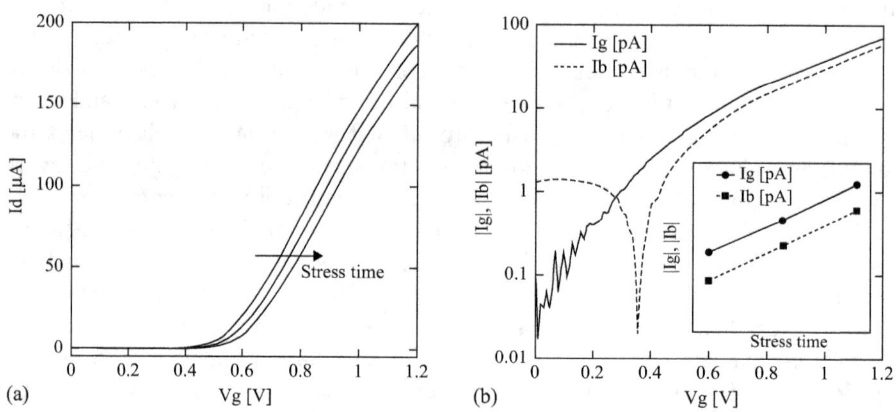

Figure 5.3 (a) Id-Vg curves with stress time as a parameter, (b) absolute value of Ig and Ib versus Vg voltage with no stress. The inset shows the logarithmic variation of Ig and Ib with stress time

the device, which accelerates the aging process in a time period in the order of 1000 seconds, which is much shorter than that of the lifetime of a product. The process is repeated several times enabling characterization engineers to study the degradation mechanism that contributes to device reliability and lifetime prediction. The shift of the threshold voltage and the reduction of the current drive capability are one of the typical consequences of device aging. A typical rule of thumb allows a maximum 10% reduction in transistor current drive capability over a timeframe of 10 years. As described in Chapter 3 the degradation of the electrical performance of an FET device is intimately linked to physical mechanisms, such as hot electrons and gate oxide and band-to-band drain-bulk tunneling leakage currents, just to mention a few. With the voltage and time stress, not only the current-voltage characteristics degrade but also the parasitic current components become worse as shown in the inset of Figure 5.3b. The increase of Ig and Ib with time contributes to the parasitic energy consumption, which then impact in a negative way the device energy efficiency and ultimately the performance of the end-user products. An excellent analysis of the different aging or accelerating test conditions is given in [5.15], where degradation can happen at the device or interconnects level as shown in Table 5.1.

Under normal operation an integrated circuit is exposed to DC and AC electrical signals as well as temperature stress. Depending on the kind of stress, DC, AC, or temperature, the degradation occurs at the device or interconnect level or at both. At the device level, the major degradation mechanisms are those related to hot carriers, gate oxide dielectric breakdown, and bias temperature instability (BTI). When the electrical signal goes out of the device, it circulates through the metal lines that interconnect different devices (Figure 5.4), which then causes open and shorts in the metal lines, or electromigration. Both degradation mechanisms may result in a catastrophic failure at the circuit or system level. The interconnect system is an array of metal lines placed at different layers, which are surrounded by a rigid insulator material as shown in Figure 5.5.

The interconnect line is composed of metal grains with an average size of 42 nm, but it ranges from 10 to 100 nm as seen from Figure 5.6 [5.15].

Table 5.1 Examples of accelerating conditions for device and interconnect degradation

Level	DC voltage	AC voltage	Temperature
Device	• Dielectric breakdown • Hot carriers • Temperature bias instability	• Conducting hot carrier mechanism	• Dielectric breakdown • Hot carriers • Temperature bias instability
Interconnect	• Open and shorts		• Electromigration • Open and shorts

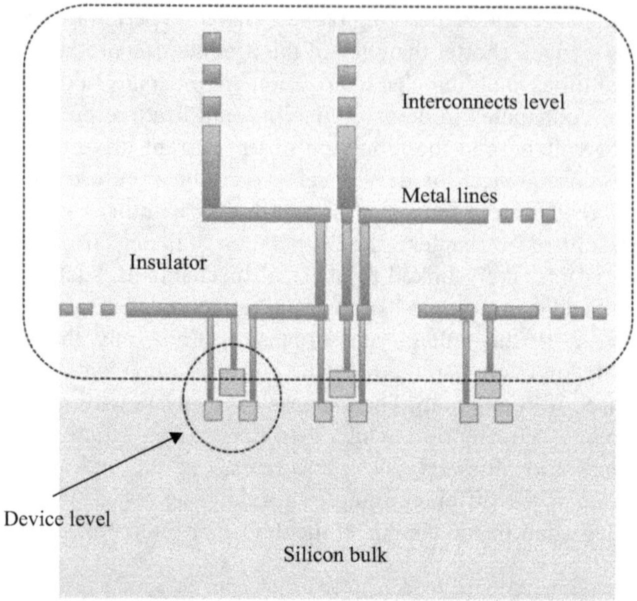

Figure 5.4 Schematic representation of devices and interconnection levels of an integrated circuit

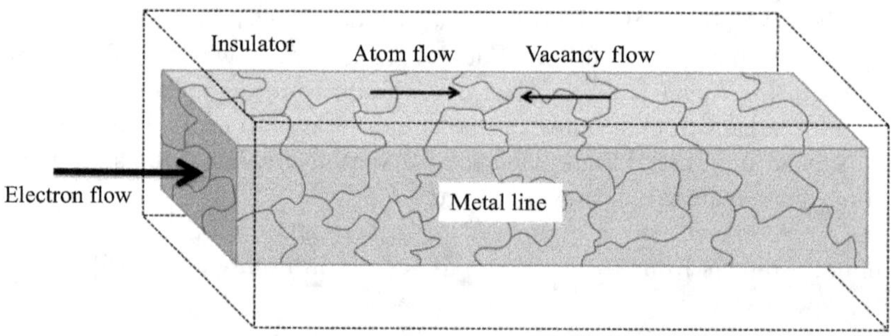

Figure 5.5 Metal interconnect line surrounded by an insulator material. Metal grains are shown

For copper dual-damascene interconnects, two main electromigration failures can happen, the late (strong) mode and the early (weak) mode [5.16]. The late mode is characterized by the growth of a void which spans the line cross section. The void grows by edge displacement in the direction of the electron flow, so that the line resistance gradually increases, as the current has to flow through a larger portion of the line. A typical reliability criterion allows one failure in 10^9 hours of device operation [5.17]. This implies that interconnect reliability is determined by the early failures, and then modeling and understanding of the early failure mode

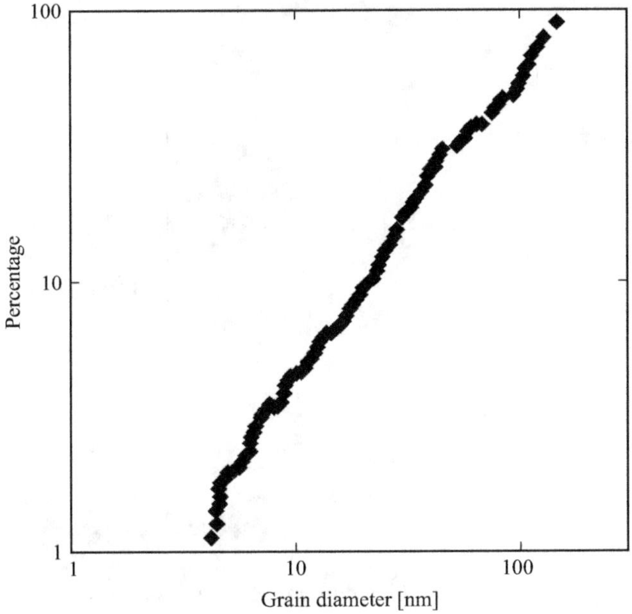

Figure 5.6 Grain size distribution for Cu lines

becomes crucial for a precise interconnects reliability assessment. As referred in Reference 5.16 the void nucleation is expected to occur at the higher hydrostatic stress magnitudes which develop at the copper/capping/barrier intersection adjacent to the via that interconnects two metal layers. A stress of a magnitude of up to 80 MPa can be built at the copper/capping/barrier layer intersection. In advanced interconnects systems the number of metal layers is larger than 10, which introduces many vias and layer intersections as shown in Figure 5.7, where a 9-metal layer interconnects system is shown. The stack of metal layers, in this particular case, has a height of 10 μm, which is quite large compared to the nano-scaled transistor and its metal contacts. The thicker top metal layer is regularly used for power delivery and input/output signal routing. Then the upper metal layer is the one that carries the higher current density and thus the one with the larger heat dissipation. A high temperature contributes to metal atom and lattice vacancies diffusion, which combined with the hydrostatic stress increases the chances for interconnect failure. A model for prediction of early failures, as the one introduced in Reference 5.16 should be very beneficial for reliability. The simulation results of the early cumulative failure percentile as a function of time are shown in Figure 5.8.

The simulation model accounts for void nucleation and simple slit void growth under the via. The simulations indicate that the void formation time and the void growth time are of the same order of magnitude. In addition to potential failures by

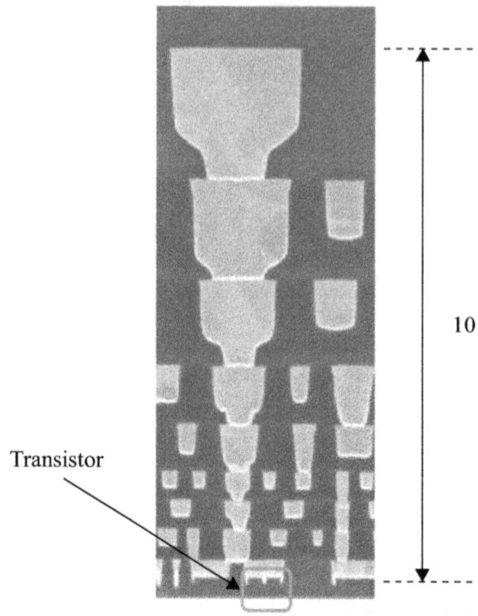

Figure 5.7 Interconnect metal lines compared to transistor size plug vias

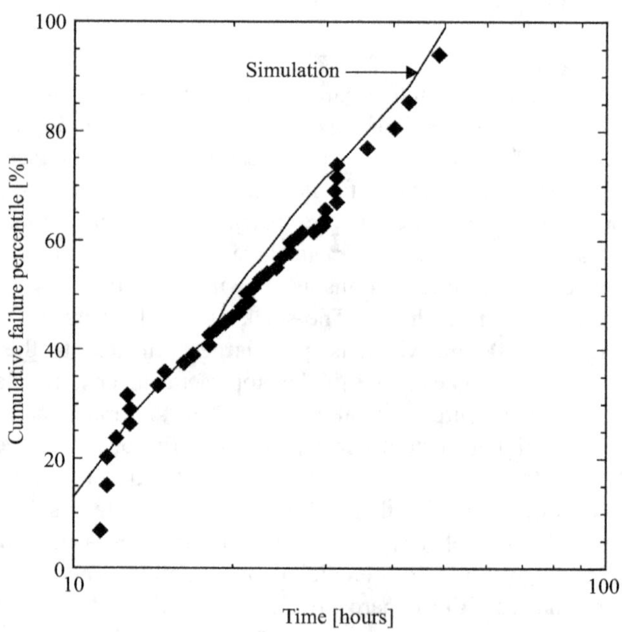

Figure 5.8 Early electromigration lifetime in a metal level 1/via intersection. Experimental results are indicated by symbols

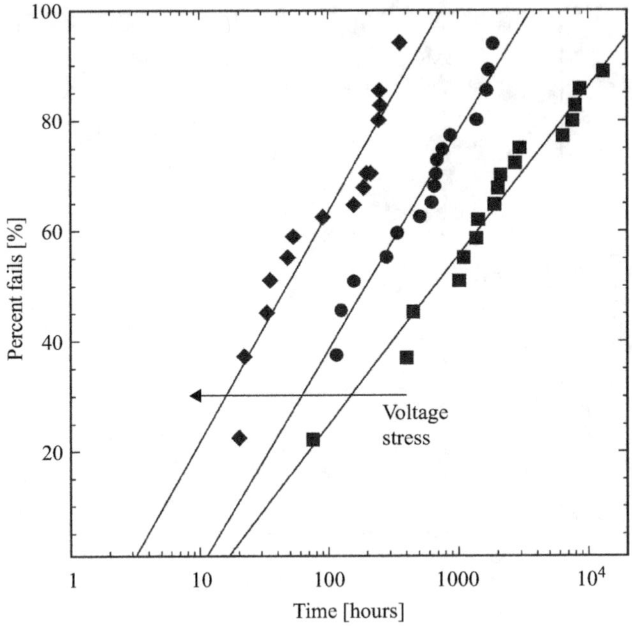

Figure 5.9 Time-dependent insulator breakdown for three different voltage stresses

electromigration, also there is the probable failure to occur in the insulator region between metal layers as shown in Figure 5.9.

As shown in Figure 5.9 the insulator breaks down earlier as the stress voltage applied between metal layers is larger. On the other hand, experiments of electromigration of Cu interconnects under AC and DC test conditions [5.18] reveal that only net DC time at test has a significant effect on time to fail. AC stressing for 1 week at 2.5 MA/cm^2 prior to DC stressing to electromigration failure had no effect on the electromigration performance. A similar result was obtained with various DC and thermal cycling conditions. The thermal history of the electromigration test samples, and in particular cooling to room temperature, has no effect on electromigration lifetime. The experiments lead to the conclusion that the standard test methodologies, using accelerated DC stress conditions at elevated temperatures, provide a good predictor for lifetime expectations under operating conditions.

At the device level, the potential failure has to do with hot carriers, gate dielectric breakdown, and BTI. However, the natural granularity inherent to the high-k/metal gates used in advanced FET devices can also introduce threshold voltage variability as reported in Reference 5.19. Figure 5.10 shows the simulated results of the standard deviation of threshold voltage versus the average metal gate grain size.

As the grain size approaches that of the 32-nm gate n-type metal–oxide–semiconductor field-effect transistor (nMOSFET), the standard variation of the

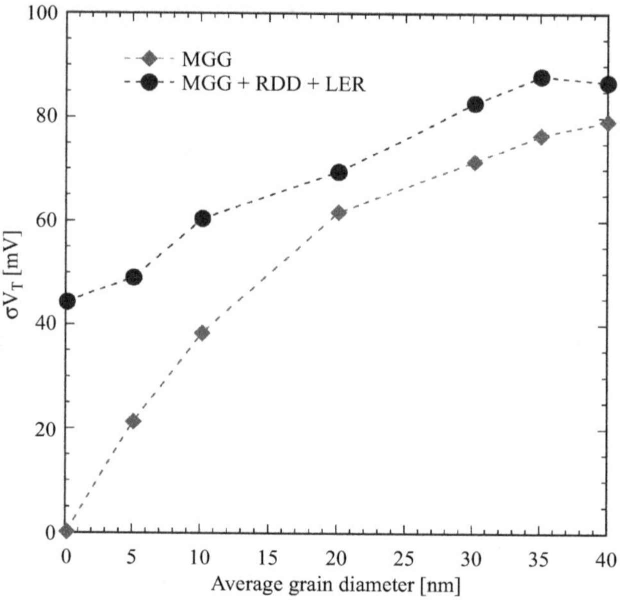

Figure 5.10 Simulated dependence of threshold voltage (V_T) variability versus average metal gate grain diameter. MGG stands for metal gate granularity, RDD for random discrete dopants, and LER for line-edge roughness

threshold voltages saturates. The distribution of the V_T voltage is bounded by the highest and lowest work function determined by the size of the grains. When the RDD and LER effects are added to the MGG, the V_T variability increases but keeps the same shape with a trend to saturate at high grain size values.

In regards to the device reliability, this can be classified as a function of the electric field intensity as shown in Figure 5.11. At low electric fields the BTI is the first degradation mechanism to show up; then as the electric field increases the random telegraph noise (RTN) shows up; and then lifetime and failure mechanisms, HCs, and finally at the largest electric field time-dependent oxide breakdown. Although there may be some overlap in between the degradation mechanisms, the progression from low to high electric fields is shown in Figure 5.11.

The degradation mechanisms are a function of the electric field and the energy gained by the carriers inside the transistor channel. The electric field and energy distribution in the channel of the transistor are shown in Figure 5.12. As both the electric field and the energy are a function of the position, the degradations mechanisms are also position dependent. At low Vd voltages (Vd_0 in Figure 5.12(a)), the electric field strength is small and confined to a narrow portion of the channel length. At high Vd voltages (Vd_f in Figure 5.12(a)), the peak electric field not only increases in magnitude but also stretches through a longer portion of the channel.

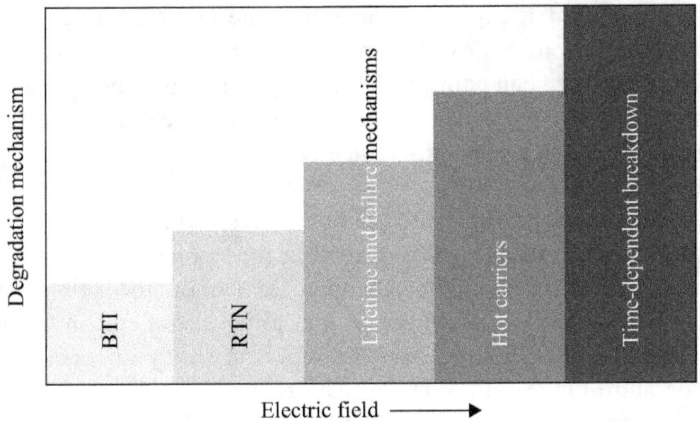

Figure 5.11 Progressive appearance of degradation effects versus the electric field

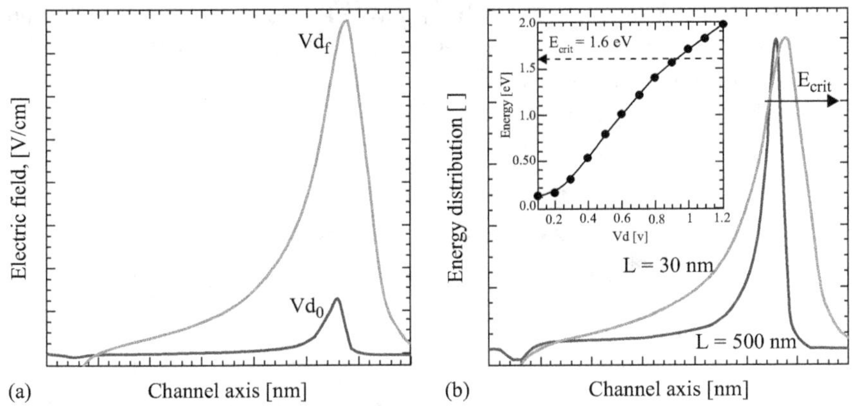

Figure 5.12 (a) Electric field versus channel axis for low (Vd_0) and high (Vd_f) drain voltages and (b) normalized energy distribution versus channel axis for two transistors with channel length L = 30 and 500 nm, respectively. The channel axis is normalized to its largest value. The energy distribution is also normalized to its maximum value. The inset in (b) shows the maximum peak energy value versus the drain voltage Vd

For short transistors, as shown in Figure 5.12(b), the carrier energy also stretches out through a longer portion of the transistor channel. This indicates that the channel of a short transistor is more energized than its longer counterpart. Therefore a nano-scaled transistor should be more prone to degradation mechanisms that a longer transistor. The inset of Figure 5.12b shows how the carrier energy increases with the Vd voltage. There is a Vd critical value from which the carrier energy overpasses that required for impact ionization. The average critical value for the ionization energy is about 1.6 eV. When impact ionization happens, HCs are generated, and some of them can gain enough energy to get its way into the oxide. This is the beginning of the gate oxide degradation process. Before going into the hot electron regime, the device may experience BTI first.

When the MOSFET is biased at low vertical gate oxide fields in the inversion regime and if it is exposed to elevated temperatures (typically from 25 to 200°C), there is a generation of interface states (ΔNit) at the semiconductor–oxide interface and a build-up of positive charges in the semiconductor side. This wearout mechanism is known as BTI. With the downscaling in dimension the self-heating exacerbates, and thus the internal device temperature may go easily into the BTI damage temperature range. The BTI causes a reduction of the MOSFET electrical performance measured as a degradation of the threshold voltage, channel mobility, transconductance, linear and saturation currents, and subthreshold slope. When dealing with negative gate voltage applied to pMOSFETs, we talk of negative bias temperature instability (NBTI), and in the case of positive gate voltage applied to nMOSFETs, we talk of positive PBTI. The NBTI used to be the most dominant degradation mechanisms observed only on pMOSFETs. However, since the introduction of high-K oxides and metal gates, not only the NBTI has been exacerbated but also the magnitude of PBTI is approaching that of NBTI as shown in Figure 5.13

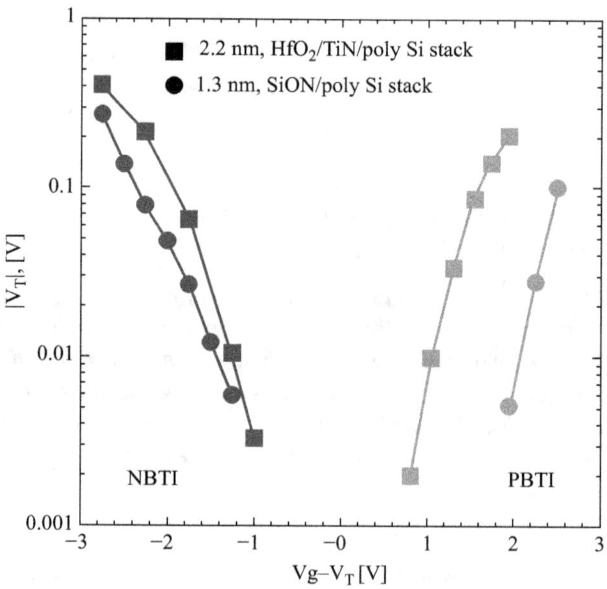

Figure 5.13 BTI-induced threshold voltage shift at a fixed stress time of 100 seconds for conventional poly-Si/SION and advanced metal gate/HfO_2 p- and n-type MOSFET devices

[5.20]. The experimental data plotted in Figure 5.13 corresponds to a short stress time of only 100 seconds, but even the PBTI for the metal-gate high-k is considerably high. The BTI in general is strongly dependent on the gate stack structure and processing conditions. The carrier mobility is also affected by BTI as can be seen in Figure 5.14. The carrier mobility degradation is larger at low gate-overdrive voltages because of Coulomb scattering due to interface traps. As shown in Figure 5.15 the current in the saturation regime degrades more than when the transistor operates in the linear regime.

The gate-to-drain Cgd capacitance also shows a similar degradation to that of the threshold voltage V_T (Figure 5.16). Because of the Miller effect [5.22], the Cgd degradation translates into degradation for digital and analog circuits. According to Reference 5.21, a 2% Idsat degradation causes a 1% reduction in the output frequency of a ring oscillator (RO). The experimental results reported in Reference 5.21 belong to a 130 nm node technology with dual gate oxide (2.7 nm SiO2 operated at 1.5 V and 7 nm SiO2 operated at 3.3 V). More recently the BTI degradation in a high-K metal-gate 40-nm RO has been reported in Reference 5.23. The oscillation frequency degradation is shown in Figure 5.17.

The three ROs are identical, but the feedback of the 50% activity RO is controlled in a way that the RO oscillates at 50% of the full speed (≈1 GHz). In the case of 0.01% activity, the feedback is inhibited most of the time in a way that the oscillation frequency is 100 kHz. Thus, under the 50% activity condition the RO is

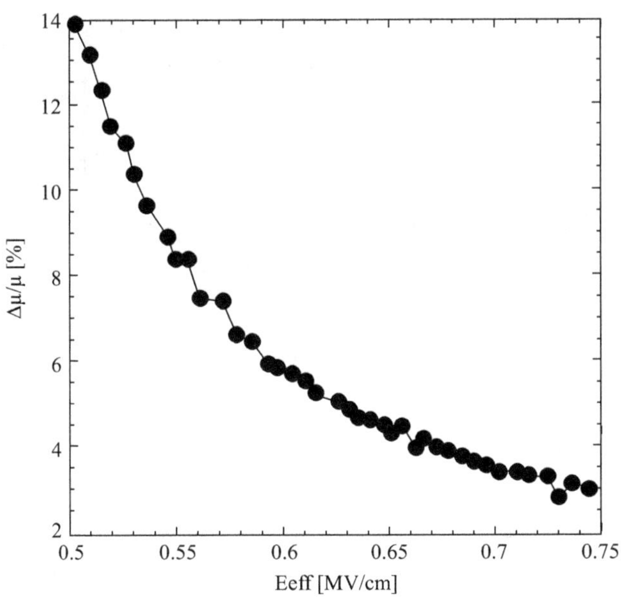

Figure 5.14 Measured mobility degradation versus effective electric field for a 2.7-nm SiO2

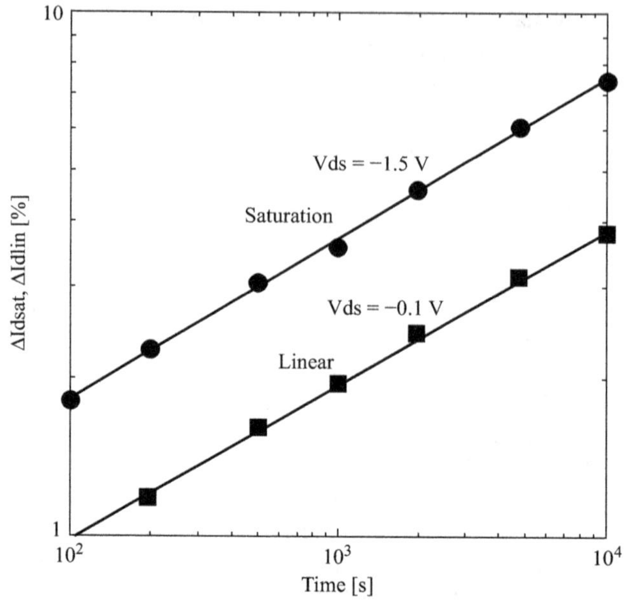

Figure 5.15 Percentage change in saturation and linear currents versus stress time

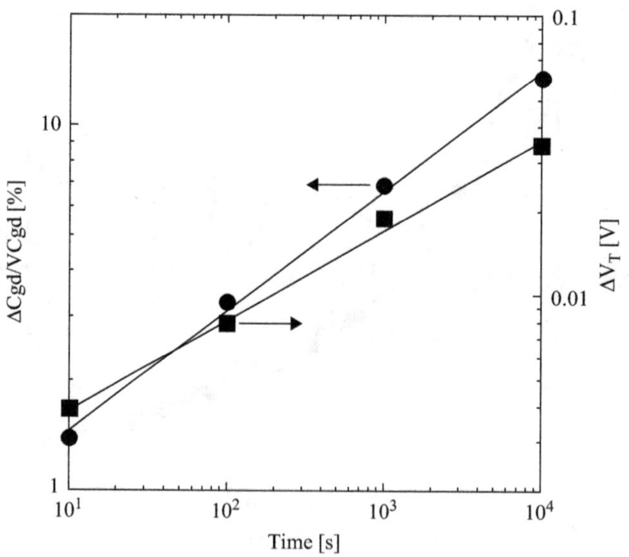

Figure 5.16 Percentage change of gate-to-drain capacitance and threshold voltage shift versus stress time

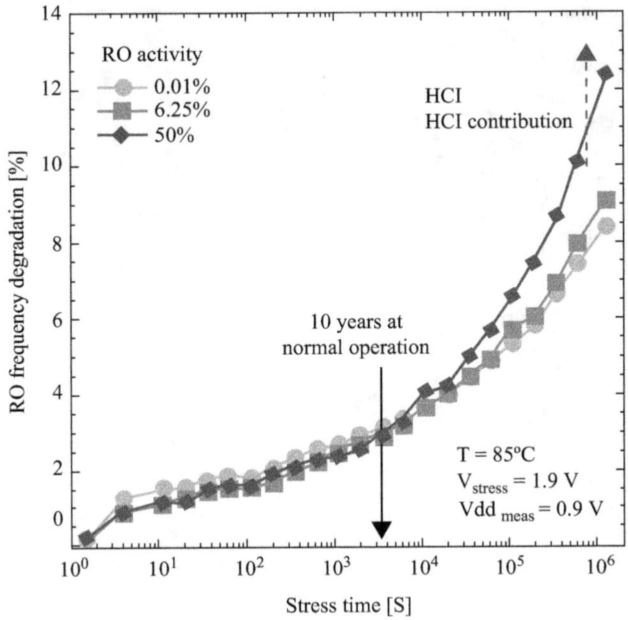

Figure 5.17 Measured BTI degradation of the oscillation frequency for three 40-nm CMOS RO stressed with the same amount of stress but different amounts of hot carrier injection (HCI) stress. The RO activity of 0.01% is toggled at very low speed and thus HCI is neglected, and only BTI degradation acts on the RO. At the 50% RO activity, which is in full speed operation, HCI adds to BTI degradation

subjected to HCI, while for the other two activity conditions does not. As shown in Figure 5.17 for the given stress voltage of 1.6 times the supply Vdd voltage, a stress time of 4000 seconds corresponds to a 10-year actual operation at the nominal Vdd supply voltage. The HCI contribution to degradation is only observed after a stress time longer than 10^4 seconds. From this condition, it is clear that the device lifetime CMOS combinational logic is only limited by BTI, and that HCI can be neglected. Also, degradation recovery during AC stress shows a benefit. The lifetime increases by a typical factor of 100, which for combinational logic applications is a huge benefit that can be translated into a significant performance by increasing the nominal value of the Vdd supply voltage. The BTI recovery, as demonstrated in Reference 5.24, is illustrated with a test in which the gate voltage is cycled as shown in Figure 5.18.

The device is first stressed for 100 seconds, and then the gate voltage is reduced to a lower value near the transistor threshold voltage for 100 seconds. This

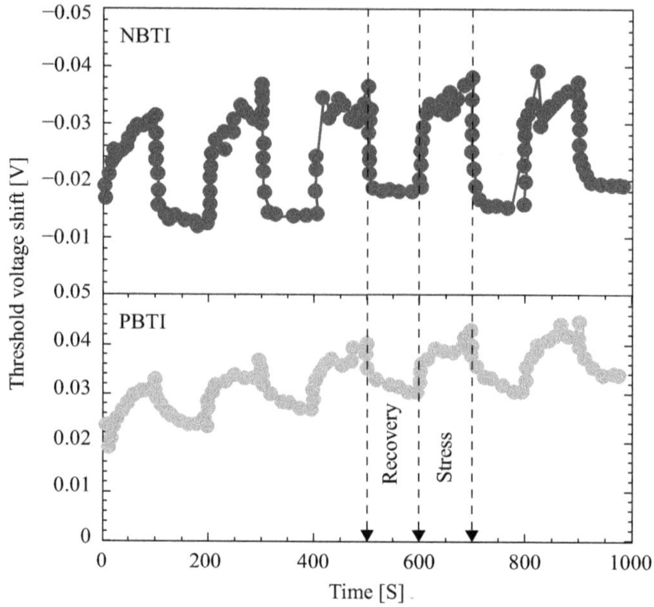

Figure 5.18 Threshold voltage shift in n- and p-type MOSFETs. The gate voltage is switched between off-state and inversion modes. When the gate voltage stress is removed, the BTI threshold voltage shift recovers for both types of transistors

process is repeated several times while the threshold voltage is monitored. Both NBTI and PBTI show qualitatively similar recovery features. The differences in the physical processes causing NBTI and PBTI are the origin for the different magnitude and time dependence of the recovery process. The interface or near-interface processes originate the NBTI, which leads to faster recovery than for PBTI, where trapping centers are situated far away from the interface. The pMOSFETs show roughly a 50% NBTI recovery, while the nMOSFETs show only a rough 25% PBTI recovery.

RTN [5.25] is a charge trapping/detrapping mechanism that contributes to degradation as well. In devices with very small areas, it is possible to have only one oxide trap in the vicinity of the surface Fermi level over the entire channel. The capture and emission of a carrier by the single trap result in a discrete modulation of the channel current that resembles that of a random telegraph signal as shown in Figure 5.19.

For device shrinking down to the range of 10 nm or sub-10 nm the granularity of matter and the discreteness of charge obliges to an atomistic approach of the device analysis [5.26], which results in a large device parameter variability. This large device parameter variability reflects also in a reliability variability that

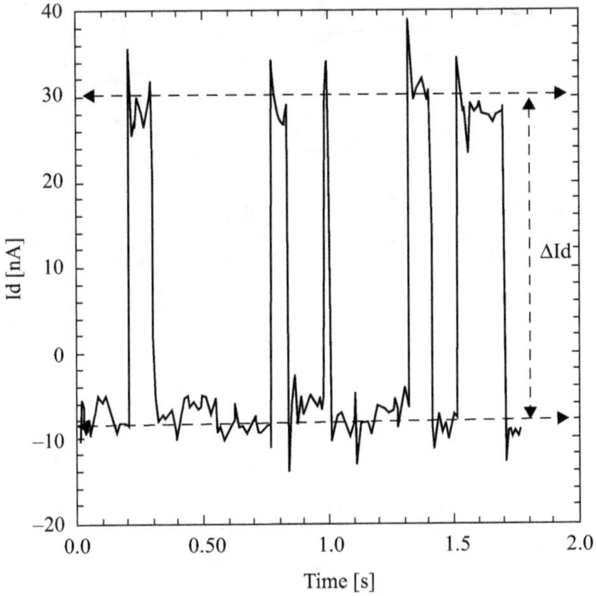

Figure 5.19 Typical measured drain current Id fluctuations at a determined Vg and Vd bias condition

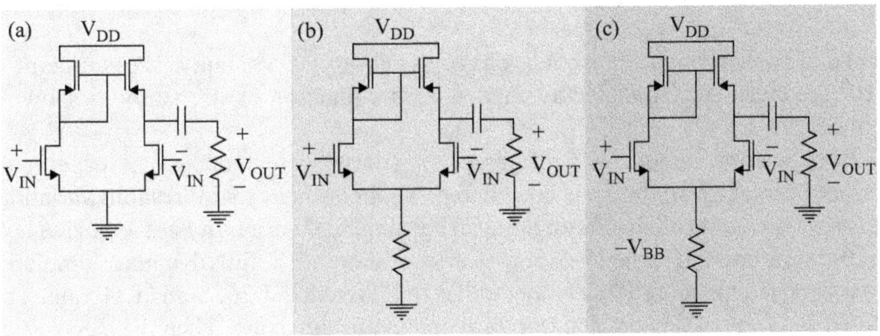

Figure 5.20 Three amplifier configurations simulated under a threshold variability of 5% and a stress time of up to 10^5 seconds

strongly affects circuits. The gate oxide aging is strongly related to RTN as shown in Reference 5.27, by the observation of exponentially distributed threshold voltage shifts corresponding to single-carrier discharges in NBTI transients in pMOSFET devices. The time-dependence observed in device reliability through the RTN mechanism adds time dependence to device mismatch as well, which in turn impacts the performance of CMOS differential amplifiers as demonstrated in Reference 5.28. Three different CMOS differential amplifier configurations (Figure 5.20) were

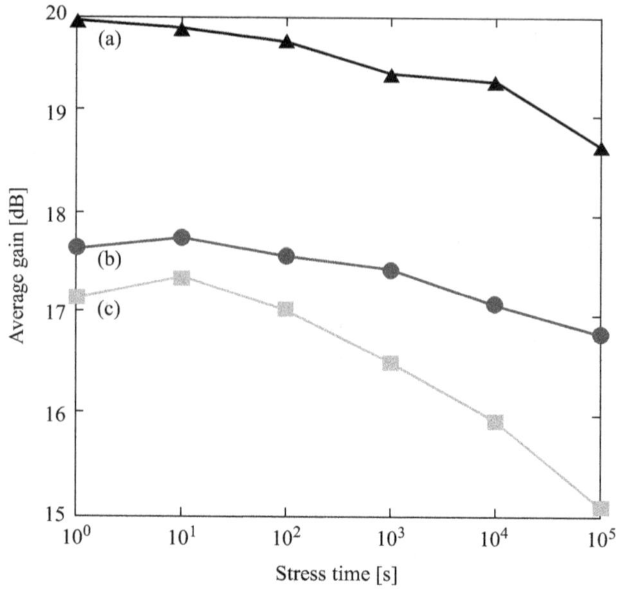

Figure 5.21 Average gain as a function of stress time for the three amplifier configuration in Figure 5.20

simulated considering a threshold voltage variability of 5% and a stress time of up to 10^5 seconds. The simulated average gain as a function of stress time is shown in Figure 5.21.

For the three amplifier configurations a large gain reduction is observed if threshold voltage variability is considered. This confirms that threshold variability plays an important role in circuit aging. The statistical variability at the device level can be simulated if kinetic Monte Carlo engine in a full dynamic simulation framework is considered as demonstrated in Reference 5.26. Statistical values are defined for a given device, and traps are created in the oxide. Then the electrostatic of the device is solved, providing the local carrier concentration and potential barrier profile required to evaluate the average time constants of each traps. Upon calculation of these rates, the kinetic Monte Carlo engine randomly selects which trap will capture or emit a charge. Under this simulation approach the effect of statistical variability and traps properties lead to different RTN sensitivity and BTI behavior as depicted in Figure 5.22.

The behavior shown in Figure 5.22 is explained as a percolative conduction regime in the presence of random dopants. The random dopant distribution is different for different devices, thus both the capture and emission time constants and the threshold voltage amplitudes are dispersed from device to device.

The first attempt to incorporate device HC degradation into circuit reliability was reported in Reference 5.29. A 0.7-µm LDD CMOS technology was used to test

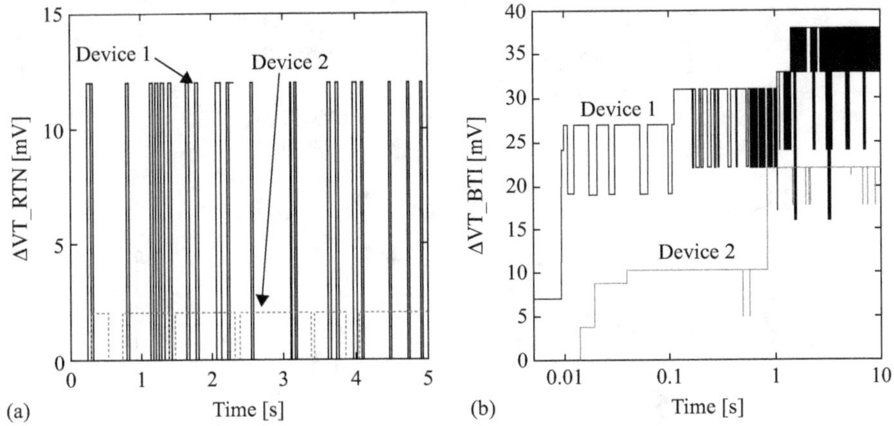

Figure 5.22 *(a). RTN signal at $V_G = V_T$ for two different devices with one trap, and (b). BTI trace at $V_G = 1$ V for two different devices with trap configurations*

the reliability of a 91-stage RO. The RO frequency degradation as a function of stress time is shown in Figure 5.23. The BERT (Berkeley Reliability Tools) is used to simulate the impact of HC at the device level on circuit reliability. The program includes both nMOSFET and pMOSFET HC models [5.30]. The simulator incorporates a parameter "Age," which relates to the amount of stress experienced by each device, and it is used to quantify device degradation during circuit operation. During circuit simulation, the Age is calculated for each device at each time step and then integrated to obtain the total Age of the SPICE analysis. After the age of each transistor in the circuit is calculated by this quasi-static method, the aged process files corresponding to the individual transistors are then used to simulate the actual circuit degradation for a specific period of time. The RO and test transistors fabricated in the 17.5-nm gate oxide thickness 0.7 μm LDD technology, were stressed at a Vdd voltage of 8.0 V, under peak substrate current peak condition for the nMOSFET and peak gate current condition for the pMOSFET. A Vdd voltage of 5 V was used for all measurements, while parameter degradation was monitored at Vgs = 5 V and Vds = 0.1 V.

The good correlation between simulated and measured results for the RO confirms that circuit degradation can be predicted using the quasi-static method for a 0.7-μm LDD CMOS technology. For advanced CMOS technologies, such as the 28-nm high-K metal gate, BTI should be used in combination of HC if an appropriate circuit reliability prediction is required. In Reference 5.31 the coexistence and weight of coupling of the two aging mechanisms, BTI and HC, is quantified.

So far it has been demonstrated that advanced CMOS technologies result in significant circuit performance, but, on the other hand, the reliability window reduces, which jeopardizes device reliability and increases the circuit failure risk at an early

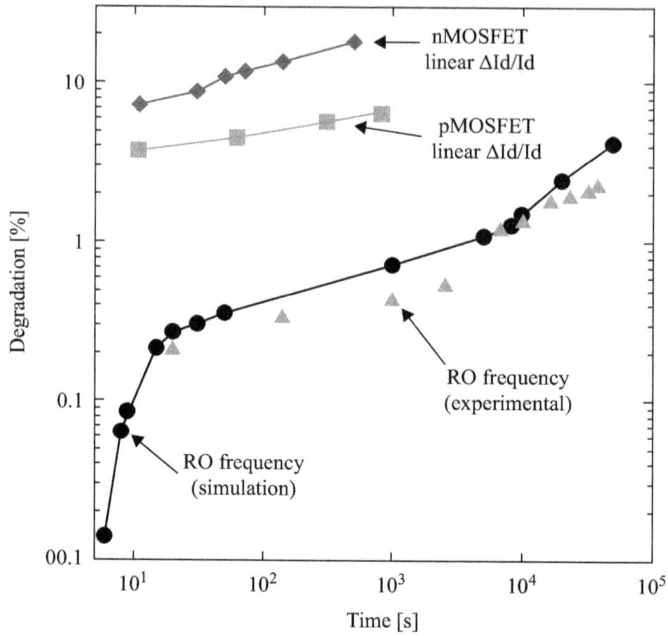

Figure 5.23 Simulated and experimental results of a 91-stage RO frequency degradation. The measured linear drain current degradation for both p- and n-MOSFET devices is also shown in the upper side of the graph

operation stage. Therefore, the appropriate incorporation of the coupled BTI–HC degradation mechanisms is crucial for an accurate circuit design and reliability prediction. Most of the device-level model assumes that BTI and HC are cumulative and independent. However, they interact with each other in a complex way, where HC degradation is affected by BTI degradation through coupling factors. The aging simulation flow has two steps. First, stimuli are analyzed for each individual device and a defect build-up quantity is computed taking into account the waveform in a digital representative time. Second, defect build-up induced degradation is performed by a SPICE parameters change. Once this is done, several ROs based on inverters of different gate lengths are stressed at wafer level. The oscillating frequency at both 25°C and 125°C are measured as a function of stress time. The monitored 1-GHz frequency at a nominal Vdd supply voltage is shown in Figure 5.24.

The standard models that do not account for BTI–HC coupling overestimate the frequency drift by a factor greater than 2. The different weights of mechanisms involved in the degradation are depicted in Figure 5.25. The two models, the one with HC and BTI decoupled (standard) and the HC+BTI coupled, are considered for stress from −40°C to 125°C. The measured HC degradation increases with

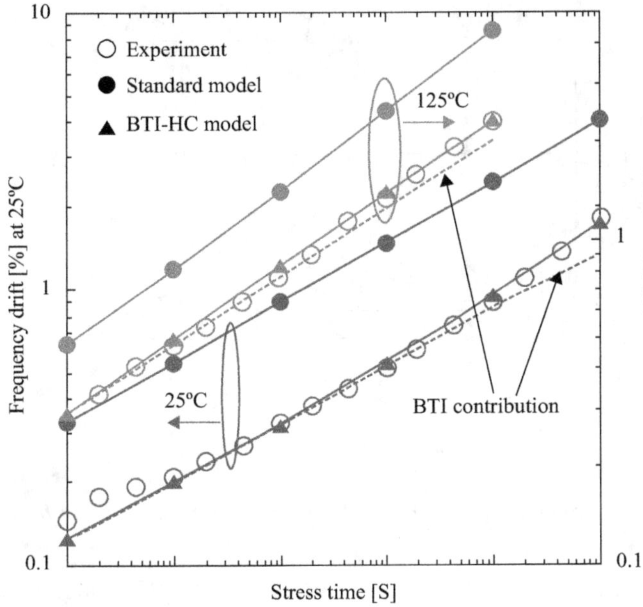

Figure 5.24 Measured and simulated RO frequency drift with BTI and HC contributions for T = 25°C (left axis) and 125°C (right axis)

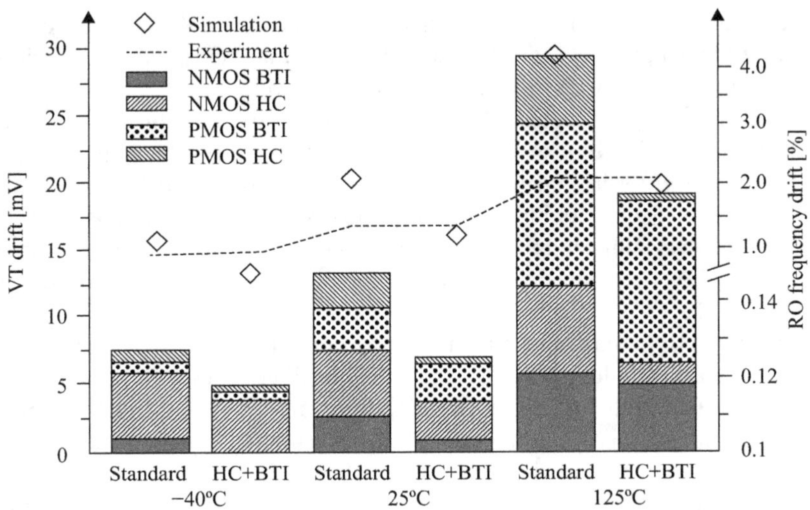

Figure 5.25 Different contributors of V_T shift (left axis) and RO frequency (right axis) drift at three different temperatures

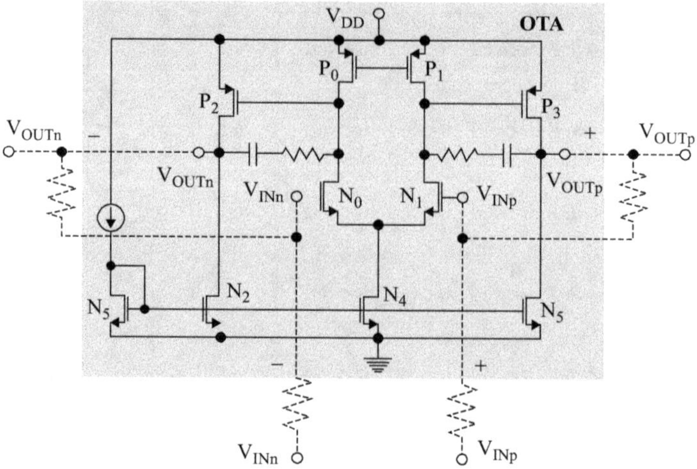

Figure 5.26 Schematic of the OTA circuit in closed loop configuration

temperature by more than 50% for pMOSFET. Only HC-BTI coupling can explain that the obtained total degradation on RO is moderated. HC in pMOSFET is one of the major contributors in the standard model at 125°C, but it totally vanishes in the HC-BTI coupled model due to the presence of NBTI. At low temperature, BTI has a small contribution, which explains the difference observed between the two models at −40°C and 25°C. At 125°C the degradation is driven by BTI. At 125°C the HC contribution is weak.

The results shown in Figures 5.24 and 5.25 demonstrate that a compact model accounting for coupled HC-BTI is required for accurate and predictive design-in-reliability simulations at IP (Intellectual Property) level.

In the analog/mixed-signal (AMS) circuit arena, special care has to be taken with the basic building blocks, such as current mirrors, operational amplifiers, and bandgap reference, for instance. In these cases, the precision and accuracy strongly depends on transistor matching. When compared to digital circuits, the AMS circuits are always subjected to a constant DC power supply, which in addition to self-heating effects recreates a voltage and temperature stress condition that may vary from place to place inside of the integrated circuit. NBTI is a significant concern for analog circuits since many analog operations require device matching [5.32], and NBTI could induce mismatching leading to circuit failure. Analog circuit designers need to anticipate to device aging effects by designing and running aging simulations. This requires special attention as the design window shrinks with the aging-induced device parameters drift. The analysis of an OTA in closed-loop configuration has been done in Reference 5.33. A Miller OTA implemented in a 32-nm metal-gate high-k CMOS technology is used as a test vehicle as shown in Figure 5.26. The closed-loop OTA is subjected to aging simulation with an equivalent stress time of 4 years at a $V_{DD} = 1.155$ V and $T = 85°C$ (Figure 5.27). Under closed-loop condition the input transistors are not subjected to

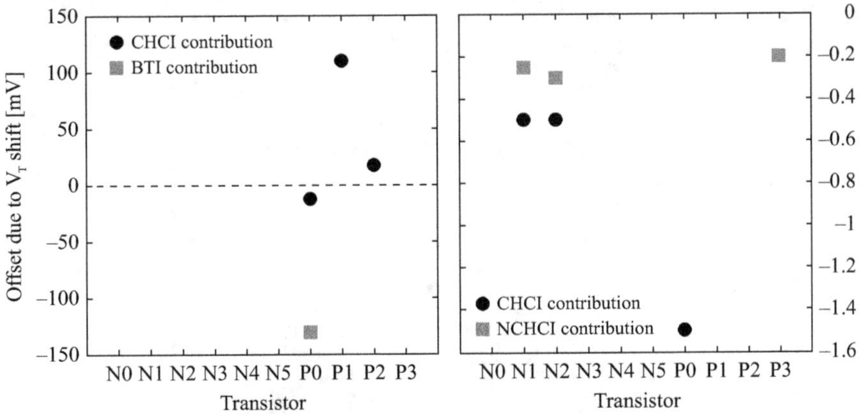

Figure 5.27 Output offset due to V_T shift (left axis) and Id shift (right axis) after aging simulations with a 4-year stress at $V_{DD} = 1.155$ V and $T = 85^oC$. CHC refers to conducting HC, while NCHC to nonconducting HC. Data taken from reference 5.33

high stress voltage because they operate near common mode voltage. On the other hand, the output transistors can be subjected to full supply voltage swing. The sensitivity of the offset is small for the parameter drifts for the output transistors, which results in a small aging-induced offset. The dominant contribution to aging-induced output-referred offset due to V_T shift comes from NBTI from the output P_2 transistor. NBTI degradation from transistors P_0 and P_1 compensates each other because they are subjected to the same gate to source voltage.

The dielectric breakdown of the gate oxide may cause partial or total loss of the transistor function as well as an undesirable increase in standby power consumption, which also leads to unpredictable consequences for the functionality at a circuit level. The precise location where dielectric breakdown occurs in the transistor determines the level of failure. For instance, soft breakdown that occurs exclusively in the transistor channel [5.34] may not lead to fatal failure. However, the hardest circuit-killing breakdowns occur above the source and drain extension regions. A gate oxide breakdown position, along the channel axis, determination technique has been developed in Reference 5.34, where an effective resistance Reff of the breakdown path is defined as Reff = (Vg = 1.5 V)/(Ig at Vg = 1.5 V) at Vs = Vd = 0 V. The soft breakdown range is 10^5–$10^9\Omega$, while the hard breakdown range is 10^2–$10^4\Omega$. The post-breakdown Reff as a function of the position x along the channel is shown in Figure 5.28.

Reff rises with x until a constant value of about 20 kΩ is reached at about 40 nm away from the source gate edge. The same happens at the drain side. The lower Reff values occur in breakdown spots located over the source/drain extension regions. Circuit failure occurs only if Reff is smaller than a critical value Reffcrit.

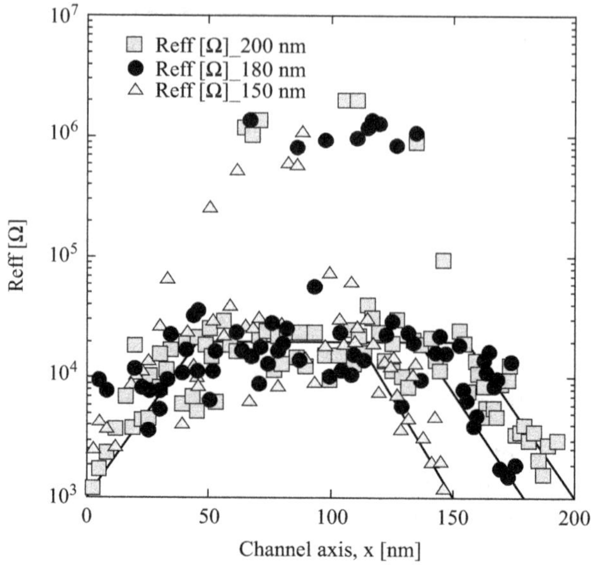

Figure 5.28 Measured Reff as a function of the breakdown position along the channel for three different channel length MOSFETs. The eye-guidelines assume a linear drop of Reff over a 40-nm wide interval at both source and drain sides

The technical description of the physics, modeling, characterization, simulation, energy-efficiency, and degradation and lifetime of semiconductor electron devices has been covered, through Chapters 2 and 3. This has been done with the aim of blending that information at the applications level and not to the level of a reliability textbook. This enables an integral view to our reader about the physics of nano-scaled semiconductor devices; the way they are fabricated and their electrical performance, reliability, and lifetime can be extrapolated to the energy-efficiency of systems used in our daily life. These systems, which find wide utilization in medicine, research, industry, food production, health care, and many other sectors, have a relevant societal impact for mankind. They have been reviewed in Chapter 4 and shown to have a profound impact for instance on optimizing the milk and meat production and by reducing the irrigation used in crop fields, to mention just two examples. These examples clearly illustrate the emblematic societal impact that at the same time reduce greenhouse gasses by the use of wireless network sensors/actuators in conjunction with signal processors to monitor and reduce the methane emission produced by cows. Adjusting the cow diet so the pH and temperature variables in the cow's rumen stay in a safe window where milk quality improves and methane emission are reduced to accomplish this win-win scenario.

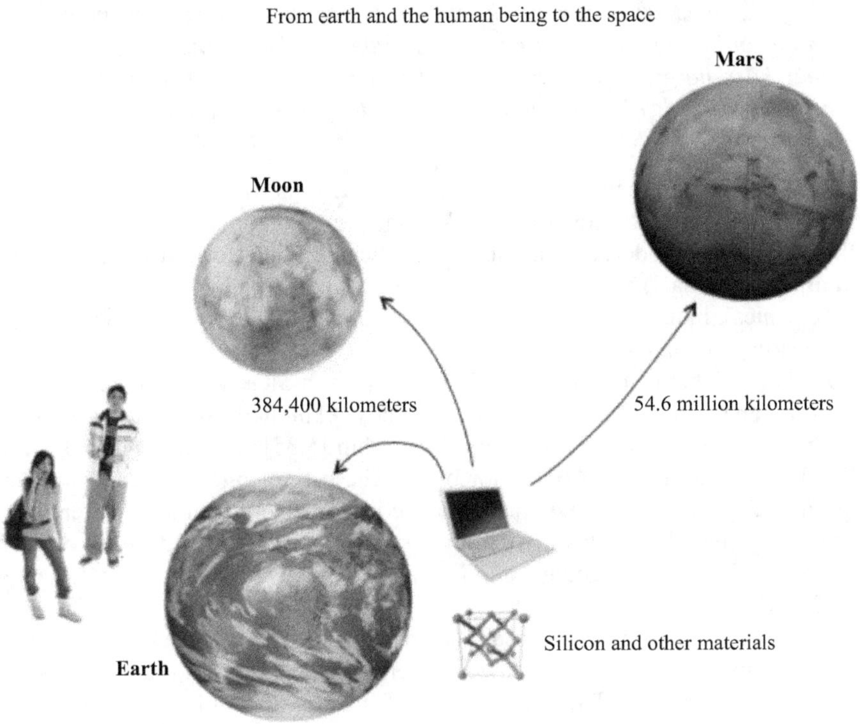

Figure 5.29 Silicon has already enabled multiple electronic applications used on earth for the benefit of human beings

Beyond agricultural, urban ("smart city," "IoT"), industry, research, and other sector applications, we find the medical and health care applications, which have a direct societal impact on population. They help people monitor health conditions and diminish the risks of unattended medical crisis, like heart attacks or diabetes events for instance. So far silicon and its associated devices, circuits, and systems have planted the starting seed for the development and production of a wide array of electronic instruments that we see in our daily life. But this is not the end of the story; the electronic applications have traveled outside of our planet. In 1969 the first flight computer, using integrated circuits, was used for man to land on the moon. In 2012 the Curiosity (rover) landed on Mars. The expedition to Mars is a 54.6 million kilometer journey (Figure 5.29). The outer space environmental conditions require special considerations for the electronics to be reliable, and thus send true and relevant information back to earth.

Electronic applications, with their advanced nano-scaled semiconductor devices, surrounded by computing capabilities, communications, and the ability, through sensors and actuators, to interact with the environment and the human being have become a ubiquitous element and inherent attribute in our society. They will continue pervading our lives at any moment and any time

> *The pace at which electronics keep permeating our lives will depend only on our ability to incorporate new materials with better energy efficiency and the synergistic way in which they are used to save energy, thus reducing worldwide pollution, and providing a greener and more sustainable world.*

Among the alternative and additional materials beyond silicon, we can generate a long list, starting from III-V compounds, germanium (Ge) alloys for high-speed and photodetector applications, silicon carbide for power applications, gallium nitride (GaN) for optoelectronic applications, graphene for display, solar cells, chemical, industrial, and various medical applications, ferroelectrics for high-density storage data, etc.

Among the list of optional or Si-based alloys is SiGe with unique electronic transport capabilities that have been recently exploited by IBM and Global Foundries to fabricate the world's first 7-nm chip [5.35]. This is a FinFET with a SiGe channel and a 30-nm transistor pitch. These semiconductor companies are targeting a 50%, at least, power/performance improvement for the next generation of systems when shrinking down from 10 to 7 nm. This would be an amazing achievement in terms of energy efficiency, and thus would contribute to energy saving.

Even with the use of compounds such as InGaAs, as the channel of FET-based transistors, the fabrication process would benefit from the well-established scaled process of Si technology. The incorporation of InGaAs on SOI has been experimentally demonstrated in Reference 5.36, where an electron mobility of 5400 cm^2/Vs on an InAs nanowire cross-junction has been obtained. An MuG-FET fabricated with this process exhibits a 660 μA/μm on current and a peak transconductance of 1.0 mS/μm at Vds = 0.5 V. This is promising for the next generation of III-V CMOS an opto-electronics on Si. The commercial fabrication of true 3D integrated circuits is now a reality; it will create a new paradigm that will open the door to a new and exciting host of multiple applications for the near future.

5.2 Conclusion

A conclusion drawn out of the scientific work reported in 2015 is that silicon will keep being the primary seed material for the next generations of electron devices. Nature and the existing manufacturing infrastructure continue to favor silicon!

References

[5.1] http://home.web.cern.ch/topics/higgs-boson.
[5.2] http://www.nasa.gov/mission_pages/apollo/apollo11.html.
[5.3] http://home.web.cern.ch/topics/large-hadron-collider.
[5.4] https://www.nasa.gov/jpl/msl/nasa-mars-rover-moves-onward-after-marias-pass-studies.

[5.5] http://spectrum.ieee.org/automaton/robotics/industrial-robots/earthbound-robots-today-need-to-take-flight.

[5.6] Snoeys, W., "How chips helped discover the Higgs boson at CERN", IEEE ESSDERC 2014 Conference, Venice, Italy, pp. 9-19.

[5.7] http://mars.nasa.gov/mer/technology/bb_avionics.html.

[5.8] Gutiérrez-D., E. A., Deen, M. J., and Claeys, C. L., Low Temperature Electronics: Physics, Devices, Circuits, and Applications, Academic Press, 2000.

[5.9] http://www.avinc.com/nano.

[5.10] https://www.microdrones.com/en/references/case-studies-and-use-cases/.

[5.11] http://robotsinsider.com/tlaloc-ii-remote-controlled-robot-uncovers-hidden-chambers-in-mexico/.

[5.12] http://www.jornada.unam.mx/2013/04/23/cultura/a05n1cul.

[5.13] http://3dprintingfromscratch.com/common/history-of-3d-printing/.

[5.14] http://spectrum.ieee.org/tech-talk/biomedical/bionics/powered-prosthetic-legs-work-better-by-tracking-emg.

[5.15] Cao, L., Ganesh, K. J., Zhang, L., *et al.*, Grain structure analysis and effect on electromigration reliability in nanoscale Cu interconnects, Applied Physics Letter, 102, 13, 2013, pp. 131907-1-131907-4.

[5.16] de Orio, R. L., Ceric, H., and Selberherr, S., A compact model for early electromigration failures of copper dual-damascene interconnects, Microelectronics Reliability, 51, 2011, pp. 1573–1577.

[5.17] Gall, M., Capasso, C., Jawarani, D., Hernandez, R., Kawasaki, H., and Ho, P. S., Statistical analysis of early failures in electromigration, Journal of Applied Physics, 90, 2, 2001, pp. 732–740.

[5.18] Shaviv, R., Harm, G. J., Kumari, S., Keller, R. R., and Read, D. T., Electromigration of Cu interconnects under AC and DC test conditions, Microelectronic Engineering, 92, 2012, pp. 111–114.

[5.19] Brown, A. R., Idris, N. M., Watling, J. R., and Asenov, A., Impact of metal gate granularity on threshold voltage variability: a full-scale three-dimensional statistical simulation study, IEEE Electron Device Letters, 31, 11, 2010, pp. 1199–1201.

[5.20] Kerber, A., Maitra, K., Majumdar, A., Hargrove, M., Carter, R. J., and Cartier, E. A., Characterization of fast relaxation during BTI stress in conventional and advanced CMOS devices with HfO2/TiN gate stacks, IEEE Transaction on Electron Devices, 55, 11, 2008, pp. 3175–3183.

[5.21] Krishnan, A. T., Reddy, V., Chakravarthi, S., Rodriguez, J., John, S., and Krishnan, S., NBTI impact on transistor & circuit: models, mechanisms, and scaling effects, Proceedings of the IEDM, 2003, pp. 349–352.

[5.22] Miller, J. M., Dependence of the input impedance of a three-electrode vacuum tube upon the load in the plate circuit, Scientific Papers of the Bureau of Standards, 15, 351, 1920, pp. 367–385.

[5.23] Reisinger, H., Grasser, T., Hofmann, K., Gustin, W., and Schlünder, C., The impact of recovery on BTI reliability assessment, IEEE International Integrated Reliability Workshop Final Report (IRW), 2010, pp. 12–16.

[5.24] Grasser, T., Bias Temperature Instability for Devices and Circuits, Springer, New York, 2014. Chapter 1, pp. 7–9.

[5.25] Hung, K. K., Ko, P. K., Hu, C., and Cheng, Y.-C., Random telegraph noise of deep-submicrometer MOSFETs, IEEE Electron Device Letters, 11, 2, 1990, pp. 90–92.

[5.26] Gerrer, I., Ding, J., Amoroso, S. M., *et al.*, Modelling RTN and BTI in nanoscale MOSFETs from device to circuit: a review, Microelectronics Reliability, 54, 2014, pp. 682–697.

[5.27] Kaczer, B., Grasser, T., Roussel, J. P., *et al.*, Origin of NBTI variability in deeply scaled pFETs, Proceedings of Reliability Physics Symposium (IRPS) 2010, pp. 26–32.

[5.28] Martin-Martinez, J., Rodríguez, R., Nafria, M., and Aymerich, X., Time-dependent variability related to BTI effects in MOSFETs: impact on CMOS differential amplifiers, IEEE Transactions on Device and Materials Reliability, 9, 2, 2009, pp. 305–310.

[5.29] Tu, R. H., Rosenbaum, E., Chan, W. Y., *et al.*, Berkeley reliability tools-BERT, IEEE Transactions on Computer-Aided Design of Integrated Circuits and Systems, 12, 10, 1993, pp. 1524–1534.

[5.30] Hu, C., Tam, S., Hsu, F.-C., *et al.*, Hot-electron induced MOSFET degradation-model, monitor, improvement, IEEE Transactions on Electron Devices, 32, 1985, pp. 375–385.

[5.31] Cacho, F., Mora, P., Arfaoui, W., Federspiel, X., and Huard, V., HCI/BTI coupled model: the path for accurate and predictive reliability simulations, IEEE International Reliability Physics Symposium, IRPS 2014, pp. 5D.4.1–5D.4.5.

[5.32] Latif, M., Ali, N., and Hussin, F., A case study for reliability-aware in SoC analog circuit design. In: International Conference on Intelligent and Advanced Systems (ICIAS), June 2010, pp. 1–6.

[5.33] More, S., Aging degradation and countermeasures in deep-submicrometer analog and mixed signal integrated circuits, PhD Dissertation, Technical University of Munich, Germany, November, 2011.

[5.34] Degraeve, R., Kaczer, B., De Keersgieter, A., and Groeseneken, G., Relation between breakdown mode and location in short-channel nMOSFTEs and its impact on reliability specifications, IEEE Transactions on Device and Materials Reliability, 1, 3, 2001, pp. 163–169.

[5.35] http://arstechnica.co.uk/gadgets/2015/07/ibm-unveils-industrys-first-7nm-chip-moving-beyond-silicon/.

[5.36] Schmid, H., Bor, M., Moselund, K., *et al.*, Template-assisted selective epitaxy of III-V nanoscale devices for co-planar heterogeneous integration with Si, Applied Physics Letters, 106, 2015, pp. 233101-1–233101-5.

[5.37] Jan, C.-H., Agostinelli, M., Deshpande, H., *et al.*, RF CMOS technology scaling in High-k/metal gate era for RF SoC (system-on-chip) applications, IEEE IEDM Proceedings 2010, San Francisco CA, USA, 2010, pp. 27.2.1–27.2.4.

Further reading

http://www.dailymail.co.uk/news/article-2313312/Robot-discovers-unexplored-passages-2-000-year-old-tunnel-near-Pyramid-Sun-Mexico.html.

http://www.jornada.unam.mx/2013/04/23/cultura/a05n1cul.

"Silicon-unit-cell-3D-balls" by Ben Mills (Own Work). Licensed under Public Domain via Commons. Available at: https://commons.wikimedia.org/wiki/File:Silicon-unit-cell-3D-balls.png#/media/File:Silicon-unit-cell-3D-balls.png.

Mars image taken from http://mars.nasa.gov/allaboutmars/extreme/.

Moon image from http://www.google.com.mx/imgres?imgurl=https://upload.wikimedia.org/wikipedia/commons/e/e1/FullMoon2010.jpg&imgrefurl=https://en.wikipedia.org/wiki/Moon&h=2452&w=2580&tbnid=e0MOONwtisYikM:&tbnh=186&tbnw=195&usg=__p4dxcQ47BaqiYUSFG3qI5abvjOI=&docid=rqZWM4HaPWtFJM&itg=1.

Index

www.ingramcontent.com/pod-product-compliance
Lightning Source LLC
LaVergne TN
LVHW011901200926
848484LV00002B/502